Current Developments in Mathematics 2002

Edited by

David Jerison
Barry Mazur
Tomasz Mrowka
Wilfried Schmid
Richard P. Stanley
Shing-Tung Yau

International Press
www.intlpress.com

Current Developments in Mathematics, 2002

ISBN 978-1-57146-346-3

Paperback reissue 2017. Originally published in 2003 under ISBN 1-57146-102-7 (casebound). The table of contents and papers herein are reproduced in facsimile from the original publication.

Printed in the United States of America.

Contents

Current Developments in Mathematics, 2002
Pages 1 – 37

One Dimensional Hyperbolic Systems of Conservation Laws

Alberto Bressan

1. Introduction

These notes are meant to provide a survey of some recent results and techniques in the theory of conservation laws. In one space dimension, a system of conservation laws can be written as

$$u_t + f(u)_x = 0\,. \tag{1.1}$$

Here $u = (u_1, \ldots, u_n)$ is the vector of ***conserved quantities*** while the components of $f = (f_1, \ldots, f_n)$ are called the ***fluxes***. Integrating (1.1) over the interval $[a, b]$ one obtains

$$\begin{aligned}\frac{d}{dt} \int_a^b u(t,x)\,dx &= \int_a^b u_t(t,x)\,dx = -\int_a^b f\big(u(t,x)\big)_x\,dx \\ &= f\big(u(t,a)\big) - f\big(u(t,b)\big) = [\text{inflow at } a] - [\text{outflow at } b]\,.\end{aligned} \tag{1.2}$$

In other words, each component of the vector u represents a quantity which is neither created nor destroyed: its total amount inside any given interval $[a, b]$ can change only because of the flow across boundary points.

Systems of the form (1.1) can be used to express the fundamental balance laws of continuum physics, when small viscosity or dissipation effects are neglected [D2]. A primary example is provided by the Euler equations describing the evolution of a compressible, non viscous fluid:

$$\begin{cases} \rho_t + (\rho v)_x = 0 & \text{(conservation of mass)} \\ (\rho v)_t + (\rho v^2 + p)_x = 0 & \text{(conservation of momentum)} \\ (\rho E)_t + (\rho E v + p v)_x = 0 & \text{(conservation of energy)} \end{cases}$$

Here ρ is the mass density, v is the velocity while $E = e + v^2/2$ is the energy density per unit mass. The system is closed by a ***constitutive relation*** of the form $p = p(\rho, e)$, determining the pressure as a function of the density and the internal energy. The particular values of p depends on the gas under consideration.

Using the chain rule, (1.1) can be written in the ***quasilinear*** form

$$u_t + A(u) u_x = 0\,, \tag{1.3}$$

where $A(u) \doteq Df(u)$ is the Jacobian matrix of first order partial derivatives of f. For smooth solutions, the two equations (1.1) and (1.3) are entirely equivalent. However, if u has a jump at a point x_0, the left hand side of (1.3) will contain the product of the discontinuous function $x \mapsto A(u(x))$ with the distributional derivative u_x, which in this case contains a Dirac mass at the point x_0. In general, such a product is not well defined. Hence (1.3) is meaningful only within a class of continuous functions. On the other hand, working with the equation in divergence form (1.1) allows us to consider discontinuous solutions as well, interpreted in distributional sense. More precisely, a locally integrable function $u = u(t,x)$ is a ***weak solution*** of (1.1) provided that

$$\iint \{u\phi_t + f(u)\phi_x\}\, dxdt = 0 \tag{1.4}$$

for every differentiable function with compact support $\phi \in C_c^1$.

We say that the above system is ***strictly hyperbolic*** if every matrix $A(u)$ has n real, distinct eigenvalues, say $\lambda_1(u) < \cdots < \lambda_n(u)$. In this case, one can find dual bases of left and right eigenvectors of $A(u)$, denoted by $l_1(u), \ldots, l_n(u)$ and $r_1(u), \ldots, r_n(u)$, normalized according to

$$|r_i| = 1, \qquad l_i \cdot r_j = \begin{cases} 1 & if \quad i = j, \\ 0 & if \quad i \neq j. \end{cases} \tag{1.5}$$

To appreciate the effect of the non-linearity, consider first the case of a linear system with constant coefficients

$$u_t + Au_x = 0. \tag{1.6}$$

Call $\lambda_1 < \cdots < \lambda_n$ the eigenvalues of the matrix A, and let l_i, r_i be the corresponding left and right eigenvectors as in (1.5). One can then write the general solution of (1.6) as a superposition of independent linear waves:

$$u(t,x) = \sum_i \phi_i(x - \lambda_i t) r_i\,, \qquad \phi_i(y) \doteq l_i \cdot u(0,y)\,.$$

Notice that here the solution is completely decoupled along the eigenspaces of A, and each component travels with constant speed, given by the corresponding eigenvalue of A.

In the nonlinear case (1.3) where the matrix A depends on the state u, new features will appear in the solutions.

(i) Since the eigenvalues λ_i now depend on u, the shape of the various components in the solution will vary in time (fig. 1). In particular, rarefaction waves will decay, and compression waves will become steeper, possibly leading to shock formation in finite time.

(ii) Since the eigenvectors r_i also depend on u, nontrivial interactions between different waves will occur (fig. 2). In particular, the strength of the interacting waves may change, and new waves of different families can be created, as a result of the interaction.

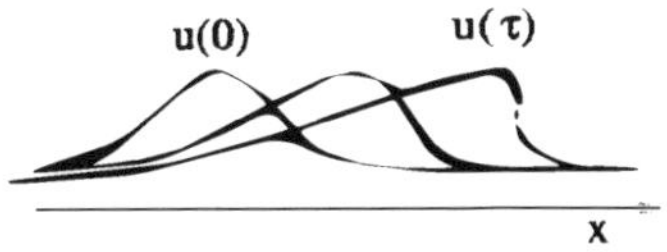

figure 1

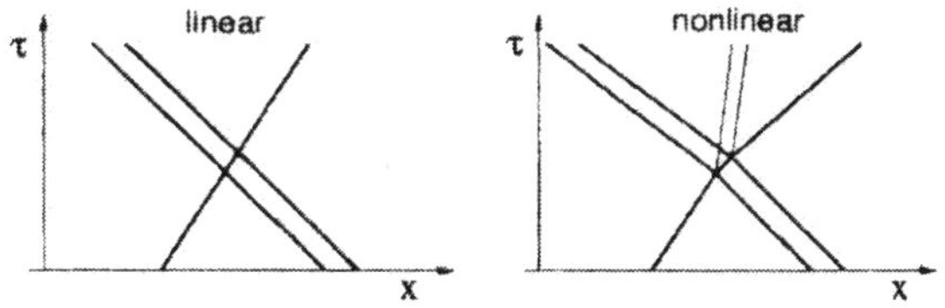

figure 2

Of all these effects, the most important for the mathematical analysis is the possible loss of regularity. This lack of regularity (also due to the absence of second order terms that could provide a smoothing effect), together with the strong nonlinearity of the equations, is the main source of the difficulties encountered in a rigorous mathematical analysis of this subject. Indeed, most of the powerful techniques of functional analysis do not apply in this context. Solutions cannot be represented as fixed points of continuous transformations, or in variational form, as critical points of suitable functionals. Dealing with vector valued functions, comparison principles based on upper or lower solutions cannot be used. Similarly, the theory of accretive operators and contractive nonlinear semigroups works well in the scalar case [C], but does not apply to systems. For the above reasons, the theory of hyperbolic conservation laws has largely developed by *ad hoc* methods, along two main lines.

1. The BV setting, considered by Glimm [G]. Solutions are here constructed within a space of functions with bounded variation, controlling the BV norm by a wave interaction functional.

2. The $\mathbf{L}^\infty$ setting, introduced by Tartar and DiPerna [DP2], based on weak convergence and a compensated compactness argument.

Both approaches yield results on the global existence of weak solutions. However, the method of compensated compactness appears to be suitable only for 2×2 systems. Moreover, it is only in the BV setting that the well-posedness of the Cauchy problem could recently be proved, as well as the stability and convergence of vanishing viscosity approximations. In the remainder of this paper we thus concentrate on the theory of BV solutions, referring to [DP2] or [Se] for the alternative approach based on compensated compactness.

Since the pioneering work of Glimm, the basic building block toward the construction and the analysis of more general solutions has been provided by the ***Riemann problem***, i.e. the initial value problem with piecewise constant data

$$u(0,x) = \bar{u}(x) = \begin{cases} u^- & if \quad x < 0, \\ u^+ & if \quad x > 0. \end{cases} \tag{1.7}$$

This was first introduced by B. Riemann (1860) in the context of isentropic gas dynamics. A century later, P. Lax [Lx] solved the Riemann problem for a general class of $n \times n$ systems. Solutions are self-similar, having the form

$$u(t,x) = U(x/t).$$

The central position taken by the Riemann problem is related to a symmetry of the equations (1.1). If $u = u(t,x)$ is a solution of (1.1), then for any $\theta > 0$ the function

$$u^\theta(t,x) \doteq u(\theta t, \theta x)$$

provides another solution. The solutions which are invariant under these rescalings of the independent variables are precisely those which correspond to some Riemann data.

Both the Glimm scheme and the method of front tracking yield approximate solutions of a general Cauchy problem by piecing together a large number of Riemann solutions. The approach is successful because one can provide an a priori bound on the total amount of new waves produced by nonlinear interactions, and hence on the total variation of the solution. It is safe to say that, in the context of weak solutions with small total variation, nearly all results on the existence, uniqueness, continuous dependence and qualitative behavior have relied on a careful analysis of the Riemann problem.

Very recently, in [BB] a substantially different perspective has emerged from the study of vanishing viscosity approximations. Solutions of (1.1) are sought as limits for $\varepsilon \to 0$ of solutions to the parabolic problems

$$u^\varepsilon_t + A(u^\varepsilon)u^\varepsilon_x = \varepsilon u^\varepsilon_{xx}\,. \tag{1.8}$$

Here $A(u) \doteq Df(u)$. This approach is very natural, and has been considered since the 1950's. However, it was only in the scalar case [O], [K], that complete results could be obtained. For general $n \times n$ systems, the main difficulty lies in establishing the compactness of the approximating sequence. We observe that $u^\varepsilon(t,x)$ solves (1.8) if and only if $u^\varepsilon(t,x) = u(t/\varepsilon,\ x/\varepsilon)$ for some function u which satisfies

$$u_t + A(u)u_x = u_{xx}\,. \tag{1.9}$$

In the analysis of vanishing viscosity approximations, the key step is to derive a priori estimates on the total variation and on the stability of solutions of (1.9). For this parabolic system, the rescaling $(t,x) \mapsto (\theta t,\ \theta x)$ no longer determines a symmetry. Hence the Riemann data no longer hold a privileged position. The role of basic building block is now taken by the *viscous travelling profiles*, i.e. the solutions of the form

$$u(t,x) = U(x - \lambda t).$$

Of course, the function U must then satisfy the second order O.D.E.

$$U'' = \big(A(U) - \lambda\big)U'.$$

The new point of view consists in decomposing the gradient u_x locally as a sum of gradients of viscous travelling waves. More precisely, let a smooth function $u : \mathbb{R} \mapsto \mathbb{R}^n$ be given. At each point x, looking at the third order jet (u, u_x, u_{xx}) we seek travelling profiles $U_1, \ldots, U_n$ such that

$$U_i'' = \big(A(U_i) - \sigma_i\big)U_i' \tag{1.10}$$

for some speed σ_i close to the characteristic speed λ_i, and moreover

$$U_i(x) = u(x) \qquad\qquad i = 1, \ldots, n, \tag{1.11}$$

$$\sum_i U_i'(x) = u_x(x)\,, \qquad\qquad \sum_i U_i''(x) = u_{xx}(x)\,. \tag{1.12}$$

It turns out that this decomposition is unique provided that the travelling profiles are chosen within suitable center manifolds. We let $\tilde{r}_i$ be the unit vector parallel to U_i', so that $U_i' = v_i\tilde{r}_i$ for some scalar v_i. One can show that $\tilde{r}_i$ remains close to the eigenvector $r_i(u)$ of the Jacobian matrix $A(u) \doteq Df(u)$, but $\tilde{r}_i \neq r_i(u)$ in general. The first equation in (1.12) now yields the decomposition

$$u_x = \sum_i v_i\tilde{r}_i\,. \tag{1.13}$$

If $u = u(t,x)$ is a solution of (1.9), we can think of v_i as the ***density of i-waves*** in u. The remarkable fact is that these components satisfy a system of evolution equations

$$v_{i,t} + (\tilde{\lambda}_i v_i)_x - v_{i,xx} = \phi_i \qquad\qquad i = 1,\ldots,n\,, \tag{1.14}$$

where the source terms ϕ_i on the right hand side are INTEGRABLE over the whole domain $\{x \in \mathbb{R}\,,\ t > 0\}$. Indeed, we can think of the sources ϕ_i as the new waves produced by interactions between viscous waves. Their total strength is controlled by means of ***viscous interaction functionals***, somewhat similar to the one introduced by Glimm in the hyperbolic case. Since the left hand side of (1.14) is in conservation form and the vectors $\tilde{r}_i$ have unit length, we obtain the bound

$$\big\|u_x(\tau)\big\|_{\mathbf{L}^1} \leq \sum_i \big\|v_i(\tau)\big\|_{\mathbf{L}^1} \leq \sum_i \left(\big\|v_i(0)\big\| + \int_0^T\!\!\int \big|\phi_i(t,x)\big|\,dxdt\right). \tag{1.15}$$

This line of argument yields global BV bounds and stability estimates for viscous solutions. In turn, letting $\varepsilon \to 0$ in (1.8), a standard compactness argument yields the convergence of u^ε to a weak solution u of (1.1).

The plan of these notes is as follows. In Section 2 we review some basic facts about shock and rarefaction waves, and describe the solution of the Riemann problem. Section 3 is concerned with solutions of the more general Cauchy problem, obtained as limits of Glimm or front tracking approximations. Stability and uniqueness issues are discussed in the next two sections. The remaining sections 6 to 8 describe the vanishing viscosity approach, based on decomposition along gradients of viscous travelling waves and the new Lyapunov functionals controlling the interactions among viscous waves. Some further research directions are discussed in Section 9.

2. The Riemann problem

Here and in the next section, we shall consider a strictly hyperbolic system of conservation laws satisfying the additional hypothesis

(H) For each $i = 1,\ldots,n$, the i-th field is either ***genuinely nonlinear***, so that $D\lambda_i(u)\cdot r_i(u) > 0$ for all u, or ***linearly degenerate***, with $D\lambda_i(u)\cdot r_i(u) = 0$ for all u.

Notice that, in the genuinely nonlinear case, the i-th eigenvalue λ_i is required to be strictly increasing along each integral curve of the corresponding field of eigenvectors r_i. In the linearly degenerate case, on the other hand, the eigenvalue λ_i is constant along each such curve.

Shocks and Admissibility Conditions.

The simplest type of discontinuous solution is given by a *shock* :

$$U(t,x) = \begin{cases} u^+ & if \quad x > \lambda t, \\ u^- & if \quad x < \lambda t, \end{cases} \tag{2.1}$$

for some left and right states $u^-, u^+ \in I\!R^n$ and a speed $\lambda \in I\!R$. Using the divergence theorem, one checks that the identity (1.4) is satisfied if and only if the following Rankine-Hugoniot conditions hold:

$$\lambda\,(u^+ - u^-) = f(u^+) - f(u^-)\,. \tag{2.2}$$

Denote by $A(u) = Df(u)$ the $n \times n$ Jacobian matrix of f at u. For any u, v, define the averaged matrix

$$A(u,v) \doteq \int_0^1 A\big(\theta u + (1-\theta)v\big)\; d\theta$$

and call $\lambda_i(u,v)$, $i = 1,\ldots,n$ its eigenvalues. We can then write (2.2) in the equivalent form

$$\lambda\,(u^+ - u^-) = \int_0^1 Df\big(\theta u^+ + (1-\theta)u^-\big)\cdot(u^+ - u^-)\; d\theta = A(u^+,u^-)\cdot(u^+ - u^-). \tag{2.3}$$

In other words, the Rankine-Hugoniot conditions hold if and only if the jump $u^+ - u^-$ is an eigenvector of the averaged matrix $A(u^+,u^-)$ and the speed λ coincides with the corresponding eigenvalue.

Definition 1. A function $u = u(t,x)$ has an *approximate jump discontinuity* at the point (τ,ξ) if there exists vectors $u^+ \neq u^-$ and a speed λ such that

$$\lim_{r\to 0+} \frac{1}{r^2}\int_{-r}^{r}\int_{-r}^{r} \big|u(\tau+t,\,\xi+x) - U(t,x)\big|\, dxdt = 0\,,$$

with U as in (2.1). We say that u is *approximately continuous* at the point (τ,ξ) if the above relations hold with $u^+ = u^-$ (and λ arbitrary).

More generally, if u is now a solution of the system of conservation laws (1.1) having an approximate jump, one can prove that the states u^-, u^+ and the speed λ again satisfy the Rankine-Hugoniot conditions.

In order to achieve uniqueness of solutions to initial value problems, it is customary to supplement the conservation equations (1.1) with additional admissibility conditions, to be satisfied at points of jump. We recall here two basic approaches.

Definition 2. A continuously differentiable function $\eta : I\!R^n \mapsto I\!R$ is called an *entropy* for the system of conservation laws (1.1), with *entropy flux* $q : I\!R^n \mapsto I\!R$, if it satisfies the identity

$$D\eta(u)\cdot Df(u) = Dq(u). \tag{2.4}$$

An immediate consequence of (2.4) is that, if $u = u(t,x)$ is a C^1 solution of (2.1), then

$$\eta(u)_t + q(u)_x = 0. \tag{2.5}$$

In other words, for a smooth solution u, not only the quantities $u_1, \ldots, u_n$ are conserved, but the additional conservation law (2.5) holds as well. However one should be aware that, for a discontinuous solution u, the quantity $\eta(u)$ may not be conserved.

A standard admissibility condition for weak solutions can now be formulated as follows.

Definition 3. Let η be a convex entropy for the system (1.1), with entropy flux q. A weak solution u is ***entropy-admissible*** if

$$\eta(u)_t + q(u)_x \leq 0 \tag{2.6}$$

in distribution sense, i.e.

$$\iint \{\eta(u)\varphi_t + q(u)\varphi_x\}\, dxdt \geq 0 \tag{2.7}$$

for every function $\varphi \geq 0$, continuously differentiable with compact support.

In analogy with (2.3), if u is an entropy admissible solution, at every point of approximate jump one can show that

$$\lambda\big[\eta(u^+) - \eta(u^-)\big] \geq q(u^+) - q(u^-). \tag{2.8}$$

An alternative admissibility condition, due to Lax [Lx], is particularly useful because it can be applied to any system and has a simple geometrical interpretation. We recall that, at a point of approximate jump, the speed λ must be an eigenvalue of the averaged matrix $A(u^-, u^+)$ at (2.3), i.e. $\lambda = \lambda_i(u^-, u^+)$ for some $i \in \{1, \ldots, n\}$.

Definition 4. A solution $u = u(t,x)$ of (1.1) satisfies the *Lax admissibility condition* if, at each point of approximate jump, the left and right states u^-, u^+ and the speed $\lambda = \lambda_i(u^-, u^+)$ of the jump satisfy

$$\lambda_i(u^-) \geq \lambda \geq \lambda_i(u^+). \tag{2.9}$$

To appreciate the geometric meaning of this condition, consider a piecewise smooth solution, having a discontinuity along the line $x = \gamma(t)$, where the solution jumps from a left state u^- to a right state u^+. According to (2.3), this discontinuity must travel with a speed $\lambda \doteq \dot{\gamma} = \lambda_i(u^-, u^+)$ equal to an eigenvalue of the averaged matrix $A(u^-, u^+)$. If we now look at the *i-characteristics*, i.e. at the solutions of the O.D.E.

$$\dot{x} = \lambda_i\big(u(t,x)\big),$$

we see that the Lax condition (2.9) requires that these lines run into the shock, from both sides.

Centered Rarefaction Waves.

Next, we consider another special type of solutions, in the form of *centered rarefaction waves*. Fix a state $u_0 \in \mathbb{R}^n$ and an index $i \in \{1, \ldots, n\}$. As before, let $r_i(u)$ be an i-eigenvector of the Jacobian matrix $A(u) = Df(u)$. The integral curve of the vector field r_i through the point u_0 is called the ***i-rarefaction curve*** through u_0. It is obtained by solving the Cauchy problem in state space:

$$\frac{du}{d\sigma} = r_i(u), \qquad u(0) = u_0. \tag{2.10}$$

We shall denote this curve as

$$\sigma \mapsto R_i(\sigma)(u_0).$$

Clearly, the parametrization depends on the choice of the eigenvectors r_i. In particular, if we impose the normalization $|r_i(u)| \equiv 1$, then the rarefaction curve R_i will be parametrized by arc-length.

Let the i-th field be genuinely nonlinear, and assume that u^+ lies on the positive i-rarefaction curve through u^-, i.e. $u^+ = R_i(\sigma)(u^-)$ for some $\sigma > 0$. For each $s \in [0, \sigma]$, define

$$\lambda_i(s) = \lambda_i\big(R_i(s)(u^-)\big).$$

Observe that, by genuine nonlinearity, the map $s \mapsto \lambda_i(s)$ is strictly increasing. Hence, for every $\lambda \in [\lambda_i(u^-),\ \lambda_i(u^+)]$, there is a unique value $s \in [0, \sigma]$ such that $\lambda = \lambda_i(s)$. We claim that, for $t > 0$, the function

$$u(t,x) = \begin{cases} u^- & if \quad x/t < \lambda_i(u^-), \\ R_i(s)(u^-) & if \quad x/t = \lambda_i(s) \in [\lambda_i(u^-),\ \lambda_i(u^+)], \\ u^+ & if \quad \mathrm{x/t} > \lambda_i(u^+), \end{cases} \tag{2.11}$$

is a continuous solution of the Riemann problem (1.1), (1.7). Indeed, by construction it follows

$$\lim_{t \to 0+} \|u(t,\cdot) - \bar{u}\|_{\mathbf{L}^1} = 0.$$

Moreover, the equation (1.3) is trivially satisfied in the sectors where $x/t < \lambda_i(u^-)$ or $x/t > \lambda_i(u^+)$, since here $u_t = u_x = 0$. Next, assume $x/t = \lambda_i(s)$ for some $s \in]0, \sigma[$. Since u is constant along each ray through the origin $\{x/t = c\}$, we have

$$u_t(t,x) + \frac{x}{t} u_x(t,x) = 0.$$

We now observe that the definition (2.11) implies $x/t = \lambda_i\big(u(t,x)\big)$. Moreover, u_x is parallel to $r_i(u)$, hence it is an eigenvector of the Jacobian matrix $A(u)$ with eigenvalue $\lambda_i(u)$. On the sector $\{\lambda_i(u^-) < x/t < \lambda_i(u^+)\}$ we thus have $u_t + A(u)u_x = 0$, proving our claim. Notice that the assumption $\sigma > 0$ is essential for the validity of this construction. In the opposite case $\sigma < 0$, the definition (2.11) would yield a triple-valued function in the region where $x/t \in \big]\lambda_i(u^+),\ \lambda_i(u^-)\big[$.

Solution of the Riemann Problem.

Relying on the previous analysis, the solution of the general Riemann problem

$$u_t + f(u)_x = 0, \qquad u(0,x) = \begin{cases} u^- & if \quad x < 0, \\ u^+ & if \quad x > 0. \end{cases} \tag{2.12}$$

can be obtained by finding intermediate states $\omega_0 = u^-, \omega_1, \ldots, \omega_n = u^+$ such that each pair of adiacent states ω_{i-1}, ω_i can be connected by an i-shock, or by a centered rarefaction i-wave. By the implicit function theorem, this can always be done

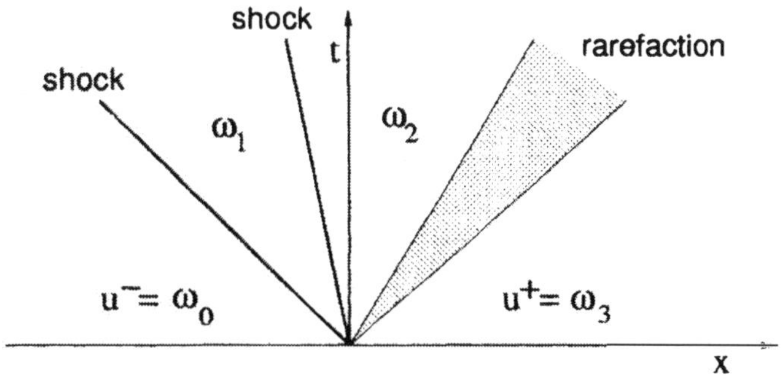

figure 3

provided that the two states u^-, u^+ are sufficiently close. The complete solution is now obtained by piecing together the solutions of the n Riemann problems

$$u_t + f(u)_x = 0, \qquad u(0,x) = \begin{cases} \omega_{i-1} & if \quad x < 0, \\ \omega_i & if \quad x > 0, \end{cases} \tag{2.13}$$

on different sectors of the t-x plane. A typical solution is illustrated in fig. 3.

3. Glimm and front tracking approximations

Approximate solutions to a more general Cauchy problem can be constructed by patching together solutions of several Riemann problems. In the Glimm scheme (fig. 4), one works with a fixed grid in the t-x plane, with mesh sizes Δt, Δx. At time $t = 0$ the initial data is approximated by a piecewise constant function, with jumps at grid points. Solving the corresponding Riemann problems, a solution is constructed up to a time Δt sufficiently small so that waves emerging from different nodes do not interact. At time $t_1 \doteq \Delta t$, we replace the solution $u(\Delta t, \cdot)$ by a piecewise constant function having jumps exactly at grid points. Solving the new Riemann problems at every one of these points, one can prolong the solution to the next time interval $[\Delta t,\ 2\Delta t]$. At time $t_2 \doteq 2\Delta t$, the solution is a gain approximated by a piecewise constant functions with jumps exactly at grid points, etc... A key ingredient of the Glimm scheme is the restarting procedure. At each time $t_j \doteq j\,\Delta t$, a natural way to approximate a BV function with a piecewise constant one is by taking its average value on each subinterval $J_i \doteq [x_{i-1}, x_i]$. However, this procedure may generate an arbitrarily large amount of oscillations. Instead, the Glimm scheme is based on random sampling: a point y_i is selected at random inside each interval J_i and the value $u(t_j-, y_i)$ is taken as the new value of $u(t_j, x)$ for all $x \in J_i$. An excellent introduction to the Glimm scheme can be found in the book by J. Smoller [Sm].

An alternative technique for contructing approximate solutions is by wave-front tracking (fig. 5). This method was introduced by Dafermos [D1] in the scalar case and later developed by various authors [DP1], [B1], [Ri], [BJ]. It now provides an

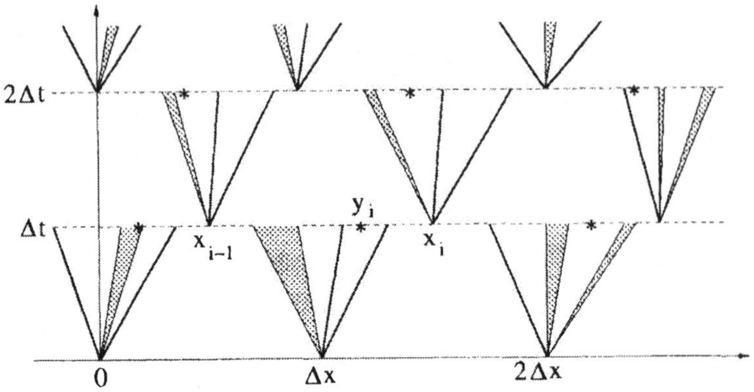

figure 4

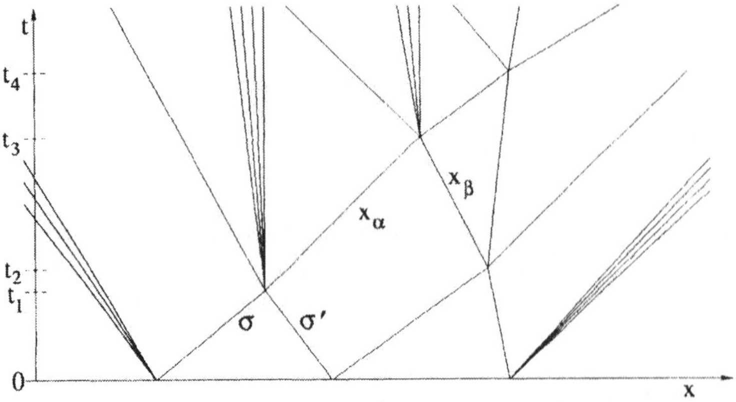

figure 5

efficient tool in the study of general $n \times n$ systems of conservation laws, both for theoretical and numerical purposes [B3], [HR].

The initial data is here approximated with a piecewise constant function, and each Riemann problem is solved approximately, within the class of piecewise constant functions. In particular, if the exact solution contains a centered rarefaction, this must be approximated by a *rarefaction fan*, containing several small jumps. At the first time t_1 where two fronts interact, the new Riemann problem is again approximately solved by a piecewise constant function. The solution is then prolonged up to the second interaction time t_2, where the new Riemann problem is solved, etc... The main difference is that in the Glimm scheme one specifies a priori the nodal points where the the Riemann problems are to be solved. On the other hand, in a solution constructed by wave-front tracking the locations of the jumps and of the interaction points depend on the solution itself. No restarting procedure is needed and the map $t \mapsto u(t, \cdot)$ is thus continuous with values in $\mathbf{L}^1_{loc}$.

In the end, both algorithms produce a sequence of approximate solutions, whose convergence relies on a compactness argument based on uniform bounds on the total variation. We sketch the main idea involved in these a priori BV bounds. Consider a piecewise constant function $u : \mathbb{R} \mapsto \mathbb{R}^n$, say with jumps at points $x_1 < x_2 < \cdots < x_N$. Call σ_α the amplitude of the jump at x_α. The ***total strength of waves*** is then defined as

$$V(u) \doteq \sum_\alpha |\sigma_\alpha|, \tag{3.1}$$

Clearly, this is an equivalent way to measure the total variation. Along a solution $u = u(t, x)$ constructed by front tracking, the quantity $V(t) = V(u(t, \cdot))$ may well increase at interaction times. To provide global a priori bounds, following [G] one introduces a ***wave interaction potential***, defined as

$$Q(u) = \sum_{(\alpha,\beta)\in\mathcal{A}} |\sigma_\alpha \sigma_\beta|, \tag{3.2}$$

where the summation runs over the set $\mathcal{A}$ of all couples of approaching waves. Roughly speaking, we say that two wave-fronts located at $x_\alpha < x_\beta$ are ***approaching*** if the one at x_α has a faster speed than the one at x_β (hence the two fronts are expected to collide at a future time). Now consider a time τ where two incoming wave-fronts interact, say with strengths σ, σ' (for example, take $\tau = t_1$ in fig. 5). The difference between the outgoing waves emerging from the interaction and the two incoming waves σ, σ' is of magnitude $\mathcal{O}(1) \cdot |\sigma\sigma'|$. On the other hand, after time τ the two incoming waves are no longer approaching. This accounts for the decrease of the functional Q in (3.2) by the amount $|\sigma\sigma'|$. Observing that the new waves generated by the interaction could approach all other fronts, the change in the functionals V, Q across the interaction time τ is estimated as

$$\Delta V(\tau) = \mathcal{O}(1) \cdot |\sigma\sigma'|, \qquad \Delta Q(\tau) = -|\sigma\sigma'| + \mathcal{O}(1) \cdot |\sigma\sigma'|\, V(\tau-).$$

If the initial data has small total variation, for a suitable constant C_0 the quantity

$$\Upsilon(t) \doteq V(u(t,\cdot)) + C_0\, Q(u(t,\cdot))$$

is monotone decreasing in time. This argument provides the uniform BV bounds on all approximate solutions. Using Helly's compactness theorem, one obtains the convergence of a subsequence of approximate solutions, and hence the global existence of a weak solution.

Theorem 1. *Let the system (1.1) be strictly hyperbolic and satisfy the assumptions (H). Then, for a sufficiently small $\delta > 0$ the following holds. For every initial condition $\bar u$ with*

$$\|\bar u\|_{\mathbf{L}^\infty} < \delta, \qquad \text{Tot. Var.}\{\bar u\} < \delta, \tag{3.3}$$

the Cauchy problem has a weak solution, defined for all times $t \geq 0$.

The proof is based on careful analysis of solutions of the Riemann problem and on the use of a quadratic interaction functional (3.2) to control the creation of new waves. These techniques also provided the basis for subsequent investigations of Glimm and Lax [GL] and Liu [L2], [L3] on the asymptotic behavior of weak solutions as $t \to \infty$.

4. Stability

The previous existence result relied on a compactness argument which, by itself, does not provide informations on the uniqueness of solutions. A first understanding of the dependence of weak solutions on the initial data was provided by the analysis of front tracking approximations. The idea is to perturb the initial data by shifting the position of one of the jumps, say from x to a nearby point x'. By carefully estimating the corresponding shifts in the positions of all wave-fronts at a later time t, one obtains a bound on the $\mathbf{L}^1$ distance between the original and the perturbed approximate solution. After much technical work, this approach yielded a proof of the Lipschitz continuous dependence of solutions on the initial data, first in [BC1] for 2×2 systems, then in [BCP] for general $n \times n$ systems.

Theorem 2. *Let the system (1.1) be strictly hyperbolic and satisfy the assumptions (H). Then, for every initial data $\bar{u}$ satisfying (3.3) the weak solution obtained as limit of front tracking approximations is unique and depends Lipschitz continuously on the initial data, in the $\mathbf{L}^1$ distance.*

These weak solutions can thus be written in the form $u(t,\cdot) = S_t\bar{u}$, as trajectories of a semigroup $S : \mathcal{D} \times [0,\infty[\mapsto \mathcal{D}$ on some domain $\mathcal{D}$ containing all functions with sufficiently small total variation. For some Lipschitz constants L, L' one has

$$\left\|S_t\bar{u} - S_s\bar{v}\right\|_{\mathbf{L}^1} \le L\,\|\bar{u} - \bar{v}\|_{\mathbf{L}^1} + L'|t-s|\,, \tag{4.1}$$

for all $t, s \ge 0$ and initial data $\bar{u}, \bar{v} \in \mathcal{D}$.

A more transparent proof of Theorem 2 was later achieved by a technique introduced by Liu and Yang in [LY] and developed in [BLY] into its final form. The heart of the matter is to construct a nonlinear functional, equivalent to the $\mathbf{L}^1$ distance, which is decreasing in time along every pair of solutions. We thus seek $\Phi = \Phi(u,v)$ and a constant C such that

$$\frac{1}{C}\cdot\left\|v-u\right\|_{\mathbf{L}^1} \le \Phi(u,v) \le C\cdot\left\|v-u\right\|_{\mathbf{L}^1}\,, \tag{4.2}$$

$$\frac{d}{dt}\Phi\big(u(t),\ v(t)\big) \le 0. \tag{4.3}$$

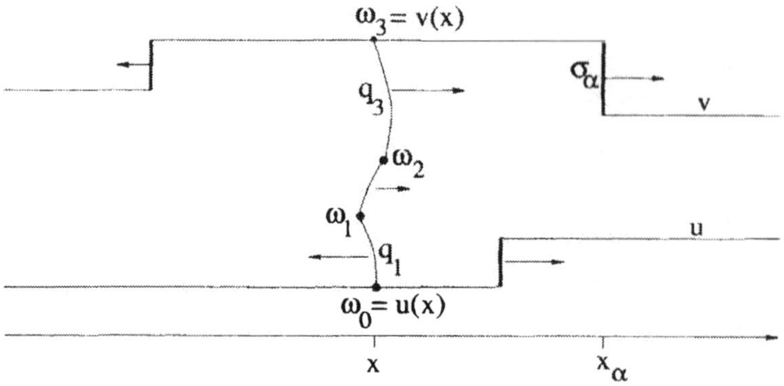

figure 6

Throughout the following, it will be convenient to measure the strength of wave-fronts of a genuinely nonlinear family in terms of the change in the characteristic speed. If u has a jump in the k-th family at the point x_α, in the genuinely nonlinear case we thus define the strength as

$$\sigma_\alpha \doteq \lambda_k\big(u(x_\alpha+)\big) - \lambda_k\big(u(x_\alpha-)\big). \tag{4.4}$$

In connection with piecewise constant functions $u, v : \mathbb{R} \mapsto \mathbb{R}^n$ generated by a front tracking algorithm, the functional Φ can be defined as follows (fig. 6). At each point x, we connect the states $u(x)$, $v(x)$ by means of n shock curves. In other words, we construct intermediate states $\omega_0 = u(x), \omega_1, \ldots, \omega_n = v(x)$ such that each pair ω_{i-1}, ω_i is connected by an i-shock. These states can be uniquely determined by the implicit function theorem. Call $q_1, \ldots, q_n$, the strengths of these shocks. We regard $q_i(x)$ as the i-th scalar component of the jump $\big(u(x),\ v(x)\big)$. For some constant C', one clearly has

$$\frac{1}{C'} \cdot \big|v(x) - u(x)\big| \le \sum_{i=1}^{n} \big|q_i(x)\big| \le C' \cdot \big|v(x) - u(x)\big|. \tag{4.5}$$

The functional Φ is now defined as

$$\Phi(u, v) \doteq \sum_{i=1}^{n} \int_{-\infty}^{\infty} W_i(x)\, \big|q_i(x)\big|\, dx, \tag{4.6}$$

where the weights W_i take the form

$$\begin{aligned} W_i(x) &\doteq 1 \quad +\kappa_1 \cdot \big[\text{total strength of waves in } u \text{ and in } v \text{ which approach the } i\text{-wave } q_i(x)\big] \\ &\qquad\quad +\kappa_2 \cdot \big[\text{wave interaction potentials of } u \text{ and of } v\big] \\ &\doteq 1 \quad +\kappa_1 V_i(x) + \kappa_2\big[Q(u) + Q(v)\big] \end{aligned} \tag{4.7}$$

for suitable constants κ_1, κ_2. Notice that, by construction, $q_i(x)$ represents the strength of a fictitious shock wave located at x, travelling with a speed $\lambda_i(x)$ determined by the Rankine-Hugoniot equations. In (4.7), it is thus meaningful to consider the quantity

$$V_i(x) \doteq \sum_{\alpha \in \mathcal{A}_i(x)} |\sigma_\alpha|,$$

where the summation extends to all wave-fronts σ_α in u and in v which are *approaching* the i-shock $q_i(x)$. By (4.5) and the boundedness of the weights W_i, it is clear that (4.2) holds.

Next, we will sketch an argument showing why the functional Φ should decrease along pairs of solutions. For a detailed proof, see [BLY] or [B3]. We first observe that the maps $t \mapsto q_i(t, \cdot)$ are Lipschitz continuous as maps with values into $\mathbf{L}^1$. At a time τ where an interaction occurs in u or in v, if the coefficient κ_2 is sufficiently large the decrease of the functionals $Q(u)$ or $Q(v)$ implies

$$W_i(\tau+,\ x) < W_i(\tau-,\ x),$$

for every $i \in \{1, \ldots, n\}$ and almost every $x \in \mathbb{R}$. Hence Φ is decreasing at each interaction time.

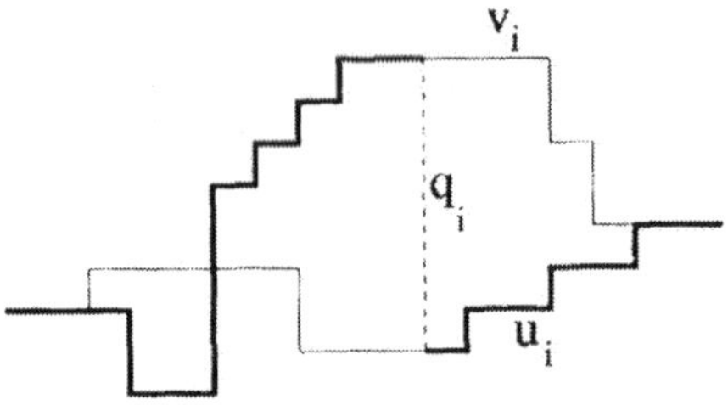

figure 7

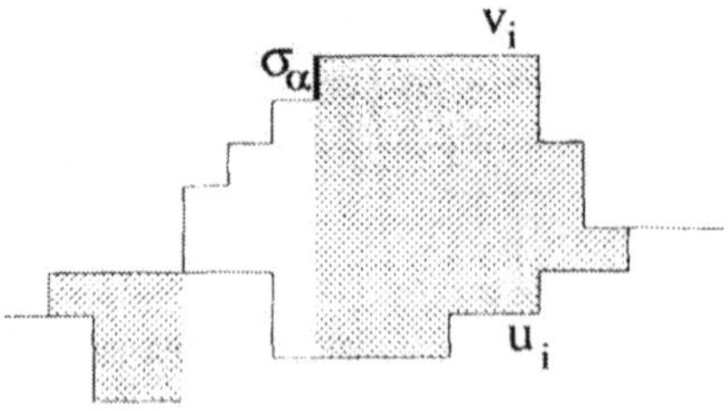

figure 8

Interchanging the order of the summation and the integral, the functional Φ in (4.6)-(4.7) can be written in the equivalent form

$$\begin{aligned}\Phi(u,v) &= \big[1+\kappa_2 Q(u)+\kappa_2 Q(v)\big]\cdot\textstyle\sum_{i=1}^{n}\int_{-\infty}^{\infty}\big|q_i(x)\big|\,dx \\ &\quad+\kappa_1\cdot\textstyle\sum_{\sigma_\alpha\in\mathcal{J}(u)\cup\mathcal{J}(v)}|\sigma_\alpha|\cdot\sum_i\int_{q_i(x)\text{ approaches }\sigma_\alpha}\big|q_i(x)\big|\,dx\,.\end{aligned}\tag{4.8}$$

For our further analysis, the following construction will be useful. Given two piecewise constant functions u,v with compact support, for $i=1,\ldots,n$ we can define the scalar components u_i,v_i by induction on the jump points of u,v. We start by setting $u_i(-\infty)=v_i(-\infty)=0$. If $x_\alpha\in\mathcal{J}(u)$ is a jump point of u, then we let v_i be constant across x_α and set

$$u_i(x_\alpha+)\doteq u_i(x_\alpha-)-\big[q_i(x_\alpha+)-q_i(x_\alpha-)\big].$$

On the other hand, if $x_\alpha\in\mathcal{J}(v)$ is a jump point of v, then we let u_i be constant across x_α and set

$$v_i(x_\alpha+)\doteq v_i(x_\alpha-)+\big[q_i(x_\alpha+)-q_i(x_\alpha-)\big].$$

These definitions trivially imply

$$q_i(x)=v_i(x)-u_i(x)\qquad\text{for all } x\in\mathbb{R},\ i=1,\ldots,n.$$

By definition, the i-waves in u and v which approach $q_i(x)$ are those located within the thick portions of the graphs of u_i, v_i in fig. 7. Viceversa, for a given i-wave σ_α located at x_α, the regions where the jumps $q_i(x)$ approach σ_α are represented by the shaded areas in in fig. 8.

To understand why the expression (4.8) decreases in time, assume first that v has a wave-front at x_α with strength σ_α, say in the k-th family. In connection with this front (fig. 9), for every $i<k$ the functional $\Phi(u,v)$ contains a term of the form

$$\Phi_{\alpha,i}\doteq\kappa_1\cdot|\sigma_\alpha|\cdot\big[\text{area of the region between the graphs of } u_i \text{ and } v_i\text{, to the right of } x_\alpha\big].$$

By strict hyperbolicity, the i-th and k-th characteristic speeds are strictly separated, say $\lambda_k-\lambda_i\geq c>0$. We thus expect

$$\frac{d\Phi_{\alpha,i}}{dt}\approx-\kappa_1|\sigma_\alpha|\big|q_i^\alpha\big|(\dot{x}_\alpha-\lambda_i^\alpha)\leq-c\kappa_1|\sigma_\alpha|\big|q_i^\alpha\big|,\tag{4.9}$$

up to higher order terms. Here $\lambda_i^\alpha\doteq\lambda_i\big(\omega_{i-1}(x_\alpha),\ \omega_i(x_\alpha)\big)$ is the speed of the i-shock q_i^α. In addition, for every $i>k$ the functional $\Phi(u,v)$ contains a term of the form

$$\Phi_{\alpha,i}\doteq\kappa_1\cdot|\sigma_\alpha|\cdot\big[\text{area of the region between the graphs of } u_i \text{ and } v_i\text{, to the left of } x_\alpha\big].$$

Entirely similar estimates can be proved also for these terms.

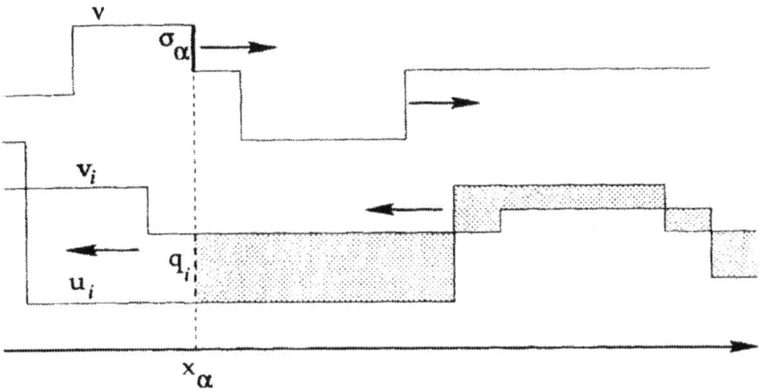

figure 9

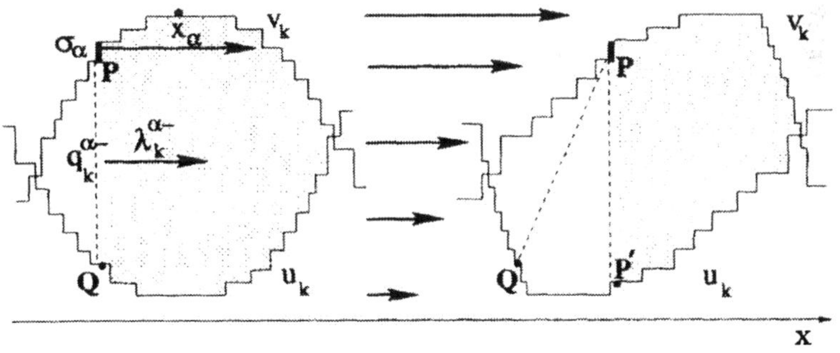

figure 10

Next, if the k-th field is genuinely nonlinear, the functional $\Phi(u,v)$ also contains a term of the form (fig. 10)

$$\Phi_{\alpha,k} \doteq \kappa_1 \cdot |\sigma_\alpha| \cdot \big[\text{area of the region between the graphs of } u_k \text{ and } v_k, \text{ to the right of } x_\alpha\big].$$

Notice that, because of genuine nonlinearity, the points on the graphs of u_k and v_k move with different speeds. As a consequence, the shape of these graphs changes in time. In particular, while the area enclosed by the two graphs may remain constant, the portion of this area located to the right of a given front σ_α will decrease. In figure 10, the two points P, Q lie initially on the same vertical line. At a later time this is no longer true. The area of the shaded region, enclosed by the graphs of u_k and v_k and by a vertical line through P, has decreased by an amount roughly equal to the area of the triangular region with vertices P, P', Q.

The difference between the speed $\dot{x}_\alpha$ of the jump σ_α at P and the average speed $\lambda_k^{\alpha-}$ of points between P and Q is estimated by

$$\dot{x}_\alpha - \lambda_k^{\alpha-} \approx \frac{1}{2}\left|q_k^{\alpha-} + \sigma_\alpha\right|.$$

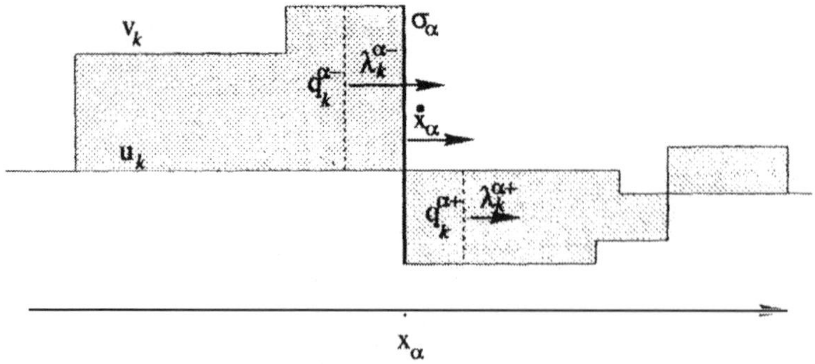

figure 11

Hence, up to higher order terms, the time derivative of $\Phi_{\alpha,k}$ satisfies

$$\frac{d\Phi_{\alpha,k}}{dt} \approx -\kappa_1 |\sigma_\alpha| |q_k^{\alpha-}| (\dot{x}_\alpha - \lambda_k^{\alpha-}) \approx -\kappa_1 |\sigma_\alpha| |q_k^{\alpha-}| \cdot \frac{1}{2} |q_k^{\alpha-} + \sigma_\alpha|. \tag{4.10}$$

Yet another kind of estimate is needed in the case where the jump in v_k crosses the graph of u_k, say $v_k(x_\alpha+) < u_k(x_\alpha) < v_k(x_\alpha-)$. In this case, the estimates (4.9) remain valid. In connection with the k-th field, the functional Φ contains a term of the form

$$\Phi_k^* \doteq 1 \cdot \big[\text{area of the region between the graphs of } u_k \text{ and } v_k\big],$$

where the above area includes points both on the right and on the left of x_α.

For convenience, call $\bar{\lambda}_k \doteq \lambda_k\big(v(x_\alpha+)\big)$. Since $q_k^{\alpha+} \approx q_k^{\alpha-} + \sigma_\alpha$, the genuine nonlinearity of the k-th characteristic field and the parametrization (4.4) yield

$$\lambda_k^{\alpha-} \approx \bar{\lambda}_k + |q_k^{\alpha+}| + \frac{1}{2}|q_k^{\alpha-}| \qquad \lambda_k^{\alpha+} \approx \bar{\lambda}_k + \frac{1}{2}|q_k^{\alpha+}|,$$

$$\dot{x}_\alpha \approx \bar{\lambda}_k + \frac{1}{2}\big(|q_k^{\alpha+}| + |q_k^{\alpha-}|\big).$$

Neglegting higher order terms, the shaded area in fig. 11 thus decreases at the rate

$$\frac{d\Phi_k^*}{dt} \approx |q_k^{\alpha-}|(\lambda_k^{\alpha-} - \dot{x}_\alpha) + |q_k^{\alpha+}|(\dot{x}_\alpha - \lambda_k^{\alpha+}) \approx |q_k^{\alpha-}||q_k^{\alpha+}| \approx |q_k^{\alpha-}|\,|\sigma_\alpha + q_k^{\alpha-}|. \tag{4.11}$$

Notice that in this last case the decrease of the functional Φ is simply due to the decrease of the $\mathbf{L}^1$ distance between the k-th components of u and v. In this occasion, the weights W_k play no role. For all details we always refer to [B3].

For general $n \times n$ systems, in (4.1) one finds a Lipschitz constant $L > 1$. Indeed, it is only in the scalar case that the semigroup is contractive and the theory of accretive operators and abstract evolution equations in Banach spaces can be applied, see [K], [C]. We refer to the flow generated by a system of conservation laws as a *Riemann semigroup*, because it is entirely determined by specifying how Riemann problems are solved. As proved in [B2], if two semigroups S, S' yield the same solutions to all Riemann problems, then they coincide, up to the choice of their domains.

From (4.1) one can deduce the error bound

$$\|w(T) - S_T w(0)\|_{\mathbf{L}^1} \le L \cdot \int_0^T \left\{ \liminf_{h \to 0+} \frac{\|w(t+h) - S_h w(t)\|_{\mathbf{L}^1}}{h} \right\} dt, \tag{4.12}$$

valid for every Lipschitz continuous map $w : [0,T] \mapsto \mathcal{D}$ taking values inside the domain of the semigroup. We can think of $t \mapsto w(t)$ as an approximate solution of (1.1), while $t \mapsto S_t w(0)$ is the exact solution having the same initial data. According to (4.12), the distance at time T is bounded by the integral of an *instantaneous error rate*, amplified by the Lipschitz constant L of the semigroup.

Using (4.12), one can estimate the distance between a front tracking approximation and the corresponding exact solution. For approximate solutions constructed by the Glimm scheme, a direct application of this same formula is not possible because of the additional errors introduced by the restarting procedures at times $t_k \doteq k\,\Delta t$. However, relying on a careful analysis of Liu [L4], one can construct a front tracking approximate solution having the same initial and terminal values as the Glimm solution. By this technique, in [BM] the authors proved the estimate

$$\lim_{\Delta x \to 0} \frac{\|u^{\text{Glimm}}(\tau,\cdot) - u^{\text{exact}}(\tau,\cdot)\|_{\mathbf{L}^1}}{\sqrt{\Delta x} \cdot |\ln \Delta x|} = 0 \tag{4.13}$$

for every $\tau > 0$. In other words, letting the mesh sizes $\Delta x, \Delta t \to 0$ while keeping their ratio $\Delta x/\Delta t$ constant, the $\mathbf{L}^1$ norm of the error in the Glimm approximate solution tends to zero at a rate slightly slower than $\sqrt{\Delta x}$.

5. Uniqueness

The uniqueness and stability results stated in Theorem 2 refer to a special class of weak solutions: those obtained as limits of Glimm or front tracking approximations. For several applications, it is desirable to have a uniqueness theorem valid for general weak solutions, without reference to any particular constructive procedure. Results in this direction were proved in, [BLF], [BG], [BLw]. They are all based on the error formula (4.12). In the proofs, one considers a weak solution $u = u(t,x)$ of the Cauchy problem

$$u_t + f(u)_x = 0, \qquad u(0,x) = \bar{u}(x). \tag{5.1}$$

Assuming that u satisfies suitable entropy and regularity conditions, one shows that

$$\liminf_{h \to 0+} \frac{\|u(t+h) - S_h u(t)\|_{\mathbf{L}^1}}{h} = 0 \tag{5.2}$$

at almost every time t. By (4.12), u thus coincides with the semigroup trajectory $t \mapsto S_t u(0) = S_t \bar{u}$. Of course, this implies uniqueness. As an example, we state below the result of [BLw]. Consider the following assumptions:

(A1) (Conservation Equations) The function $u = u(t,x)$ is a weak solution of the Cauchy problem (5.1) taking values within the domain $\mathcal{D}$ of the semigroup S.

More precisely, $u : [0,T] \mapsto \mathcal{D}$ is continuous w.r.t. the $\mathbf{L}^1$ distance. The initial condition $u(0,x) = \bar{u}(x)$ holds, together with

$$\iint \left[u\,\phi_t + f(u)\,\phi_x\right] dxdt = 0$$

for every $\mathcal{C}^1$ function ϕ with compact support contained inside the open strip $]0,T[\times\mathbb{R}$.

(A2) (Lax Entropy Condition) Let u have an approximate jump discontinuity at some point $(\tau,\xi) \in]0,T[\times\mathbb{R}$. In other words, assume that there exists states $u^-, u^+ \in \Omega$ and a speed $\lambda \in \mathbb{R}$ such that, calling

$$U(t,x) \doteq \begin{cases} u^- & if \quad x < \xi + \lambda(t-\tau), \\ u^+ & if \quad x > \xi + \lambda(t-\tau), \end{cases} \tag{5.3}$$

there holds

$$\lim_{\rho\to 0+} \frac{1}{\rho^2} \int_{\tau-\rho}^{\tau+\rho} \int_{\xi-\rho}^{\xi+\rho} \left|u(t,x) - U(t,x)\right| dxdt = 0. \tag{5.4}$$

Then, for some $i \in \{1,\ldots,n\}$, one has the entropy inequality:

$$\lambda_i(u^-) \geq \lambda \geq \lambda_i(u^+). \tag{5.5}$$

(A3) (Bounded Variation Condition) For some small constant $\delta > 0$, the function $x \mapsto u\big(\tau(x),x\big)$ has bounded variation along every Lipschitz continuous space-like curve $\big\{t = \tau(x)\big\}$, such that $\big|\tau(x) - \tau(x')\big| \leq \delta|x - x'|$ for all x, x'.

Theorem 3. *Let $u = u(t,x)$ be a weak solution of the Cauchy problem (5.1), satisfying the assumptions (A1), (A2) and (A3). Then*

$$u(t,\cdot) = S_t\bar{u} \tag{5.6}$$

for all t. In particular, the solution that satisfies the three above conditions is unique.

An additional characterization of these solutions, based on local integral estimates, was given in [B2]. The underlying idea is as follows. In a forward neighborhood of a point (τ,ξ) where u has a jump, the weak solution u should behave much in the same way as the solution of the corresponding Riemann problem. On the other hand, on a region where its total variation is small, our solution u should be accurately approximated by the solution of a linear hyperbolic system with constant coefficients.

To state the result more precisely, we introduce some notations. Given a function $u = u(t,x)$ and a point (τ,ξ), we denote by $U^\sharp_{(u;\tau,\xi)}$ the solution of the Riemann problem with initial data

$$u^- = \lim_{x\to\xi-} u(\tau,x), \qquad u^+ = \lim_{x\to\xi+} u(\tau,x). \tag{5.7}$$

In addition, we define $U^\flat_{(u;\tau,\xi)}$ as the solution of the linear hyperbolic Cauchy problem with constant coefficients

$$w_t + \widehat{A}w_x = 0, \qquad w(0,x) = u(\tau,x). \tag{5.8}$$

Here $\widehat{A} \doteq A(u(\tau,\xi))$. Observe that (5.8) is obtained from the quasilinear system

$$u_t + A(u)u_x = 0 \qquad (A = Df) \tag{5.9}$$

by "freezing" the coefficients of the matrix $A(u)$ at the point (τ,ξ) and choosing $u(\tau)$ as initial data. A new notion of "good solution" can now be introduced, by locally comparing a function u with the self-similar solution of a Riemann problem and with the solution of a linear hyperbolic system with constant coefficients. More precisely, we say that a function $u = u(t,x)$ is a **viscosity solution** of the system (1.1) if $t \mapsto u(t,\cdot)$ is continuous as a map with values into $\mathbf{L}^1_{loc}$, and moreover the following integral estimates hold.

(i) At every point (τ,ξ), for every $\beta' > 0$ one has

$$\lim_{h\to 0+} \frac{1}{h}\int_{\xi-\beta' h}^{\xi+\beta' h} \Big| u(\tau+h,\, x) - U^\sharp_{(u;\tau,\xi)}(h,\, x-\xi)\Big|\, dx = 0. \tag{5.10}$$

(ii) There exist constants $C,\beta > 0$ such that, for every $\tau \geq 0$ and $a < \xi < b$, one has

$$\limsup_{h\to 0+} \frac{1}{h}\int_{a+\beta h}^{b-\beta h} \Big| u(\tau+h,\, x) - U^b_{(u;\tau,\xi)}(h,x)\Big|\, dx \leq C\cdot \Big(\text{Tot.Var.}\{u(\tau);\ ]a,\, b[\ \}\Big)^2. \tag{5.11}$$

As proved in [B2], this concept of viscosity solution completely characterizes semigroup trajectories.

Theorem 4. *Let $S : \mathcal{D}\times[0,\infty[\times\mathcal{D}$ be a semigroup generated by the system of conservation laws (1.1). A function $u : [0,T] \mapsto \mathcal{D}$ is a viscosity solution of (1.1) if and only if $u(t) = S_t u(0)$ for all $t \in [0,T]$.*

6. Vanishing viscosity approximations

It is natural to expect that the entropy weak solutions of the hyperbolic system (1.1) actually coincide with the limits of solutions to the parabolic system

$$u^\varepsilon_t + f(u^\varepsilon)_x = \varepsilon\, u^\varepsilon_{xx} \tag{6.1}$$

letting the viscosity coefficient $\varepsilon \to 0$. For smooth solutions, this convergence result is easy to show. However, one should keep in mind that a weak solution of the hyperbolic system (1.1) in general is only a function with bounded variation, possibly with a countable number of discontinuities. In this case, as the smooth functions u^ε approach the discontinuous solution u, near points of jump their gradients u^ε_x must tend to infinity (fig. 12), while their second derivatives u^ε_{xx} become even more singular. Therefore, the convergence $u^\varepsilon \to u$ of the vanishing viscosity limit is a highly nontrivial matter.

In earlier literature, results in this direction relied on three different approaches:

1 - Comparison principles for parabolic equations. For a scalar conservation law, the existence, uniqueness and global stability of vanishing viscosity solutions

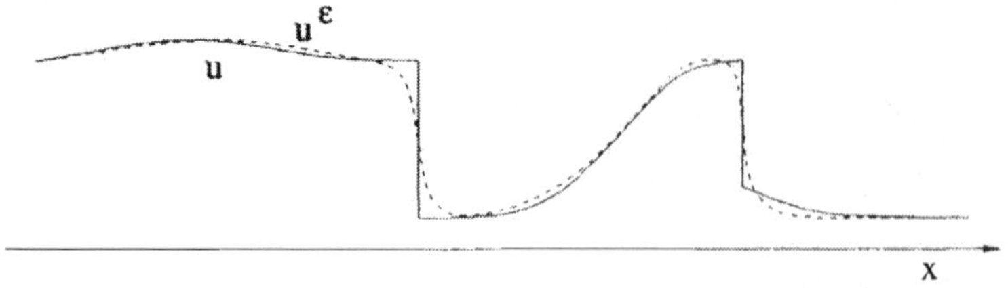

figure 12

was first established by Oleinik [O] in one space dimension. The famous paper by Kruzhkov [K] covers the more general class of $\mathbf{L}^\infty$ solutions and is also valid in several space dimensions.

2 - Singular perturbations. This technique was developed by Goodman and Xin [GX], and covers the case where the limit solution u is piecewise smooth, with a finite number of non-interacting, entropy admissible shocks. See also [Y] and [Ro], for further results in this direction.

3 - Compensated compactness. With this approach, introduced by Tartar and DiPerna [DP2], one first considers a weakly convergent subsequence $u^\varepsilon \rightharpoonup u$. For a class of 2×2 systems, one can show that this weak limit u actually provides a distributional solution to the nonlinear system (1.1). The proof relies on a compensated compactness argument, based on the representation of the weak limit in terms of Young measures, which must reduce to a Dirac mass due to the presence of a large family of entropies.

In view of recent results on the well-posedness of the hyperbolic Cauchy problem within the space of BV functions, it is natural to seek uniform BV bounds also on the viscous approximations u^ε in (1.3). This is indeed the main goal accomplished in [BB]. As soon as these BV bounds are established, the existence of a vanishing viscosity limit follows by a standard compactness argument. The uniqueness of the limit can then be deduced from the uniqueness theorem in [BG]. By further analysis, one can also prove the continuous dependence on the initial data for the viscous approximations u^ε, in the $\mathbf{L}^1$ norm. These results are valid, more generally, for hyperbolic systems not necessarily in conservation form. In this case, however, the unique limits of vanishing viscosity approximations cannot be interpreted as weak solutions in a distributional sense. Their characterization thus requires more care. The main results in [BB] can be stated as follows.

Theorem 5. *Consider the Cauchy problem for the hyperbolic system with viscosity*

$$u_t^\varepsilon + A(u^\varepsilon)u_x^\varepsilon = \varepsilon\, u_{xx}^\varepsilon \qquad\qquad u^\varepsilon(0,x) = \bar{u}(x)\,. \tag{6.2}$$

Assume that the matrices $A(u)$ are strictly hyperbolic, smoothly depending on u in a neighborhood of the origin. Then there exist constants C, L, L' and $\delta > 0$ such that the following holds. If

$$\|\bar{u}\|_{\mathbf{L}^\infty} < \delta\,, \qquad\qquad Tot.Var.\{\bar{u}\} < \delta\,,$$

then for every $\varepsilon > 0$ the Cauchy problem (6.2) has a unique solution u^ε, defined for all $t \geq 0$. Adopting a semigroup notation, this will be written as $t \mapsto u^\varepsilon(t,\cdot) \doteq S_t^\varepsilon \bar{u}$. In addition, one has:

BV bounds :
$$Tot. Var.\{S_t^\varepsilon \bar{u}\} \leq C \; Tot. Var.\{\bar{u}\}. \tag{6.3}$$

L^1 stability :
$$\left\|S_t^\varepsilon \bar{u} - S_t^\varepsilon \bar{v}\right\|_{L^1} \leq L \left\|\bar{u} - \bar{v}\right\|_{L^1}, \tag{6.4}$$

$$\left\|S_t^\varepsilon \bar{u} - S_s^\varepsilon \bar{u}\right\|_{L^1} \leq L' \left(|t-s| + \left|\sqrt{\varepsilon t} - \sqrt{\varepsilon s}\right|\right). \tag{6.5}$$

Convergence: *As $\varepsilon \to 0+$, the solutions u^ε converge to the trajectories of a semigroup S such that*

$$\left\|S_t \bar{u} - S_s \bar{v}\right\|_{L^1} \leq L \left\|\bar{u} - \bar{v}\right\|_{L^1} + L' \, |t-s|. \tag{6.6}$$

These vanishing viscosity limits can be regarded as the unique **vanishing viscosity solutions** *of the hyperbolic Cauchy problem*

$$u_t + A(u) u_x = 0, \qquad u(0,x) = \bar{u}(x). \tag{6.7}$$

In the conservative case $A(u) = Df(u)$, every vanishing viscosity solution is a weak solution of

$$u_t + f(u)_x = 0, \qquad u(0,x) = \bar{u}(x), \tag{6.8}$$

satisfying the Liu admissibility conditions [L1].

We outline here the main ideas in the proof of Theorem 5.

1. The rescaling $t \mapsto t/\varepsilon$, $x \mapsto x/\varepsilon$ transforms the problem (6.2) into

$$u_t + A(u) u_x = u_{xx} \qquad u(0,x) = \bar{u}(\varepsilon x). \tag{6.9}$$

Clearly, the total variation of a solution u is not affected by this rescaling. Notice however that the values of a solution u^ε of (6.2) on a time interval $[0,T]$ correspond to the value of a solution u of (6.9) on the much longer interval $[0, \; T/\varepsilon]$. To obtain BV estimates uniformly valid for all $\varepsilon > 0$, we thus need to prove similar estimates for the problem

$$u_t + A(u) u_x = u_{xx} \tag{6.10}$$

uniformly valid for all $t \in [0, \infty[$ and all initial data $u(0,\cdot)$ with suitably small total variation.

2. From now on we thus consider the parabolic system (6.10). On an initial time interval $[0,T]$, the solution can be represented as

$$u(t) = G(t) * u(0) - \int_0^t G(t-s) * A(u(s)) u_x(s) \, ds, \tag{6.11}$$

in terms of convolutions with the standard heat kernel $G(t,x) \doteq (4\pi t)^{-1/2} e^{-x^2/4t}$. Even if the initial data is discontinuous, the presence of viscosity has an immediate regularizing effect and the solution becomes smooth for all times $t > 0$. The smoothing effect of the parabolic equation is summarized in fig. 13. Let

$$\delta_0 \doteq \text{Tot.Var.}\{\bar{u}\}$$

be the total variation of the initial data. Then, on an initial interval $t \in]0,T]$ with $T = \mathcal{O}(1) \cdot \delta_0^{-2}$ one can show that

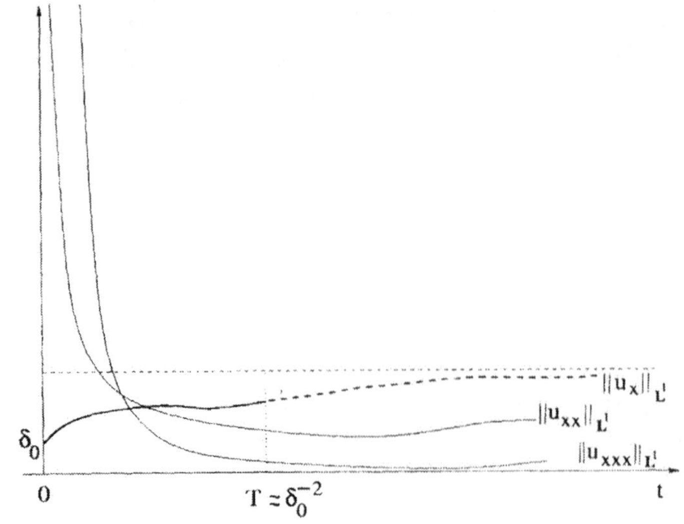

figure 13

- The total variation of $u(t,\cdot)$ remains bounded, i.e.

$$\big\|u_x(t,\cdot)\big\|_{\mathbf{L}^1} = \mathcal{O}(1)\cdot\delta_0\,. \tag{6.12}$$

- The norms of all higher order derivatives decay at algebraic rates. In particular

$$\begin{aligned} \big\|u_x(t,\cdot)\big\|_{\mathbf{L}^\infty} &\leq \big\|u_{xx}(t,\cdot)\big\|_{\mathbf{L}^1} = \mathcal{O}(1)\cdot\delta_0 t^{-1/2},\\ \big\|u_{xx}(t,\cdot)\big\|_{\mathbf{L}^\infty} &\leq \big\|u_{xxx}(t,\cdot)\big\|_{\mathbf{L}^1} = \mathcal{O}(1)\cdot\delta_0 t^{-1}. \end{aligned} \tag{6.13}$$

All these bounds can be derived by looking at integral representations of the form (6.11).

For $t\in[T,\,\infty[$, these parabolic type estimates yield only partial informations. What can be proved is the following: as long as the bound (6.12) on the total variation remains valid, also the norms of all higher derivatives remain small. Namely

$$\begin{aligned} \big\|u_x(t,\cdot)\big\|_{\mathbf{L}^\infty} &\leq \big\|u_{xx}(t,\cdot)\big\|_{\mathbf{L}^1} = \mathcal{O}(1)\cdot\delta_0^2\,,\\ \big\|u_{xx}(t,\cdot)\big\|_{\mathbf{L}^\infty} &\leq \big\|u_{xxx}(t,\cdot)\big\|_{\mathbf{L}^1} = \mathcal{O}(1)\cdot\delta_0^3\,. \end{aligned} \tag{6.14}$$

3. The main part of the proof is to establish the uniform bound (6.12) on the total variation, for all times $t\in[0,\infty[$. By the previous step, we can assume that the bounds (6.14) on higher order derivatives already hold. As mentioned in the Introduction, the basic approach is as follows. At each point (t,x) we decompose the gradient along a suitable basis of unit vectors $\tilde r_i$, say

$$u_x = \sum v_i\tilde r_i\,. \tag{6.15}$$

Then we derive an evolution equation for these gradient components, and show that their $\mathbf{L}^1$ norm remains uniformly bounded.

Toward this decomposition, the first step is to select n families of travelling wave profiles U_i, each depending on $n+2$ scalar parameters. This will allows us to solve the system (1.11)-(1.12) in unique way, depending continuously on the data (u,u_x,u_{xx}).

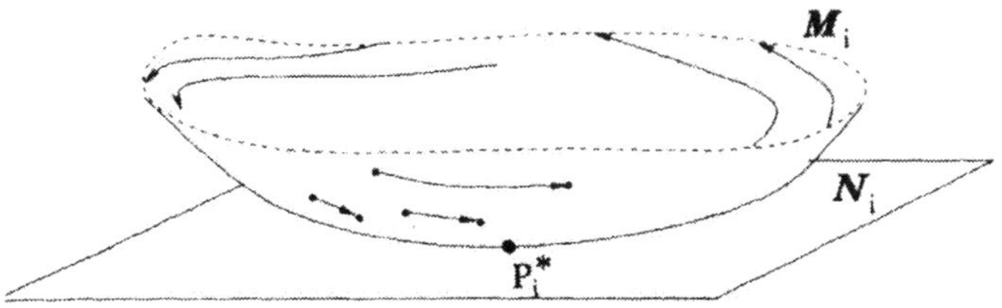

figure 14

4. A viscous travelling i-wave profile is a solution of (1.10), where the speed σ is close to the i-th eigenvalue λ_i of the hyperbolic matrix A. This second order O.D.E. can be rewritten as a first order system:

$$\begin{cases} \dot{u} &= v\,, \\ \dot{v} &= \big(A(u) - \sigma\big)v\,, \\ \dot{\sigma} &= 0\,. \end{cases} \tag{6.16}$$

Linearizing at a point $P_i^* = \big(u^*,\, 0,\, \lambda_i(u^*)\big)$ we obtain the linear system with constant coefficients

$$\begin{cases} \dot{u} &= v\,, \\ \dot{v} &= \big(A(u^*) - \lambda_i(u^*)\big)v\,, \\ \dot{\sigma} &= 0\,. \end{cases} \tag{6.17}$$

This system admits a center subspace $\mathcal{N}_i$ of dimension $n+2$. By the center manifold theorem [V], there exists a manifold $\mathcal{M}_i$, tangent to $\mathcal{N}_i$ at the point P_i^* (see fig. 14), locally invariant for the flow of the nonlinear system (6.16). Instead of looking at all viscous travelling profiles, we can now restrict ourselves to only those profiles corresponding to a trajectory on the center manifold $\mathcal{M}_i$. Since this manifold is $(n+2)$-dimensional, through each state $u \in \mathbb{R}^n$ there is a 2-parameter family of such profiles, say $U_i(u, v_i, \sigma_i)$. Here v_i denotes the scalar amplitude of the wave, and σ_i is the speed. Calling $\tilde{r}_i(u, v_i, \sigma_i)$ the unit vector parallel to U_i', this profile U_i is characterized by the equations

$$U_i'' = \big(A(U_i) - \sigma_i\big)U_i', \qquad U_i(0) = u\,, \qquad U_i'(0) = v_i\tilde{r}_i\,. \tag{6.18}$$

The manifold can then be defined as

$$\mathcal{M}_i = \big\{(u, v, \sigma_i) \in \mathbb{R}^{n+n+1}\,;\ v = v_i\tilde{r}_i(u, v_i, \sigma_i)\big\}\,. \tag{6.19}$$

We can now try to solve the system (1.11)-(1.12) using not all travelling wave profiles in (1.10), but only those U_i which correspond to a trajectory on the center manifold $\mathcal{M}_i$. Notice that in this case we get the correct number of parameters: in all, the n profiles $U_1, \ldots, U_n$ through the point $u(x)$ depend on the $2n$ scalar parameters $v_1, \sigma_1, \ldots, v_n, \sigma_n$. Exactly the right number to fit the data $\big(u_x(x),\, u_{xx}(x)\big) \in \mathbb{R}^{n+n}$.

To better understand how our decomposition works, consider first the special case where u is precisely the profile of a viscous travelling wave of the j-th family. In this case, the decomposition should clearly contain one single component:

$$u_x = v_j\tilde{r}_j(u, v_j, \sigma_j)\,.$$

It is easy to guess what v_j and σ_j should be. Indeed, since by construction $|\tilde{r}_j| = 1$, the quantity

$$v_j = \pm|u_x|$$

is the signed strength of the wave. We also notice that, for a travelling wave, the vectors u_x and u_t are always parallel. Indeed, $u_t = -\sigma_j u_x$, where σ_j is the speed of the wave. We can thus write

$$u_t = u_{xx} - A(u)u_x = \omega_j \tilde{r}_j(u, v_j, \sigma_j)$$

for some scalar ω_j. The speed of the wave is now obtained as $\sigma_j = -\omega_j/v_j$.

Motivated by the previous analysis, as a first attempt we define our decomposition in terms of the equations

$$\begin{cases} u_x &= \sum_i v_i \tilde{r}_i(u, v_i, \sigma_i), \\ u_t &= \sum_i \omega_i \tilde{r}_i(u, v_i, \sigma_i), \end{cases} \qquad \sigma_i = -\frac{\omega_i}{v_i}. \tag{6.20}$$

The trouble with (6.20) is that the vectors $\tilde{r}_i$ are defined only for speeds σ_i close to the i-th characteristic speed $\lambda_i^* \doteq \lambda_i(u^*)$. However, when $u_x \approx 0$ one has $v_i \approx 0$ and the ratio ω_i/v_i may become arbitrarily large. To overcome this problem, we introduce a smooth cutoff function θ such that

$$\theta(s) = \begin{cases} s & if \quad |s| \leq \delta_1 \\ 0 & if \quad |s| \geq 3\delta_1 \end{cases} \tag{6.21}$$

for some small constant $\delta_1 > 0$. We now rewrite (6.20) in terms of the new variable w_i, related to ω_i by $\omega_i \doteq w_i - \lambda_i^* v_i$. We require that σ_i coincides with $-\omega_i/v_i$ only when this ratio is sufficiently close to $\lambda_i^* \doteq \lambda_i(u^*)$. Our basic equations thus take the form

$$\begin{cases} u_x &= \sum_i v_i \tilde{r}_i(u, v_i, \sigma_i), \\ u_t &= \sum_i (w_i - \lambda_i^* v_i)\, \tilde{r}_i(u, v_i, \sigma_i), \end{cases} \tag{6.22}$$

where

$$u_t = u_{xx} - A(u)u_x\,, \qquad \sigma_i = \lambda_i^* - \theta\left(\frac{w_i}{v_i}\right). \tag{6.23}$$

The system (6.22)-(6.23) can be written in the more compact form

$$\begin{pmatrix} u_x \\ u_t \end{pmatrix} = \Lambda(u, v, w). \tag{6.24}$$

Using the implicit function theorem one can prove that, for each given state $u \approx u^*$ and u_x, u_t sufficiently small, there exists unique $v = (v_1, \ldots, v_n)$ and $w = (w_1, \ldots, w_n)$ such that (6.22)-(6.23) hold. Furthermore, the map $(u, u_x, u_t) \mapsto (v, w)$ is C^1 with Lipschitz continuous first derivative. This achieves the desired decomposition of the gradient u_x.

We observe that above equations are closely related to (1.12). Indeed, recalling (6.18), from the first equation in (6.22) it follows

$$u_x(x) = \sum_i U_i'(x).$$

If $\sigma_i = \lambda_i^* - w_i/v_i$ for all $i = 1, \ldots, n$, i.e. if none of the cutoff functions is active, then

$$\begin{aligned} u_{xx}(x) &= u_t + A(u)u_x = \textstyle\sum_i (w_i - \lambda_i^* v_i)\tilde{r}_i + A(u)\sum_i v_i \tilde{r}_i \\ &= \textstyle\sum_i \big(A(u) - \sigma_i\big) v_i \tilde{r}_i = \sum_i U_i''(x). \end{aligned}$$

In this case, both of the equalities in (1.12) hold. Notice however that the second equality in (1.12) may fail if $|w_i/v_i| > \delta_1$ for some i, so that $\theta(w_i/v_i) \neq w_i/v_i$.

Remark 1. For each $i = 1, \dots, n$, in addition to the eigenvalue $\lambda_i(u)$ of the matrix $A(u)$, one can define the *viscous eigenvalue* $\tilde\lambda_i(u, v_i, \sigma_i)$ by setting

$$\tilde\lambda_i \doteq \langle \tilde r_i\,,\ A(u)\tilde r_i\rangle\,. \tag{6.25}$$

Instead of the standard relation

$$\big(A(u) - \lambda_i\big) r_i = 0\,,$$

a direct computation shows that these viscous eigenvectors satisfy

$$\big(A(u) - \tilde\lambda_i\big)\tilde r_i = v_i\big(\tilde r_{i,u}\tilde r_i + \tilde r_{i,v}(\tilde\lambda_i - \sigma_i)\big)\,. \tag{6.26}$$

Here $\tilde r_{i,u} = \partial\tilde r_i/\partial u$ and $\tilde r_{i,v} = \partial\tilde r_i/\partial v_i$ denote partial derivatives. In the special case $v_i = 0$, an easy consequence of (6.26) is the identity

$$\tilde r_i(u, 0, \sigma_i) = r_i(u).$$

The presence of non-zero terms on the right hand side of (6.26) is of fundamental importance. Indeed, in the evolution equation for the component v_i, these terms achieve a crucial cancellation with other terms that would otherwise not be integrable.

5. We now consider a smooth solution $u = u(t,x)$ of the parabolic system (6.10). By the previous step, we can define the scalar functions $v_i = v_i(t,x)$, $w_i = w_i(t,x)$, according to the decomposition (6.22)-(6.23). The equations governing the evolution of these $2n$ components can be written in the form

$$\begin{cases} v_{i,t} + (\tilde\lambda_i v_i)_x - v_{i,xx} &= \phi_i\,,\\ w_{i,t} + (\tilde\lambda_i w_i)_x - w_{i,xx} &= \psi_i\,. \end{cases} \tag{6.27}$$

We recall that $\tilde\lambda_i$ is the speed defined at (6.25). The source terms ϕ_i, ψ_i can be computed by inserting (6.24) in the evolution equation

$$\begin{pmatrix} u_x\\ u_t\end{pmatrix}_t + \left(\begin{bmatrix} A(u) & 0\\ 0 & A(u)\end{bmatrix}\begin{pmatrix} u_x\\ u_t\end{pmatrix}\right)_x - \begin{pmatrix} u_x\\ u_t\end{pmatrix}_{xx} = \begin{pmatrix} 0\\ (DA(u)\cdot u_x)u_t - (DA(u)\cdot u_t)u_x\end{pmatrix}$$

and differentiating various times w.r.t. t, x, using the fundamental relation (6.26). For all details we refer to [BB]. Rather than the exact form of the source terms, what matters here is their magnitude. This can be estimated as follows.

Lemma 1. *The source terms in (6.27) satisfy the bounds*

$$\begin{aligned}\phi_i,\ \psi_i = \quad & \mathcal{O}(1)\cdot\textstyle\sum_j\big(|v_{j,x}| + |w_{j,x}|\big)\cdot|w_j - \theta_j v_j| \qquad \textbf{(cutoff error)}\\ & +\mathcal{O}(1)\cdot\textstyle\sum_{j\neq k}\big(|v_jv_k| + |v_{j,x}v_k| + |v_jw_k| + |v_{j,x}w_k| + |v_jw_{k,x}| + |w_jw_k|\big)\\ & \qquad\qquad\qquad \textbf{(interaction of waves of different families)}\\ & + \mathcal{O}(1)\cdot\textstyle\sum_j|v_{j,x}w_j - v_jw_{j,x}| \ + \ \mathcal{O}(1)\cdot\sum_j\left|v_j\left(\frac{w_j}{v_j}\right)_x\right|^2\cdot\chi_{\{|w_j/v_j|<3\delta_1\}}\end{aligned} \tag{6.28}$$

Here δ_1 is the constant in the cutoff function at (6.21). A rigorous proof of Lemma 1 requires lengthy calculations. Here we shall only provide an intuitive explanation of how the various terms arise. Consider first the special case where u is precisely one of the travelling wave profiles on the center manifold (fig. 15a), say $u(t,x) = U_j(x - \sigma_j t)$. In this case one has

$$u_x = v_j\tilde r_j\,, \qquad u_t = (w_j - \lambda_j^* v_j)\tilde r_j\,, \qquad v_i = w_i = 0 \quad \text{for } i \neq j\,,$$

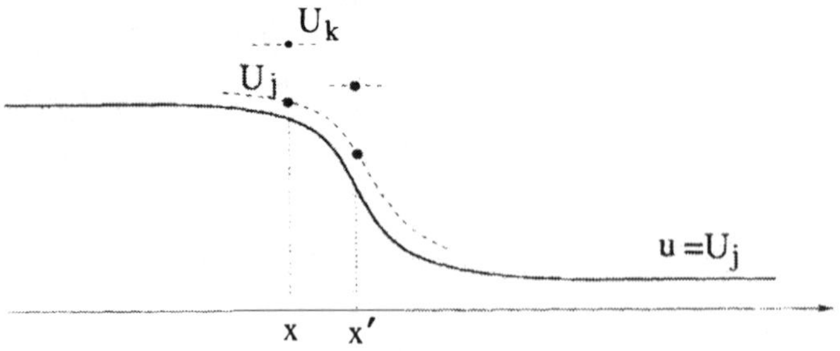

figure 15a

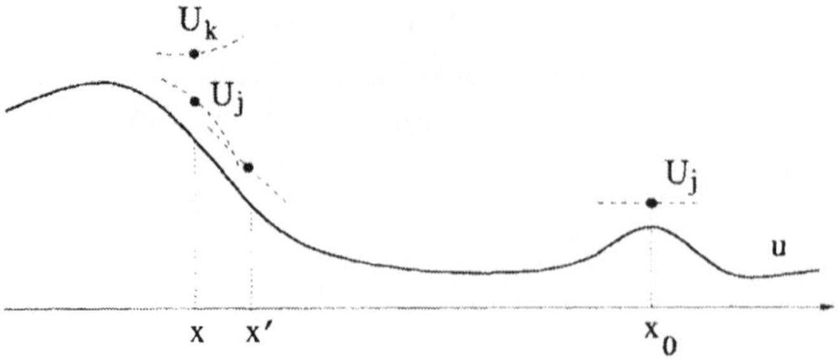

figure 15b

and all source terms ϕ_i, ψ_i vanish identically.

Next, consider the case of a general solution $u = u(t,x)$ of (6.10). The sources on the right hand sides of (6.27) arise for three different reasons (fig. 15b).

(i) The ratio $|w_j/v_j|$ is large and hence the cutoff function $\theta_j \doteq \theta(w_j/v_j)$ is active. Typically, this will happen near a point x_0 where $u_x = 0$ but $u_t = u_{xx} \neq 0$. In this case the second identity in (1.12) fails because of a "wrong" choice of the speed: $\sigma_j \neq \lambda_j^* - (w_j/v_j)$.

(ii) Waves of two different families $j \neq k$ are present at a given point x. These will produce quadratic source terms, due to transversal interactions.

(iii) Since the decomposition (6.22) is defined pointwise, it may well happen that the travelling j-wave profile U_j at a point x is not the same as the profile U_j at a nearby point x'. Indeed, these two travelling waves may have slightly different speeds. It is the rate of change in this speed: $\sigma_{j,x} = -(w_j/v_j)_x$, that determines the infinitesimal interaction between nearby waves of the same family. A detailed analysis shows that the corresponding source terms can only be linear or quadratic w.r.t. $\sigma_{j,x}$, with the square of the strength of the wave always appearing as a factor. These terms can thus be estimated as $\mathcal{O}(1) \cdot v_j^2 \sigma_{j,x} + \mathcal{O}(1) \cdot v_j^2 \sigma_{j,x}^2$.

To complete the proof of the BV bounds, it remains to show that all terms on the right hand side of (6.28) are integrable over the domain $\{t > T,\ x \in \mathbb{R}\}$. For this purpose, we shall introduce four types of Lyapunov functionals, called ***energy, transversal interaction, curve length*** and ***area functional***, monotonically decreasing

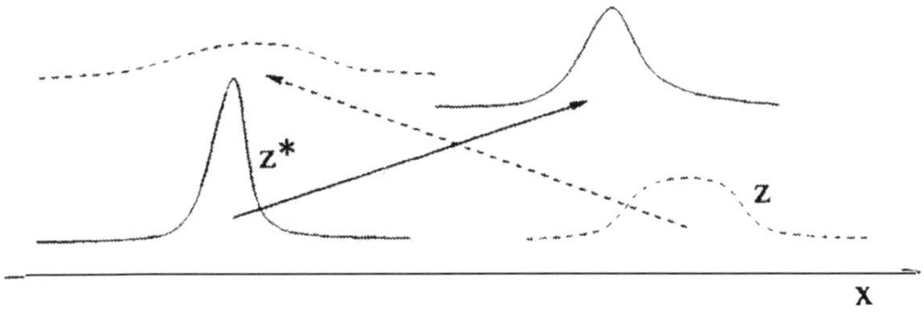

figure 16

along solutions. Each of the terms that we need to estimate will be dominated by the rate of decrease of one of these functionals.

7. Lyapunov functionals

We begin by describing a functional which controls the amount of interaction between viscous waves of different families.

Consider two independent, scalar diffusion equations with strictly different drifts (fig. 16):

$$\begin{cases} z_t + \left[\lambda(t,x)z\right]_x - z_{xx} &= 0\,, \\ z^*_t + \left[\lambda^*(t,x)z^*\right]_x - z^*_{xx} &= 0\,, \end{cases}$$

assuming that

$$\inf_{t,x} \lambda^*(t,x) - \sup_{t,x} \lambda(t,x) \geq c > 0\,.$$

We can think z as the density of waves with a slow speed λ and z^* as the density of waves with a fast speed λ^*. We seek an a priori bound on the total amount of interaction

$$\int_0^\infty \int_{\mathbb{R}} \left|z(t,x)\right| \cdot \left|z^*(t,x)\right| dxdt\,. \tag{7.1}$$

This can be achieved by means of a ***transversal interaction potential***, defined by

$$Q(z,z^*) \doteq \frac{1}{c} \iint_{\mathbb{R}^2} K(x_2 - x_1)\left|z(x_1)\right| \left|z^*(x_2)\right| dx_1 dx_2\,, \tag{7.2}$$

$$K(y) \doteq \begin{cases} e^{-cy/2} & if \quad y¿0\,, \\ 1 & if \quad y \leq 0\,. \end{cases} \tag{7.3}$$

Notice that, to the product of a slow wave located at x_1 and a fast wave at x_2, in (7.2) we assign weight $K(x_2 - x_1) = 1$ if the two waves are approaching (i.e. if $x_1 > x_2$), and an exponentially small weight if they are moving away from each other (i.e. if $x_1 < x_2$). It turns out that this functional is monotonically decreasing along every couple of solutions z, z^*. Moreover, the rate of decrease controls the

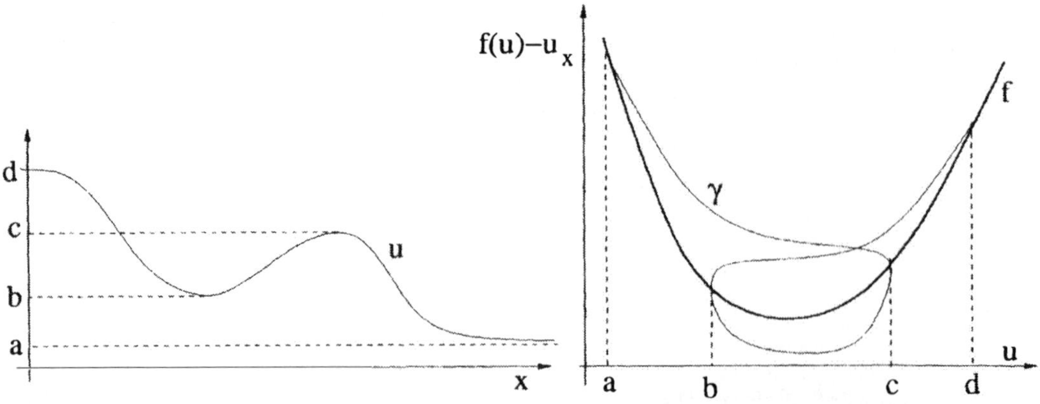

figure 17

integrand in (7.1). Namely

$$\int_0^\infty \int_{\mathbb{R}} |z(t,x)|\,|z^*(t,x)|\,dxdt \le -\int_0^\infty \left[\frac{d}{dt} Q\big(z(t),\, z^*(t)\big)\right] dt$$

$$\le Q\big(z(0),\, z^*(0)\big) \le \frac{1}{c}\int_{\mathbb{R}} |z(0,x)|\,dx \cdot \int_{\mathbb{R}} |z^*(0,x)|\,dx\,.$$

By introducing Lyapunov functionals of this type, we can bound the integrals of all terms of the form

$$|v_j v_k|\ |v_{j,x} v_k|\,,\ |v_j w_k|\,,\ |v_{j,x} w_k|\,,\ |w_j w_k|$$

with $j \neq k$, due to the interaction of waves of different families.

Next, we seek functionals which control the interactions between waves of the same family. Toward this goal, we first consider Lyapunov functionals which are decreasing along every solution of the scalar viscous conservation law

$$u_t + f(u)_x = u_{xx}\,. \tag{7.4}$$

Given a scalar function $x \mapsto u(x)$, define the curve (fig. 17)

$$\gamma \doteq \begin{pmatrix} u \\ f(u) - u_x \end{pmatrix} = \begin{pmatrix} \text{conserved quantity} \\ \text{flux} \end{pmatrix}. \tag{7.5}$$

In connection with a solution $u = u(t,x)$ of (7.4), the curve γ evolves according to

$$\gamma_t + f'(u)\gamma_x = \gamma_{xx}\,. \tag{7.6}$$

Notice that the vector $f'(u)\gamma_x$ is parallel to γ, hence the presence of this term in (7.6) only amounts to a reparametrization of the curve, and does not affect its shape. The curve thus evolves in the direction of curvature. An obvious Lyapunov functional is thus the *length* of the curve. In terms of the variables

$$\gamma_x = \begin{pmatrix} v \\ w \end{pmatrix} \doteq \begin{pmatrix} u_x \\ -u_t \end{pmatrix}, \tag{7.7}$$

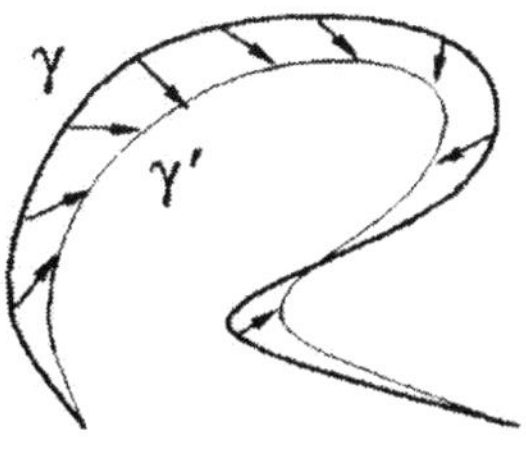

figure 18a

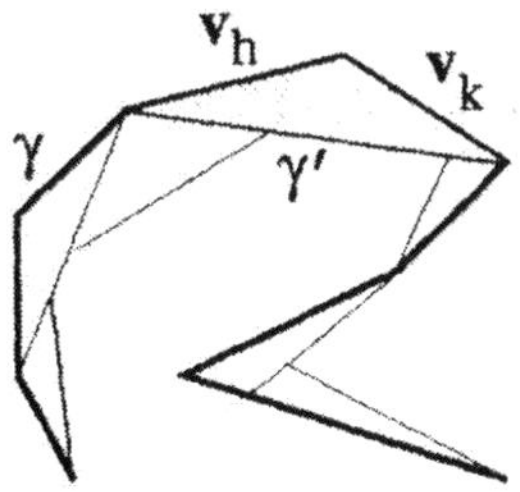

figure 18b

this length is given by

$$L(\gamma) \doteq \int |\gamma_x|\, dx = \int \sqrt{v^2 + w^2}\, dx\,. \tag{7.8}$$

We can estimate the rate of decrease of the length as

$$-\frac{d}{dt}L(\gamma(t)) = \int_{\mathbb{R}} \frac{|v|\,\big[(w/v)_x\big]^2}{\big(1+(w/v)^2\big)^{3/2}}\, dx \;\geq\; \frac{1}{(1+\delta^2)^{3/2}} \int_{|w/v|\leq\delta} |v|\,\big[(w/v)_x\big]^2\, dx\,,$$

for any given constant $\delta > 0$. Therefore

$$\int_0^\infty \int_{|w/v|\leq\delta_1} |v|\,\big[(w/v)_x\big]^2\, dx\, dt = \mathcal{O}(1)\cdot \int_0^\infty \left|\frac{d}{dt}L(\gamma(t))\right| dt = \mathcal{O}(1)\cdot L(\gamma(0)).$$

This argument allows us to estimate all source terms in (6.28) of the form

$$\mathcal{O}(1)\cdot \left| v_j \left(\frac{w_j}{v_j}\right)_x \right|^2 \cdot \chi_{\{|w_j/v_j|<3\delta_1\}}\,.$$

In connection with the same curve γ in (7.5), we now introduce a second functional, defined in terms of a wedge product.

$$Q(\gamma) \doteq \frac{1}{2} \iint_{x<x'} \big|\gamma_x(x) \wedge \gamma_x(x')\big|\, dx\, dx'\,. \tag{7.9}$$

For any curve that moves in the plane in the direction of curvature (fig. 18a), this functional is monotone decreasing and its decrease controls the area swept by the curve: $|dA| \leq -dQ$. An intuitive proof of this fact can be obtained by looking at polygonal approximations. In the special case (fig. 18b) where γ is a polygonal with edges $\mathbf{v}_1, \ldots, \mathbf{v}_m$, the functional in (7.9) reduces to

$$Q(\gamma) = \frac{1}{2}\sum_{i<j} |\mathbf{v}_i \wedge \mathbf{v}_j|\,.$$

Consider now a second polygonal γ', obtained from γ by replacing two consecutive edges $\mathbf{v}_h, \mathbf{v}_k$ by a single one. The area of the triangle between γ and γ' is then

$$\text{Area} = \frac{1}{2}|\mathbf{v}_h \wedge \mathbf{v}_k| \leq Q(\gamma) - Q(\gamma').$$

Observing that every continuous curve moving in the direction of curvature can be approximated by a suitable sequence of polygonals, the relation $|dA| \leq -dQ$ is obtained in the limit.

Using (7.6)-(7.7) we now compute

$$-\frac{dQ}{dt} \geq \left|\frac{dA}{dt}\right| = \int |\gamma_t \wedge \gamma_x|\, dx = \int |\gamma_{xx} \wedge \gamma_x|\, dx = \int |v_x w - v w_x|\, dx\,.$$

Therefore, integrating w.r.t. time one finds

$$\int_0^\infty \int |v_x w - v w_x|\, dx\, dt \;\leq\; \int_0^\infty \left|\frac{d}{dt} Q\big(\gamma(t)\big)\right| \,dt \;\leq\; Q\big(\gamma(0)\big)\,.$$

This argument allows us to estimate all source terms in (6.28) of the form

$$\mathcal{O}(1)\cdot \big|v_{j,x} w_j - v_j w_{j,x}\big|\,.$$

Remark 2. Defining the speed of a scalar viscous wave as $\sigma(x) \doteq -u_t(x)/u_x(x)$, the area functional $Q(\gamma)$ in (7.9) can be written as

$$\begin{aligned} Q(\gamma) &= \tfrac{1}{2} \iint_{x<y} \big|u_x(x) u_t(y) - u_t(x) u_x(y)\big|\, dxdy \\ &= \tfrac{1}{2} \iint_{x<y} \big|u_x(x)\, dx\big| \cdot \big|u_x(y)\, dy\big| \cdot \big|\sigma(x) - \sigma(y)\big| \\ &= \tfrac{1}{2} \iint_{x<y} [\text{wave at } x] \times [\text{wave at } y] \times [\text{difference in speeds}]\,. \end{aligned}$$

It now becomes clear that the area functional can be regarded as an interaction potential between waves of the same family. Indeed, it closely resembles the functional used by T. P. Liu in [L5].

Remark 3. In the case where u is precisely a viscous travelling wave, the curve γ reduces to a segment. Assume now that the flux f is genuinely nonlinear, say with $f'' \geq c > 0$. Consider a solution u which initially consists of two viscous travelling waves, far apart from each other (fig. 19). To fix the ideas, let the first wave join a left state a with a middle state b, and the second wave join the middle state b with a right state c, with $a > b > c$. The strength of the two waves can be measured as $s = a - b$, $s' = b - c$. The corresponding curve γ is approximately given by two segments, joining the points $P \doteq \big(a, f(a)\big)$, $Q \doteq \big(b, f(b)\big)$, $P' \doteq \big(c, f(c)\big)$. After a long time $\tau >> 0$, the two shocks will interact, merging into one single viscous shock. The curve $\gamma(\tau)$ thus reduces to one single segment, joining P with P'. The area swept by the curve is approximately the area of the triangle PQP'. The assumption of genuine nonlinearity implies

$$\text{area swept} \;=\; \mathcal{O}(1)\cdot |ss'|\big(|s| + |s'|\big).$$

In this case, the decrease in the area functional is of cubic order w.r.t. the strengths s, s' of the interacting waves. This is indeed the correct order of magnitude needed to control the strength of new waves generated by the interaction in the genuinely nonlinear case. It is remarkable that this area functional gives the correct order of magnitude of waves generated by an interaction also for a general flux function f, not necessarily convex.

Finally, we describe two functionals which can be used to control the source terms in (6.28) due to the cutoff errors.

Recalling the constant δ_1 in (6.21), we define a second cutoff function η such that

$$\eta(s) = \begin{cases} 0 & \text{if} \quad |s| \leq 3\delta_1/5 \\ 1 & \text{if} \quad |s| \geq 4\delta_1/5. \end{cases}$$

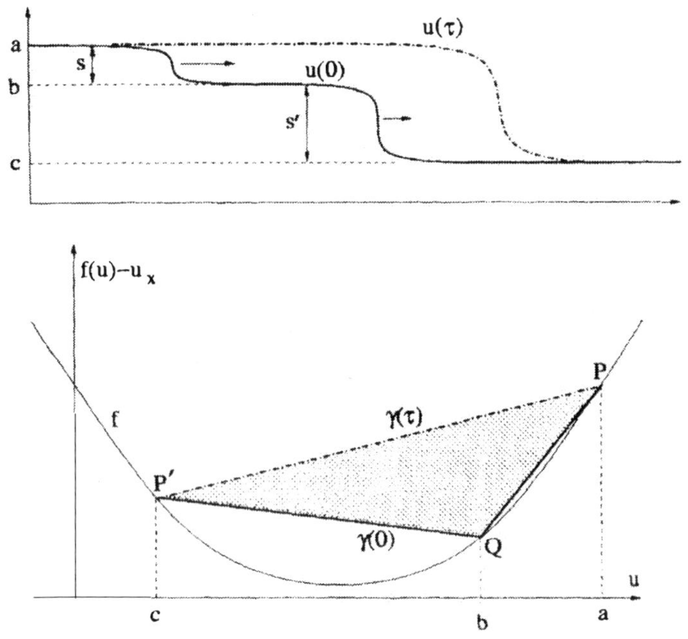

figure 19

In connection with a solution (v_i, w_i) of (6.27) we define $\eta_i \doteq \eta(w_i/v_i)$. The integral of the first terms on the right hand side of (6.28), due to the cutoff error, is then controlled by the decrease of the ***weighted energy functionals***

$$E_i(u) \doteq \int \eta_i \cdot v_{i,x}^2 \, dx \,, \qquad \widetilde{E}_i(u) \doteq \int \eta_i \cdot w_{i,x}^2 \, dx \,. \tag{7.10}$$

For all details we always refer to [BB].

Remark 4. Intuitively, we expect that the cutoff function θ will be active when $|u_x| << |u_{xx}|$. In these regions, the diffusion term u_{xx} dominates over the convection term $A(u)u_x$ and the system (6.10) essentially reduces to the heat equation. It is thus natural to estimate the error due to cutoff in terms of an energy functional, similar to the one commonly used for the heat equation.

By means of the above functionals we obtain the $\mathbf{L}^1$ estimates on the source terms ϕ_i, ψ_i in (6.27), proving the uniform bounds on the total variation of a solution u of (6.10).

8. Stability and convergence of viscous approximations

To prove the uniform stability of all solutions of the parabolic system (6.10) having small total variation, we consider the linearized system describing the evolution of a first order variation. Inserting the formal expansion $u = u_0 + \epsilon z + O(\epsilon^2)$ in (1.9), we obtain

$$z_t + \big[DA(u)\cdot z\big]u_x + A(u)z_x = z_{xx}\,. \tag{8.1}$$

A basic step in the proof is to establish the a priori bound

$$\big\|z(t)\big\|_{\mathbf{L}^1} \le L\,\big\|z(0)\big\|_{\mathbf{L}^1}\,, \tag{8.2}$$

for some constant L and all $t \ge 0$ and every solution z of the linearized system (8.1). By a standard homotopy argument, from (8.2) one easily deduces the Lipschitz continuity of the solution of (6.10) on the initial data. Namely, for every couple of solutions $u, \tilde u$ with small total variation one has

$$\big\|u(t) - \tilde u(t)\big\|_{\mathbf{L}^1} \le L\,\big\|u(0) - \tilde u(0)\big\|_{\mathbf{L}^1}\,. \tag{8.3}$$

To see this, let any two solutions $u, \tilde u$ be given. We can connect them by a smooth path of solutions u^θ, whose initial data satisfy

$$u^\theta(0,x) \doteq \theta u(0,x) + (1-\theta)\tilde u(0,x)\,, \qquad\qquad \theta \in [0,1]\,.$$

The distance $\big\|u(t,\cdot) - \tilde u(t,\cdot)\big\|_{\mathbf{L}^1}$ at any later time $t > 0$ is clearly bounded by the length of the path $\theta \mapsto u^\theta(t)$. In turn, this can be computed by integrating the norm of a tangent vector. Calling $z^\theta \doteq du^\theta/d\theta$, each vector z^θ is a solution of the corresponding equation (8.1), with u replaced by u^θ. Using (8.2) we thus obtain

$$\begin{aligned}\big\|u(t\ ,\cdot) - \tilde u(t,\cdot)\big\|_{\mathbf{L}^1} &\le \int_0^1 \Big\|\tfrac{d}{d\theta}u^\theta(t)\Big\|_{\mathbf{L}^1}\,d\theta \;=\; \int_0^1 \big\|z^\theta(t)\big\|_{\mathbf{L}^1}\,d\theta\\ &\le L\int_0^1 \big\|z^\theta(0)\big\|_{\mathbf{L}^1}\,d\theta \;=\; L\,\big\|u(0,\cdot) - \tilde u(0,\cdot)\big\|_{\mathbf{L}^1}\,.\end{aligned}$$

By the simple rescaling of coordinates $t \mapsto \varepsilon t$, $x \mapsto \varepsilon x$, all of the above estimates remain valid for solutions u^ε of the system (1.8). By a compactness argument, these BV bounds imply the existence of a strong limit $u^{\varepsilon_m} \to u$ in $\mathbf{L}^1_{\text{loc}}$, at least for some subsequence $\varepsilon_m \to 0$. In the conservative case where $A = Df$, it is now easy to show that this limit u provides a weak solution to the Cauchy problem (1.1).

To establish (8.2) one follows the same strategy as in the proof of the BV bounds. We decompose the vector z as a sum of scalar components:

$$z = \sum_i h_i \tilde r_i\,,$$

write an evolution equation for these components and prove that the source terms are integrable on the domain $\{t > 0\,, x \in \mathbb{R}\}$.

Calling $\Upsilon \doteq z_x - A(u)z$, the decomposition takes the form

$$\begin{cases} z &= \sum_i h_i \tilde r_i\big(u, v_i, \lambda_i^* - \theta(g_i/h_i)\big)\,,\\ \Upsilon &= \sum_i (g_i - \lambda_i^* h_i)\tilde r_i\big(u, v_i, \lambda_i^* - \theta(g_i/h_i)\big)\,,\end{cases} \tag{8.4}$$

where θ is the cutoff function introduced at (6.21). These components h_i, g_i satisfy the system of parabolic equations

$$\begin{cases} h_{i,t} + (\tilde\lambda_i h_i)_x - h_{i,xx} &= \hat\phi_i\,,\\ g_{i,t} + (\tilde\lambda_i g_i)_x - g_{i,xx} &= \hat\psi_i\,.\end{cases} \tag{8.5}$$

By similar arguments as in the previous sections, one obtains uniform bounds on all source terms $\hat{\phi}_i, \hat{\psi}_i$. This proves (8.2) and hence the stability of all solutions of (6.10) with small total variation.

Remark 5. The evolution equations (8.1) for a first order perturbation z are linear. It thus seems natural to use a decomposition $(z, z_x) \mapsto (h, g)$ where the components h_i, g_i are linear functions of z, z_x. This is the case if in (8.4) one chooses the same vectors $\tilde{r}_i = \tilde{r}_i(u, v_i, \sigma_i)$ as in (6.22), depending only on the reference solution u and not on the perturbation z. Unfortunately, this linear decomposition does not seem to work, since it leads to a system of the form (8.6) with possibly non-integrable source terms. On the contrary, our decomposition (8.5) is positive homogeneous of degree one, but not linear. Indeed, the unit vectors $\tilde{r}_i$ depend on the reference solution u through the amplitude v_i, but also on the perturbation z implicitly through the speed $\sigma_i = \lambda_i^* - \theta(g_i/h_i)$.

For every initial data $u(0,\cdot) = \bar{u}$ with small total variation, the previous steps yield the existence of a unique global solution to the parabolic system (6.10), depending Lipschitz continuously on the initial data, in the $\mathbf{L}^1$ norm. Performing the rescaling $t \mapsto t/\varepsilon$, $x \mapsto x/\varepsilon$, we immediately obtain the same results for the Cauchy problem

$$u_t^\varepsilon + A(u^\varepsilon)u_x^\varepsilon = \varepsilon\, u_{xx}^\varepsilon, \qquad\qquad u^\varepsilon(0,x) = \bar{u}(x).$$

Adopting a semigroup notation, this solution will be written as $u^\varepsilon(t,\cdot) = S_t^\varepsilon \bar{u}$. Thanks to the uniform bounds on the total variation, a compactness argument yields the existence of a strong limit in $\mathbf{L}^1_{\mathrm{loc}}$

$$u = \lim_{\varepsilon_m \to 0} u^{\varepsilon_m}, \tag{8.6}$$

at least for some subsequence $\varepsilon_m \to 0$. Since the u^ε depend continuously on the initial data, with a uniform Lipschitz constant, the same is true of $u(t,\cdot) = S_t \bar{u}$. In the conservative case where $A(u) = Df(u)$, it is not difficult to show that this limit u actually provides a weak solution to the Cauchy problem (6.8).

The only remaining issue is to show that the limit in (8.6) is unique, i.e. it does not depend on the subsequence $\{\varepsilon_m\}$. In the standard conservative case, this fact can already be deduced from the uniqueness criterion in [BG]. In the general case, uniqueness is proved in two steps. First we show that, in the special case of a Riemann problem, the solution obtained as vanishing viscosity limit is unique and can be completely characterized. To conclude the proof, we then rely on the same general argument as in [B2]: if two Lipschitz semigroups S, S' provide the same solutions to all Riemann problems, then they must coincide.

9. Concluding remarks

It remains an open problem to establish the convergence of vanishing viscosity approximations of the form

$$u_t + A(u)u_x = \varepsilon\big(B(u)u_x\big)_x \tag{9.1}$$

for more general viscosity matrices B, as in [MP]. In the present paper we are exclusively concerned with the case where B is the identity matrix. For systems which are not in conservative form, we expect that the limit of solutions of (**9**.1), as $\varepsilon \to 0$, will be heavily dependent on the choice of the matrix B. Studying the case where the viscosity matrix $B(u)$ is only semidefinite is particularly important: for applications to more realistic physical models and also toward the understanding of general relaxation approximations [N].

In the present survey we only considered initial data with small total variation. This is a convenient setting, adopted in much of the current literature, which guarantees the global existence of BV solutions of (1.1) and captures the main features of the problem. In particular, a single solution can exhibit a rich structure, including shock formation, decay of genuinely nonlinear rarefaction waves, interaction and cancellation phenomena. On the other hand, from the point of view of dynamical systems, the global flow generated by a hyperbolic systems looks rather trivial: for large times, all small BV solutions asymptotically converge to a limit described by the solution of a Riemann problem. In particular, we are not yet in a position to add a large source term on the right hand side of (1.1) and study the resulting flow from a topological point of view, say determining periodic or heteroclinic orbits. The reason is, quite simply, that this rich dynamics can only be observed in solutions with large data. But in this case, unfortunately, no global existence theorem is yet known, except for very special systems.

A recent example of Jenssen [J] shows that, for initial data with large total variation, the $\mathbf{L}^\infty$ norm of the solution can blow up in finite time. In this more general setting, one can only expect that the existence and uniqueness of weak solutions, together with the convergence of vanishing viscosity approximations, will hold locally in time as long as the total variation remains bounded. For the hyperbolic system (1.1), results on the local existence and stability of solutions with large BV data can be found in [Sc] and [BC2].

We remark that all of the blow up examples given so far arise from hyperbolic systems that do not admit any strictly convex entropy. This allows one to hope that, for physical systems, finite time blow up of solutions should not occur. At present, the role of entropies in preventing this finite time blow-up (as well as other pathological behavior) remains one of the major open problems in the one-dimensional theory of hyerbolic conservation laws.

References

[1] P. Baiti and H. K. Jenssen, On the front tracking algorithm, *J. Math. Anal. Appl.* **217** (1998), 395-404.

[2] S. Bianchini and A. Bressan, Vanishing viscosity solutions to nonlinear hyperbolic systems, *Annals of Mathematics*, to appear.

[3] A. Bressan, Global solutions to systems of conservation laws by wave-front tracking, *J. Math. Anal. Appl.* **170** (1992), 414-432.

[4] A. Bressan, The unique limit of the Glimm scheme, *Arch. Rational Mech. Anal.* **130** (1995), 205-230.

[5] A. Bressan, *Hyperbolic Systems of Conservation Laws. The One Dimensional Cauchy Problem.* Oxford University Press, 2000.

[6] A. Bressan and R. M. Colombo, The semigroup generated by 2×2 conservation laws, *Arch. Rational Mech. Anal.* **133** (1995), 1-75.

[7] A. Bressan and R. M. Colombo, Unique solutions of 2×2 conservation laws with large data, *Indiana Univ. Math. J.* **44** (1995), 677-725.

[8] A. Bressan, G. Crasta and B. Piccoli, Well posedness of the Cauchy problem for $n \times n$ conservation laws, *Amer. Math. Soc. Memoir* **694** (2000).

[9] A. Bressan and P. Goatin, Oleinik type estimates and uniqueness for $n \times n$ conservation laws, *J. Diff. Equat.* **156** (1999), 26-49.

[10] A. Bressan and P. LeFloch, Uniqueness of weak solutions to systems of conservation laws, *Arch. Rat. Mech. Anal.* **140** (1997), 301-317.

[11] A. Bressan and M. Lewicka, A uniqueness condition for hyperbolic systems of conservation laws, *Discr. Cont. Dynam. Syst.* **6** (2000), 673-682.

[12] A. Bressan, T. P. Liu and T. Yang, L^1 stability estimates for $n \times n$ conservation laws, *Arch. Rational Mech. Anal.* **149** (1999), 1-22.

[13] A. Bressan and A. Marson, Error bounds for a deterministic version of the Glimm scheme, *Arch. Rational Mech. Anal.* **142** (1998), 155-176.

[14] M. Crandall, The semigroup approach to first-order quasilinear equations in several space variables, *Israel J. Math.* **12** (1972), 108-132.

[15] C. Dafermos, Polygonal approximations of solutions of the initial value problem for a conservation law, *J. Math. Anal. Appl.* **38** (1972), 33-41.

[16] C. Dafermos, *Hyperbolic Conservation Laws in Continuum Physics*, Springer-Verlag, Berlin 1999.

[17] R. DiPerna, Global existence of solutions to nonlinear hyperbolic systems of conservation laws, *J. Diff. Equat.* **20** (1976), 187-212.

[18] R. DiPerna, Convergence of approximate solutions to conservation laws, *Arch. Rational Mech. Anal.* **82** (1983), 27-70.

[19] J. Glimm, Solutions in the large for nonlinear hyperbolic systems of equations, *Comm. Pure Appl. Math.* **18** (1965), 697-715.

[20] J. Glimm and P. Lax, Decay of solutions of systems of nonlinear hyperbolic conservation laws, *Amer. Math. Soc. Memoir* **101** (1970).

[21] J. Goodman and Z. Xin, Viscous limits for piecewise smooth solutions to systems of conservation laws, *Arch. Rational Mech. Anal.* **121** (1992), 235-265.

[22] H. Holden and N. H. Risebro, *Front Tracking for Hyperbolic Conservation Laws*, Springer Verlag, New York 2002.

[23] S. Kruzhkov, First order quasilinear equations with several space variables, *Math. USSR Sbornik* **10** (1970), 217-243.

[24] H. K. Jenssen, Blowup for systems of conservation laws, *SIAM J. Math. Anal.* **31** (2000), 894-908.

[25] P. Lax, Hyperbolic systems of conservation laws II, *Comm. Pure Appl. Math.* **10** (1957), 537-566.

[26] T. P. Liu, The entropy condition and the admissibility of shocks, *J. Math. Anal. Appl.* **53** (1976), 78-88.

[27] T. P. Liu, Decay to N-waves of solutions of general systems of nonlinear hyperbolic conservation laws, *Comm. Pure Appl. Math.* **30** (1977), 585-610.

[28] T. P. Liu, Linear and nonlinear large-time behavior of solutions of general systems of hyperbolic conservation laws, *Comm. Pure Appl. Math.* **30** (1977), 767-796.

[29] T. P. Liu, The deterministic version of the Glimm scheme, *Comm. Math. Phys.* **57** (1977), 135-148.

[30] T. P. Liu, Admissible solutions of hyperbolic conservation laws, *Amer. Math. Soc. Memoir* **240** (1981).

[31] T. P. Liu and T. Yang, L^1 stability for 2×2 systems of hyperbolic conservation laws, *J. Amer. Math. Soc.* **12** (1999), 729-774.

[32] A. Majda and R. Pego, Stable viscosity matrices for systems of conservation laws, *J. Differential Equations* **56** (1985), 229-262.

[33] R. Natalini, Recent results on hyperbolic relaxation problems, in *Analysis of Systems of Conservation Laws*, H. Freisthüler Ed., Chapman & Hall/CRC, 1998, pp.128-198.

[34] O. Oleinik, Discontinuous solutions of nonlinear differential equations (1957), *Amer. Math. Soc. Translations* **26**, 95-172.

[35] N. H. Risebro, A front-tracking alternative to the random choice method, *Proc. Amer. Math. Soc.* **117** (1993), 1125-1139.

[36] F. Rousset, Viscous approximation of strong shocks of systems of conservation laws, *SIAM J. Math.Anal.*, to appear.

[37] S. Schochet, Sufficient conditions for local existence via Glimm's scheme for large BV data, *J. Differential Equations* **89** (1991), 317-354.

[38] D. Serre, *Systems of Conservation Laws I, II*, Cambridge University Press, 2000.

[39] J. Smoller, *Shock Waves and Reaction-Diffusion Equations*, Springer-Verlag, New York, 1983.

[40] A. Vanderbauwhede, Centre manifolds, normal forms and elementary bifurcations, *Dynamics Reported, Vol. 2* (1989), 89-169.

[41] S. H. Yu, Zero-dissipation limit of solutions with shocks for systems of hyperbolic conservation laws, *Arch. Rational Mech. Anal.* **146** (1999), 275-370.

S.I.S.S.A., TRIESTE 34014, ITALY.

Current Developments in Mathematics, 2002
Pages 39 - 111

Combinatorics, symmetric functions, and Hilbert schemes

Mark Haiman

ABSTRACT. We survey the proof of a series of conjectures in combinatorics using new results on the geometry of Hilbert schemes. The combinatorial results include the positivity conjecture for Macdonald's symmetric functions, and the "$n!$" and "$(n+1)^{n-1}$" conjectures relating Macdonald polynomials to the characters of doubly-graded S_n modules. To make the treatment self-contained, we include background material from combinatorics, symmetric function theory, representation theory and geometry. At the end we discuss future directions, new conjectures and related work of Ginzburg, Kumar and Thomsen, Gordon, and Haglund and Loehr.

CONTENTS

1. Introduction

About ten years ago, Garsia and I [**22**] began to investigate certain doubly-graded S_n-modules whose characters appeared to be related to the new class of symmetric functions then recently introduced by Macdonald. Our modules are "doubled" analogs of well-known singly-graded modules arising in geometry and representation theory connected with the flag variety and Springer correspondence for GL_n. From the outset, we hoped to use them to prove the ***positivity conjecture*** for certain coefficients $K_{\lambda\mu}(q,t)$ occurring in Macdonald's theory, but by the time this was achieved [**39**], it had become clear that the solution of the positivity problem was just one chapter in a longer story. As the investigation advanced, we and others noticed that the characters of our doubled modules seemed to be related not only

Work supported in part by N.S.F. grants DMS-0070772, DMS-0296203.

to Macdonald's symmetric functions but also to a number of classical "q-analogs" familiar to combinatorialists. Among these are q-enumerations of rooted forests and parking functions, two q-analogs of the Catalan number $C_n = \frac{1}{n+1}\binom{2n}{n}$, and a q-analog of Lagrange inversion discovered independently by Garsia and Gessel. Eventually these various observations were formulated as a series of conjectures which have come to be known by the nicknames $n!$ and $(n+1)^{n-1}$ conjectures.

It develops that the doubled modules are associated naturally with the Hilbert scheme H_n of points in the plane. Understanding this geometric context has led to proofs of the $n!$ and $(n+1)^{n-1}$ conjectures, along lines first suggested by Procesi. As it turns out, the full explanation depends on properties of the Hilbert scheme that were not known before, and had to be established from scratch in order to complete the picture. One might say, then, that the main results are not the $n!$ and $(n+1)^{n-1}$ theorems, but new theorems in algebraic geometry. In a nutshell, the new theorems are, first, that the Hilbert scheme H_n is isomorphic to another kind of Hilbert scheme, which parametrizes orbits of the symmetric group S_n on $\mathbb{C}^n \oplus \mathbb{C}^n$, and second, a cohomology vanishing theorem for vector bundles on H_n provided by the first theorem.

I think it is fair to say that such results were unexpected, and became plausible to conjecture only because of evidence accumulated from the combinatorial study. Certainly it was this study that provided the incentive to prove them. One thing I wish to emphasize in these notes and in the accompanying lectures is how important the combinatorial origins of the problem were for all the subsequent developments.

It is perfectly possible to state the $n!$ and $(n+1)^{n-1}$ conjectures without reference to anything but elementary algebra, but to properly appreciate their content and the context in which they were discovered, it is helpful to understand some concepts which are familiar to combinatorialists and experts on special functions, but not so well known to a larger public. To this end I give in §2 and §3 a review of background material from combinatorics and the theory of Hall-Littlewood and Macdonald polynomials. To the same end, I will finish this introduction with a capsule history of some earlier developments which inspired the current results.

In the theory of Hall-Littlewood symmetric functions one meets q-analogs $K_{\lambda\mu}(q)$ of the *Kostka numbers* $K_{\lambda\mu}$, which count semi-standard Young tableaux of shape λ and content μ, or equivalently the weight multiplicity of μ in the irreducible representation of GL_n with highest weight λ. The $K_{\lambda\mu}(q)$ are called *Kostka-Foulkes polynomials*, and it is an important theorem that their coefficients are positive integers. A major development in the combinatorial theory of symmetric functions was Lascoux and Schützenberger's interpretation of $K_{\lambda\mu}(q)$ as q-enumerating Young tableaux according to a numerical statistic called *charge*. This gives one of two proofs of the positivity theorem for Kostka-Foulkes polynomials; the other being that of Hotta and Springer, who interpreted $K_{\lambda\mu}(q)$ in terms of characters of cohomology rings of Springer fibers for GL_n.

In a 1987 preprint [**62**], Macdonald unified the theory of Hall-Littlewood symmetric functions with that of spherical functions on symmetric spaces, introducing the symmetric functions now known as Macdonald polynomials. For root systems of type A, they are symmetric functions in the classical sense, but with coefficients that depend on two parameters q and t. In this new context there are bivariate analogs $K_{\lambda\mu}(q,t)$ of the Kostka-Foulkes polynomials, specializing at $q=0$ to $K_{\lambda\mu}(t)$. Macdonald conjectured in [**60**] that these more general *Kostka-Macdonald polynomials* should also have positive integer coefficients. (Actually, the $K_{\lambda\mu}(q,t)$ were defined as rational functions and their being polynomials was part of the conjecture.) In light of Lascoux and Schützenberger's work, it was natural to try to prove Macdonald's positivity conjecture by suitably generalizing the definition of charge. Indeed, the proof of the positivity conjecture notwithstanding, it remains an open problem to give an explicit combinatorial proof along such lines.

As a first step toward the positivity conjecture, Garsia and Procesi [**26**] revisited and simplified the Hotta-Springer proof of the positivity theorem for Kostka-Foulkes polynomials. Beginning with an elementary description due to Tanisaki of the cohomology ring R_μ of a Springer fiber, they derived the formula relating $K_{\lambda\mu}(q)$ to the character of R_μ directly, without invoking geometric machinery. Their work led Garsia and me to the $n!$ conjecture, which similarly relates $K_{\lambda\mu}(q,t)$ to the character of a doubled analog of R_μ, and so implies Macdonald's positivity conjecture.

The spaces figuring in the $n!$ conjecture are quotients of the ring of coinvariants for the diagonal action of S_n on $\mathbb{C}^n \oplus \mathbb{C}^n$, and so we decided also to investigate the characters of the full coinvariant ring. That precipitated the key event leading to the discoveries recounted here: the recognition by us and others, especially Gessel and Stanley, of striking combinatorial patterns among characters of diagonal coinvariants. The space of coinvariants apparently had dimension equal to $(n+1)^{n-1}$, a combinatorially interesting number. Paying closer attention to the grading and the S_n action revealed known q-analogs of the number $(n+1)^{n-1}$ and the Catalan numbers C_n in the data. A menagerie of things studied earlier by combinatorialists for their own sake thus turned up unexpectedly in this new context.

It was Procesi who suggested, upon learning of their remarkable behavior, that the diagonal coinvariants might be interpreted as sections of a vector bundle on the Hilbert scheme H_n. Then it might be possible to compute their dimension and character using the Lefshetz formula of Atiyah and Bott, and so explain the phenomena. It soon became clear that the existence of Procesi's alleged vector bundle was in fact equivalent to the $n!$ conjecture. By 1994, we had managed to fully develop Procesi's idea, explaining the observations on the diagonal coinvariants by combining known combinatorial properties of Macdonald polynomials with a character formula predicated on the assumption that the geometric theorems alluded to earlier on the Hilbert scheme would hold. This explanation was successful enough to persuade us that the theorems surely must be true, although it took quite some time after that before the proof was finally complete.

2. Background from combinatorics

This section is a review of combinatorial facts needed later on. In part, the purpose here is to fix notation and terminology, but we will also introduce several fundamental concepts.

The ***jeu-de-taquin*** operations on Young tableaux have become central to the combinatorial theory of symmetric functions, and lead to the definition of *charge*, the basis of Lascoux and Schützenberger's combinatorial interpretation of the Kostka-Foulkes polynomials, discussed in more detail in §3.4.5. The positivity theorem for Kostka-Foulkes polynomials was the progenitor of Macdonald's positivity conjecture. The beautiful and subtle combinatorics associated with the earlier theorem was a major reason why Macdonald's conjecture seized the attention of combinatorialists.

Next we will touch on a few aspects of q-enumeration, involving Catalan numbers, trees and parking functions. These particular q-enumerations have especially appealing and well studied theories of their own, apart from their surprising connection with the diagonal coinvariants, which is our main concern here.

Finally, we outline Garsia and Gessel's theory of q-Lagrange inversion, which, as it happens, nicely ties together precisely the q-enumerations we have just discussed. This fact is hardly accidental, and indeed q-Lagrange inversion turns out to be indispensable to the full development of the theory relating the combinatorial phenomena to the Hilbert scheme. The reason for this is that to get specific predictions from the geometry, such as the dimension $(n+1)^{n-1}$ for the diagonal coinvariants, is not a trivial exercise. What makes it possible is a reformulation of q-Lagrange inversion in terms of symmetric function operators arising in the theory of Macdonald polynomials, as will be explained in §3.5.7.

2.1. Partitions, tableaux and jeu-de-taquin. We write the parts of an integer partition as usual in decreasing order, as

$$\lambda = (\lambda_1, \ldots, \lambda_l), \quad \text{where} \quad \lambda_1 \geq \lambda_2 \geq \cdots \geq \lambda_l.$$

Its *length*, denoted $l(\lambda)$, is the number of parts, and its *size*, denoted $|\lambda|$, is the sum of the parts. The *diagram* of λ, sometimes called its *Young diagram* or its *Ferrers diagram*, is the array of lattice points

$$\{(i,j) \in \mathbb{N} \times \mathbb{N} : i+1 \leq \lambda_{j+1}\}.$$

It is drawn as an array of square cells, as for example

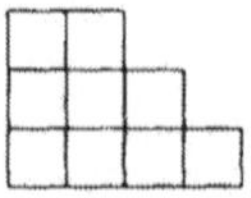

for the partition $(4,3,2)$. Note that our conventions follow the French style, and that the lower-left box is $(0,0)$, not $(1,1)$. We often abuse notation by identifying λ with its diagram, writing for instance $x \in \lambda$ to mean that $x = (i,j)$ is a cell in the diagram of λ. The *conjugate partition*, denoted λ', is the partition obtained by transposing the diagram of λ.

The *dominance order* is the partial order on partitions of a given size defined by

$$(1) \qquad \lambda \geq \mu \quad \text{if} \quad \lambda_1 + \cdots + \lambda_k \geq \mu_1 + \cdots + \mu_k \quad \text{for all } k.$$

It is a slightly non-trivial exercise to prove that

$$\lambda \leq \mu \quad \Leftrightarrow \quad \mu' \leq \lambda'.$$

The *arm* of a cell $x \in \lambda$ is the set of cells to the right of x in its row; the *leg* is the set of cells above x in its column. The *hook* of x consists of x together with its arm and leg. We denote by

$$a(x), \quad l(x), \quad h(x) = 1 + a(x) + l(x)$$

the sizes of the arm, leg and hook of x. Thus in the example

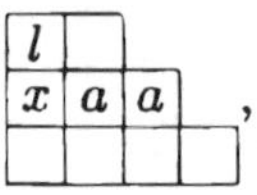

,

we have $a(x) = 2$, $l(x) = 1$ and $h(x) = 4$.

A ***standard Young tableau*** of shape λ, where $|\lambda| = n$, is a bijective map $T\colon \lambda \to \{1, \dots, n\}$ such that T is increasing along each row and column of λ (here already we are identifying λ with its diagram). A ***semi-standard Young tableau*** of shape λ is a map $T\colon \lambda \to \mathbb{N}$ which is weakly increasing along each row of λ and strictly increasing along each column. The number of standard Young tableaux of shape λ is given by the ***hook formula*** of Frame, Robinson and Thrall [**19**]:

$$|SYT(\lambda)| = \frac{n!}{\prod_{x\in\lambda} h(x)}. \tag{2}$$

An important numerical statistic associated with λ is

$$n(\lambda) \underset{\text{def}}{=} \sum_i (i-1)\lambda_i = \sum_i \binom{\lambda'_i}{2}.$$

This number has the following representation-theoretical significance. Put $|\lambda| = n$, and let T be any standard Young tableau of shape λ. Let $S_T \cong S_{\lambda'_1} \times \cdots \times S_{\lambda'_l}$ be the subgroup of S_n consisting of elements that only permute the numbers within columns of T. The ***Garnir polynomial*** $g_T(x_1, \dots, x_n)$ is defined to be the product of the Vandermonde determinants $\Delta(X_{C_0}) \cdots \Delta(X_{C_k})$ in the subsets X_{C_i} of the variables indexed by the entries C_i in column i of T, as for example

$$T = \begin{array}{|c|c|c|c|}\hline 3 & 6 & & \\ \hline 2 & 5 & 8 & \\ \hline 1 & 4 & 7 & 9 \\ \hline\end{array} \quad : \quad g_T = \Delta(x_1, x_2 x_3)\Delta(x_4, x_5, x_6)\Delta(x_7, x_8).$$

Thus g_T is the essentially unique homogeneous polynomial of minimal degree that is antisymmetric with respect to S_T. Its degree is $n(\lambda)$. As T ranges over standard tableaux of shape λ, the Garnir polynomials g_T form a basis of an S_n-invariant subspace of $\mathbb{C}[x_1, \dots, x_n]$, which affords the irreducible representation V^λ of S_n whose character is denoted χ^λ in the standard indexing. In particular, its degree $\chi^\lambda(1)$ is the number of standard Young tableaux of shape λ.

A ***skew shape*** λ/ν, where $\nu \subseteq \lambda$, is the array of cells in the difference between the diagrams, as for example

$(4, 3, 2)/(2, 1) =$ 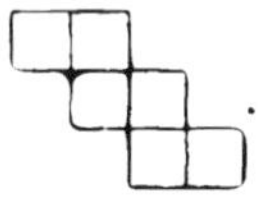 .

Standard and semistandard Young tableaux of skew shape are defined by the same rules as for straight shapes. Suppose given a semistandard tableau T of shape λ/ν and a cell x outside λ/ν but on its lower boundary, so that $\{x\} \cup (\lambda/\nu)$ is again a (skew) shape. There is a unique process, called a ***forward slide***, by which an entry of T is moved from an adjacent cell above or to the right of x, then another entry into the cell thus vacated, and so on, until finally a cell is left vacant along the upper boundary, in such a way that the result is again a tableau. This is best illustrated by an example:

$$\begin{array}{cccc} \boxed{1} & \boxed{6} & & \\ & \boxed{2} & \boxed{5} & \\ & \boxed{x} & \boxed{3} & \boxed{4} \end{array} \quad \to \quad \begin{array}{cccc} \boxed{1} & \boxed{6} & & \\ & \boxed{5} & \boxed{y} & \\ & \boxed{2} & \boxed{3} & \boxed{4} \end{array}.$$

Here y indicates the final vacated cell, and the tableau resulting from the slide consists of the other cells in the diagram on the right. The opposite process, in which tableau entries move up and to the right while the vacant cell moves down and to the left, is a ***reverse slide***. It is easy to see that following a slide with an opposite slide into the cell just vacated undoes the the first slide. In particular, the figure above with the arrow reversed gives an example of a reverse slide. The general name for manipulation of tableaux by slides is *jeu-de-taquin.* The basic properties of jeu-de-taquin were established by Lascoux and Schützenberger [**56, 77**].

PROPOSITION 2.1.1. *(1) When a skew tableau T is brought to a straight shape by jeu-de-taquin, the result depends only on T and not on the choice of slides used.*

(2) The number of tableaux T of a given skew shape λ/ν that go by jeu-de-taquin to a given tableau S of straight shape μ is the Littlewood-Richardson coefficient $c^{\lambda}_{\mu\nu}$, independent of S.

A semistandard tableau T has *content* μ if its entries are μ_1 1's, μ_2 2's, and so on. If $\mu_1 \geq \mu_2 \geq \cdots$, we say that T has *partition content.*

DEFINITION 2.1.2. The *Kostka number* $K_{\lambda\mu}$ is the number of semistandard Young tableaux of shape λ and content μ. In particular, $K_{\lambda\mu} = 0$ unless $\lambda \geq \mu$, and $K_{\lambda\lambda} = 1$.

Lascoux and Schützenberger defined a numerical invariant called *charge* on tableaux with partition content and arbitrary skew shape. It is easiest to define it in terms of a complementary invariant, called *cocharge.*

DEFINITION 2.1.3. The *cocharge* of a tableaux T with partition content μ is the integer $cc(T)$ uniquely characterized by the following properties:

(i) Cocharge is invariant under jeu-de-taquin slides.
(ii) Suppose the shape of T is disconnected, say $T = X \cup Y$, with X above and left of Y, and no entry of X is equal to 1. Let $S = Y \cup X$ be a tableau obtained by swapping X and Y. Then $cc(T) = cc(S) + |X|$.
(iii) If T is a single row, then $cc(T) = 0$.

The *charge* of T is defined as $c(T) = n(\mu) - cc(T)$.

The existence of an invariant $cc(T)$ with properties (i)-(iii) is of course a theorem. To compute $cc(T)$ for a tableau T of straight shape λ with more than one row, one may first put it in the form $X \cup Y$ in (ii) using jeu-de-taquin, by sliding the bottom row to the right until it detaches from the rest of the shape. Then swapping the detached row to the top and normalizing again to straight shape by jeu-de-taquin, we obtain a new tableau S with $cc(S) = cc(T) - |\lambda| + \lambda_1$. This process is called *catabolism.* Since it diminishes the cocharge, repeated catabolism eventually produces a tableau with one row and cocharge zero.

2.2. Catalan numbers and q-analogs. The *Catalan numbers* are given by the formula

$$C_n = \frac{1}{n+1}\binom{2n}{n}.$$

They enumerate a wide range of interesting combinatorial objects. For example, C_n is the number of binary trees with n vertices, the number of ordered rooted trees with $n+1$ vertices, the number of standard Young tableaux of shape (n,n), and the number of partitions λ whose diagram is contained inside that of the partition $\delta_n = (n-1, n-2, \ldots, 1)$. For literally dozens of other combinatorial interpretations of C_n, one may consult the book of Stanley [**82**].

The formulation most relevant here is that C_n is the number of partitions $\lambda \subseteq \delta_n$, listed below for $n = 3$.

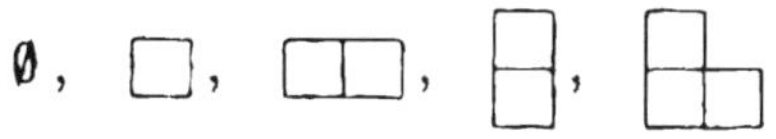

The list exhibits that $C_3 = 5$. The Carlitz-Riordan q-analog of the Catalan number [**12**] is the polynomial defined by

$$C_n(q) = \sum_{\lambda \subseteq \delta_n} q^{\binom{n}{2} - |\lambda|}. \tag{3}$$

From the list of partitions for $n = 3$, we see that

$$C_3(q) = q^3 + q^2 + 2q + 1.$$

The Carlitz-Riordan q-Catalan numbers satisfy the recurrence

$$C_n(q) = \sum_{k=0}^{n-1} q^k C_k(q) C_{n-1-k}(q), \tag{4}$$

which is the q-analog of a classical recurrence for C_n. The recurrence (4) is easily proven by taking k to be the number of consecutive cells along the outermost diagonal of δ_n, beginning at the top left, that do not belong to λ, as illustrated here with $n = 6$ and $\lambda = (4, 4, 1)$, so $k = 3$.

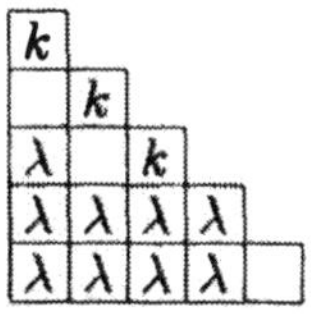

For fixed k, the top k rows of δ_n/λ may be chosen independently of the bottom $n - 1 - k$, with the choice in the top k rows contributing a factor $q^k C_k(q)$ and the choice in the remaining rows contributing $C_{n-1-k}(q)$.

The ordinary generating function for Catalan numbers is

$$C(x) \underset{\text{def}}{=} \sum_{n=0}^{\infty} C_n x^n = \frac{1 - \sqrt{1 - 4x}}{2x}. \tag{5}$$

Thus $C(x)$ is the solution of a quadratic equation, which we will find it convenient to write in the form

$$xC(x)(1 - xC(x)) = x. \tag{6}$$

In other words, $xC(x)$ is the compositional inverse of the function $F(x) = x(1 - x)$.

2.3. Tree enumeration and q-analogs. Among the things enumerated by Catalan numbers are various types of unlabelled trees. It is also classical in combinatorics (going back to Cayley) to enumerate trees and forests with vertices labelled by the integers $\{1, \ldots, n\}$. A *tree* is a connected graph without cycles. A *rooted tree* is a tree with one vertex distinguished as the root. A *forest* is a possibly non-connected graph without cycles. Every forest is the disjoint union of its connected components, which are trees. A *rooted forest* is a forest in which each component is assigned a distinguished root.

Labelled rooted forests on the vertex set $\{1, \ldots, n\}$ are in natural bijective correspondence with trees on the vertex set $\{0, 1, \ldots, n\}$, the trees being thought of either as unrooted or as always rooted at vertex 0. The tree corresponding to a forest F is constructed by adding the vertex 0 and attaching it by an edge to the root vertex of each component of F. It is of little consequence whether we enumerate trees or forests; for the present discussion I prefer forests.

PROPOSITION 2.3.1. *The number of rooted forests on* n *labelled vertices is* $(n+1)^{n-1}$.

There are many ways to prove this classical theorem, none of them entirely trivial. Perhaps the most direct is to consider the exponential generating function for trees

$$T(x) = \sum_{n=1}^{\infty} t_n \frac{x^n}{n!},$$

where t_n is the number of rooted trees on n labelled vertices. On the one hand, it follows from general principles that the exponential generating function $F(x)$ for rooted forests is given by $F(x) = e^{T(x)}$, because a forest is a disjoint union of trees. On the other hand, viewing a rooted tree as composed of a root vertex and an attached forest on the other vertices, one obtains $T(x) = xF(x)$. Hence we have $T(x) = xe^{T(x)}$, or better,

$$T(x)e^{-T(x)} = x, \tag{7}$$

so $T(x)$ is the compositional inverse of xe^{-x}. Now we apply the Lagrange inversion formula.

PROPOSITION 2.3.2. *Let* $xK(x)$ *be the compositional inverse of the formal power series* $x/E(x)$. *Then the coefficients of* $K(x)$ *are given by*

$$k_n \underset{\text{def}}{=} [x^n]K(x) = [x^n]\frac{E(x)^{n+1}}{n+1}. \tag{8}$$

In our case, we take $E(x) = e^x$, so $xK(x) = T(x)$ and $K(x) = F(x)$. Then $k_n = f_n/n!$, where f_n is our desired number of forests. Proposition 2.3.2 yields $k_n = (n+1)^{n-1}/n!$ and hence $f_n = (n+1)^{n-1}$.

DEFINITION 2.3.3. An *inversion* in a rooted forest F on vertices labelled $\{1,\dots,n\}$ is a pair of vertices $i < j$ such that vertex j is on the unique path from vertex i to the root of its component in F.

As an example, the forest

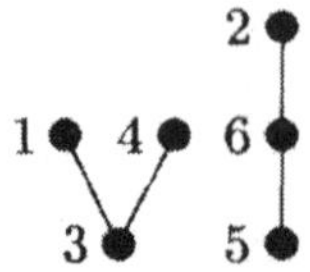

(drawn with the roots at the bottom) has 3 inversions: $(1,3)$, $(2,6)$ and $(2,5)$. This definition provides a combinatorial q-analog of the number $(n+1)^{n-1}$, the *inversion enumerator for forests*

$$J_n(q) \underset{\text{def}}{=} \sum_F q^{i(F)}, \tag{9}$$

where the sum is over rooted forests with vertices labelled $\{1,\dots,n\}$, and $i(F)$ denotes the number of inversions in the forest F. By listing the forests, one computes for example

$$J_3(q) = q^3 + 3q^2 + 6q + 6.$$

Mallows and Riordan [**63**] determined $J_n(q)$ by a generating function identity.

PROPOSITION 2.3.4 ([**63**]). *We have*

(10) $$\sum_{n=1}^{\infty} q^{n-1} J_{n-1}(q+1)\frac{x^n}{n!} = \log \sum_{n=0}^{\infty}(q+1)^{\binom{n}{2}}\frac{x^n}{n!}.$$

A pleasant proof of this result was given by Gessel and Wang [**32**]. The right-hand side of (10) is the exponential generating function for connected graphs with labelled vertices, enumerated with weight $q^{e(G)}$ for a graph G with $e(G)$ edges. Gessel and Wang described a search algorithm that selects a distinguished subtree in each graph. All graphs with a given distinguished tree T are gotten by adding to T some subset of certain optional edges. The number of optional edges is the number of inversions in T, regarded as rooted at vertex 1. Since vertex 1 never participates in an inversion of T, this is the same as the number of inversions in the rooted forest on the other vertices. Taking account of the $n-1$ fixed edges in T, together with the additional optional edges, and summing over all trees T on n vertices, gives the coefficient $q^{n-1}J_{n-1}(q+1)$ of $x^n/n!$ on the left-hand side.

2.4. q-Lagrange inversion. A q-analog of Lagrange inversion was defined independently by Garsia [**30**] and Gessel [**31**]. I will follow the approach of Garsia, who defined a q-analog of functional composition of formal power series by the identity

$$F \circ_q G(x) \underset{\text{def}}{=} \sum_n f_n G(x)G(qx)\cdots G(q^{n-1}x), \quad \text{where } F(x) = \sum_n f_n x^n.$$

He proved the following basic result, which shows that it is a good definition.

PROPOSITION 2.4.1 ([**30**]). *We have*

(11) $$F \circ_q G(x) = x \quad \textit{if and only if} \quad G \circ_{1/q} F(x) = x$$

More generally, when they hold, we have for any $\phi(x)$ *and* $\psi(x)$

(12) $$\psi(x) = \phi \circ_q G(x) \quad \textit{if and only if} \quad \phi(x) = \psi \circ_{1/q} F(x).$$

Garsia also obtained a q-analog of the Lagrange inversion formula (8) for the q-compositional inverse $G(x)$ of $F(x) = x/E(x)$, as defined by (11). His formula involves operations called *roofing* and *starring*, which I will not go into here. Garsia and I later reformulated it more combinatorially as follows.

PROPOSITION 2.4.2 ([**24, 30**]). *Let* $E(x) = \sum_{n=0}^{\infty} e_n x^n$, *with* $e_0 = 1$, *and let* $G(x)$ *be the q-compositional inverse of* $F(x) = x/E(x)$ *in the sense of* (11). *Then* $G(x) = xK(qx)$, *where* $K(x) = \sum_{n=0}^{\infty} k_n(q)x^n$, *with* $k_n(q)$ *given by*

$$k_n(q) = \sum_{\lambda \subseteq \delta_n} q^{\binom{n}{2}-|\lambda|} e_{\alpha_0(\lambda)} e_{\alpha_1(\lambda)} \cdots e_{\alpha_{n-1}(\lambda)}.$$

Here $\alpha_i(\lambda)$ *is the number of parts equal to* i *in* λ, *with* $\alpha_0(\lambda)$ *defined to make* $\sum_{i=0}^{n-1} \alpha_i(\lambda) = n$.

One sees immediately that when $E(x) = 1/(1-x)$, so $e_n = 1$ for all n, then $k_n(q) = C_n(q)$, the Carlitz-Riordan q-Catalan number defined by (3). In other words the generating function for q-Catalan numbers

$$C(x;q) = \sum_{n=0}^{\infty} C_n(q)x^n$$

has the property that $G(x) = xC(qx;q)$ is the q-compositional inverse of $F(x) = x/E(x) = x(1-x)$. This is the the q-analog of equation (6).

It turns out that equation (7) also has a q-analog, which involves the inversion enumerators for forests.

PROPOSITION 2.4.3 (Gessel [**31**]). *Let*

$$J(x;q) = \sum_{n=0}^{\infty} J_n(q)\frac{x^n}{n!}$$

be the exponential generating function for the forest inversion enumerators $J_n(q)$ *in* (9). *Then*

$$xe^{-x} \circ_q xJ(qx;q) = x. \tag{13}$$

This is proved using Proposition 2.3.4. Proposition 2.4.3 has an alternate interpretation that is worth mentioning. Combining (13) with Proposition 2.4.2, we see that $J_n(q)$ is equal to the specialization of $n!k_n(q)$ at $e_k = 1/k!$, or in symbols,

$$J_n(q) = \sum_{\lambda\subseteq\delta_n} q^{\binom{n}{2}-|\lambda|}\binom{n}{\alpha_0(\lambda),\alpha_1(\lambda),\dots,\alpha_{n-1}(\lambda)}. \tag{14}$$

Consider the diagram of λ with its border extended by a line segment along the vertical axis ending at $(0,n)$. The numbers $\alpha_i(\lambda)$, including $\alpha_0(\lambda)$, are then the heights of the vertical segments along this extended border. The multinomial coefficient $\binom{n}{\alpha_0(\lambda),\alpha_1(\lambda),\dots,\alpha_{n-1}(\lambda)}$ is the number of ways to place labels $\{1,\dots,n\}$ to the right of the n vertical steps on the extended border of λ, so that the labels increase along each contiguous vertical segment. Here is an example with $n = 6$ and $\lambda = (4,4,1)$.

(15)

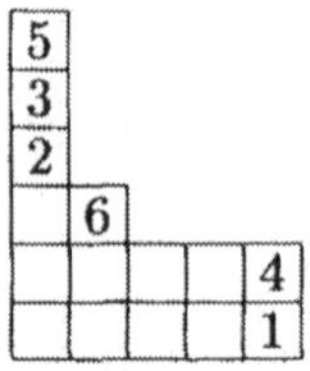

Such a labelled diagram is specified completely by giving the function $f\colon \{1,\dots,n\} \to \{1,\dots,n\}$ in which $f(i) - 1$ is the column occupied by label i. Not every function is admissible, because of the condition $\lambda \subseteq \delta_n$. However, it is not hard to see that this condition on λ is precisely equivalent to the condition on f that

$$|f^{-1}(\{1,\dots,k\})| \geq k \quad \text{for all } k = 1,\dots,n. \tag{16}$$

DEFINITION 2.4.4. A function $f : \{1,\dots,n\} \to \{1,\dots,n\}$ satisfying (16) is a *parking function*. The *weight* of f is the quantity $\binom{n+1}{2} - \sum_i f(i)$.

Note that the parts of λ are one less than the values of the corresponding parking function f, so the weight of f is simply $\binom{n}{2} - |\lambda|$. Hence Propositions 2.4.2 and 2.4.3 have the following corollary, which was also proved by Kreweras [**51**] using a combinatorial bijection.

COROLLARY 2.4.5. *The number of parking functions of weight d on* $\{1,\dots,n\}$ *is equal to the number of rooted forests on* $\{1,\dots,n\}$ *with d inversions.*

There is a natural action of S_n on parking functions defined by

$$w(f) = f \circ w^{-1}, \quad \text{for } w \in S_n.$$

This action is well-defined since it leaves condition (16) invariant; and it preserves the weight. We have the following nice result.

PROPOSITION 2.4.6. *The permutation action of the symmetric group* S_n *on parking functions is isomorphic to its action on the finite Abelian group*

$$Q/(n+1)Q,$$

where $Q = \mathbb{Z}^n/\mathbb{Z}\cdot(1,1,\dots,1)$ *and* S_n *acts on* Q *by permuting coordinates.*

PROOF. Fix the integers $\{0,\dots,n\}$ as representatives of the residue classes modulo $n+1$, and identify $(\mathbb{Z}/(n+1)\mathbb{Z})^n$ with the set of functions $f\colon \{1,..,n\} \to \{0,\dots,n\}$ in the obvious way. It is not hard to show that in every coset of $(\mathbb{Z}/(n+1)\mathbb{Z})\cdot(1,1,\dots,1)$ there is exactly one f that is a parking function. This gives an S_n-equivariant bijection between parking functions and the elements of $Q/(n+1)Q$. □

Note that the proposition gives in particular an easy proof that there are exactly $(n+1)^{n-1}$ parking functions. The Abelian group Q and its S_n action may be identified with the weight lattice for $\mathfrak{sl}_n$ and its Weyl group action.

3. Background from symmetric function theory

3.1. Generalities. We work with formal symmetric functions in infinitely many variables $z = z_1, z_2, \dots$, with coefficients implicitly assumed to be rational functions of any parameters q, t, etc. under discussion. I'll use the standard notation from Macdonald's book [61] for the classical families of symmetric functions: e_k for the k-th elementary symmetric function, h_k for the complete homogeneous symmetric function of degree k, p_k for the power-sum $z_1^k + z_2^k + \cdots$, and e_λ, h_λ, p_λ for products of these. The monomial symmetric functions are denoted m_λ, and the Schur functions s_λ.

The ***Hall inner product*** makes the Schur functions orthonormal, or equivalently, makes the monomial symmetric functions m_λ dual to the complete homogeneous symmetric functions h_λ:

$$\langle s_\lambda, s_\mu\rangle = \langle m_\lambda, h_\mu\rangle = \delta_{\lambda\mu}. \tag{17}$$

Recall that the power-sum symmetric functions are orthogonal, but not orthonormal; their norms are given by

$$\langle p_\lambda, p_\lambda\rangle = z_\lambda \underset{\text{def}}{=} \prod_i \alpha_i!\, i^{\alpha_i}, \quad \text{where } \lambda = (1^{\alpha_1}, 2^{\alpha_2}, \dots). \tag{18}$$

The Kostka numbers $K_{\lambda\mu}$ of Definition 2.1.2 are related to symmetric functions by the identity

$$K_{\lambda\mu} = \langle s_\lambda, h_\mu\rangle, \tag{19}$$

which in light of (17) is equivalent to either of the expansions

$$s_\lambda = \sum_\mu K_{\lambda\mu} m_\mu \tag{20}$$

$$h_\mu = \sum_\lambda K_{\lambda\mu} s_\lambda. \tag{21}$$

3.2. The Frobenius map. A classical theorem of Frobenius expresses the irreducible characters of the symmetric groups in terms of symmetric functions.

PROPOSITION 3.2.1. *Let λ be a partition of n, and $w \in S_n$. The value of the irreducible character χ^λ of S_n at w is given by*

$$\chi^\lambda(w) = \langle s_\lambda, p_{\tau(w)} \rangle, \tag{22}$$

where $\tau(w)$ is the partition whose parts are the lengths of the disjoint cycles of the permutation w.

The number of permutations $w \in S_n$ with given cycle-type $\tau(w) = \lambda$ is equal to $n!/z_\lambda$, where $z_\lambda = \langle p_\lambda, p_\lambda \rangle$, as in (18). Hence Proposition 3.2.1 can also be written as the identity

$$s_\lambda = \frac{1}{n!} \sum_{w \in S_n} \chi^\lambda(w) p_{\tau(w)}.$$

This suggests the following definition.

DEFINITION 3.2.2. The *Frobenius characteristic map* is the the linear map from class functions on S_n to symmetric functions homogeneous of degree n given by

$$F\chi \underset{\text{def}}{=} \frac{1}{n!} \sum_{w \in S_n} \chi(w) p_{\tau(w)},$$

or equivalently, the unique linear map sending the irreducible character χ^λ to the Schur function s_λ.

Note that F is an isometry of the usual inner product on characters onto the Hall inner product on symmetric functions.

We will often deal with graded, or doubly-graded, S_n-modules. In this context it is useful to extend the Frobenius map to a generating function for their graded characters.

DEFINITION 3.2.3. Let $A = \bigoplus_r A_r$ be a graded S_n-module (with each A_r finite-dimensional). The *Frobenius series* of A is the generating function

$$F_A(z; t) = \sum_r t^r F \operatorname{char} A_r.$$

If $A = \bigoplus_{r,s} A_{r,s}$ is doubly graded, its Frobenius series is the bivariate generating function

$$F_A(z; q, t) = \sum_{r,s} t^r q^s F \operatorname{char} A_{r,s}.$$

Note that the degree of χ^λ is equal to $\chi^\lambda(1) = \langle s_\lambda, p_1^n \rangle$. It follows that the dimension of any finite-dimensional S_n-module V is given by $\langle p_1^n, F \operatorname{char} V \rangle$. Hence the Frobenius series of a graded or doubly graded module A determines its Hilbert series by the formula

$$H_A(t) = \langle p_1^n, F_A(z;t)\rangle \quad \text{or} \quad H_A(q,t) = \langle p_1^n, F_A(z;q,t)\rangle.$$

Of course the Frobenius series of any module is *a fortiori* Schur-positive, that is, the coefficients of its expansion in Schur functions are polynomials or formal series in t and/or q with positive integer coefficients. Our basic tool for establishing the Schur positivity of symmetric functions will be to interpret them as Frobenius series for suitable graded modules.

3.3. Plethystic or λ-ring notation. The algebra of symmetric functions $\Lambda_{\mathbb{F}}$ over a coefficient field $\mathbb{F}$ containing $\mathbb{Q}$ is freely generated by the power-sums p_k, that is,

$$\Lambda_{\mathbb{F}} \cong \mathbb{F}[p_1, p_2, \ldots].$$

Hence we may specify arbitrary values for the p_k's in any algebra $\mathcal{A}$ over $\mathbb{F}$ and extend uniquely to an $\mathbb{F}$-algebra homomorphism $\Lambda_{\mathbb{F}} \to \mathcal{A}$.

Now let A be a formal Laurent series with rational coefficients in indeterminates $a_1, a_2, \ldots$ (possibly including parameters q, t from the coefficient field). We define $p_k[A]$ to be the result of replacing each indeterminate a_i in A by a_i^k. Then for any $f \in \Lambda_{\mathbb{F}}$, the ***plethystic substitution*** of A into f, denoted $f[A]$, is the image of f under the homomorphism sending p_k to $p_k[A]$.

If A is a sum of indeterminates, $A = a_1 + \cdots + a_n$, then $p_k[A] = p_k(a_1, a_2, \ldots, a_n)$, and hence for every f we have $f[A] = f(a_1, a_2, \ldots, a_n)$. This is why we view this operation as a kind of substitution. In particular, when dealing with symmetric functions in an alphabet z, we always denote the sum of the variables by

$$Z = z_1 + z_2 + \cdots,$$

whence

$$f[Z] = f(z)$$

for all f. More generally, if A has a series expansion as a sum of monomials, then $f[A]$ is f evaluated on these monomials, for example

$$f[Z/(1-t)] = f(z_1, z_2, \ldots, tz_1, tz_2, \ldots, t^2z_1, t^2z_2, \ldots).$$

Among the virtues of this notation is that the substitution $Z \to Z/(1-t)$ as above has an explicit inverse, namely the substitution $Z \to Z(1-t)$.

One caution that must be observed with plethystic notation is that indeterminates must always be treated as formal symbols, never as variable numeric quantities. For instance, if f is homogeneous of degree d then it is true that

$$f[tZ] = t^d f[Z],$$

but it is *false* that $f[-Z] = (-1)^d f[Z]$, that is, we cannot set $t = -1$ in the equation above. Actually, $f[-Z]$ is an interesting quantity: it is equal to $(-1)^d \omega f(z)$, where ω is the classical involution on symmetric functions defined by $\omega p_k = (-1)^{k+1} p_k$, which interchanges the elementary and complete symmetric functions e_k and h_k, and more generally exchanges the Schur function s_λ with $s_{\lambda'}$.

It is convenient when using plethystic notation to define

$$\Omega = \exp(\sum_{k=1}^{\infty} p_k/k). \tag{23}$$

Then since $p_k[A+B] = p_k[A] + p_k[B]$ and $p_k[-A] = -p_k[A]$ we have

$$\Omega[A+B] = \Omega[A]\Omega[B], \quad \Omega[-A] = 1/\Omega[A]. \tag{24}$$

From this and the single-variable evaluation $\Omega[x] = \exp(\sum_{k\geq 1} x^k/k) = 1/(1-x)$ we obtain

$$\Omega[Z] = \prod_i \frac{1}{1-z_i} = \sum_{n=0}^{\infty} h_n(z) \tag{25}$$

$$\Omega[-Z] = \prod_i (1-z_i) = \sum_{n=0}^{\infty} (-1)^n e_n(z). \tag{26}$$

The plethystic substitutions $Z \to Z(1-t)$ and $Z \to Z/(1-t)$ have an important representation-theoretical interpretation.

PROPOSITION 3.3.1. *Let S^kV and $\wedge^kV$ denote the symmetric and exterior powers repectively of the defining representation $V = \mathbb{C}^n$ of S_n, and let $f(z) = F_A(z;t)$ be the Frobenius series of a graded S_n-module A. Then we have*

$$f[Z(1-t)] = \sum_k (-1)^k t^k F_{A\otimes\wedge^k V}(z;t), \tag{27}$$

$$f[\frac{Z}{1-t}] = \sum_k t^k F_{A\otimes S^k V}(z;t). \tag{28}$$

The proof is in two steps. First, a direct computation gives

$$h_n[Z(1-t)] = \sum_k (-1)^k t^k F_{\wedge^k V}(z;t),$$

$$h_n[\frac{Z}{1-t}] = \sum_k t^k F_{S^k V}(z;t).$$

Second, an easy exercise shows that if ϕ is a virtual character whose Frobenius image has the form $F\phi = h_n[ZQ]$ for some Q, then $F(\phi\otimes\chi) = (F\chi)[ZQ]$ for all χ, from which the result follws.

In §3.5 we will need the following classical generalization of the hook formula (2).

PROPOSITION 3.3.2. *The Schur function specialization $s_\lambda(1,t,t^2,\dots) = s_\lambda[1/(1-t)]$ is given by*

$$s_\lambda[1/(1-t)] = \frac{t^{n(\lambda)}}{\prod_{x\in\lambda}(1-t^{h(x)})}, \tag{29}$$

where $h(x)$ is the hook-length of the cell $x \in \lambda$.

3.4. Hall-Littlewood polynomials. Hall-Littlewood polynomials are symmetric functions with coefficients depending on a parameter q. They play an important role in the representation theory of $GL_n(\mathbb{F}_q)$ and in the geometry associated with it, involving character sheaves and the Springer correspondence (see §3.4.4). Together with Jack's symmetric functions, Hall-Littlewood polynomials were the precursors of Macdonald's symmetric functions.

The connecting coefficients between Hall-Littlewood polynomials and Schur functions are the *Kostka-Foulkes polynomials* $K_{\lambda\mu}(q)$. The positivity theorem for Kostka-Foulkes polynomials was the precursor to Macdonald's positivity conjecture, and the combinatorial proof of that theorem by Lascoux and Schützenberger was the inspiration for our work on Macdonald's conjecture.

3.4.1. *Definition; Kostka-Foulkes polynomials.* We begin with the classical definition of Hall-Littlewood polynomials, as in Macdonald [**61**]. We use the conventional q-notation (with t instead of q)

$$
\begin{aligned}
[k]_t &\underset{\text{def}}{=} \frac{1-t^k}{1-t} = t^{k-1} + t^{k-2} + \cdots + 1; \\
[k]_t! &\underset{\text{def}}{=} [k]_t[k-1]_t \cdots [1]_t.
\end{aligned}
\tag{30}
$$

DEFINITION 3.4.1. The *Hall-Littlewood polynomial* $P_\lambda(z;t)$ is defined in $n \geq l(\lambda)$ variables $z = z_1, \ldots, z_n$ by the formula

$$
P_\lambda(z;t) = \frac{1}{\prod_{i\geq 0}[\alpha_i]_t!} \sum_{w\in S_n} w\left(z^\lambda \frac{\prod_{i<j}(1-tz_j/z_i)}{\prod_{i<j}(1-z_j/z_i)} \right).
\tag{31}
$$

Here $\lambda = (1^{\alpha_1}, 2^{\alpha_2}, \ldots)$, with α_0 defined so as to make $\sum_i \alpha_i = n$, and z^λ is shorthand for $z_1^{\lambda_1} z_2^{\lambda_2} \cdots z_l^{\lambda_l}$.

We will see below that the definition is stable with respect to changing the number of variables, so $P_\lambda(z;t)$ makes sense formally in infinitely many variables. At $t = 0$, the denominator $\prod_{i\geq 0}[\alpha_i]_t!$ disappears and (31) reduces to the classical formula for Schur functions (equation (36) below), so we have

$$
P_\lambda(z;0) = s_\lambda(z).
\tag{32}
$$

At $t = 1$, the products inside the sum cancel, and $\prod_{i\geq 0}[\alpha_i]_t!$ becomes the number of permutations $w \in S_n$ that stabilize z^λ, so

$$
P_\lambda(z;1) = m_\lambda(z).
\tag{33}
$$

DEFINITION 3.4.2. The *Kostka-Foulkes polynomials* $K_{\lambda\mu}(t)$ are the coefficients in the expansion

$$
s_\lambda(z) = \sum_\mu K_{\lambda\mu}(t) P_\mu(z;t).
$$

In particular, by (32) and (33), we have

$$
\begin{aligned}
K_{\lambda\mu}(1) &= K_{\lambda\mu} \\
K_{\lambda\mu}(0) &= \delta_{\lambda\mu}.
\end{aligned}
$$

It is not yet obvious that $K_{\lambda\mu}(t)$ is in fact a polynomial, but this will become clear below.

3.4.2. *Jing's operators and transformed Hall-Littlewood polynomials.* There is another way of defining Hall-Littlewood polynomials which gives easier access to many of their properties. We begin with "vertex operators" due to Bernstein (see Macdonald [**61**] or Zelevinsky [**86**]) that have the effect of adding a part to a Schur function.

DEFINITION 3.4.3. The *Bernstein operators* are the coefficients $S^0_m = [u^m]S^0(u)$ of the operator generating function $S^0(u)$ defined by

$$S^0(u)f = f[Z - u^{-1}]\Omega[uZ]. \tag{34}$$

PROPOSITION 3.4.4. *The Bernstein operators add a part to the indexing partition of a Schur function, that is, for $m \geq \lambda_1$, we have $S^0_m s_\lambda(z) = s_{(m,\lambda)}(z)$.*

PROOF. It suffices to work in finitely many variables $Z = z_1 + \cdots + z_n$. Then we have the partial fraction expansion

$$\Omega[uZ] = \prod_i \frac{1}{1 - uz_i} = \sum_i \frac{1}{1 - uz_i} \prod_{j \neq i} \frac{1}{1 - z_j/z_i}. \tag{35}$$

The classical formula in n variables for a Schur function as a ratio of determinants, or equivalently the Weyl character formula for GL_n, can be written

$$s_\lambda(z) = \sum_{w \in S_n} w\left(\frac{z^\lambda}{\prod_{i<j}(1 - z_j/z_i)}\right). \tag{36}$$

Now observe that for any polynomial f, we have

$$[u^m]f(u^{-1})\frac{1}{1 - uz} = [u^0]u^{-m}f(u^{-1})\frac{1}{1 - uz} = z^m f(z).$$

Combined with (34) and (35), this gives the formula for the Bernstein operator in n variables

$$S^0_m f(z) = \sum_i z_i^m \frac{f(Z - z_i)}{\prod_{j \neq i}(1 - z_j/z_i)}.$$

The desired result now follows easily by induction using (36). □

It is customary to define s_λ by (36) when λ is any integer sequence, not necessarily a partition. Then, setting $\delta = (n-1, n-2, \ldots, 1, 0)$, we have $s_\lambda = 0$ if $\lambda + \delta$ does not have distinct parts, and otherwise $s_\lambda = \varepsilon(w)s_\nu$, where $w(\lambda + \delta) = \nu + \delta$ with ν a partition. With these conventions, the identity $S^0_m s_\lambda(z) = s_{(m,\lambda)}(z)$ holds for all $m \geq 0$.

We now introduce two t-deformations of Bernstein's operators that will allow us to construct Hall-Littlewood polynomials.

DEFINITION 3.4.5. The *Jing operators* [**45**] are the coefficients $S^t_m = [u^m]S^t(u)$ of the operator generating function $S^t(u)$ defined by

$$S^t(u)f = f[Z + (t-1)u^{-1}]\Omega[uZ]. \tag{37}$$

Let $\Pi_{(1-t)}$ denote the plethystic substitution operator $\Pi_{(1-t)}f(z) = f[Z(1-t)]$. The *modified Jing operators* are

$$\tilde{S}^t_m = \Pi_{(1-t)} S^t_m \Pi^{-1}_{(1-t)},$$

or equivalently, the coefficients of the generating function

$$\tilde{S}^t(u)f = f[Z - u^{-1}]\Omega[(1-t)uZ]. \tag{38}$$

The ***transformed Hall-Littlewood polynomials*** are defined by

$$H_\mu(z;t) = S^t_{\mu_1} S^t_{\mu_1} \cdots S^t_{\mu_l}(1). \tag{39}$$

We also set $Q_\mu(z;t) \underset{\text{def}}{=} H_\mu[(1-t)Z;t]$, so

$$Q_\mu(z;t) = \tilde{S}^t_{\mu_1} \tilde{S}^t_{\mu_1} \cdots \tilde{S}^t_{\mu_l}(1). \tag{40}$$

The notation $Q_\mu(z;t)$ agrees with that used in Macdonald's book.

3.4.3. *Orthogonality and triangularity*. The basic orthogonality and triangularity properties of Hall-Littlewood polynomials are readily deduced using the Rodrigues-type formulas (39) and (40).

LEMMA 3.4.6. *If $m \geq \mu_1$ and $\lambda \geq \mu$, then*

$$S^t_m s_\lambda \in \mathbb{Z}[t]\{s_\gamma : \gamma \geq (m,\mu)\}.$$

Moreover, $s_{(m,\mu)}$ occurs with coefficient 1 *in $S^t_m s_\mu$.*

PROOF. Recall the Schur function identity (*dual Pieri rule*)

$$s_\lambda[Z+a] = \sum_k a^k \sum_{\lambda/\nu \in H_k} s_\nu,$$

where the notation $\lambda/\nu \in H_k$ means that the skew shape λ/ν is a ***horizontal strip*** of size k, that is, it has at most one cell in each column. Write the Jing operator in terms of the Bernstein operator: $S^t(u)f = S^0(u)f[Z+tu^{-1}]$. Taking the coefficient of u^m and applying the Pieri rule, we get

$$S^t_m s_\lambda = \sum_k t^k \sum_{\lambda/\nu \in H_k} s_{(m+k,\nu)}, \tag{41}$$

with the conventions discussed above if $(m+k,\nu)$ is not a partition. It is convenient to extend the dominance partial order to sequences that may not be partitions by maintaining the same definition (1). In particular, rewriting $s_{(m+k,\nu)}$, if nonzero, as $\pm s_\gamma$ where γ is a partition, we have

$$\gamma \geq (m+k,\nu) \geq (m,\lambda) \geq (m,\mu).$$

Equality can only occur for $k=0$, and does occur then if $\lambda = \mu$. □

COROLLARY 3.4.7. *We have*

$$H_\mu(z;t) = \sum_{\lambda \geq \mu} C_{\lambda\mu}(t) s_\lambda(z)$$

for suitable coefficients $C_{\lambda\mu}(t) \in \mathbb{Z}[t]$, with $C_{\mu\mu}(t) = 1$.

We will see presently that in fact $C_{\lambda\mu}(t) = K_{\lambda\mu}(t)$. Now let us use our second Rodrigues formula (40) to establish an opposite triangularity for $Q_\mu(z;t)$.

LEMMA 3.4.8. *If $m \geq \lambda_1$, then*

$$\tilde{S}^t_m s_\lambda \in \mathbb{Z}[t]\{s_\gamma : \gamma \leq (m,\lambda)\},$$

and the coefficient of $s_{(m,\lambda)}$ in $\tilde{S}^t_m s_\lambda$ is equal to $1-t^\alpha$, where α is the multiplicity of m as a part of (m,λ).

PROOF. In n variables, we can get an explicit formula for $\tilde{S}^t_m$, much as we did for S^0_m in the proof of Proposition 3.4.4, by using the partial fraction expansion

$$\Omega[(1-t)uZ] = \prod_i \frac{1-tuz_i}{1-uz_i} = t^n + (1-t)\sum_i \frac{1}{1-uz_i}\prod_{j\neq i}\frac{1-tz_j/z_i}{1-z_j/z_i}. \tag{42}$$

The resulting formula is

$$\tilde{S}^t_m f = \delta_{m,0}t^n f + (1-t)\sum_i z_i^m f[Z-z_i]\frac{\prod_{j\neq i}(1-tz_j/z_i)}{\prod_{j\neq i}(1-z_j/z_i)} \tag{43}$$

$$= \delta_{m,0}t^n f + (1-t)\sum_i\sum_k (-t)^k z_i^{m-k}\frac{e_k[Z-z_i]f[Z-z_i]}{\prod_{j\neq i}(1-z_j/z_i)}. \tag{44}$$

Taking $f = s_\lambda$ and applying the usual Pieri rule for multiplication by e_k, we obtain for $m > 0$

$$\tilde{S}^t_m s_\lambda = (1-t)\sum_k (-t)^k \sum_{\gamma/\lambda \in V_k} s_{(m-k,\gamma)}.$$

The worst case γ for each k, that is, the maximal one in the dominance order, is $\lambda + (1^k)$. For this case, $s_{(m-k,\gamma)}$, if non-zero, is $\pm s_\nu$ with ν having one of the forms

$$\nu = (\lambda_1, \dots, \lambda_i, m-k+i, \lambda_{i+1}+1, \dots, \lambda_k+1, \lambda_{k+1}, \dots) \quad \text{and } i \leq k, \text{ or}$$
$$\nu = (\lambda_1, \dots, \lambda_k, \lambda_{k+1}-1, \dots, \lambda_i - 1, m-k+i, \lambda_i, \dots) \quad \text{and } i > k.$$

In either case, $\nu \leq (m,\lambda)$ provided $m \geq \lambda_1$. Equality occurs when $m = \lambda_1 = \cdots = \lambda_i$ and $k = i$. Hence the coefficient of $s_{(m,\lambda)}$ in $\tilde{S}^t_m s_\lambda$ is equal to $(1-t)[\alpha]_t = (1-t^\alpha)$. $\square$

COROLLARY 3.4.9. *We have*

$$Q_\mu(z;t) = \sum_{\lambda \leq \mu} B_{\lambda\mu}(t) s_\lambda$$

for suitable coefficients $B_{\lambda\mu}(t) \in \mathbb{Z}[t]$, *and* $B_{\mu\mu} = (1-t)^{l(\mu)}\prod_i [\alpha_i]_t!$, *where* $\mu = (1^{\alpha_1}, 2^{\alpha_2}, \dots,)$.

The operator $\Pi_{(1-t)}$ is self-adjoint for the Hall inner product, that is, $\langle f, g[(1-t)Z]\rangle = \langle f[(1-t)Z], g\rangle$. More generally, for any A there holds the identity

$$\langle f, g[AZ]\rangle = \langle f[AZ], g\rangle.$$

This is easily seen using the Cauchy formula: homogeneous bases $\{u_\lambda\}$ and $\{v_\lambda\}$ are dual with respect to $\langle -,- \rangle$ if and only if

$$\sum_\lambda u_\lambda[Y]v_\lambda[Z] = \Omega[YZ]. \tag{45}$$

Hence $\{u_\lambda\}$ and $\{v_\lambda[AZ]\}$ are dual bases if and only if $\sum_\lambda u_\lambda[Y]v_\lambda[Z] = \Omega[YZ/A]$, and this condition is symmetric between $\{u_\lambda\}$ and $\{v_\lambda\}$.

By Corollaries 3.4.7 and 3.4.9, if $\langle H_\mu, H_\nu[(1-t)Z]\rangle \neq 0$, we must have $\mu \leq \nu$. By symmetry, we must also have $\nu \leq \mu$, so $\mu = \nu$. Together with the leading terms determined in Corollaries 3.4.7 and 3.4.9, this gives the following result.

COROLLARY 3.4.10. *The transformed Hall-Littlewood polynomials are orthogonal with respect to the inner product $\langle f, g[(1-t)Z]\rangle$, and their self-inner-products are given by*

$$\langle H_\mu, H_\mu[(1-t)Z]\rangle = (1-t)^{l(\mu)} \prod_i [\alpha_i]_t!, \quad \mu = (1^{\alpha_1}, 2^{\alpha_2}, \ldots). \tag{46}$$

The polymomials H_μ are uniquely characterized by any two of the three properties expressed by Corollaries 3.4.7, 3.4.9, and 3.4.10.

Equation (43) yields by induction an explicit formula for $Q_\lambda(z;t)$ in n variables, analogous to the classical formula (36) for Schur functions. In this way we can recover the classical formula (31) for Hall-Littlewood polynomials.

PROPOSITION 3.4.11. *The symmetric functions $Q_\lambda(z;t) = H_\lambda[(1-t)Z;t]$ are given in n variables $z_1, \ldots, z_n$ by*

$$Q_\lambda(z;t) = (1-t)^{l(\lambda)}[n-l(\lambda)]_t! \sum_{w \in S_n} w\left(z^\lambda \frac{\prod_{i<j}(1-tz_j/z_i)}{\prod_{i<j}(1-z_j/z_i)}\right).$$

Hence the classical Hall-Littlewood polynomials $P_\lambda(z;t)$ are equal to

$$P_\lambda(z;t) = \frac{Q_\lambda(z;t)}{(1-t)^{l(\lambda)}\prod_i [\alpha_i]_t!}, \quad \lambda = (1^{\alpha_1}, 2^{\alpha_2}, \ldots). \tag{47}$$

In the classical theory of Hall-Littlewood polynomials, one uses the t-inner product given in our language by

$$\langle f, g\rangle_t = \langle f, g[Z/(1-t)]\rangle.$$

Corollary 3.4.10 and equation (47) imply that $\langle P_\lambda, Q_\mu\rangle_t = \delta_{\lambda\mu}$, that is, $\{P_\lambda\}$ and $\{Q_\lambda\}$ are orthogonal (but not orthonormal) bases for $\langle -,-\rangle_t$, dual to each other. Hence

$$K_{\lambda\mu}(t) = \langle s_\lambda, Q_\mu\rangle_t = \langle s_\lambda, H_\mu\rangle.$$

This shows, as claimed earlier, that the coefficients $C_{\lambda\mu}(t)$ in Corollary 3.4.7 are equal to the $K_{\lambda\mu}(t)$.

To keep things organized, let me summarize in one place our conclusions so far.

COROLLARY 3.4.12. *The transformed Hall-Littlewood polynomials H_μ are related to the classical Hall-Littlewood polynomials by*

$$H_\mu[(1-t)Z;t] = Q_\mu(z;t) = (1-t)^{l(\mu)} \prod_{i=1}^{\mu_1} [\alpha_i(\mu)]_t! P_\mu(z;t). \tag{48}$$

They are uniquely characterized by the following properties.

(i) $H_\mu(z;t) \in \mathbb{Z}[t]\cdot\{s_\lambda : \lambda \geq \mu\}$,
(ii) $H_\mu[(1-t)Z;t] \in \mathbb{Z}[t]\cdot\{s_\lambda : \lambda \leq \mu\}$
(iii) $\langle s_\mu, H_\mu\rangle = 1$.

The Kostka-Foulkes polynomials $K_{\lambda\mu}(t)$ can be defined through the Schur function expansion

$$H_\mu(z;t) = \sum_{\lambda\mu} K_{\lambda\mu}(t)s_\lambda(z),$$

and enjoy the following properties.

(iv) $K_{\lambda\mu}(t) \in \mathbb{Z}[t]$,

(v) $K_{\lambda\mu}(t) = 0$ *unless* $\lambda \geq \mu$, *and* $K_{\mu\mu}(t) = 1$,
(vi) $K_{\lambda\mu}(0) = \delta_{\lambda\mu}$, *that is,* $H_\mu(z;0) = s_\mu(z)$,
(vii) $K_{\lambda\mu}(1) = K_{\lambda\mu}$, *that is,* $H_\mu(z;1) = h_\mu(z)$.

3.4.4. *Interpretation in geometry and representation theory.* The geometric interpretation of the Kostka-Foulkes polynomials that I wish to discuss is summarized in the next proposition, which combines results of Hotta, Lusztig and Springer [**42, 58, 81**]. To state it we need a bit of terminology. Let $\mathcal{N}$ be the variety of $n \times n$ nilpotent matrices, that is, the ***nilpotent variety*** in the Lie algebra $\mathfrak{gl}_n$. Via the exponential map, we can identify $\mathcal{N}$ with the variety of unipotent elements in GL_n. The adjoint action of GL_n on $\mathcal{N}$ is by similarities $x \mapsto gxg^{-1}$, so the GL_n-orbit of an element $x \in \mathcal{N}$ is given by its Jordan canonical form. For each partition λ of n, let O_λ denote the orbit whose elements have Jordan block sizes $\lambda_1, \ldots, \lambda_l$.

The ***flag variety*** $\mathcal{B} = GL_n/B$ may be identified concretely either with the variety of Borel subalgebras $\mathfrak{b} \subseteq \mathfrak{gl}_n$ or with the variety of flags of subspaces

$$0 \subset F_1 \subset F_2 \subset \cdots \subset F_{n-1} \subset \mathbb{C}^n, \quad \dim F_i = i. \tag{49}$$

A flag F is ***compatible*** with $x \in \mathcal{N}$ if $xF_i \subseteq F_{i-1}$ for each i, or equivalently, F is fixed by the unipotent element $\exp(x) \in GL_n$, or yet again equivalently, we have $x \in \mathfrak{b}$ for the corresponding Borel $\mathfrak{b}$, the Lie algebra of the stabilizer of F. The variety

$$Z_0 = \{(x, \mathfrak{b}) \in \mathcal{N} \times \mathcal{B} : x \in \mathfrak{b}\} \tag{50}$$

is a vector bundle over $\mathcal{B}$, hence nonsingular. The *Grothendieck resolution* is the proper and birational map

$$g\colon Z_0 \to \mathcal{N}$$

given by projection on the first factor. Its fiber over any element $x \in O_\lambda$ is called a *Springer fiber*, and denoted $\mathcal{B}_\lambda$. Thus $\mathcal{B}_\lambda \subseteq \mathcal{B}$ is just the variety of flags fixed by a unipotent matrix of Jordan type λ. The dimension of $\mathcal{B}_\lambda$, which is equal to one-half the codimension of O_λ in $\mathcal{N}$, is the statistic $n(\lambda)$ discussed in §2.1.

Finally, let $GL_n(q)$ be the linear group over the finite field $\mathbb{F}_q$ with q elements. Identifying O_λ with a unipotent orbit in GL_n, its $\mathbb{F}_q$-rational points form a conjugacy class $O_\lambda(q) \subseteq GL_n(q)$ (here life is simpler for GL_n than for other semisimple Lie groups G; in general a unipotent orbit may break up into several conjugacy classes of the corresponding finite Chevalley group). The finite group $GL_n(q)$ acts on the finite set $\mathcal{B}(q)$ of $\mathbb{F}_q$-rational points of the flag variety, which in our case is just the set of flags as in (49), but in the finite vector space $\mathbb{F}_q^n$ instead of $\mathbb{C}^n$.

The characters χ_λ of those irreducible representations V_λ of $GL_n(q)$ that occur in the permutation representation on $\mathbb{C}\cdot\mathcal{B}(q)$ are called ***unipotent characters.*** Their indexing by partitions λ comes about as follows. Fix the Borel subgroup $B \subseteq GL_n(q)$ of upper triangular matrices. The ***Hecke algebra*** $H_n(q)$ is the subalgebra $H_n(q) \subseteq \mathbb{C}\,GL_n(q)$ of elements $\sum_g a_g g$ with coefficients a_g constant on double cosets BwB. As is well-known, $H_n(q)$ is the specialization at the integer q of a generic Hecke algebra in which q is an indeterminate, and its specialization at $q = 1$ is the group algebra of the Weyl group, that is, $H_n(1) = \mathbb{C}S_n$ for $G = GL_n$. Both the generic Hecke algebra and these specializations are semi-simple, so the irreducible representations V^λ of $H_n(q)$ are naturally identified with those of $\mathbb{C}S_n$. By general principles for double coset algebras, $H_n(q) \times GL_n(q)$ acts on $\mathbb{C} \cdot \mathcal{B}(q)$ and this representation decomposes into irreducibles as

$$\mathbb{C} \cdot B(q) \cong \bigoplus_{\lambda} V^{\lambda} \otimes V_{\lambda}.$$

Thus there is a natural correspondence between the unipotent characters χ_λ of $GL_n(q)$ and the irreducible characters χ^λ of S_n.

DEFINITION 3.4.13. The "cocharge" Kostka-Foulkes polynomials are

$$\tilde{K}_{\lambda\mu}(q) \underset{\text{def}}{=} q^{n}(\mu) K_{\lambda\mu}(q^{-1}).$$

It is not hard to see—from the proof of Lemma 3.4.6, for instance—that the degree of $K_{\lambda\mu}(t)$ is at most $n(\mu)$, so $\tilde{K}_{\lambda\mu}(q)$ is in fact a polynomial.

PROPOSITION 3.4.14. *The Kostka-Foulkes polynomials $K_{\lambda\mu}(q)$ and their cocharge variants $\tilde{K}_{\lambda\mu}(q)$ have the following interpretations.*

(i) *The value of a unipotent character of $GL_n(q)$ on a unipotent conjugacy class is given by $\chi_\lambda(u) = \tilde{K}_{\lambda\mu}(q)$, for $u \in O_\mu(q)$.*

(ii) *The Poincaré series of the local intersection homology of the closure $\overline{O}_\lambda$ at any point $x \in O_\mu$ is given by $\sum_k q^k\, IH^{2k}_x(\overline{O}_\lambda) = q^{n(\mu)-n(\lambda)} K_{\lambda\mu}(q)$.*

(iii) *For the Springer action* [**81**] *of the Weyl group S_n on the cohomology ring $H^*(\mathcal{B}_\mu)$ of a Springer fiber, we have $\sum_q q^k \langle \chi^\lambda, \operatorname{char} H^{2k}(\mathcal{B}_\mu)\rangle = \tilde{K}_{\lambda\mu}(q)$. In other words, the Frobenius series of $H^*(\mathcal{B}_\mu)$ is given by $F_{H^*(\mathcal{B}_\mu)}(z;q) = q^{n(\mu)} H_\mu(z;q^{-1})$.*

It is somewhat difficult to gain an understanding of these results from the original papers, because the theory of perverse sheaves, which simplifies and clarifies the proofs, was developed later. I will not discuss the proofs in any detail, but will only mention a few points to make the present discussion self-contained. For further information I recommmend the excellent exposition by Shoji [**78**].

One point that deserves attention is the definition of the Springer action referred to in part (iii) of the Proposition. The variety Z_0 in (50) is part of a larger bundle over the flag variety,

$$Z = \{(x, \mathfrak{b}) \in \mathfrak{gl}_n \times \mathcal{B} : x \in \mathfrak{b}\},$$

and projection on the first factor again yields a proper map

$$f\colon Z \to \mathfrak{gl}_n.$$

This map f is not birational, but it is generically finite. Specifically, the preimage of the set $(\mathfrak{gl}_n)_{\mathrm{rss}}$ of regular semi-simple elements has a natural structure of principal S_n-bundle. Let $\mathbb{C}_Z$ be the trivial constant sheaf on Z. The fundamental *decomposition theorem* of Beilinson-Bernstein-Deligne-Gabber [**4**], together with an easy dimension argument, implies that the object $Rf_*\mathbb{C}_Z$ in the derived category of constructible sheaves on $\mathfrak{gl}_n$ is a perverse sheaf, and furthermore, it is the perverse extension of its restriction to $(\mathfrak{gl}_n)_{\mathrm{rss}}$. Since S_n acts naturally on this restriction, it acts on $Rf_*\mathbb{C}_Z$. But the cohomology ring $H^*(\mathcal{B}_\mu)$ is just the stalk of $Rf_*\mathbb{C}_Z$ at $x \in O_\mu$. The action of S_n on the stalk is the Springer action.

The equivalence of interpretations (ii) and (iii) in Proposition 3.4.14 follows from similar considerations involving $Rg_*\mathbb{C}_{Z_0}$. Their further equivalence with interpretation (i) is part of Lusztig's theory of *character sheaves.*

As for the identification with $\tilde{K}_{\lambda\mu}(q)$ of the quantity described by all three interpretations, Shoji gave a procedure for computing the characters of the cohomology rings of Springer fibers (referred to in this context as *Green polynomials*) for the classical groups G, and this was extended by Lusztig to all G in [**59**]. Shoji and Lusztig characterize the Green polynomials by triangularity conditions, which in the case of GL_n, when translated into symmetric function language using the Frobenius series and Proposition 3.3.1, amount to conditions (i)-(iii) in Corollary 3.4.12.

Finally, a word is in order here about the situation for a general semisimple Lie group G. The things described by the three parts of Proposition 3.4.14 again coincide for general G, with certain adjustments.

First of all, the cohomology rings $H^*(\mathcal{B}_u)$ depend, strictly speaking, on the choice of u in a nilpotent orbit O. On a fixed orbit they are all isomorphic, but the locally constant sheaf with stalk $H^*(\mathcal{B}_u)$ at u is usually not trivial. Various pairs (O, L) consisting of a nilpotent orbit O and an irreducible local system L on it arise in this way. These pairs, rather than the orbits themselves, are what correspond to unipotent conjugacy classes in the finite Chevalley group $G(q)$. To formulate the Proposition correctly, in (iii) one should consider separately the summands of $H^*(\mathcal{B}_u)$ corresponding to different local systems L on O. Also (ii) has to be rephrased to take account of the local systems.

Secondly, part (i) is only "almost" correct for general G. The unipotent characters χ_λ have to be replaced with certain linear combinations, called *almost characters*. This phenomenon occurs because Lusztig's character sheaves are imperfect geometric analogs of characters. The linear transformations required involve only a few characters at a time, and are given by certain small matrices which have been explicitly determined by Lusztig.

3.4.5. *Combinatorial interpretations.* The following theorem was discovered by Lascoux and Schützenberger [**57**] and its proof completed by Butler [**11**].

THEOREM 3.4.15. *The Kostka-Foulkes polynomial $K_{\lambda\mu}(t)$ is given in terms of the charge statistic defined in §2.1, evaluated on semistandard tableaux T of shape λ and content μ, by*

$$K_{\lambda\mu}(t) = \sum_T t^{c(T)}.$$

Equivalently, the cocharge Kostka-Foulkes polynomials are given in terms of cocharge by

$$\tilde{K}_{\lambda\mu}(t) = \sum_T t^{cc(T)}.$$

An alternative combinatorial interpretation of $K_{\lambda\mu}(t)$ was given by Kirillov and Reshetikhin [**47, 48**]. For any partition ν, let

$$s_k(\lambda) = \lambda_1 + \cdots + \lambda_k$$

denote the sum of the first k parts of ν. A (λ, μ) *configuration* is a sequence of partitions

$$\nu = (\nu^0 = \mu', \nu^1, \ldots, \nu^{l(\lambda)})$$

of sizes

$$|\nu^k| = |\lambda| - s_k(\lambda).$$

Note that $\nu^{l(\lambda)}$ is empty by definition. The configuration ν is said to be ***admissible*** if the numbers

$$p_j^k(\nu) \underset{\text{def}}{=} s_j(\nu^{k-1}) - 2s_j(\nu^k) + s_j(\nu^{k+1}), \quad \text{for } i = 1, \dots, l(\lambda) - 1 \text{ and all } j$$

are non-negative. Note that for j sufficiently large, $p_j^k(\nu) = \lambda_k - \lambda_{k+1}$, which is necessarily non-negative.

THEOREM 3.4.16 (Kirillov-Reshetikhin [**48**]). *The Kostka-Foulkes polynomial $K_{\lambda\mu}(t)$ is given by the sum*

$$K_{\lambda\mu}(t) = \sum_{\nu} t^{n(\mu) - \sum_{j\geq 1} \mu_j' \nu_j^1 + \sum_{k,j\geq 1} \nu_j^k (\nu_j^k - \nu_j^{k+1})} \prod_{k,j\geq 1} \begin{bmatrix} p_j^k + \nu_j^k - \nu_{j+1}^k \\ \nu_j^k - \nu_{j+1}^k \end{bmatrix}_t \tag{51}$$

over all (λ, μ)-admissible configurations ν, where

$$\begin{bmatrix} n \\ k \end{bmatrix}_t \underset{\text{def}}{=} \frac{[n]_t!}{[k]_t![n-k]_t!}$$

are the Gauss binomial coefficients.

The theorem is proved by attaching to each configuration ν some additional data, called a ***rigging***, so that the term corresponding to ν in (51) is a weighted enumeration of the possible riggings. Then Kirillov and Reshetikhin give a bijection between rigged configurations and semistandard tableaux in which the weight of the configuration corresponds to the charge of the tableau. This reduces Theorem 3.4.16 to Theorem 3.4.15.

Theorem 3.4.16 has an interesting origin. A technique from mathematical physics known as the *Bethe ansatz* enables one to produce highest weight vectors (called ***Bethe vectors*** in this context) for the irreducible constituents in tensor products of GL_n modules $V_{\mu_1} \otimes \cdots \otimes V_{\mu_r}$, where the μ_i are rectangular partitions. In this particular application of the Bethe ansatz it turns out that the resulting system of Bethe vectors is complete, and that they are naturally indexed by rigged configurations. The weight of the rigged configuration has a physical interpretation as a quantum number of the state described by the corresponding Bethe vector. When the rectangles μ_i are the rows of μ, the relevant configurations are the admissible configurations in Theorem 3.4.16. Thus the theorem says that $K_{\lambda\mu}(t)$ enumerates Bethe vectors in $V_{(\mu_1)} \otimes \cdots \otimes V_{(\mu_r)}$ by quantum number.

3.4.6. *The method of Garsia and Procesi.* The proof of Theorem 3.4.15 is complicated and not particularly illuminating, while the proof of Proposition 3.4.14 requires heavy intersection cohomology machinery. A simpler route than either of these to the positivity theorem for the Kostka-Foulkes polynomials was found by Garsia and Procesi [**26**], with some improvements by N. Bergergon and Garsia [**8**]. I will outline their approach.

The basic idea is to describe the cohomology ring $R_\mu = H^*(\mathcal{B}_\mu)$ in elementary terms, without reference to its geometric origin. In order to motivate the description it is helpful first to recall some geometrical facts about R_μ, although they are not actually needed for the construction. For any semisimple Lie group G, the cohomology ring $H^*(\mathcal{B})$ of the whole flag variety is isomorphic to the ring of ***coinvariants*** for the Weyl group W acting on a Cartan subalgebra $\mathfrak{h}$. The coinvariant ring is by definition $\mathbb{C}[\mathfrak{h}]/I$, where $I = (\mathbb{C}[\mathfrak{h}]_+^W)$ is generated by the W-invariant polynomials without constant term. The isomorphism $H^*(\mathcal{B}) \cong \mathbb{C}[\mathfrak{h}]/I$ is W-equivariant for the Springer action on $H^*(\mathcal{B})$ and the tautological W action on $\mathbb{C}[\mathfrak{h}]/I$.

For $G = GL_n$, the ring $\mathbb{C}[\mathfrak{h}]$ is a polynomial ring $\mathbb{C}[\mathbf{x}] = \mathbb{C}[x_1, \dots, x_n]$ in n variables, the Weyl group W is S_n, and $I = (e_1, \dots, e_n)$ is generated by the elementary symmetric functions. The homomorphism $H^*(\mathcal{B}) \to H^*(\mathcal{B}_\mu)$ induced by the inclusion $\mathcal{B}_\mu \subseteq \mathcal{B}$ is always W-equivariant. For $G = GL_n$ these homomorphims are surjective, so R_μ is $\mathbb{C}[\mathbf{x}]/I_\mu$ for a homogeneous S_n-invariant ideal $I_\mu \supseteq I$. The highest degree in which R_μ is nonzero is $\dim(\mathcal{B}_\mu) = n(\mu)$, and by the Springer correspondence, this top degree component of R_μ affords the irreducible representation V_μ of S_n.

Recall from §2.1 that $\mathbb{C}[\mathbf{x}]$ contains a unique copy of V_μ in degree $n(\mu)$, spanned by the Garnir polynomials $g_T(\mathbf{x})$ for standard tableaux T of shape μ. It develops that

> *I_μ is the unique largest homogeneous S_n-invariant ideal having zero intersection with the unique copy of V^μ in degree $n(\mu)$.*

So, following the Garsia-Procesi approach, let us forget about $H^*(\mathcal{B}_\mu)$, define I_μ and $R_\mu = \mathbb{C}[\mathbf{x}]/I_\mu$ by the above characterization, and proceed to study the graded character of R_μ. The first lemma is an easy general consequence of the definition of I_μ.

LEMMA 3.4.17 (Bergeron-Garsia [8]). *Let $\mathbb{C}[\partial] = \mathbb{C}[\partial/\partial x_1, \dots, \partial/\partial x_n]$ act on $\mathbb{C}[\mathbf{x}]$ by differentiation, and let $G_\mu \subseteq \mathbb{C}[\mathbf{x}]$ be the $\mathbb{C}[\partial]$-module generated by the Garnir polynomials g_T for T of shape μ.*

(i) *A polynomial $f \in \mathbb{C}[\mathbf{x}]$ belongs to I_μ if and only if $f(\partial)g_T = 0$ for all T.*
(ii) *The canonical projection of $G_\mu \subseteq \mathbb{C}[\mathbf{x}]$ on $R_\mu = \mathbb{C}[\mathbf{x}]/I_\mu$ is an isomorphism.*

Let $\mathbf{x}_S$ denote the subset of the variables $x_1, \dots, x_n$ indexed by a subset $S \subseteq \{1, \dots, n\}$. Using Lemma 3.4.17 one can prove that the elementary symmetric function $e_k(\mathbf{x}_S)$ belongs to I_μ whenever

$$k > |S| - n + \mu'_1 + \mu'_2 + \cdots \mu'_{n-|S|}. \tag{52}$$

By considering the leading terms of derivatives of the Garnir polynomials, Bergeron and Garsia showed that G_μ and R_μ have dimension at least $\binom{n}{\mu_1, \dots, \mu_l}$. On the other hand Garsia and Procesi constructed a set of $\binom{n}{\mu_1, \dots, \mu_l}$ monomials which span $\mathbb{C}[\mathbf{x}]$ modulo the ideal generated by the functions $e_k(\mathbf{x}_S)$ with k and S satisfying (52). This proves the following result.

PROPOSITION 3.4.18 (Garsia-Procesi [26]). *The* Tanisaki generators $e_k(\mathbf{x}_S)$ *for k and S satisfying* (52) *generate the ideal I_μ.*

With this established, Garsia and Procesi showed that the Frobenius series of R_μ satisfies a simple recurrence that also characterizes the "cocharge" version of the transformed Hall-Littlewood polynomials, thereby recovering the result in Proposition 3.4.14 (iii) by direct, elementary means.

THEOREM 3.4.19 (Garsia-Procesi [26]). *The Frobenius series of R_μ is given by*

$$F_{R_\mu}(z;t) = t^{n(\mu)} H_\mu(z;t^{-1}) = \sum_\lambda \tilde{K}_{\lambda\mu}(t) s_\lambda.$$

In particular, $\tilde{K}_{\lambda\mu}(t)$ is a polynomial with non-negative integer coefficients.

Another aspect of Garsia and Procesi's approach is worth mentioning here. Fix distinct complex numbers $\alpha_1, \dots, \alpha_l$, and let $a \in \mathbb{C}^n$ be a point with μ_1 coordinates equal to α_1, μ_2 equal to α_2, and so on. Let $I_a \subseteq \mathbb{C}[\mathbf{x}]$ be the ideal of polynomials vanishing on the orbit $S_n \cdot a$, and let $\operatorname{gr} I_a$ denote its ideal of leading forms. Clearly $\dim \mathbb{C}[\mathbf{x}]/\operatorname{gr} I_a = \dim \mathbb{C}[\mathbf{x}]/I_a = |S_n \cdot a| = \binom{n}{\mu_1,\ldots,\mu_l}$. It is not hard to show, as Garsia and Procesi do in [**26**], that the Tanisaki generators $e_k(\mathbf{x}_S)$ belong to $\operatorname{gr} I_a$, and hence the dimension count implies that $\operatorname{gr} I_a = I_\mu$. In particular, R_μ affords the induced representation $1\uparrow_{S_\mu}^{S_n}$, where $S_\mu \cong S_{\mu_1} \times \cdots \times S_{\mu_l}$ is the stabilizer of a. Written in terms of the Frobenius characteristic, this says

$$F_{R_\mu}(z;1) = h_\mu(z).$$

Since the top degree part of R_μ is V^μ, it follows that $t^{n(\mu)}F_{R_\mu}(z;t^{-1})$ has properties (i) and (iii) in the characterization of $H_\mu(z;t)$ in Corollary 3.4.12. The remaining property (ii) is a consequence of Proposition 3.3.1 and the following proposition.

PROPOSITION 3.4.20. *The S_n-modules $\operatorname{Tor}_i^{\mathbb{C}[\mathbf{x}]}(R_\mu, \mathbb{C})$ contain only irreducible representations V_λ with $\lambda \geq \mu'$. In particular, R_μ has an S_n-equivariant graded free resolution over $\mathbb{C}[\mathbf{x}]$ whose terms are generated by S_n-modules containing only those irreducibles.*

This proposition follows from the geometric theorems around the $n!$ conjecture, but I think an elementary proof should be possible. This would improve further on the results of Bergeron, Garsia and Procesi by providing an even more direct route to the identification of the character of R_μ.

3.4.7. *Characterization of the cocharge Kostka-Foulkes polynomials.* For comparison with the Macdonald polynomials discussed in the next section it will be useful to reformulate the conditions characterizing $H_\mu(z;t)$ in Corollary 3.4.12 in terms of their cocharge variant

$$\tilde{H}_\mu(z;t) \underset{\text{def}}{=} t^{n(\mu)}H_\mu(z;t^{-1}) = \sum_\lambda \tilde{K}_{\lambda\mu}(t)s_\lambda(t).$$

To do this, we note that $\tilde{K}_{(n)\mu}(t) = 1$ for all μ. This is clear from the geometric interpretation, since $\tilde{K}_{(n)\mu}(t)$ is the Hilbert series of the S_n invariants in R_μ. Alternatively, it is not difficult to deduce from (41) that $K_{(n)\mu}(t) = t^{n(\mu)}$. We also note that $\tilde{H}_\mu[(1-t)Z;t] = t^{n(\mu)}(-t)^{|\mu|}\omega H_\mu[(1-t^{-1})z;t^{-1}]$, which contains only Schur functions s_λ with $\lambda' \leq \mu$, or equivalently $\lambda \geq \mu'$. Hence the desired characterization is as follows.

COROLLARY 3.4.21. *The cocharge variant transformed Hall-Littlewood polynomials $\tilde{H}_\mu(z;t)$ are uniquely characterized by the properties*

(i) $\tilde{H}_\mu(z;t) \in \mathbb{Z}[t]\{s_\lambda : \lambda \geq \mu\}$;
(ii) $\tilde{H}_\mu[(1-t)Z;t] \in \mathbb{Z}[t]\{s_\lambda : \lambda \geq \mu'\}$;
(iii) $\tilde{H}_\mu[1;t] = \langle s_{(n)}, \tilde{H}_\mu(z;t)\rangle = 1$.

3.5. Macdonald polynomials.

3.5.1. *Definition and transformed version; Kostka-Macdonald coefficients.* We begin with Macdonald's original definition [**60, 61**] of his polynomials as deformations of the Hall-Littlewood polynomials $P_\lambda(z;t)$ with an extra parameter q. Macdonald first defines a q,t-deformation of the Hall inner product which in our notation is

$$\langle f,g\rangle_{q,t} \underset{\text{def}}{=} \langle f(z),\, g[\tfrac{1-q}{1-t}Z]\rangle.$$

DEFINITION 3.5.1. The *Macdonald symmetric functions* $P_\mu(z;q,t)$ are uniquely characterized by the orthogonality and triangularity conditions

(i) $P_\mu(z;q,t) = s_\mu + \sum_{\lambda<\mu} a_{\lambda\mu}(q,t)s_\lambda$, for suitable coefficients $a_{\lambda\mu} \in \mathbb{Q}(q,t)$;
(ii) $\langle P_\lambda, P_\mu\rangle = 0$ if $\lambda \neq \mu$.

For all purposes that we will be concerned with here it is better to work with transformed versions of the Macdonald polynomials, which are q-deformations of the cocharge variant transformed Hall-Littlewood polynomials $\tilde{H}_\mu(z;t)$.

DEFINITION 3.5.2. The *transformed Macdonald symmetric functions* $\tilde{H}_\mu(z;q,t)$ are uniquely characterized by the conditions

(i) $\tilde{H}_\mu[(1-q)Z;q,t] \in \mathbb{Q}(q,t)\{s_\lambda : \lambda \geq \mu\}$;
(ii) $\tilde{H}_\mu[(1-t)Z;q,t] \in \mathbb{Q}(q,t)\{s_\lambda : \lambda \geq \mu'\}$;
(iii) $\tilde{H}_\mu[1;q,t] = 1$.

Observe that from Corollary 3.4.21 it immediately follows that $\tilde{H}_\mu(z;0,t) = \tilde{H}_\mu(z;t)$. Of course the existence of polynomials meeting the conditions in either of the above definitions requires proof. We will prove in §3.5.2 that the $\tilde{H}_\mu(z;q,t)$ exist, so let us now see that this implies that the $P_\mu(z;q,t)$ exist. To this end, we define another q,t-inner product

$$\langle f,g\rangle_* \underset{\text{def}}{=} \langle f,\, \omega g[(1-q)(1-t)Z]\rangle = (-t)^d\langle f,\, g[(1-q)(1-t^{-1})Z]\rangle, \tag{53}$$

where the last equality is for f and g homogeneous of degree d in z. We have $\langle \tilde{H}_\mu, \tilde{H}_\nu\rangle_* = \langle \tilde{H}_\mu[(1-q)Z;q,t],\, \omega\tilde{H}_\nu[(1-t)Z;q,t]\rangle$. If this is non-zero then by (i) and (ii), we must have $\mu \leq \nu$. By symmetry, we must also have $\nu \leq \mu$, so we have

(iv) $\langle \tilde{H}_\mu, \tilde{H}_\nu\rangle_* = 0$ if $\mu \neq \nu$.

It follows that the symmetric functions

$$J_\mu(z;q,t) \underset{\text{def}}{=} t^{n(\mu)}\tilde{H}_\mu[(1-t^{-1})Z;q,t^{-1}] \tag{54}$$

have the orthogonality property (ii) of Definition 3.5.1, and they also have the trianguarity property (i), except that the leading coefficient need not be 1. But we can of course divide by the leading coefficient to obtain the required polynomials $P_\mu(z;q,t)$.

DEFINITION 3.5.3. The *Kostka-Macdonald polynomials* $\tilde{K}_{\lambda\mu}(q,t)$ are defined through the Schur function expansion

$$\tilde{H}_\mu(z;q,t) = \sum_\lambda \tilde{K}_{\lambda\mu}(q,t)s_\lambda(z).$$

From the usual proof of the existence theorem for Macdonald symmetric functions (see §3.5.2, below) it is by no means obvious that the "polynomials" $\tilde{K}_{\lambda\mu}(q,t)$ are anything more than rational functions of q and t. Their ***integrality*** property,

$$\tilde{K}_{\lambda\mu}(q,t) \in \mathbb{Z}[q,t],$$

remained unproven until circa 1995, when several proofs were independently discovered by a number of people using a variety of methods [**27, 28, 29, 49, 50, 55, 76**]. I will mention one way of proving integrality later.

Definition 3.5.3 is related to the original one of Macdonald in the following way. The symmetric functions $J_\mu(z;q,t)$ defined in (54) are the ***integral forms*** of Macdonald, who defined coefficients $K_{\lambda\mu}(q,t)$ through the expansion

$$J_\mu(z;q,t) = \sum_\lambda K_{\lambda\mu}(q,t) s_\lambda[Z/(1-t)].$$

By (54), this is equivalent to $\tilde{K}_{\lambda\mu}(q,t) = t^{n(\mu)} K_{\lambda\mu}(q,t^{-1})$. Macdonald defined the integral forms $J_\mu(z;q,t)$ to be scalar multiples of $P_\mu(z;q,t)$ by an explicit normalizing factor. To see that our J_μ is the same scalar multiple of P_μ as Macdonald's, one may compare the identity $K_{(n)\mu}(q,t) = t^{n(\mu)}$ obtained by Macdonald [**61**] with our $\tilde{K}_{(n)\mu}(q,t) = 1$, which is another way of stating part (iii) of Definition 3.5.2.

3.5.2. *Existence.* We will prove that the polynomials $\tilde{H}_\mu(z;q,t)$ meeting the conditions in Definition 3.5.2 exist by exhibiting them as eigenfunctions of the operator

$$D \underset{\text{def}}{=} \frac{1}{(1-q)(1-t)}(1-D_0), \tag{55}$$

where

$$D_0 f = [u^0] f[Z + (1-q)(1-t)u^{-1}]\Omega[-uZ]. \tag{56}$$

We shall see that the existence theorem also follows as a consequence of the geometric results to be discussed later. This is the case for various other aspects of the elementary theory of Macdonald polynomials as well, such as integrality results. I have included the elementary existence proof here for clarity and to keep this part of the discussion self-contained.

LEMMA 3.5.4. *Set* $\hat{D} = \Pi_{(1-t^{-1})} D_0 \Pi^{-1}_{(1-t^{-1})}$, *or explicitly,*

$$\hat{D} f = [u^0] f[Z - (1-q)u^{-1}]\Omega[(1-t)uZ].$$

Then $\hat{D}$ is lower-triangular with respect to the Schur basis. More precisely,

$$\hat{D} s_\mu = (1-(1-q)(1-t)B_\mu(q,t)) s_\mu + \sum_{\lambda<\mu} b_{\lambda\mu}(q,t) s_\lambda \tag{57}$$

for suitable coefficients $b_{\lambda\mu}(q,t) \in \mathbb{Z}[q,t]$, where

$$B_\mu(q,t) \underset{\text{def}}{=} \sum_{(i,j)\in\mu} q^i t^j. \tag{58}$$

PROOF. Write $\hat{D}$ in terms of the modified Jing operators as $\hat{D}f = [u^0]\tilde{S}^t(u) f[Z+qu^{-1}]$. Then using the dual Pieri rule as in the proof of Lemma 3.4.6, we have

$$\hat{D} s_\mu = \sum_m q^m \sum_{\mu/\lambda \in H_m} \tilde{S}^t_m s_\lambda.$$

As in the proof of Lemma 3.4.8, this is equal in n variables $z_1, \ldots, z_n$ to

$$t^n s_\mu + (1-t)\sum_{m,k} q^m (-t)^k \sum_{\mu/\lambda\in H_m} \sum_{\gamma/\lambda \in V_k} s_{(m-k,\gamma)}.$$

Also as in the proof of Lemma 3.4.8, the worst case is $\gamma = \lambda + (1^k)$, and then $s_{(m-k,\gamma)} = 0$ or $\pm s_\nu$ with ν of the form indicated there. The condition $\mu/\lambda \in H_m$

implies that $\mu_1 \geq \lambda_1 \geq \mu_2 \geq \lambda_2 \geq \cdots$, and it follows with little difficulty that $\nu \leq \mu$.

By carefully examining how equality can be attained, one sees that the contribution to the coefficient of s_μ from the sum over m and k is equal to $\sum_{j=1}^{n} t^{j-1} q^{\mu_j}$, with μ_j interpreted as zero for $j > l(\mu)$. But this is equal to $[n]_t - (1-q)B_\mu(q,t)$, completing the proof. □

Lemma 3.5.4 implies that $\hat{D}$ has distinct eigenvalues and that its eigenfunction $F(z;q,t)$ with eigenvalue $1-(1-q)(1-t)B_\mu(q,t)$ belongs to the space $\mathbb{Q}(q,t)\{s_\lambda : \lambda \leq \mu\}$. Then $H(z;q,t) = F[Z/(1-t^{-1});q,t]$ is an eigenfunction of D with eigenvalue $B_\mu(q,t)$, and $H[(1-t)Z;q,t]$ belongs to $\mathbb{Q}(q,t)\{s_\lambda : \lambda \geq \mu'\}$. But D is symmetric between q and t, so we have the following corollary.

COROLLARY 3.5.5. *The operator D has eigenfunctions $\tilde{H}_\mu(z;q,t)$ with eigenvalue $B_\mu(q,t)$ satisfying conditions (i) and (ii) in Definition 3.5.2.*

To see that condition (iii) can also be satisfied we only have to verify that the eigenfunctions have $\langle s_{(n)}, \tilde{H}_\mu(z;q,t)\rangle \neq 0$. But their specializations at $q=0$ are non-zero scalar multiples of the Hall-Littlewood polynomials $\tilde{H}_\mu(z;t)$, so this is clear from Corollary 3.4.21.

COROLLARY 3.5.6. *Polynomials $\tilde{H}_\mu(z;q,t)$ satisfying conditions (i)-(iii) in Definition 3.5.2 exist.*

Note that the solution of (i)-(iii) is necessarily unique, since the matrix giving any other solution in terms of the basis $\{\tilde{H}_\mu\}$ would have to be upper triangular by (i), lower triangular by (ii), and 1 on the diagonal by (iii).

3.5.3. *Specializations.* For special values of the parameters, there are simpler expressions for the Macdonald symmetric functions, as follows.

PROPOSITION 3.5.7. *The Macdonald symmetric function $\tilde{H}_\mu(z;0,t)$ at $q=0$ is equal to the Hall-Littlewood symmetric function $\tilde{H}_\mu(z;t)$. Equivalently, the Kostka-Macdonald polynomials at $q=0$ reduce to the Kostka-Foulkes polynomials:*

$$\tilde{K}_{\lambda\mu}(0,t) = \tilde{K}_{\lambda\mu}(t). \tag{59}$$

PROOF. Compare Corollary 3.4.21 with Definition 3.5.2. □

PROPOSITION 3.5.8. *The Macdonald symmetric function at $q=1$ is given by*

$$\tilde{H}_\mu(z;1,t) = (1-t)^{|\mu|}\Big(\prod_i [\mu_i']_t!\Big)\, h_{\mu'}[Z/(1-t)]. \tag{60}$$

COROLLARY 3.5.9. *The Macdonald symmetric function at $q=t=1$ is given by*

$$\tilde{H}_\mu(z;1,1) = e_1^n,$$

for every partition μ. In other words, the Kostka-Macdonald polynomials satisfy

$$\tilde{K}_{\lambda\mu}(1,1) = \chi^\lambda(1),$$

the number of standard Young tableaux of shape λ, independent of μ.

PROOF. The operator D in (55) is well-defined in the limit as $q \to 1$, and a calculation shows that $D_{q=1}$ is a derivation: $D_{q=1}(fg) = (D_{q=1}f)g + f(D_{q=1}g)$. Since $B_\mu(1,t) = \sum_i [\mu'_i]_t$, it follows that

$$\tilde{H}_\mu(z;1,t) = \prod_i \tilde{H}_{(1^{\mu'_i})}(z;1,t).$$

Therefore we only need to show that $\tilde{H}_{(1^n)}(z;1,t) = (1-t)^n [n]_t! h_n[Z/(1-t)]$. This latter fact holds even for $q \neq 1$, since condition (ii) in Definition 3.5.2 implies that $\tilde{H}_{(1^n)}(z;q,t)$ is a scalar multiple of $h_n[Z/(1-t)]$, and the identity $h_n[1/(1-t)] = 1/([n]_t!(1-t)^n)$, which is a special case of (29), fixes the scalar factor.

For the corollary, the Cauchy formula and (29) imply that

$$\begin{aligned}(1-t)^n[n]_t! h_n[Z/(1-t)] &= (1-t)^n[n]_t! \sum_{|\lambda|=n} s_\lambda(z) s_\lambda[\frac{1}{1-t}] \\ &= \sum_{|\lambda|=n} s_\lambda(z) \frac{[n]_t!}{\prod_{x\in\lambda}[h(x)]_t!}.\end{aligned}$$

Setting $t = 1$, this becomes

$$\sum_{|\lambda|} s_\lambda(z)\chi^\lambda(1),$$

by the hook formula (2), and this is finally equal to e_1^n, the Frobenius characteristic of the regular representation of S_n. □

PROPOSITION 3.5.10. *The Macdonald symmetric function $\tilde{H}_\mu(z;q,q^{-1})$ at $t = 1/q$ is given by*

$$\tilde{H}_\mu(z;q,q^{-1}) = q^{-n(\mu)} \prod_{x\in\mu} (1 - q^{h(x)}) s_\mu[Z/(1-q)], \tag{61}$$

where $h(x) = 1 + a(x) + l(x)$ denotes the hook-length of the cell x in the diagram of μ.

PROOF. One verifies immediately that $s_\mu[Z/(1-q)]$ satisfies conditions (i) and (ii) in Definition 3.5.2 when $t = 1/q$, and hence $\tilde{H}_\mu(z;q,q^{-1})$ is a scalar multiple of $s_\mu[Z/(1-q)]$. Formula (29) fixes the scalar factor. □

In addition to the specializations the transformed Macdonald polynomials obey two fundamental symmetries. The first one is obvious from the definition.

PROPOSITION 3.5.11. *For every μ we have $\tilde{H}_{\mu'}(z;q,t) = \tilde{H}_\mu(z;t,q)$, or equivalently,*

$$\tilde{K}_{\lambda\mu'}(q,t) = \tilde{K}_{\lambda\mu}(t,q).$$

PROPOSITION 3.5.12. *For every μ we have $\omega\tilde{H}_\mu(z;q,t) = t^{n(\mu)}q^{n(\mu')}\tilde{H}_\mu(z;q^{-1},t^{-1})$, or equivalently,*

$$\tilde{K}_{\lambda'\mu}(q,t) = t^{n(\mu)}q^{n(\mu')}\tilde{K}_{\lambda\mu}(q^{-1},t^{-1}).$$

PROOF. It is easy to see that $\omega\tilde{H}_\mu(z;q^{-1},t^{-1})$ satisfies conditions (i) and (ii) in Definition 3.5.2, and hence is a scalar multiple of $\tilde{H}_\mu(z;q,t)$. To fix the scalar, we need the identity

$$\tilde{K}_{(1^n),\mu} = t^{n(\mu)}q^{n(\mu')}.$$

Somewhat surprisingly, this is one of the more subtle results in the elementary theory of Macdonald polynomials. We will prove something a little more general in the next section, in Corollary 3.5.20. □

3.5.4. *The positivity problem.* From the observations in the preceding section, we see that the specializations $\tilde{K}_{\lambda\mu}(0,t)$, $\tilde{K}_{\lambda\mu}(q,0)$, $\tilde{K}_{\lambda\mu}(1,t)$ and $\tilde{K}_{\lambda\mu}(q,1)$ of the Kostka-Macdonald polynomials $\tilde{K}_{\lambda\mu}(q,t)$ have non-negative integer coefficients. We shall also see explicitly in Corollary 3.5.20 below that $\tilde{K}_{\lambda\mu}(q,t) \in \mathbb{N}[q,t]$ when λ is a hook shape partition. These facts and tables which he had computed for $n \leq 6$ led Macdonald to conjecture the following theorem, already in his 1988 paper [**60**].

THEOREM 3.5.13 (Macdonald positivity conjecture). *The Kostka-Macdonald polynomials $\tilde{K}_{\lambda\mu}(q,t)$ are polynomials with non-negative integer coefficients.*

Part of the conjecture was the *integrality* property, $\tilde{K}_{\lambda\mu}(q,t) \in \mathbb{Z}[q,t]$, which as we have mentioned has since been proved by elementary methods. The additional *positivity* property, $\tilde{K}_{\lambda\mu}(q,t) \in \mathbb{N}[q,t]$, lies deeper, as might be expected from the fact that it is a serious theorem even for the Kostka-Foulkes polynomials $\tilde{K}_{\lambda\mu}(0,t)$. The only proof known at present is the one based on the geometry of Hilbert schemes that I will describe in §5.4.

In the course of working on the positivity conjecture, Garsia and I and others were led to integrality and positivity conjectures for other quantities related to the Kostka-Macdonald polynomials. Some of these conjectures have now also been proven using geometric methods. In the rest of this section I will explain some additional aspects of the theory that lead to these further conjectures.

3.5.5. *Operators Δ and ∇ and the plethystic formula.* For any symmetric function f we define a homogeneous linear operator Δ_f on symmetric functions with coefficients in $\mathbb{Q}(q,t)$ by the formula

$$\Delta_f \tilde{H}_\mu(z;q,t) = f[B_\mu]\tilde{H}_\mu(z;q,t), \tag{62}$$

where $B_\mu = B_\mu(q,t)$ is given by (58). In particular, the operator D of §3.5.2 is $D = \Delta_{e_1}$ in this notation. In [**7**] one can find formulas similar to (56), but more complicated, for the operators Δ_f. We also define

$$\nabla \tilde{H}_\mu(z;q,t) = t^{n(\mu)}q^{n(\mu')}\tilde{H}_\mu(z;q,t), \tag{63}$$

which is equivalent to setting $\nabla f(z) = \Delta_{e_n} f(z)$ for f homogeneous of degree n. The operator ∇ plays an especially important role in the theory.

F. Bergeron, Garsia, Tesler and I studied these operators in [**7, 25**]. We proved integrality properties and we conjectured positivity properties for them, and we showed that various results previously discovered by others could be deduced easily with their aid. The next proposition, which was proved in [**7**] by elementary means, also follows (in part) from a natural geometric interpretation of the operators, as we shall see in §5.4.4.

PROPOSITION 3.5.14 ([**7**]). *The operators Δ_f and hence also ∇ are integral, in the sense that if f and F have coefficients in $\mathbb{Z}[q,t]$, then so does $\Delta_f F$. The operator ∇^{-1} is Laurent-integral, that is, $\nabla^{-1}F$ has coefficients in $\mathbb{Z}[q,t,q^{-1},t^{-1}]$*

CONJECTURE 3.5.15 ([**7**]). *When expanded on the Schur basis $\{s_\lambda\}$, the following quantities all have coefficients in $\mathbb{N}[q,t]$:*

(I)

$$(-1)^{i(\mu)}\nabla^m s_\mu(z)$$

for all μ *and all* $m \geq 1$, *where* $i(\mu) = \binom{\mu_1}{2} + \sum_{\mu'_i < i-1} (i - 1 - \mu'_i)$;

(II)

$$(-1)^{|\mu|-l(\mu)}\nabla \tilde{H}_\mu(z; 0, t)$$

for all μ;

(III)

$$\nabla\omega\tilde{H}_\mu(z; 0, t^{-1})$$

for all μ, *and also*

$$\nabla\omega\tilde{H}_\mu(z; 0, t^{-1}) - \nabla\omega\tilde{H}_\nu(z; 0, t^{-1})$$

whenever $\mu \geq \nu$;

(IV)

$$(-1)^{|\mu|-l(\mu)}\nabla m_\mu(z)$$

for all μ, *and moreover the coefficients are doubly unimodal in* q *and* t;

(V)

$$\Delta_{s_\nu} e_n(z)$$

for all ν *and* n.

All except (II) above have as special cases the quantity

$$\nabla e_n(z),$$

whose special significance will be explained below. Each of (I)–(V) has a geometric interpretation in the Hilbert scheme setting, and in principle it should eventually be possible to prove all five parts of the conjecture using geometric methods. To date, this has only been carried out for the folllowing weakened form of (V).

THEOREM 3.5.16 ([**40**]). *When expanded on the Schur basis, the quantity*

$$\Delta_{s_\nu}\nabla e_n(z)$$

has coefficients in $\mathbb{N}[q, t]$ *for all* ν *and* n.

In [**7**] we proved Proposition 3.5.14 by considering the operator D_1 defined by taking the coefficient of u^1 instead of u^0 in (56), and showing that there is an integral basis of the algebra of symmetric functions consisting of elements of the form

$$u_\lambda(z; q, t) = e_1^a D_1 e_1^{\lambda_1 - 1} D_1 e_1^{\lambda_2 - 1} \cdots D_1 e^{\lambda_k - 1},$$

where $\lambda = (\lambda_1, \ldots, \lambda_k, 1^a)$ with $\lambda_k > 1$. The proposition then follows from various commutation relations between the operators Δ_f, D_1 and multiplication by e_1. I will not go into further detail on these relations here. The following fundamental result is proved in a similar way.

PROPOSITION 3.5.17 ([**25**]). *For any symmetric function* f, *we have*

$$\langle f, \tilde{H}_\mu[Z + 1; q, t]\rangle_* = K_f[(1 - q)(1 - t)B_\mu - 1;\ -1],$$

where

$$K_f(z; u) \underset{\text{def}}{=} \nabla^{-1}(f[Z - u])$$

and $\langle -,- \rangle_*$ *is defined in* (53).

Let me mention two important consequences that follow by straightforward calculations from this proposition. The first of these is the *plethystic formula* for Kostka-Macdonald polynomials.

COROLLARY 3.5.18. *Fix a partition* γ *of size* k. *Then the Kostka-Macdonald coefficients* $\tilde{K}_{(n-k,\gamma)\mu}(q,t)$, *where* $n = |\mu| \geq k+\gamma_1$, *are given for all* μ *simultaneously by*

$$\tilde{K}_{(n-k,\gamma)\mu}(q,t) = k_\gamma[B_\mu; q, t],$$

where

$$k_\gamma[\tfrac{Z+1}{(1-q)(1-t)}; q, t] = \nabla^{-1}\omega\left(s_\lambda[\tfrac{Z+1}{(1-q)(1-t)} - 1]\right).$$

This corollary was first proved prior to Proposition 3.5.17, by Garsia and Tesler in [**28**], but without the simple formula for k_γ. Another consequence of Proposition 3.5.17 is the reciprocity formula of Koornwinder and Macdonald, which takes the following pleasant form when written in terms of the transformed Macdonald symmetric functions.

PROPOSITION 3.5.19 ([**61**]). *For all pairs of partitions* λ, μ, *we have*

$$\tilde{H}_\mu[1 - uA_\lambda; q, t]\Omega[uB_\mu] = \tilde{H}_\lambda[1 - uA_\mu; q, t]\Omega[uB_\lambda],$$

where $B_\mu = B_\mu(q,t)$ *is given by* (58) *and* $A_\mu = 1 - (1-q)(1-t)B_\mu$.

As a particular consequence, by taking $\lambda = \emptyset$ in the reciprocity formula, we obtain the specialization theorem of Macdonald [**61**].

COROLLARY 3.5.20. *Under the plethystic specialization* $Z \mapsto 1 - u$, *we have*

$$\tilde{H}_\mu[1 - u; q, t] = \Omega[-uB_\mu].$$

Equivalently, for $\lambda = (n - r, 1^r)$ *a hook shaped partition and all partitions* μ *of* n, *we have*

$$\tilde{K}_{\lambda\mu} = e_r[B_\mu - 1]. \tag{64}$$

3.5.6. *Raising operators.* Operators on Macdonald symmetric functions analogous to the Jing operators S^t_m in Definition 3.4.5 were first found by Lapointe and Vinet [**55**] and Kirillov and Noumi [**49**]. Their work provides several different families of operators B_m which add a part to the indexing partition of a Macdonald polynomial, that is,

$$B_m\tilde{H}_\mu(z; q, t) = \tilde{H}_{(m,\mu)}(z; q, t) \quad \text{for } m \geq \mu_1. \tag{65}$$

Equivalently, for every partition μ, there holds the Rodrigues formula

$$\tilde{H}_\mu(z; q, t) = B_{\mu_1}B_{\mu_2}\cdots B_{\mu_l}(1). \tag{66}$$

Note that in light of the symmetry given by Proposition 3.5.11 it is the same thing to give operators B_{1^m} with the property

$$B_{1^m}\tilde{H}_\mu(z; q, t) = \tilde{H}_{\mu+(1^m)}(z; q, t) \quad \text{for } m \geq l(\mu), \tag{67}$$

so that

$$\tilde{H}_{\mu'}(z; q, t) = B_{1^{\mu'_1}}B_{1^{\mu'_2}}\cdots B_{1^{\mu'_r}}(1). \tag{68}$$

Zabrocki subsequently discovered that such raising operators can be manufactured at will from *any* operators which have the property in (67) for the Hall-Littlewood polynomials.

THEOREM 3.5.21 ([**29**]). *For any operator V on symmetric functions, define its q-deformation V^q by the formula*

$$V^q f = V_Y(\, f[qZ + (1-q)Y]\,)\,|_{Y \mapsto Z},$$

where V_Y denotes V acting on symmetric functions in the variables y and treating the variables z as scalars (so in particular, $V^0 = V$ and $V^1 f = V(1)\cdot f$). Let T_{1^m} be any linear operators whatsoever that satisfy

$$T_{1^m} \tilde{H}_\mu(z;t) = \tilde{H}_{\mu+(1^m)}(z;t) \quad \textit{for } m \geq l(\mu). \tag{69}$$

Then

$$T^q_{1^m} \tilde{H}_\mu(z;q,t) = \tilde{H}_{\mu+(1^m)}(z;q,t) \quad \textit{for } m \geq l(\mu). \tag{70}$$

It turns out rather amazingly that this theorem is almost trivial to prove, once it is given that there exists *some* family of operators satisfying both (69) and (70). Conveniently, the operators of Kirillov and Noumi have the correct form, *i.e.*, they are the Zabrocki q-deformations of their $q = 0$ specializations, and the result follows. One nice application of Zabrocki's theorem is to give the simplest of many proofs of the integrality theorem.

COROLLARY 3.5.22 ([**29**]). *The Kostka-Macdonald polynomials are polynomials, $\tilde{K}_{\lambda\mu}(q,t) \in \mathbb{Z}[q,t]$.*

PROOF. It is easy to see that Theorem 3.5.21 is equivalent to the corresponding result for $H_\mu(z;t)$ and $H_\mu(z;q,t) \underset{\text{def}}{=} t^{n(\mu)}\tilde{H}(z;q,t^{-1})$. In this variant of the theorem, take for T_{1^m} the *trivial* operator defined by

$$T_{1^m} H_\mu(z;t) = \begin{cases} H_{\mu+(1^m)}(z;t) & \text{if } m \geq l(\mu), \\ 0 & \text{otherwise.} \end{cases}$$

Corollary 3.4.12, part (v) shows that the transformed Hall-Littlewood symmetric functions $H_\mu(z;t)$ form an integral basis for the algebra of symmetric functions with coefficients in $\mathbb{Z}[t]$. In particular, the operators T_{1^m} act on the Schur basis with coefficients in $\mathbb{Z}[t]$, and their q-deformations $T^q_{1^m}$ act with coefficients in $\mathbb{Z}[q,t]$. Then the Rodrigues formula implies that $H_\mu(z;q,t) \in \mathbb{Z}[q,t]\{s_\lambda\}$, that is, $K_{\lambda\mu}(q,t) \in \mathbb{Z}[q,t]$ for all λ and μ. By the symmetry in Proposition 3.5.12, this is equivalent to $\tilde{K}_{\lambda\mu} \in \mathbb{Z}[q,t]$ for all λ and μ. □

In is worthy of remark that the Jing operators S^t_m are just the Zabrocki t-deformations of the Bernstein operators S^0_m.

3.5.7. *Operator ∇ and q-Lagrange inversion.* The q-Catalan numbers discussed in §2.2, the inversion enumerator for forests in §2.3, and more generally the q-Lagrange inversion coefficients $k_n(q)$ in Proposition 2.4.2 turn out to have remarkable expressions involving the operator ∇ specialized at $t = 1$.

PROPOSITION 3.5.23. *The quantity*

$$\nabla_{t=1} e_n$$

is given by the formula for $k_n(q)$ in Proposition 2.4.2, when the indeterminates e_k in the formula are interpreted as elementary symmetric functions.

COROLLARY 3.5.24. *The Carlitz-Riordan q-Catalan numbers are given in terms of* ∇ *by*

$$C_n(q) = \langle e_n, \nabla_{t=1} e_n \rangle, \tag{71}$$

while the inversion enumerator for forests, or weight enumerator for parking functions, is given by

$$J_n(q) = \langle e_1^n, \nabla_{t=1} e_n \rangle. \tag{72}$$

PROOF OF THE COROLLARY. The inner product $\langle e_n, e_\nu \rangle$ is equal to 1 for all ν, so the right-hand side of (71) is the specialization of $k_n(q)$ at $e_k = 1$ for all k. We have seen in §2.4 that this is the same as $C_n(q)$.

The inner product $\langle e_1^n, e_\nu \rangle$ is equal to the multinomial coefficient $\binom{n}{\nu_1, \dots, \nu_l}$. Using (14), we obtain (72). □

Proposition 3.5.23 was proved in [**23**]. Here I will sketch a slick version of the proof. There are two key points. The first is that Garsia's original solution of the q-Lagrange inversion problem in [**30**] can be recast in symmetric function language when the coefficients e_k are understood as elementary symmetric functions, to give another formula for $k_n(q)$, different from the one in Proposition 2.4.2, namely

$$k_n(q) = \sum_{|\mu|=n} q^{n(\mu')} h_\mu[Z/(1-q)] f_\mu[1-q],$$

where the $f_\mu(z) = \omega m_\mu(z)$ are the so-called *forgotten* symmetric functions. The second is that the specialization in Proposition 3.5.8, taken with $t = 1$ instead of $q = 1$, implies that

$$\nabla_{t=1} h_\mu[Z/(1-q)] = q^{n(\mu')} h_\mu[Z/(1-q)].$$

But $\{h_\mu[Z/(1-q)]\}$ and $\{m_\mu[(1-q)Z]\}$ are dual bases with respect to the Hall-inner product, so the Cauchy formula (45) yields

$$h_n(YZ) = \sum_{|\mu|=n} h_\mu[Z/(1-q)] m_\mu[(1-q)Y].$$

Now applying ω in the Y variables and then setting $Y = 1$ yields

$$e_n(z) = \sum_{|\mu|=n} h_\mu[Z/(1-q)] f_\mu[1-q],$$

and hence

$$k_n(q) = \nabla_{t=1} e_n.$$

The $t = 1/q$ specialization of ∇e_n can be derived by similar devices.

PROPOSITION 3.5.25. *We have*

$$\nabla_{t=q^{-1}} e_n = q^{-\binom{n}{2}} \frac{1}{[n+1]_q} e_n\left[[n+1]_q Z\right].$$

Note that the above formula is a kind of naive q-analog of the classical Lagrange inversion formula (8).

The proof of Proposition 3.5.23 given above and the similar (unstated) proof of Proposition 3.5.25 evade the issue of determining ∇e_n explicitly. We did, however, find an explicit formula in [**23**]. Since we will need it later, I will review its derivation.

In order to apply ∇, we need to write e_n in terms of the transformed Macdonald polynomials. To do this, we use the fact that the $\tilde{H}_\mu(z;q,t)$ are orthogonal with respect to the inner product $\langle -,-\rangle_*$ in (53), together with the identity

$$\langle \tilde{H}_\mu, \tilde{H}_\mu\rangle_* = t^{n(\mu)}q^{n(\mu')}\prod_{x\in\mu}(1-t^{1+l(x)}q^{-a(x)})(1-t^{-l(x)}q^{1+a(x)}). \tag{73}$$

Here $a(x)$ and $l(x)$ denote the arm and leg of x, as defined in §2.1. Identity (73) follows from Macdonald's formula for the inner product $\langle P_\mu, P_\mu\rangle$. However, rather than repeat its derivation here, I prefer to appeal to the geometric proof that will be given in §5.4.3. The latter proof is more illuminating than the elementary one because it explains the meaning of the factors in (73).

The $\langle -,-\rangle_*$ orthogonality and (73) yield as an instance of the Cauchy formula

$$\omega\Omega\left[\frac{YZ}{(1-q)(1-t)}\right] = \sum_\mu \frac{t^{-n(\mu)}q^{-n(\mu')}\tilde{H}_\mu(y;q,t)\tilde{H}_\mu(z;q,t)}{\prod_{x\in\mu}(1-t^{1+l(x)}q^{-a(x)})(1-t^{-l(x)}q^{1+a(x)})}.$$

Taking the homogeneous component of degree n, setting $Y = 1-u$, and using Corollary 3.5.20, we arrive at

$$e_n\left[\frac{(1-u)Z}{(1-q)(1-t)}\right] = \sum_{|\mu|=n} \frac{t^{-n(\mu)}q^{-n(\mu')}\Omega[-uB_\mu]\tilde{H}_\mu(z;q,t)}{\prod_{x\in\mu}(1-t^{1+l(x)}q^{-a(x)})(1-t^{-l(x)}q^{1+a(x)})}.$$

Dividing both sides by $1-u$ and setting $u=1$ in what remains produces

$$(-1)^{n-1}p_n\left[\frac{Z}{(1-q)(1-t)}\right] = \sum_{|\mu|=n} \frac{t^{-n(\mu)}q^{-n(\mu')}\Pi_\mu(q,t)\tilde{H}_\mu(z;q,t)}{\prod_{x\in\mu}(1-t^{1+l(x)}q^{-a(x)})(1-t^{-l(x)}q^{1+a(x)})},$$

where

$$\Pi_\mu \underset{\text{def}}{=} \Omega[B_\mu - 1] = \prod_{\substack{(i,j)\in\mu\\ (i,j)\neq(0,0)}} (1-q^i t^j). \tag{74}$$

Finally, applying the operator $D = \Delta_{e_1}$ to both sides yields, by a calculation on the left-hand side and using Corollary 3.5.5 on the right,

$$e_n(z) = \sum_{|\mu|=n} \frac{t^{-n(\mu)}q^{-n(\mu')}(1-q)(1-t)\Pi_\mu(q,t)B_\mu(q,t)\tilde{H}_\mu(z;q,t)}{\prod_{x\in\mu}(1-t^{1+l(x)}q^{-a(x)})(1-t^{-l(x)}q^{1+a(x)})}.$$

This proves the following proposition.

PROPOSITION 3.5.26. *We have explicitly*

$$\nabla e_n(z) = \sum_{|\mu|=n} \frac{(1-q)(1-t)\Pi_\mu(q,t)B_\mu(q,t)\tilde{H}_\mu(z;q,t)}{\prod_{x\in\mu}(1-t^{1+l(x)}q^{-a(x)})(1-t^{-l(x)}q^{1+a(x)})}, \tag{75}$$

with $B_\mu(q,t)$ *as in* (58) *and* $\Pi_\mu(q,t)$ *as in* (74).

4. The $n!$ and $(n+1)^{n-1}$ conjectures

4.1. The $n!$ conjecture. The proof of the Macdonald positivity conjecture rests on the interpretation of $\tilde{H}_\mu(z;q,t)$ as the Frobenius series of a suitable doubly graded S_n module. This interpretation, which Garsia and I proposed in 1991, is what has come to be known as the *n! conjecture*. I will present it here in a way that I hope gives a flavor of how we discovered it. The results below and some of our other early results on the $n!$ conjecture were announced in [**22**] and given a fuller treatment in [**24**].

In §3.4.6, we saw that the Kostka-Foulkes polynomials $\tilde{K}_{\lambda\mu}(t)$ describe multiplicities of S_n characters χ^λ in the graded character of the Garsia-Procesi ring R_μ, which is the same thing as the cohomology ring $H^*(\mathcal{B}_\mu)$ of a springer fiber. We also saw that these rings are quotients of the ring

$$R_{1^n}(\mathbf{x}) = \mathbb{C}[\mathbf{x}]/I$$

of coinvariants for the natural action of S_n on the polynomial ring $\mathbb{C}[\mathbf{x}] = \mathbb{C}[x_1, \ldots, x_n]$.

From Propositions 3.5.7 and 3.5.11 we know that the Kostka-Macdonald polynomials $\tilde{K}_{\lambda\mu}(q,t)$ specialize to

$$\tilde{K}_{\lambda\mu}(0,t) = \tilde{K}_{\lambda\mu}(t); \quad \tilde{K}_{\lambda\mu}(q,0) = \tilde{K}_{\lambda\mu'}(q).$$

This suggests that $\tilde{H}_\mu(z;q,t)$ might be the Frobenius series of a doubly graded ring $R_\mu(\mathbf{x},\mathbf{y})$, a quotient of the polynomial ring $\mathbb{C}[\mathbf{x},\mathbf{y}] = \mathbb{C}[x_1, y_1, \ldots, x_n, y_n]$ in two sets of variables, whose components of degree zero in the $\mathbf{y}$ and $\mathbf{x}$ variables are $R_\mu(\mathbf{x})$ and $R_{\mu'}(\mathbf{y})$, respectively. In other words, we may expect $R_\mu(\mathbf{x},\mathbf{y})$ to be a suitable quotient of

$$R_\mu(\mathbf{x}) \otimes R_\mu(\mathbf{y}) \cong H^*(\mathcal{B}_\mu) \otimes H^*(\mathcal{B}_{\mu'}).$$

Other information about the $\tilde{K}_{\lambda\mu}(q,t)$ gives more clues. In particular, by Corollary 3.5.9, we want $R_\mu(\mathbf{x},\mathbf{y})$ to afford the regular representation of S_n. To this end, recall that the defining ideal of $R_\mu(\mathbf{x})$ is generated by the leading forms of polynomials $p(\mathbf{x})$ that vanish on the S_n-orbit of a point $a = (a_1, \ldots, a_n) \in \mathbb{C}^n$ with μ_i coordinates a_j equal to α_i, for distinct complex numbers α_i. Suppose now that we fix a point

$$b = (a_1, b_1, \ldots, a_n, b_n) \in \mathbb{C}^{2n} \tag{76}$$

by assigning distinct numbers α_j to each row j in the diagram of μ and β_i to each column i, and setting $a_k = \alpha_{j_k}$, $b_k = \alpha_{i_k}$, where $(i_1, j_1), \ldots, (i_n, j_n)$ is a list of the cells $(i,j) \in \mu$ in some arbitrary order. The n pairs (a_k, b_k) are all distinct, so the S_n orbit of b in $\mathbb{C}^{2n} = (\mathbb{C}^2)^n$ is a *regular* orbit, with $n!$ distinct points. Let J_b be the ideal of leading forms of polynomials $p(\mathbf{x},\mathbf{y})$ that vanish on the orbit $S_n \cdot b$, and put $R_b = \mathbb{C}[\mathbf{x},\mathbf{y}]/J_b$. Then we have automatically that R_b affords the regular representation of S_n. Projecting b on the $\mathbf{x}$ coordinates gives a point with coordinate multiplicities μ, so the subring of R_b generated by $\mathbf{x}$ is isomorphic to R_μ. The subring generated by $\mathbf{y}$ is isomorphic to $R_{\mu'}$ by symmetry. In short, the ring R_b has the properties we desire, except that it is apparently only singly, not doubly, graded.

In practice, it always turns out that the ideal J_b is doubly homogeneous. Another way to say this is that if we assign the $\mathbf{x}$ variables different weights from the $\mathbf{y}$ variables for the purpose of defining "leading forms," the resulting ideal does not depend on the weights. It also turns out that the ring R_b is always *Gorenstein*, and this property is the key to reformulating the problem indepenendtly of the choice of b. A special case of the definition will suffice for the moment.

DEFINITION 4.1.1. The *socle* of a finite-dimensional graded $\mathbb{C}$-algebra R is the set $\operatorname{soc}(R)$ of elements $x \in R$ annihilated by the maximal homogeneous ideal R_+. The algebra R is *Gorenstein* if $\dim \operatorname{soc}(R) = 1$.

Note that the highest degree homogeneous component $R_{d_{\max}}$ is always contained in $\mathrm{soc}(R)$, so if R is Gorenstein, then $R_{d_{\max}} = \mathrm{soc}(R)$. Recall also that the only one-dimensional representations of S_n are the trivial representation and the sign representation. Hence if R has an action of S_n by algebra automorphisms, and R is Gorenstein, then the socle of R must afford either the trivial or the sign representation.

Now suppose that R is a graded Gorenstein quotient of $R_\mu(\mathbf{x}) \otimes R_{\mu'}(\mathbf{y})$. The ring R_μ contains only S_n modules V_λ with $\lambda \geq \mu$, while $R_{\mu'}$ contains only V_λ with $\lambda \geq \mu'$ (see Corollary 3.4.21). However, the sign representation $\varepsilon = V_{(1^n)}$ can only occur in tensor products of the form $V_\lambda \otimes V_{\lambda'}$. Hence it occurs only once in $R_\mu \otimes R_{\mu'}$, in the top degree component $(R_\mu)_{n(\mu)} \otimes (R_{\mu'})_{n(\mu')}$. If we assume that this unique copy of the sign representation is not in the kernel of the canonical projection $R_\mu \otimes R_{\mu'} \to R$, then its image in R must be equal to $\mathrm{soc}(R)$. These considerations lead to the following result.

PROPOSITION 4.1.2. *There is a unique S_n-invariant Gorenstein (doubly) graded ideal $J_\mu \subseteq R_\mu(\mathbf{x}) \otimes R_{\mu'}(\mathbf{y})$ such that $R_\mu(\mathbf{x},\mathbf{y}) = \mathbb{C}[\mathbf{x},\mathbf{y}]/J_\mu$ contains a copy of the sign representation. A polynomial $f \in \mathbb{C}[\mathbf{x},\mathbf{y}]$ belongs to J_μ if and only if the principal ideal (f) generated by its image in $R_\mu(\mathbf{x}) \otimes R_{\mu'}(\mathbf{y})$ has zero intersection with the unique copy of the sign representation.*

PROOF. Define J_μ to be the set of polynomials f satisfying the stated criterion. It is easy to see that J_μ is a doubly homogeneous ideal and that $R_\mu(\mathbf{x},\mathbf{y}) = \mathbb{C}[\mathbf{x},\mathbf{y}]/J_\mu$ is a quotient of $R_\mu(\mathbf{x}) \otimes R_{\mu'}(\mathbf{y})$ in which the socle is the sign representation. Hence it is Gorenstein. For uniqueness, note that any ideal in a finite-dimensional graded algebra has non-zero intersection with the socle. On one hand, if R is any Gorenstein quotient R of $R_\mu(\mathbf{x}) \otimes R_{\mu'}(\mathbf{y})$ in which the sign representation survives, this implies that every $f \in J_\mu$ must vanish in R. Hence R is a quotient of $R_\mu(\mathbf{x},\mathbf{y})$. On the other hand, any proper ideal in $R_\mu(\mathbf{x},\mathbf{y})$ contains the socle, so $R = R_\mu(\mathbf{x},\mathbf{y})$. □

Henceforth we will simply write

$$R_\mu = \mathbb{C}[\mathbf{x},\mathbf{y}]/J_\mu$$

for the ring in Proposition 4.1.2, and use the notation $R_\mu(\mathbf{x})$, $R_{\mu'}(\mathbf{y})$ to distinguish the Garsia-Procesi rings in one set of variables. Before continuing further, let me give also a more elementary description of R_μ. As before let $(i_1, j_1), \ldots, (i_n, j_n)$ be a list of the cells $(i,j) \in \mu$ in some order, and set

$$\Delta_\mu(\mathbf{x},\mathbf{y}) = \det \begin{bmatrix} x_1^{j_1} y_1^{i_1} & x_2^{j_1} y_2^{i_1} & \ldots & x_n^{j_1} y_n^{i_1} \\ \vdots & & \ddots & \vdots \\ x_1^{j_n} y_1^{i_n} & x_2^{j_n} y_2^{i_n} & \ldots & x_n^{j_n} y_n^{i_n} \end{bmatrix}. \tag{77}$$

It is not hard to prove the following characterization of the defining ideal J_μ of R_μ.

PROPOSITION 4.1.3. *The ideal J_μ in Proposition 4.1.2 coincides with the ideal of polynomial differential operators that annihilate Δ_μ, that is, we have $f \in J_\mu$ if and only if*

$$f(\frac{\partial}{\partial x_1}, \frac{\partial}{\partial y_1}, \ldots, \frac{\partial}{\partial x_n}, \frac{\partial}{\partial x_n})\Delta_\mu = 0.$$

Consider once again the ideal of leading forms J_b of the vanishing ideal of an orbit $S_n \cdot b$, with b as in (76). We will prove that every $f \in J_b$ satisfies the condition in Proposition 4.1.2, so $J_b \subseteq J_\mu$. If $f \in J_b$, then by definition f coincides on $S_n \cdot b$ with some polynomial g of lower degree. There is an essentially unique function ϵ on $S_n \cdot b$ that affords the sign representation. It cannot be represented by a polynomial of degree less than $n(\mu) + n(\mu')$, because $R_\mu(\mathbf{x}) \otimes R_{\mu'}(\mathbf{y})$ has no copy of the sign representation in lower degree. Suppose now that hf is a homogeneous multiple of f of degree $n(\mu) + n(\mu')$ which coincides on $S_n \cdot b$ with ϵ. Then hg also represents ϵ and has smaller degree, a contradiction. This shows that f satisfies the condition in Proposition 4.1.2. The containment $J_b \subseteq J_\mu$, has the following consequences.

PROPOSITION 4.1.4. *We have* $\dim R_\mu \leq n!$. *If equality holds, then* $J_b = J_\mu$. *In particular,* J_b *is then doubly homogeneous and does not depend on the choice of* b, *and* R_μ *affords the regular representation of* S_n.

At this point the interpretation we have been seeking for the Macdonald symmetric function $\tilde{H}_\mu(z; q, t)$ becomes natural and plausible to conjecture.

THEOREM 4.1.5 ($n!$ conjecture). *The Frobenius series of* R_μ *is the transformed Macdonald symmetric function*

$$F_{R_\mu}(z; q, t) = \tilde{H}_\mu(z; q, t).$$

In particular R_μ *affords the regular representation of* S_n, *and hence* $\dim R_\mu = n!$.

The "in particular" follows by Corollary 3.5.9. The Macdonald positivity conjecture, Theorem 3.5.13, is a corollary to this theorem.

4.1.1. *Examples.* Let us briefly consider the simplest case of the $n!$ conjecture, when $\mu = (1^n)$. In this case Δ_μ is the Vandermonde determinant $\Delta(\mathbf{x})$ in the $\mathbf{x}$ variables, and $R_{(1^n)} = R_{(1^n)}(\mathbf{x})$ is the coinvariant ring of S_n acting on $\mathbb{C}^n$, or the cohomology ring $H(\mathcal{B})$ of the flag variety. Here we are in a classical situation, and the $n!$ conjecture is an instance of the following more general result.

PROPOSITION 4.1.6. *Let* W *be a group generated by complex reflections on a vector space* $\mathfrak{h}$. *Then the coinvariant ring* $R = \mathbb{C}[\mathfrak{h}]/W$ *affords the regular representation of* W. *It is Gorenstein, and the representation of* W *on its socle is the determinant of the representation on* $\mathfrak{h}$. *When* W *is the Weyl group of a semisimple Lie algebra* $\mathfrak{g}$, *then* R *is isomorphic to the cohomology ring of the flag variety for the corresponding Lie group* G.

It is also instructive to relate the known symmetries and specializations of the Kostka-Macdonald polynomials to Theorem 4.1.5. We have already seen that the specializations $q = 0$ and $t = 0$, and also $q = t = 1$, are in agreement with the theorem. These led us to the construction of R_μ in the first place.

The specialization $q = 1$ (and by symmetry $t = 1$) can be obtained with a bit of effort from the identification of J_μ as the ideal of leading forms J_b for an orbit $S_n \cdot b$. To do this one uses the fact noted above that J_b is independent of the choice of weights for leading forms, which allows one to take leading forms in the $\mathbf{x}$ variables first and the $\mathbf{y}$ variables afterwards.

The symmetry $\tilde{H}_{\mu'}(z; q, t) = \tilde{H}_\mu(z; q, t)$ is in obvious agreement with the theorem.

More interesting is the symmetry $\tilde{K}_{\lambda'\mu}(q, t) = t^{n(\mu)} q^{n(\mu')} \tilde{K}_{\lambda\mu}(q^{-1}, t^{-1})$. This symmetry reflects the Gorenstein property of R_μ, which implies that multiplication gives a perfect pairing of complementary degrees

$$(R_\mu)_{r,s} \otimes (R_\mu)_{n(\mu)-r,n(\mu)-s} \to (R_\mu)_{n(\mu),n(\mu')} = \operatorname{soc}(R_\mu).$$

The pairing is S_n equivariant, and it pairs complementary irreducible representations V_λ and $V_{\lambda'}$ because the socle affords the sign representation. Hence V_λ and $V_{\lambda'}$ have the same multiplicities in complementary degrees.

4.2. The $(n+1)^{n-1}$ conjecture.

4.2.1. *Diagonal harmonics and coinvariants.* The rings R_μ involved in the $n!$ conjecture are graded quotients of $\mathbb{C}[\mathbf{x},\mathbf{y}]$ that afford the regular representation of S_n. In particular, their only copy of the trivial representation is given by the constants, and so every R_μ is a quotient of the *diagonal coinvariant ring*

$$R_n = \mathbb{C}[\mathbf{x},\mathbf{y}]/I_n, \tag{78}$$

where I_n is the ideal generated by the space $\mathbb{C}[\mathbf{x},\mathbf{y}]_+^{S_n}$ of invariant polynomials without constant term. Because of this connection, Garsia and I were led during our early investigation of the rings R_μ to study also the ring R_n. The result was the discovery of its remarkable combinatorial aspects, as related in the introduction. Below I will review some of these aspects in more detail.

First we should take note of some basic facts about R_n. The first is that the ideal I_n has a well-known set of generators.

PROPOSITION 4.2.1 (Weyl [**85**]). *The ideal I_n is generated by the bivariate power-sums*

$$p_{r,s}(\mathbf{x},\mathbf{y}) \underset{\text{def}}{=} \sum_{i=1}^{n} x^r y^s, \quad 1 \le r+s \le n.$$

The corresponding result also holds for more than two sets of variables $\mathbf{x},\mathbf{y},\ldots,\mathbf{z}$.

The second fact is that as far as the study of its Frobenius series is concerned, it doesn't matter whether we work with the coinvariant ring R_n or the space of harmonics.

DEFINITION 4.2.2. The space DH_n of *diagonal harmonics* for S_n is the set of polynomials $f \in \mathbb{C}[\mathbf{x},\mathbf{y}]$ annihilated by all S_n-invariant constant coefficient differential operators without constant term, that is, by $\mathbb{C}[\partial\mathbf{x},\partial\mathbf{y}]_+^{S_n}$.

Since R_μ is a quotient of R_n, it follows from Proposition 4.1.3 that the polynomial Δ_μ is a diagonal harmonic. It at first seemed more natural in the early days to work with the space of derivatives of Δ_μ rather than with R_μ, and with DH_n rather than R_n, but in the end it is immaterial, by the following easy proposition.

PROPOSITION 4.2.3. *The canonical projection of DH_n on R_n is an isomorphism. Similarly, if $D_\mu = \mathbb{C}[\partial\mathbf{x},\partial\mathbf{y}]\Delta_\mu$, then the canonical projection of D_μ on R_μ is an isomorphism.*

4.2.2. *The initial conjectures.* In §2 we have described q-enumerations of Catalan numbers, forests and parking functions. They turn out to be connected with the Frobenius and Hilbert series of R_n.

THEOREM 4.2.4 ($(n+1)^{n-1}$ conjecture). *Let R_n denote the ring of diagonal coinvariants and R_n^ϵ its subspace of antisymmetric elements. Their characters, dimensions, and Hilbert series enjoy the following properties.*

(i)
$$\dim R_n = (n+1)^{n-1}.$$
(ii)
$$\dim R_n^\varepsilon = C_n = \frac{1}{n+1}\binom{2n}{n},$$
the n-th Catalan number.

(iii)
$$H_{R_n}(q,q^{-1}) = q^{-\binom{n}{2}}[n+1]_q^{n-1}.$$
(iv)
$$H_{R_n^\varepsilon}(q,q^{-1}) = q^{-\binom{n}{2}}\frac{1}{[n+1]_q}\begin{bmatrix}2n\\ n\end{bmatrix}_q.$$
(v)
$$H_{R_n}(q,1) = J_n(q) = \sum_F q^{i(F)},$$
the enumerator of forests on n vertices by number of inversions, or parking functions by weight.

(vi)
$$H_{R_n^\varepsilon}(q,1) = C_n(q),$$
the Carlitz-Riordan q-Catalan number.

(vii) *As an S_n-module, R_n is isomorphic to $\varepsilon \otimes PF$, where PF is the permutation representation on parking functions.*

(viii) *Ignoring the y-degrees and grading R_n only by x-degrees, the isomorphism $\varepsilon \otimes R_n \cong PF$ is homogeneous, where PF is graded by weights.*

We have listed items (i)-(viii) cumulatively for clarity, but of course some of them are redundant. Specifically, (i)-(ii) are special cases of both (iii)-(iv) and (v)-(vi), (viii) is of course stronger than (vii), and both (v) and (vi) follow from (viii).

4.2.3. *The master conjecture.* To appreciate the effect that the empiricial discovery of the facts in Theorem 4.2.4 had at the time, it must be borne in mind that the connection between q-Lagrange inversion and the operator ∇ discussed in §3.5.7 was not then known. In hindsight, however, armed with Propositions 3.5.23 and 3.5.25, we can readily recognize that everything in Theorem 4.2.4 is implied by the following master formula.

THEOREM 4.2.5. *The Frobenius series of the diagonal coinvariant ring is given by*
$$F_{R_n}(z;q,t) = \nabla e_n(z).$$

Indeed, part (viii) is equivalent to Proposition 3.5.23, and parts (v)-(vi) are its Corollary 3.5.24. Parts (iii) and (iv) follow from Proposition 3.5.25 using the symmetric function identities
$$\langle e_1^n, e_n[[n+1]_q Z]\rangle = [n+1]_q^n$$
$$\langle e_n, e_n[[n+1]_q Z]\rangle = \begin{bmatrix}2n\\ n\end{bmatrix}_q.$$

5. Hilbert scheme interpretation

Theorems 4.1.5 and 4.2.5 and the various entities associated with them—the Macdonald symmetric functions $\tilde{H}_\mu(z;q,t)$ and the operators Δ_f and ∇—can be understood geometrically in the context of Hilbert schemes of points in the plane. I will explain in some detail how this comes about in §5.4, below. First we need some background material on Hilbert schemes, and a review of classically known results and the new theorems that have made possible the solution of the $n!$ and $(n+1)^{n-1}$ conjectures.

5.1. The Hilbert scheme and isospectral Hilbert scheme.

DEFINITION 5.1.1. Let $R = \mathbb{C}[x,y] = \mathcal{O}(\mathbb{C}^2)$ denote the coordinate ring of the affine plane. The *Hilbert scheme* $H_n = \mathrm{Hilb}^n(\mathbb{C}^2)$ *of* n *points in the plane* is the algebraic variety parametrizing ideals $I \subseteq R$ such that $\dim_{\mathbb{C}} R/I = n$.

The definition of Hilbert schemes and other facts about them stated below without specific attribution are due to Grothendieck [**35**]. The term *parametrizing* in the definition has the following technical meaning. By definition, the ideals parametrized by H_n correspond one-to-one with zero-dimensional subschemes $S \subseteq \mathbb{C}^2$ of length n, so we may also think of H_n as parametrizing these subschemes. Now there is a flat family, the *universal scheme*

$$(79)\qquad \begin{array}{c} F_n \subseteq H_n \times \mathbb{C}^2 \\ \big\downarrow{\scriptstyle \pi} \\ H_n, \end{array}$$

over H_n, whose fiber over a point of H_n is the corresponding subscheme S. The structure of H_n as an algebraic variety is characterized by the universal property that any family $Y \subseteq T \times \mathbb{C}^2$, flat and finite of degree n over a scheme T, is the pullback of the univeral family F via a unique morphism $T \to H_n$.

Let us pause to see what various points of H_n might look like. In the "generic" case, the subscheme $S \subseteq \mathbb{C}^2$ is a set of n distinct points and $I = I(S)$ its vanishing ideal. The least generic case, but most important for us, is when I is a monomial ideal. In that case, there are n monomials $x^j y^i$ not in I, which form a basis of R/I, and their exponents (i,j) are the cells in the diagram of a partition μ. We index the monomial ideals by their corresponding partitions, writing

$$I = I_\mu.$$

For example, $I_{(1^n)} = (x^n, y)$ and $I_{(n)} = (x, y^n)$. The subscheme $S = V(I_\mu)$ defined by a monomial ideal is non-reduced and concentrated at the origin. It may be usefully and correctly pictured as an infinitesimal copy of the diagram of μ.

The first result on H_n is Fogarty's theorem.[1]

THEOREM 5.1.2 ([**18**]). *The Hilbert scheme of points in the plane (or any smooth surface) is irreducible and nonsingular, of dimension* $2n$.

The irreducibility part of the theorem means that the "generic" subschemes $S \subseteq \mathbb{C}^2$ consisting of n distinct points really are generic, in the sense that they are dense in H_n.

---

[1] I am told that this theorem was actually proved earlier but not published by Hartshorne.

Next we need an auxilliary variety, called the *isospectral Hilbert scheme*. To define it, we first need to introduce the *Hilbert-Chow morphism*. For each ideal $I \in H_n$, the ring R/I is isomorphic to the direct product of its local rings $(R/I)_P$ at each point $P \in V(I)$. Hence the multiplicities $m_P(I) = \dim_{\mathbb{C}}(R/I)_P$ sum to n, so that to I corresponds a zero-dimensional algebraic cycle

$$\sigma(I) = \sum_P m_P(I) \cdot P \tag{80}$$

of weight n. These cycles may be identified with the points of the the orbit variety $(\mathbb{C}^2)^n/S_n$, where S_n acts on $(\mathbb{C}^2)^n$ by permuting coordinates.

PROPOSITION 5.1.3. *The map*

$$\sigma\colon H_n \to \mathbb{C}^{2n}/S_n$$

described by (80) *is a projective and birational morphism of algebraic varieties. In particular it is a desingularization of the quotient singularity* $\mathbb{C}^{2n}/S_n$.

DEFINITION 5.1.4. The *isospectral Hilbert scheme* is the reduced fiber product

$$\begin{array}{ccc} X_n & \xrightarrow{f} & \mathbb{C}^{2n} \\ {\scriptstyle\rho}\downarrow & & \downarrow \\ H_n & \xrightarrow{\sigma} & \mathbb{C}^{2n}/S_n. \end{array} \tag{81}$$

Specifically, X_n is the underlying reduced subscheme of the scheme-theoretic fiber product (which is not reduced).

Let me explain the motivation for the term *isospectral*. For each point I of H_n, the operators of multiplication by x and y are commuting endomorphisms of the n-dimensional vector space R/I. Their joint spectrum consists of n pairs of eigenvalues $(\alpha_i, \beta_i) \in \mathbb{C}^2$, which are just the points of $\sigma(I)$, with repetitions given by the multiplicities. A point of X_n is then a tuple $(I, P_1, \dots, P_n)$ in $H_n \times (\mathbb{C}^2)^n$ where $(P_1, \dots, P_n)$ is some ordering of the joint spectrum.

The projection ρ of X_n on H_n is finite. Its degree is $n!$, since for a generic $I = I(S)$ there are $n!$ possible orderings $(P_1, \dots, P_n)$ of the n distinct points of S.

5.1.1. *Torus action.* The algebraic torus group

$$\mathbb{T}^2 = (\mathbb{C}^*)^2 \tag{82}$$

acts linearly on $\mathbb{C}^2$ as the group of 2×2 diagonal matrices

$$\tau_{t,q} = \begin{bmatrix} t^{-1} & 0 \\ 0 & q^{-1} \end{bmatrix}. \tag{83}$$

The inverse signs in (83) make $\mathbb{T}^2$ act on the coordinate ring $\mathbb{C}[x,y]$ of $\mathbb{C}^2$ by

$$\tau_{t,q}x = tx; \quad \tau_{t,q}y = qy.$$

The action of $\mathbb{T}^2$ on $\mathbb{C}^2$ induces an action on H_n, X_n, the unversal family F_n, etc., so that all relevant morphisms are equivariant. There are induced $\mathbb{T}^2$ actions on various vector spaces such as the coordinate ring $\mathbb{C}[\mathbf{x},\mathbf{y}]$ of $\mathbb{C}^{2n}$, the space of global sections of any $\mathbb{T}^2$-equivariant vector bundle, or the fiber of such a bundle at a torus-fixed point in H_n. The $\mathbb{T}^2$ action on such a space is equivalently described by a $\mathbb{Z}^2$-grading: namely, an element f is homogeneous of degree (r,s) if and only if $\tau_{t,q}f = t^r q^s f$. In most cases, the grading induced by the torus action coincides with an obvious "natural" double grading, as for example in $\mathbb{C}[\mathbf{x},\mathbf{y}]$.

PROPOSITION 5.1.5. *The $\mathbb{T}^2$-fixed points of H_n are the monomial ideals I_μ, and every $I \in H_n$ has a fixed point in the closure of its orbit.*

PROOF. An ideal $I \subseteq R = \mathbb{C}[x, y]$ is fixed if and only if it is doubly homogeneous, *i.e.*, if and only if it is a monomial ideal. The initial ideal of I with respect to any term order on R is a monomial ideal in the closure of the $\mathbb{T}^2$-orbit of I. □

5.1.2. *Zero-fiber.* We will need some basic results on the zero-fiber of the Hilbert-Chow morphism $\sigma\colon H_n \to \mathbb{C}^{2n}/S_n$ over the origin in $\mathbb{C}^{2n}/S_n$, which we denote by

$$Z_n = \sigma^{-1}(\{0\}).$$

The definition may be understood either in the set-theoretic or scheme-theoretic sense, since we will see shortly that the scheme-theoretic zero-fiber is reduced. In a naive sense, Z_n parametrizes subschemes $S \subseteq \mathbb{C}^2$ supported at the origin. One has to be a bit careful about this, however: by Skjelnes [**80**], the functor of families supported at the origin is not representable. What Z_n does represent is the restriction of this functor to families with reduced base scheme.

The first property of Z_n that we need is classical.

PROPOSITION 5.1.6 (Briançon [**9**]). *The zero-fiber Z_n is irreducible, of dimension $n-1$.*

We also need a further property of Z_n that had not been known before.

PROPOSITION 5.1.7 ([**38**]). *The zero-fiber Z_n is scheme-theoretically reduced and Cohen-Macaulay.*

This is proved by considering the zero-fiber $\tilde{Z}_n$ in the universal family F_n over H_n, that is, the fiber over the origin of the composite map

$$\sigma \circ \pi\colon F_n \to H_n \to \mathbb{C}^{2n}/S_n.$$

The universal family F_n is Cohen-Macaulay, since it is flat and finite over H_n (see the discussion following Theorem 5.2.1 below for more on this point). An explicit calculation in local coordinates on F_n, using the generators of the ideal of $\{0\}$ in $\mathbb{C}^{2n}/S_n$ given by Proposition 4.2.1, shows that $\tilde{Z}_n$ is a generically reduced local complete intersection in F_n, hence reduced and Cohen-Macaulay. It is easy to show that the projection $\pi\colon F_n \to H_n$ maps $\tilde{Z}_n$ isomorphically onto Z_n, completing the proof.

5.2. Main theorem on the isospectral Hilbert scheme. The next theorem is the key result from which all else follows.

THEOREM 5.2.1 ([**39**]). *The isospectral Hilbert scheme X_n is Cohen-Macaulay and Gorenstein. Equivalently, the projection*

$$\rho\colon X_n \to H_n$$

is flat and the coordinate ring of its scheme-theoretic fiber over each point I_μ is a finite dimensional (doubly) graded Gorenstein algebra, in the sense of Definition 4.1.1.

A few clarifying remarks may be in order here. By definition, a scheme is Cohen-Macaulay (resp. Gorenstein) if its local ring at every point is so. A local ring is *Cohen-Macaulay* if some, or equivalently every, system of parameters is a regular sequence. Phrased geometrically, this means that a scheme X is Cohen-Macaulay if some, or equivalently every, finite morphism $X \to H$ with H nonsingular, is flat. In particular this applies to the morphism ρ.

A local ring A is Gorenstein if it is Cohen-Macaulay first of all, and in addition, for some, or equivalently every, ideal $\mathcal{J}$ generated by a system of parameters, the Artin local ring $A/\mathcal{J}$ has one-dimensional socle. In the graded case this reduces to Definition 4.1.1. Since I_μ is a $\mathbb{T}^2$-fixed point, the coordinate ring $\mathcal{O}(\rho^{-1}(I_\mu))$ of its fiber is doubly graded, and it is precisely of the form $A/\mathcal{J}$, where $A = \mathcal{O}_{X_n, Q_\mu}$ is the local ring of X_n at the unique point Q_μ lying over I_μ, and $\mathcal{J}$ is generated by local coordinates on H_n at I_μ. So the rings $\mathcal{O}(\rho^{-1}(I_\mu))$ are Gorenstein if and only if X_n is Gorenstein at each point I_μ. However, the Gorenstein locus in X_n is open and $\mathbb{T}^2$-invariant, so this is equivalent to X_n being Gorenstein everywhere, by Proposition 5.1.5.

There is an important way to reformulate Theorem 5.2.1. For any finite group G of order g acting faithfully on $V = \mathbb{C}^m$, one defines the *G-Hilbert scheme*

$$V//G \tag{84}$$

as in Ito and Nakamura [**43**] to be the closure in $\mathrm{Hilb}^g(V)$ of the locus parametrizing regular G-orbits in V. The fibers of the universal family X over $V//G$ are G-invariant subschemes $S \subseteq V$ whose coordinate rings afford the regular representation of G. In particular, G acts on X, and $X/G = V//G$. The projection $X \to V$ then induces a Hilbert-Chow morphism $V//G \to V/G$.

The case of interest here is $V = \mathbb{C}^{2n}$, $G = S_n$. Because the subgroup S_{n-1} has index n in S_n, the quotient X/S_{n-1} of the universal family over $\mathbb{C}^{2n}//S_n$ is flat and finite of degree n. A simple argument based on Proposition 4.2.1 shows that the coordinates x_n and y_n on $\mathbb{C}^{2n}$ generate the S_{n-1}-invariants in the coordinate ring of each fiber of X. Hence X/S_{n-1} can be identified with a subscheme of $(\mathbb{C}^{2n}//S_n) \times \mathbb{C}^2$. The universal property of the Hilbert scheme then yields a morphism

$$\phi\colon \mathbb{C}^{2n}//S_n \to H_n,$$

and it is easy to see that it commutes with the Hilbert-Chow morphisms.

Conversely, the flatness of the projection $\rho\colon X_n \to H_n$, given by Theorem 5.2.1, means that we have a morphism $H_n \to \mathrm{Hilb}^{n!}(\mathbb{C}^{2n})$ mapping generic subschemes $S \subseteq \mathbb{C}^2$ to regular S_n-orbits in $\mathbb{C}^{2n}$. Since H_n is irreducible, this factors through a morphism

$$\eta\colon H_n \to \mathbb{C}^{2n}//S_n.$$

Because both morphisms ϕ and η commute with the Hilbert-Chow morphism, they are mutually inverse generically and hence everywhere. We have proved the following corollary.

COROLLARY 5.2.2. *The Hilbert scheme of S_n-orbits $\mathbb{C}^{2n}//S_n$ is isomorphic to the Hilbert scheme H_n of points in the plane. The isomorphism identifies the isospectral Hilbert scheme X_n with the universal family on $\mathbb{C}^{2n}//S_n$.*

We remark that Theorem 5.2.1 and Corollary 5.2.2 hold with $\mathbb{C}^2$ replaced by any smooth quasiprojective surface, since the properties in question are local in the étale topology.

5.3. Theorem of Bridgeland, King and Reid. When G is a finite subgroup of $SL(V)$, the quotient singularity V/G is Gorenstein, and in some cases it posseses a particularly nice type of desingularization.

DEFINITION 5.3.1. Let $V = \mathbb{C}^m$ and $G \subseteq SL(V)$. A resolution of singularities

$$Y \to V/G$$

is *crepant* if the sheaf ω_Y of differential m-forms is trivial, *i.e.*, $\omega \cong \mathcal{O}_Y$.

A crepant resolution is automatically minimal and has other interesting properties, among them the *generalized McKay correspondence*, conjectured by Reid [**72**] and proved by Batyrev [**2**], which says that the Betti numbers of Y enumerate the conjugacy classes of G according to certain weights. The classical McKay correspondence is its two-dimensional case.

In dimensions two and three, the G-Hilbert scheme $V//G$ turns out always to be a crepant resolution. Seeking to explain and generalize this fact, Bridgeland, King and Reid [**10**] proved the following theorem.

THEOREM 5.3.2 ([**10**]). *Let G be a finite subgroup of $SL(V)$, and consider the diagram*

$$\begin{array}{ccc} X & \xrightarrow{f} & V \\ {\scriptstyle\rho}\downarrow & & \downarrow \\ V//G & \xrightarrow{\sigma} & V/G, \end{array} \tag{85}$$

Let $D(V//G)$ be the bounded derived category of coherent sheaves on $V//G$ and $D^G(V)$ the similar derived category of G-equivariant sheaves on V. Suppose that the Hilbert-Chow morphism $\sigma\colon V//G \to V/G$ satisfies the following smallness criterion: for every d, the locus of points x with $\dim \sigma^{-1}(x) \geq d$ has codimension at least $2d-1$ in V/G. Then

(i) $V//G$ is a crepant resolution V/G;

(ii) the functor

$$\Phi = Rf_* \circ \rho^* \colon D(V//G) \to D^G(V)$$

is an equivalence of categories.

Proposition 5.1.6 implies that the Chow morphism $\sigma\colon H_n \to \mathbb{C}^{2n}/S_n$ satisfies the smallness criterion in Theorem 5.3.2. Combining this theorem with Corollary 5.2.2 we therefore obtain a powerful tool for the study of the Hilbert scheme H_n.

COROLLARY 5.3.3. *In diagram* (81), *the functor*

$$\Phi = Rf_* \circ \rho^* \colon D(H_n) \to D^{S_n}(\mathbb{C}^{2n})$$

is an equivalence of categories.

We remark that it is has long been known (see Beauville [**3**]) that H_n is a crepant resolution of $\mathbb{C}^{2n}/S_n$. Thus the interesting aspect of the Bridgeland-King-Reid theorem in this context is part (ii).

5.4. How the conjectures follow. Now let us see how Theorems 4.1.5 and 4.2.5 follow from our theorems on the Hilbert scheme. To begin, we need names for some vector bundles of geometric origin on H_n. We put

$$B = \pi_* \mathcal{O}_{F_n}, \tag{86}$$

the sheaf of regular functions on the universal family, pushed down to a sheaf on the Hilbert scheme, and

$$P = \rho_* \mathcal{O}_{X_n}, \tag{87}$$

the push-down of the sheaf of regular functions on the isospectral Hilbert scheme. Here we are identifying vector bundles with their (locally free) sheaves of sections. The sheaf P is locally free by virtue of Theorem 5.2.1.

The bundle B is the *tautological bundle* whose fiber at $I \subseteq R$ is R/I. In particular it is a bundle of algebras which are quotients of $R = \mathbb{C}[x, y]$. The bundle P is the tautological bundle on $\mathbb{C}^{2n}//S_n$, regarded as a sheaf on H_n via the isomorphism in Corollary 5.2.2. As such it is a bundle of algebras which are quotients of $\mathbb{C}[\mathbf{x}, \mathbf{y}]$ by S_n-invariant ideals, with S_n acting by the regular representaton on each fiber. The Gorenstein property of X_n in Theorem 5.2.1 means that the fibers of P are Gorenstein algebras.

We also put

$$\mathcal{O}(1) = \wedge^n B, \tag{88}$$

the *tautological line bundle* on the Hilbert scheme. In fact $\mathcal{O}(1)$ is a very ample line bundle for H_n as a variety projective over $\mathbb{C}^{2n}/S_n$, as I will explain later (see Proposition 6.1.5). As usual we also write $\mathcal{O}(k)$ with $k \in \mathbb{Z}$ for the tensor powers of $\mathcal{O}(1)$ or its dual $\mathcal{O}(-1)$.

5.4.1. *The* $n!$ *conjecture.* The first result is the "$n!$" part of the $n!$ conjecture.

PROPOSITION 5.4.1. *The ring* R_μ *in the* $n!$ *conjecture is the fiber* $P(I_\mu)$ *of the bundle* P *at the point* I_μ *in the Hilbert scheme. In particular,* $\dim R_\mu = n!$.

PROOF. Consider an orbit $S_n \cdot b \subseteq \mathbb{C}^{2n}$ with b associated to μ as in (76). Its ideal J_b is a point of $\mathbb{C}^{2n}//S_n$. The corresponding point of H_n is the ideal $I = I(S)$, where S is the set of coordinate pairs (a_i, b_i) in b. In effect, the set S is the picture in $\mathbb{C}^2$ of the diagram of μ.[2]

The ideal J of leading forms of J_b is the limit as $u \to 0$ of J_{ub}. More precisely, consider the one-parameter torus $\mathbb{C}^* = \{\tau_{u^{-1}, u^{-1}}\} \subseteq \mathbb{T}^2$. The $\mathbb{C}^*$ orbit of J_b consists of the ideals J_{ub} for $u \neq 0$. This orbit extends to an affine line $\mathbb{C}^1 \hookrightarrow \mathbb{C}^{2n}//S_n$, with the origin mapping to J.

The point of H_n corresponding to J is similarly the limit as $u \to 0$ of $I(uS)$, that is, the ideal of leading forms of $I(S)$. For (r, s) outside the diagram of μ, the polynomial

$$f(x, y) = \prod_{j<s} (x - \alpha_j) \prod_{i<r} (y - \beta_i)$$

belongs to $I(S)$. The leading form of f is $x^s y^r$, so the ideal of leading forms of $I(S)$ is I_μ. Hence J is the defining ideal of the fiber $P(I_\mu) = \mathbb{C}[\mathbf{x}, \mathbf{y}]/J$, and in particular $\mathbb{C}[\mathbf{x}, \mathbf{y}]/J$ is Gorenstein.

---

[2]For this to come out right we should think of the x-axis as vertical and the y-axis as horizontal. Unfortunately neither the English nor the French convention for Young diagrams matches the Cartesian convention for axes in the plane.

Recall from §4.1 that R_μ is uniquely characterized as the Gorenstein quotient of $\mathbb{C}[\mathbf{x},\mathbf{y}]/J$ in which the sign representation is not killed. But $\mathbb{C}[\mathbf{x},\mathbf{y}]/J$ is already Gorenstein, so it is equal to R_μ. □

To identify the Frobenius series of R_μ we can use the functorial equivalence Φ in Corollary 5.3.3. The method I will outline is a bit different from the one in [**39**].

PROPOSITION 5.4.2. *The S_n-modules $\operatorname{Tor}_i^{\mathbb{C}[\mathbf{x}]}(R_\mu, \mathbb{C})$ contain only irreducible representations V^λ of S_n with $\lambda \geq \mu'$.*

Before I indicate the proof of this proposition, let us take note of its consequences. For one thing it implies Proposition 3.4.20. Just as Proposition 3.4.20 provides a triangularity condition on the Frobenius series of the Garsia-Procesi ring, so does Proposition 5.4.2 for its doubled analog R_μ. Specifically, using Proposition 3.3.1, it implies

$$F_{R_\mu}[(1-t)Z; q, t] \in \mathbb{Z}[q,t] \cdot \{s_\lambda : \lambda \geq \mu'\},$$

which is condition (ii) in Definition 3.5.2. Taking $\mathbf{y}$ in place of $\mathbf{x}$ we see by symmetry that $F_{R_\mu}(z; q, t)$ also satisfies condition (i) in the definition of $\tilde{H}_\mu(z; q, t)$. Finally, the trivial representation of S_n occurs uniquely in degree $(0, 0)$ in R_μ, so $\langle s_{(n)}, F_{R_\mu}(z; q, t)\rangle = 1$, which is another way of stating condition (iii) in the definition. Hence we have

$$F_{R_\mu}(z; q, t) = \tilde{H}_\mu(z; q, t).$$

Now let us turn to the proof of Proposition 5.4.2. We first need to say something about the locus $W_y \subseteq H_n$ parametrizing subschemes S with support on the y-axis $V(x) \subseteq \mathbb{C}^2$. The locus W_y has a well-known explicit description. Given a partition μ of n, consider the set of points $S \in W_y$ for which the multiplicities of the points of S are the parts of μ. For every μ this set has dimension n. First there are $l(\mu)$ degrees of freedom to choose the support of S on the y-axis. For each point of the support, the choice of a non-reduced scheme structure there is locally (in the analytic topology) the same as the choice of a point in the zero-fiber of the Hilbert-Chow morphism for H_{μ_i}. By Briançon's theorem (Proposition 5.1.6) this contributes $\mu_i - 1$ degrees of freedom for each i, for a total of

$$l(\mu) + \sum_i (\mu_i - 1) = n.$$

The locus W_y is the union of these n-dimensional locally closed subsets $W_{y,\mu}$, whose closures are its irreducible components.

The functor Φ in Corollary 5.3.3 has the explicit inverse

$$\Psi = \mathcal{O}(-1) \otimes (\rho_* \circ Lf^*(-))^\epsilon, \tag{89}$$

where the notation $(\cdot)^\epsilon$ stands for the subsheaf of S_n-antisymmetric elements. Note that $D^{S_n}(\mathbb{C}^{2n})$ is just the derived category of finitely generated S_n-equivariant $\mathbb{C}[\mathbf{x},\mathbf{y}]$-modules. The key point in proving Proposition 5.4.2 is to identify the objects

$$\Psi(V^\lambda \otimes \mathbb{C}[\mathbf{y}]) = \Psi(V^\lambda \otimes \mathbb{C}[\mathbf{x},\mathbf{y}]/(\mathbf{x}))$$

in the derived category $D(H_n)$. To do this we observe that the coordinates $\mathbf{x} = x_1, x_2, \ldots, x_n$ form a regular sequence in $\mathcal{O}_{X_n}$ at every point where they vanish. The reason for this is that their vanishing locus $V(\mathbf{x}) \subseteq X_n$ is finite over W_y, hence has dimension n, so it is a complete intersection in the Cohen-Macaulay scheme

X_n. Moreover, a simple calculation at a generic point in each component of W_y shows that $V(\mathbf{x})$ is scheme-theoretically reduced, generically and hence everywhere. These facts imply that $Lf^*(V^\lambda \otimes \mathbb{C}[\mathbf{y}]) = V^\lambda \otimes \mathcal{O}_{V(\mathbf{x})}$. By (89), we have

$$\Psi(V^\lambda \otimes \mathbb{C}[\mathbf{y}]) = \mathcal{O}(-1) \otimes \operatorname{Hom}_{S_n}(V^{\lambda'}, \rho_* \mathcal{O}_{V(\mathbf{x})}).$$

Now $\rho_*\mathcal{O}_{V(\mathbf{x})}$ is supported on W_y and its S_n action is induced from that on X_n. Over the subset $W_{y,\mu}$ of W_y, the fiber of $V(\mathbf{x})$ consists of $\binom{n}{\mu_1,\ldots,\mu_l}$ reduced points, and S_n permutes them transitively, with the stabilizer of each one having the form $S_{\mu_1} \times \cdots \times S_{\mu_l}$. From this one sees that $\operatorname{Hom}_{S_n}(V^{\lambda'}, \rho_*\mathcal{O}_{V(\mathbf{x})})$ is supported on the components $\overline{W}_{y,\mu}$ with $\mu \leq \lambda'$. In turn one deduces that $\operatorname{Hom}_{S_n}(V^{\lambda'}, \rho_*\mathcal{O}_{V(\mathbf{x})})$ is locally zero at I_μ unless $\mu' \leq \lambda'$.

To complete the proof, Proposition 5.4.2 is equivalent to

$$\text{(90)} \qquad \operatorname{Ext}^i(V_\lambda \otimes \mathbb{C}[\mathbf{y}], R_\mu) = 0 \quad \text{for all } i, \text{ unless } \lambda' \geq \mu'$$

in the derived category $D^{S_n}(\mathbb{C}^{2n})$. But Proposition 5.4.1, stated in the language of Corollary 5.3.3, says simply that $\Phi k_{I_\mu} = R_\mu$, where $k_{I_\mu} = \mathcal{O}_{H_n, I_\mu}/\mathfrak{m}_{I_\mu} \in D(H_n)$ is the "skyscraper sheaf" of the point I_μ. Hence by Corollary 5.3.3, the Ext groups in (90) are equal to the Ext groups

$$\operatorname{Ext}^i(\Psi(V^\lambda \otimes \mathbb{C}[\mathbf{y}]), k_{I_\mu}).$$

We have just seen that these are zero unless $\lambda' \geq \mu'$.

5.4.2. *The $(n+1)^{n-1}$ conjecture.* To prove Theorem 4.2.5, we need stronger consequences of Corollary 5.3.3 than the ones we used in the preceding section. The next result is the second of our two main theorems on H_n (the first being Theorem 5.2.1).

THEOREM 5.4.3 ([**40**]). *With the notation B and P as in* (86), (87) *for the tautological vector bundles on the Hilbert scheme, we have*

$$\text{(91)} \qquad \begin{aligned} H^i(H_n, P \otimes B^{\otimes l}) &= 0 \quad \textit{for all } i > 0 \textit{ and all } l; \textit{ and} \\ H^0(H_n, P \otimes B^{\otimes l}) &= R(n,l), \end{aligned}$$

where $R(n,l)$ is the coordinate ring of a certain union of linear subspaces $Z(n,l) \subseteq \mathbb{C}^{2n+2l}$ (the polygraph*). Moreover, we have the similar statements on the zero-fiber*

$$\text{(92)} \qquad \begin{aligned} H^i(Z_n, P \otimes B^{\otimes l}) &= 0 \quad \textit{for all } i > 0 \textit{ and all } l; \textit{ and} \\ H^0(Z_n, P \otimes B^{\otimes l}) &= R(n,l)/\mathfrak{m}R(n,l), \end{aligned}$$

where $\mathfrak{m} = \mathbb{C}[\mathbf{x},\mathbf{y}]_+^{S_n}$ is the homogeneous maximal ideal in the ring of invariants $\mathbb{C}[\mathbf{x},\mathbf{y}]^{S_n}$, that is, the ideal of $\{0\}$ in $\mathbb{C}^{2n}/S_n$.

The definition of the polygraph $Z(n,l)$ will be given in §6, and need not concern us here except when $l = 0$. In that case, $Z(n,0) = \mathbb{C}^{2n}$ and $R(n,0) = \mathbb{C}[\mathbf{x},\mathbf{y}]$. In particular, the ring $R(n,l)/\mathfrak{m}R(n,l)$ on the right-hand side in (92) reduces for $l = 0$ to the diagonal coinvariant ring R_n. We can use Theorem 5.4.3 to calculate the Frobenius series of R_n in terms of Macdonald symmetric functions, with the aid of the following auxilliary result, which is a corollary to the proof of Proposition 5.1.7.

PROPOSITION 5.4.4. *On H_n, we have a $\mathbb{T}^2$-equivariant locally free resolution of the coordinate sheaf $\mathcal{O}_{Z_n}$ of the zero-fiber,*

(93)
$$\cdots \to B \otimes \wedge^k (B' \oplus \mathcal{O}_t \oplus \mathcal{O}_q) \to \cdots \to B \otimes (B' \oplus \mathcal{O}_t \oplus \mathcal{O}_q) \to B \to \mathcal{O}_{Z_n} \to 0,$$

where B' is a canonical direct summand of the tautological bundle B,

$$B = \mathcal{O} \oplus B',$$

and $\mathcal{O}_t$, $\mathcal{O}_q$ denote the twistings $\mathcal{O} \otimes \mathbb{C}_t$ and $\mathcal{O} \otimes \mathbb{C}_q$ by one-dimensional representations of $\mathbb{T}^2$ in which $\tau_{t,q}$ acts as t or q, respectively.

This given, we can make the desired calculation using a suitable version [**83**] of the classical Lefschetz formula of Atiyah and Bott [**1**]. What the formula gives us is actually the Frobenius series Euler characteristic

(94)
$$\sum_i (-1)^i F_{H^i(Z_n, P)}(z; q, t).$$

But from (92) we see that the only nonzero term here is $i = 0$, and that term is just $F_{R_n}(z; q, t)$. The formula gives (94) as a sum of local data for each $\mathbb{T}^2$-fixed point I_μ on the Hilbert scheme. The required local datum at I_μ is the Frobenius series of the fiber of $\mathcal{O}_{Z_n} \otimes P$ there, divided by the determinant of the $\mathbb{T}^2$ action on the cotangent space at I_μ.

From Proposition 5.4.1 and Theorem 4.1.5, we know that the Frobenius series of $P(I_\mu)$ is $\tilde{H}_\mu(z; q, t)$. We have to multiply this by a Hilbert series Euler characteristic for the fiber at I_μ of the resolution of $\mathcal{O}_{Z_n}$ in (93). Note that the Hilbert series of $B(I_\mu) = R/I_\mu$ is none other than the expression $B_\mu(q,t)$ in (58). From this we calculate the factor attributable to (93) as

$$(1-q)(1-t)\Pi_\mu(q,t)B_\mu(q,t),$$

with $\Pi_\mu(q,t)$ defined as in (74).

For the denominator factor we need the following lemma, which is a reformulation of a classical result of Ellingsrud and Strömme [**17**].

LEMMA 5.4.5 ([**38**]). *The eigenvalues of $\tau_{t,q} \in \mathbb{T}^2$ on the cotangent space of the Hilbert scheme at the fixed point I_μ are given by the $2n$ monomials*

$$t^{1+l(x)}q^{-a(x)},\ t^{-l(x)}q^{1+a(x)} \,:\, x \in \mu,$$

where $a(x)$ and $l(x)$ are the arm and leg of the cell x in the diagram of μ.

Putting all this together yields the formula

$$F_{R_n}(z; q, t) = \sum_{|\mu|=n} \frac{(1-q)(1-t)\Pi_\mu(q,t)B_\mu(q,t)\tilde{H}_\mu(z;q,t)}{\prod_{x \in \mu}(1 - t^{1+l(x)}q^{-a(x)})(1 - t^{-l(x)}q^{1+a(x)})}.$$

Comparing this with Proposition 3.5.26 we see that we have proved Theorem 4.2.5.

5.4.3. *Orthogonality of the Macdonald symmetric functions.* As promised earlier, there is a nice geometric proof of the identity (73), along with the orthogonality of the Macdonald symmetric functions $\tilde{H}_\mu(z; q, t)$ with respect to the inner product $\langle -, - \rangle_*$. The ideas that go into it will also enable us to deduce other results.

The equivalence of categories in Corollary 5.3.3 induces an isomorphism of Grothendieck groups

$$\Phi \colon K^0_{\mathbb{T}}(H_n) \cong K^0_{S_n \times \mathbb{T}}(\mathbb{C}^{2n}).$$

Here $K^0_{\mathbb{T}}(H_n)$ is the Grothendieck group of torus-equivariant coherent sheaves on the Hilbert scheme, and $K^0_{S_n\times\mathbb{T}}(\mathbb{C}^{2n})$ is the Grothendieck group of finitely generated S_n-equivariant doubly-graded $\mathbb{C}[\mathbf{x},\mathbf{y}]$-modules. These Grothendieck groups are modules over the representation ring

$$\mathbb{Z}[q,t,q^{-1},t^{-1}]$$

of the torus $\mathbb{T}^2$. More specifically, $K^0_{S_n\times\mathbb{T}}(\mathbb{C}^{2n})$ is freely generated as a $\mathbb{Z}[q,t,q^{-1},t^{-1}]$ module by the free $\mathbb{C}[\mathbf{x},\mathbf{y}]$-modules $V^\lambda\otimes\mathbb{C}[\mathbf{x},\mathbf{y}]$. By Proposition 3.3.1, the Frobenius series of $V^\lambda\otimes\mathbb{C}[\mathbf{x},\mathbf{y}]$ is

$$F_{V^\lambda\otimes\mathbb{C}[\mathbf{x},\mathbf{y}]}(z;q,t)=s_\lambda\left[\frac{Z}{(1-q)(1-t)}\right].$$

An object $A\in D(H_n)$ is supported on the zero-fiber if and only if ΦA is supported at 0, and a finitely-generated graded module is supported at 0 if and only if it is finite-dimensional. The Grothendieck group of finite-dimensional S_n-equivariant $\mathbb{C}[\mathbf{x},\mathbf{y}]$-modules is freely generated by the irreducible S_n-modules V^λ, regarded as $\mathbb{C}[\mathbf{x},\mathbf{y}]$-modules annihilated by $(\mathbf{x},\mathbf{y})$. We can summarize these observations as follows.

PROPOSITION 5.4.6. *The Frobenius series composed with the functor Φ in Corollary 5.3.3 gives an isomorphism of the Grothendieck group $K^0_{\mathbb{T}^2}(H_n)$ onto the algebra of symmetric functions f with the property that $f[(1-q)(1-t)Z]$ has coefficients in $\mathbb{Z}[q,t,q^{-1},t^{-1}]$. Under this isomorphism, the subgroup of objects supported on the zero-fiber corresponds to the subalgebra of symmetric functions f which already have coefficients in $\mathbb{Z}[q,t,q^{-1},t^{-1}]$.*

To make this more explicit, if $[I_\mu]\in K^0_{\mathbb{T}^2}(H_n)$ denotes the class of the skyscraper sheaf k_{I_μ}, then Proposition 5.4.1 and Theorem 4.1.5 give

$$\Phi[I_\mu]=\tilde{H}_\mu(z;q,t). \tag{95}$$

Also, the formula for the inverse map in (89) implies that

$$\Psi(V^\lambda\otimes\mathbb{C}[\mathbf{x},\mathbf{y}])=\operatorname{Hom}_{S_n}(V^{\lambda'},\mathcal{O}(-1)\otimes P).$$

With $\lambda=(n)$, this shows in particular that the line bundle of antisymmetric elements P^ε is isomorphic to $\mathcal{O}(1)$. Recall that fibers of P are Gorenstein algebras, and the fibers of P^ε are their socles. The Gorenstein property implies that there is a perfect pairing

$$P\otimes P\to P^\varepsilon\cong\mathcal{O}(1), \tag{96}$$

so $\mathcal{O}(-1)\otimes\varepsilon\otimes P\cong P^*$. Now define *character bundles*

$$P_\lambda \underset{\text{def}}{=} \operatorname{Hom}_{S_n}(V^\lambda,P).$$

Then the above considerations yield

$$\Phi P^*_\lambda=s_\lambda[Z/(1-q)(1-t)]. \tag{97}$$

COROLLARY 5.4.7. *For $\mathbb{T}^2$ equivariant objects A, B in $D(H_n)$, the Hilbert series Euler characteristic of their Tor groups is given by*

$$\sum_i(-1)^i H_{\operatorname{Tor}_i(A,B)}(q,t)=\langle\Phi[A],\Phi[B]\rangle_*,$$

where

$$\langle f, g\rangle_{.} \underset{\text{def}}{=} \langle \nabla^{-1} f, g\rangle_{*}$$

PROOF. Both sides depend bilinearly on the classes $[A]$ and $[B]$ in the Grothendieck group. Evaluating at $A = P^*_\lambda$ and $B = k_{I_\mu}$, we have $\tilde{K}_{\lambda\mu}(q^{-1}, t^{-1})$ on the left-hand side. On the right-hand side, using (95) and (97), we have $\langle \tilde{H}_\mu, s_\lambda[Z/(1-q)(1-t)]\rangle_{.}$, which is also equal to $\tilde{K}_{\lambda\mu}(q^{-1}, t^{-1})$. □

Now for the promised proof of (73).

COROLLARY 5.4.8. *We have* $\langle \tilde{H}_\mu, \tilde{H}_\nu\rangle_* = 0$ *if* $\mu \neq \nu$, *and*

$$\langle \tilde{H}_\mu, \tilde{H}_\mu\rangle_* = t^{n(\mu)} q^{n(\mu')} \prod_{x\in\mu} (1 - t^{1+l(x)} q^{-a(x)})(1 - t^{-l(x)} q^{1+a(x)}).$$

PROOF. From the definition of ∇ we have $\langle \tilde{H}_\mu, \tilde{H}_\nu\rangle_* = t^{n(\mu)} q^{n(\mu')} \langle \tilde{H}_\mu, \tilde{H}_\nu\rangle_{.}$. The result now follows from Corollary 5.4.7 and Lemma 5.4.5, since $\mathrm{Tor}_i(k_{I_\mu}, k_{I_\mu})$ is the i-th exterior power of the cotangent space to H_n at I_μ. □

5.4.4. *Integrality and positivity for the operators* Δ_f. The operators Δ_f defined in (62) have a simple interpretation in terms of the isomorphism of Grothendieck groups in Proposition 5.4.6. To explain it we need to recall that the *Schur functor* S^λ corresponding to a partition λ of n is defined by

$$S^\lambda(W) = \mathrm{Hom}_{S_n}(V^\lambda, W^{\otimes n}).$$

If W is (doubly) graded, then so is $S^\lambda(W)$, and its Hilbert series is given by

$$H_{S^\lambda(W)}(q,t) = s_\lambda[H_W(q,t)].$$

In particular, the Hilbert series of the fiber at I_μ of the vector bundle $S^\lambda(B)$, where B is the tautological bundle on the Hilbert scheme, is given by $s_\lambda[B_\mu(q,t)]$. Together with (95), this implies the following result.

PROPOSITION 5.4.9. *The operator* Δ_{s_λ} *on symmetric functions corresponds under the isomorphism in Proposition 5.4.6 to the operator on the Grothendieck group of the Hilbert scheme induced by the functor*

$$(S^\lambda(B) \otimes -).$$

In particular, the operators ∇ *and* ∇^{-1} *correspond to the functors*

$$(\mathcal{O}(1) \otimes -),\ (\mathcal{O}(-1) \otimes -).$$

This gives immediately a partial proof of Proposition 3.5.14.

COROLLARY 5.4.10. *The operators* Δ_{s_λ}, ∇ *and* ∇^{-1} *have the Laurent-integrality property in Property 3.5.14, that is, they map the algebra of symmetric functions with coefficients in* $\mathbb{Z}[q,t,q^{-1},t^{-1}]$ *into itself.*

Finally let us turn to the proof of Theorem 3.5.16. In the language of Proposition 5.4.6, Theorem 4.2.5 tells us that

$$\Phi\mathcal{O}_{Z_n} = \nabla e_n(z),$$

and therefore we also have

$$\Phi(S^\lambda(B) \otimes \mathcal{O}_{Z_n}) = \Delta_{s_\lambda} \nabla e_n(z).$$

To establish that this has coefficients in $\mathbb{N}[q,t]$, it suffices to show that

$$H^i(Z_n, P \otimes S^\lambda(B)) = 0, \quad \text{for } i > 0$$

and that the graded $\mathbb{C}[\mathbf{x},\mathbf{y}]$-module $H^0(Z_n, P \otimes S^\lambda(B))$ is zero in negative degrees. But $S^\lambda(B)$ is a direct summand of $B^{\otimes l}$, where $l = |\lambda|$, so both assertions follow from Theorem 5.4.3.

6. Discussion of proofs of the main theorems

In this section I will outline the proofs of Theorem 5.2.1 (the Cohen-Macaulay and Gorenstein properties of X_n) and Theorem 5.4.3 (the cohomology vanishing theorem). Procesi has also given a nice synopsis of the proof of Theorem 5.2.1 in his review article [**71**], and I have followed here his way of organizing the logic.

6.1. Theorem on the isospectral Hilbert scheme. We are to prove that X_n is Cohen-Macaulay and Gorenstein. We proceed by induction on n. Over a point I of the Hilbert scheme whose support $\sigma(I)$ is not of the form $n \cdot P$, we have locally a product structure

$$X_n \cong X_k \times X_l. \tag{98}$$

So we can assume by induction that X_n is locally Cohen-Macaulay and Gorenstein where the n points do not all coincide. In order to handle the most degenerate points, we make use of the *nested Hilbert scheme* $H_{n-1,n}$ and its corresponding isospectral scheme $X_{n-1,n}$. The nested Hilbert scheme parametrizes pairs of ideals

$$I_n \subseteq I_{n-1} \subseteq R$$

such that $\dim R/I_n = n$ and $\dim R/I_{n-1} = n-1$. Parallel to Fogarty's theorem we have the theorem of Tikhomirov and Cheah [**13**] that $H_{n-1,n}$ is irreducible and nonsingular. Now all these schemes for $n-1$ and n fit into a big commutative diagram

$$\begin{array}{ccccc} X_{n-1,n} & & \xrightarrow{g} & & X_n \\ & & \searrow & & \downarrow \\ \downarrow & & & H_{n-1,n} & \longrightarrow H_n \\ & & & \downarrow & \\ X_{n-1} & \longrightarrow & & H_{n-1}. & \end{array} \tag{99}$$

By induction, the bottom horizontal arrow is flat of degree $(n-1)!$ with Gorenstein fibers. From this fact it is easy to deduce that the square involving the bottom arrow is a scheme-theoretic fiber square. Consequently $X_{n-1,n}$ is Cohen-Macaulay and Gorenstein, since $H_{n-1,n}$ is nonsingular and the diagonal arrow is flat and finite with Gorenstein fibers.

For the next step we need to calculate the canonical line bundles on some of the schemes in (99). On $H_{n-1,n}$, and also on $X_{n-1,n}$ via pullback, we define

$$\mathcal{O}(k,l) = p_{n-1}^* \mathcal{O}_{H_{n-1}}(k) \otimes p_n^* \mathcal{O}_{H_n}(l),$$

where p_{n-1}, p_n are the projections on H_{n-1}, H_n and $\mathcal{O}(1)$ denotes the highest exterior power of the tautological bundle, as in (88). The canonical bundle on H_{n-1} is trivial, and the pairing in (96), which holds on H_{n-1} by induction, implies that the canonical bundle on X_{n-1} is $\mathcal{O}(-1)$. It follows that the relative canonical bundle of $X_{n-1,n}$ over $H_{n-1,n}$ in our fiber square is $\mathcal{O}(-1,0)$ in the notation above.

LEMMA 6.1.1 ([**39**]). *The canonical sheaf on $H_{n-1,n}$ is $\omega_{H_{n-1,n}} = \mathcal{O}(1,-1)$.*

This is proved by direct computation in local coordinates. We conclude that the canonical bundle on $X_{n-1,n}$ is

$$\omega_{X_{n-1,n}} = \mathcal{O}(0,-1).$$

The important point here is that $\omega_{X_{n-1,n}}$ is the pullback of a line bundle on X_n through the map g in diagram (99).

CLAIM 6.1.2. *For the map g in (99) we have $Rg_*\mathcal{O}_{X_{n-1,n}} = \mathcal{O}_{X_n}$.*

Suppose we prove the claim. Since $\omega_{X_{n-1,n}} = g^*\mathcal{O}_{X_n}(-1)$ we shall also have $Rg_*\omega_{X_{n-1,n}} = \mathcal{O}_{X_n}(-1)$. By duality theory it follows that $\mathcal{O}_{X_n}(-1)$ is the dualizing complex on X_n. But this means exactly that X_n is Cohen-Macaulay and Gorenstein, with canonical bundle $\mathcal{O}(-1)$, so our theorem will be proved.

As to the claim, the nested isospectral Hilbert scheme, like X_n, has a local product structure

$$X_{n-1,n} \cong X_{k,k-1} \times X_l$$

where the the n points do not all coincide. The map g in diagram (99) factors locally as

$$g_k \times 1_l \colon X_{k,k-1} \times X_l \to X_k \times X_l.$$

We can assume that Claim 6.1.2 holds for g_k as part of the induction. Then the claim holds locally away from the most degenerate points. In particular, it holds outside the locus defined by the equations

$$y_1 = y_2 = \ldots = y_n.$$

The following lemma is proved by a standard local cohomology argument.

LEMMA 6.1.3. *Let $g\colon Y \to X$ be a proper morphism of algebraic varieties. Suppose given m global regular functions $z_1, \ldots, z_m$ on X, and let Z be the subvariety of X where they vanish; $U = X \setminus Z$ its complement. Assume the following conditions hold.*

(i) *The z_i form a regular sequence in every local ring $\mathcal{O}_{X,x}$, $x \in Z$.*
(ii) *The z_i form a regular sequence in every local ring $\mathcal{O}_{Y,y}$, $y \in g^{-1}(Z)$.*
(iii) *Every fiber of g has dimension $< m-1$.*
(iv) *On the open set U, the canonical homomorphism $\mathcal{O}_X \to Rg_*\mathcal{O}_Y$ is an isomorphism.*

Then the canonical homomorphism $\mathcal{O}_X \to Rg_*\mathcal{O}_Y$ is an isomorphism everywhere.

To prove Claim 6.1.2 we apply this to the morphism $g\colon X_{n-1,n} \to X_n$, taking $z_i = y_i - y_{i+1}$, $i = 1, \ldots, n-1$. We need to verify assumptions (i)-(iv).

We have already seen that we can assume (iv) by induction.

For (iii) we have to analyze the fibers of the map $H_{n-1,n} \to H_n$. The fiber over a point I in H_n is just a projective space of dimension $\dim \operatorname{soc}(R/I) - 1$. This is always maximized at a fixed point $I = I_\mu$, and then a simple calculation shows that $\dim \operatorname{soc}(R/I) < n-2$ when $n > 3$. For $n \leq 3$ the induction hypotheses have to be verified directly.

Both (i) and (ii) reduce to the corresponding problem for the sequence $y_1, \dots, y_n$. For (ii), the local rings of $X_{n-1,n}$ are Cohen-Macaulay, so we only need to show that the locus $\mathbf{y} = 0$ in $X_{n-1,n}$ has dimension n. This follows from the cell decomposition of $H_{n-1,n}$ in Cheah [**13**].

The crucial part is (i), which amounts to the following statement.

PROPOSITION 6.1.4. *The morphism $X_n \to \mathbb{C}^n$ given by projection on the y coordinates is flat.*

At this point we have pushed our geometric induction argument as far as it will go, and must approach the proof of Proposition 6.1.4 head-on. The first step is construct H_n and X_n as blowups.

PROPOSITION 6.1.5 ([**38**]). *Let $A = \mathbb{C}[\mathbf{x}, \mathbf{y}]^\epsilon$ be the subspace of antisymmetric elements and let $J = \mathbb{C}[\mathbf{x}, \mathbf{y}]A$ be the ideal it generates. Define graded Rees algebras*

$$S = \bigoplus_{d \geq 0} A^d; \quad T = \mathbb{C}[\mathbf{x}, \mathbf{y}][tJ] = \bigoplus_{d \geq 0} J^d,$$

with the convention that $A^0 = \mathbb{C}[\mathbf{x}, \mathbf{y}]^{S_n}$. Then $H_n = \operatorname{Proj} S$ and $X_n = \operatorname{Proj} T$.

PROOF. The result for X_n follows easily from the one for H_n. For H_n, observe that the fibers of the tautological bundle B are quotients of $R = \mathbb{C}[x, y]$, so the fibers of $\mathcal{O}(1) = \wedge^n B$ are quotients of $\wedge^n R = A$. So we get an algebra homomorphism $\phi \colon S \to \bigoplus_d H^0(H_n, \mathcal{O}(d))$, and the sections of $\mathcal{O}(1)$ coming from S do not vanish simultaneously anywhere on H_n. Hence ϕ is induced by a morphism $f : H_n \to \operatorname{Proj} S$, and $\mathcal{O}_{H_n}(1)$ is the pullback $f^*\mathcal{O}(1)$ of the twisting sheaf on $\operatorname{Proj} S$. Since f is a birational morphism of schemes projective over $\mathbb{C}^{2n}/S_n$, it is projective and hence surjective.

Consider the section α of $\mathcal{O}_{H_n}(1)$ given by the Vandermonde determinant $\Delta(\mathbf{x}) \in A$. The open set U_α where $\alpha \neq 0$ consists of the ideals I such that $\{1, x, \dots, x^{n-1}\}$ is a basis of R/I. One can see with explicit coordinates that $U_\alpha \cong \mathbb{C}^{2n}$. In fact U_α is the open cell in the cell decomposition of H_n given by Ellingsrud and Strömme [**17**]. From the cell decomposition one also sees that the divisor $Z = H_n \setminus U$ is irreducible. These facts imply that the Picard group of H_n is isomorphic to $\mathbb{Z}$, and $\mathcal{O}_{H_n}(1) = f^*\mathcal{O}(1)$ is ample. It follows that f is injective. □

To prove Proposition 6.1.4 it suffices to prove that for every d the ideal J^d is a free $\mathbb{C}[\mathbf{y}]$-module. To do this we introduce an auxilliary object.

DEFINITION 6.1.6. Let $[l]$ stand for the set $\{1, \dots, l\}$. To every function $f \colon [l] \to [n]$ associate the map $\pi_f \colon (\mathbb{C}^2)^n \to (\mathbb{C}^2)^l$ sending $(P_1, \dots, P_n)$ to $(P_{f(1)}, \dots, P_{f(l)})$ and let $W_f \subseteq \mathbb{C}^{2n+2l}$ be its graph. The ***polygraph*** $Z(n, l)$ is the union of the linear subspaces $W_f \subseteq \mathbb{C}^{2n+2l}$ over all functions $f \colon [l] \to [n]$.

We denote the coordinates on $\mathbb{C}^{2n}$ by x_i, y_i and those on $\mathbb{C}^{2l}$ by a_i, b_i, so the coordinate ring $R(n, l)$ of $Z(n, l)$ is the quotient

$$R(n, l) = \mathbb{C}[\mathbf{x}, \mathbf{y}, \mathbf{a}, \mathbf{b}]/(\bigcap_f I_f),$$

where

$$I_f = \sum_{i=1}^{l} (a_i - x_{f(i)}, b_i - y_{f(i)}).$$

The connection with the ideals J^d is as follows.

PROPOSITION 6.1.7. *Put $l = dn$ and let S_n^d act on $R(n,l) = R(n,dn)$ by permuting the dn coordinate pairs a_i, b_i in blocks of size n. Then J^d is isomorphic as a $\mathbb{C}[\mathbf{x},\mathbf{y}]$-module to the space $R(n,dn)^\epsilon$ of antisymmetric elements with respect to S_n^d.*

PROOF. Fix the function $f_0\colon [dn] \to [n]$ defined by $f_0(kn+i) = i$ for $i \in [n]$ and $0 \le k < d$. Let W_{f_0} be the corresponding component of $Z(n,l)$. Restriction of functions from $Z(n,l)$ to W_{f_0} is a homomorphism of $\mathbb{C}[\mathbf{x},\mathbf{y}]$-algebras $\phi\colon R(n,l) \to \mathbb{C}[\mathbf{x},\mathbf{y}]$ mapping a_{kn+i}, b_{kn+i} to x_i, y_i. It is easy to see that ϕ maps $R(n,l)^\epsilon$ surjectively onto J^d.

We need to prove that ϕ is injective on $R(n,l)^\epsilon$. Every antisymmetric function p vanishes on W_f if $f(kn+i) = f(kn+j)$ for some $0 \le k < d$ and some $i,j \in [n]$ with $i \ne j$. But if $f(kn+1), \ldots, f(kn+n)$ are distinct for each k, then f is in the S_n^d-orbit of f_0 and so p is determined on W_f by its restriction to W_{f_0}. □

Hence our problem is reduced to a special case of the following theorem.

THEOREM 6.1.8. *The coordinate ring $R(n,l)$ of the polygraph is a free module over the polynomial ring $\mathbb{C}[\mathbf{y}]$.*

This theorem is proved in [**39**] using induction and some commutative algebra to produce a basis of $R(n,l)$ as a free module. The proof is quite constructive and one can extract from it an algorithm to generate the basis elements in any given degree. More usefully, one can deduce that that the basis elements are indexed by some simple combinatorial data. Unfortunately, the proof is also extremely complicated, and I will be the first to admit that it is rather unsatisfactory from a conceptual point of view.

I should point out, however, that the polygraph is not merely an artifice to make the proof of Proposition 6.1.4 more combinatorial. It is also the object whose coordinate ring $R(n,l)$ appears in Theorem 5.4.3 as the space of global sections of the tensor product of tautological bundles $P \otimes B^{\otimes l}$. As such it is a geometrically natural entity. From this point of view we can see for example that the identification of J^d with $R(n,dn)^\epsilon$ in Proposition 6.1.7 is secretly the representation of the line bundle $\mathcal{O}(d)$ as the summand $(\wedge^n B)^{\otimes d}$ of $B^{\otimes dn}$.

6.2. Vanishing theorem. Now let us see how the vanishing theorem, Theorem 5.4.3, comes about. The first point is to restate the theorem using the Bridgeland-King-Reid functor as the pair of identities

$$(100)\qquad \begin{aligned} \Phi(B^{\otimes l}) &= R(n,l), \\ \Phi(\mathcal{O}_{Z_n} \otimes B^{\otimes l}) &= R(n,l)/\mathfrak{m}R(n,l). \end{aligned}$$

The next point is that the second identity can be deduced from the first in a straightforward way, using the resolution of $\mathcal{O}_{Z_n}$ in Proposition 5.4.4. So we only have to prove the first identity, which we may rewrite using the Bridgeland-King-Reid theorem as

$$\Psi R(n,l) = B^{\otimes l}.$$

To put it a bit more precisely, the fiber product over the Hilbert scheme $X_n \times F_n^l/H_n$ projects on $\mathbb{C}^{2n+2l}$ with image $Z(n,l)$. Functions on $Z(n,l)$ pull back to global functions on $X_n \times F_n^l/H_n$, that is, to global sections of $P \otimes B^{\otimes l}$, and this induces a map $R(n,l) \to \Phi(B^{\otimes l})$. Applying Ψ we have a canonical map $\Psi R(n,l) \to B^{\otimes l}$ and we are to prove it is an isomorphism.

I wish to point out here the full strength of this seemingly trivial substitution. The functor Φ involves sheaf cohomology, in the guise of the derived functor Rf_*. The inverse functor Ψ, however, involves only commutative algebra: to compute ΨA for any object A you only need an S_n-equivariant free resolution of A.

It develops that Theorem 6.1.8 is exactly what we need in order to calculate $\Psi R(n,l)$. The theorem implies immediately that $R(n,l)$ has a free resolution over $\mathbb{C}[\mathbf{x},\mathbf{y}]$ of length at most n, but we can do a little better using the translation invariance of the polygraph with respect to $\mathbb{C}^2$. This yields that $R(n,l)$ is actually a free $\mathbb{C}[x_1,\mathbf{y}]$-module and hence has a free resolution of length $n-1$.

Next we observe that when not all n points coincide, so we have the local product structure in (98), both $R(n,l)$ and $B^{\otimes l}$ decompose into smaller pieces, in a manner compatible with the map $\Psi R(n,l) \to B^{\otimes l}$. Therefore we can assume by induction that our map is locally an isomorphism away from the most degenerate points.

This much turns out to be almost enough to prove the theorem. Consider an exact triangle

$$C[-1] \to \Psi R(n,l) \to B^{\otimes l} \to C$$

in the derived category $D(H_n)$. We are to prove that $C=0$. We have just shown that it has a locally free resolution of length n (from the resolution of $R(n,l)$ plus the extra term $B^{\otimes l}$), and that it vanishes on an open set whose complement has codimension $n-1$. If the codimension were $n+1$ instead, it would imply that $C=0$ by the intersection theorem of Peskine, Szpiro and Roberts [**70, 74, 75**]. All we need to do is kill off C on a slightly bigger open set.

Let U_x be the open set, called U_α in the proof of Proposition 6.1.5, where the monomials $1, x, \dots, x^{n-1}$ are independent modulo I, and let U_y be the same with y in place of x. The theorem follows from two more facts.

LEMMA 6.2.1. *Let W be the locus where not all n points coincide. Then the complement of the open set $W \cup U_x \cup U_y$ in the Hilbert scheme has codimension $n+1$.*

LEMMA 6.2.2. *On U_x, the canonical map $\Psi R(n,l) \to B^{\otimes l}$ is an isomorphism.*

Lemma 6.2.1 is an easy application of the Ellingsrud-Strömme cell decomposition.

Lemma 6.2.2 is a computation in explicit coordinates on U_x. Its proof contains two delicate points, which I will mention without giving details. First we have to check that the free resolution of $R(n,l)$ remains exact when pulled back to the open set in X_n lying over U_x. This is done using Theorem 6.1.8 again. Second, we have to check that on U_x the canonical map $\mathcal{O}(-1) \otimes (\rho_* f^* R(n,l))^\epsilon \to B^{\otimes l}$ is an isomorphism (recall the formula for Ψ in (89)). Once written out in coordinates, it is easy to see that this map is surjective. For injectivity one has to show that certain elements that are not obviously zero do in fact vanish in $f^*R(n,l)$, which requires a bit of care.

7. Current developments

7.1. Combinatorial advances. It remains an open problem to find a combinatorial formula for the Kostka-Macdonald polynomials $\tilde{K}_{\lambda\mu}(q,t)$ along the lines of Theorem 3.4.15 or 3.4.16, but there has recently been encouraging progress on the corresponding problem for the coefficients of the Frobenius series of diagonal coinvariants, $\nabla e_n(z)$.

In view of (71) there exist t-analogs of the Carlitz-Riordan q-Catalan numbers

$$C_n(q,t) \underset{\text{def}}{=} \langle e_n, \nabla e_n\rangle.$$

By Theorem 4.2.5, this ***q, t-Catalan polynomial*** is the Hilbert series of the antisymmetric part R_n^ϵ of the diagonal coinvariant ring. Hence it is a polynomial with positive integer coefficients. Unlike most of the other quantities that can be shown to be positive by geometric methods, this one has a known combinatorial interpretation, discovered by Haglund and proved by him and Garsia. I will state a variant of their theorem that lends itself better to generalization.

THEOREM 7.1.1 ([**20, 21**]). *Denote by $b(\lambda)$ the number of cells x in the diagram of λ for which the arm and leg of x are related by $l(x) \leq a(x) \leq l(x)+1$. Let δ_n be the staircase partition $(n-1, n-2, \dots, 1)$. Then the q,t-Catalan polynomial is the sum over partitions λ with diagram contained inside the diagram of δ_n*

$$C_n(q,t) = \sum_{\lambda\subseteq\delta_n} q^{\binom{n}{2}-|\lambda|}t^{b(\lambda)}. \tag{101}$$

To prove this, define more generally $S_{n,k}(q,t)$ to be the same sum taken only over those λ that contain precisely k cells on the outermost diagonal of δ_n. Garsia and Haglund find a recurrence for $S_{n,k}(q,t)$ and an expression in terms of symmetric functions that satisfies the same recurrence. Their expression reduces to $\langle e_n, \nabla e_n\rangle$ for $S_{n+1,0}(q,t)$, which is the same as the sum in (101).

Interestingly, there is a way of rewriting the combinatorial formula (101) as a sum of products of Gauss binomial coefficients, formally resembling the Kirillov-Reshetikhin formula (51) for the Kosta-Foulkes polynomials.

COROLLARY 7.1.2. *The q,t-Catalan polynomial can be written as*

$$C_n(q,t) = \sum_r \sum_{k_1+\cdots+k_r=n} q^{\sum_i (i-1)k_i}\, t^{\sum_i \binom{k_i}{2}} \prod_{i=1}^{r-1} \begin{bmatrix} k_i+k_{i+1}-1 \\ k_i-1 \end{bmatrix}_t, \tag{102}$$

the sum ranging over compositions of n into positive integers $k_1, \dots, k_r$.

PROOF. Encode each $\lambda \subseteq \delta_n$ by the sequence $(e_1, e_2, \dots, e_n)$ with $e_i = n-i-\lambda_i$ and $e_n = 0$. The constraints on the possible sequences are $0 \leq e_i \leq e_{i+1}+1$ for each i. Let k_m be the number of e_i's equal to $m-1$. If $r = 1+\max(e_1, \dots, e_n)$, the constraints imply that $k_1, \dots, k_r$ are all non-zero.

The exponent $\binom{n}{2} - |\lambda|$ of q in (101) is $\sum_j e_j = \sum_i (i-1)k_i$. The exponent $b(\lambda)$ of t is the number of pairs $i<j$ with $e_j = e_i$, contributing $\sum_i \binom{k_i}{2}$, plus the number of pairs with $e_j = e_i+1$. The constraints imply that for each m, the subsequence formed by the e_i's in $\{m-1, m\}$ ends with an $m-1$. The others may be interspersed in any order, contributing the factor $\left[\begin{smallmatrix} k_m+k_{m+1}-1 \\ k_m-1 \end{smallmatrix}\right]_t$ as we sum over all possibilities. One can show that the choices of interspersing order for each m may be made independently, and together they determine the sequence. □

Curiously, the symmetry $C_n(q,t) = C_n(t,q)$, which is obvious from the definition, appears as quite a surprising property of formula (101) or (102).

The Garsia-Haglund theorem has two conjectured extensions that have been verified computationally up to reasonably large values of n. The first conjecture I will discuss is mine; the second is due to Haglund and Loehr [**36**]. Define "higher" q,t-Catalan polynomials by the formula

$$C_n^{(m)}(q,t) = \langle e_n, \nabla^m e_n \rangle.$$

They are polynomials with positive coefficients by Theorem 3.5.16, and the following specializations can be derived similarly to the case $m = 1$:

$$C_n^{(m)}(q,1) = \sum_{\lambda \subseteq m\delta_n} q^{m\binom{n}{2} - |\lambda|};$$

$$C_n^{(m)}(q,q^{-1}) = q^{-m\binom{n}{2}} \frac{1}{[mn+1]_q} \begin{bmatrix} (m+1)n \\ n \end{bmatrix}_q.$$

From the discussion in §5.4.4 we see that $C_n^{(m)}(q,t)$ has a natural geometric interpretation as the Hilbert series of the space of sections of the line bunle $\mathcal{O}(m)$ on the zero-fiber Z_n of the Hilbert scheme. More concretely, this space of sections is $J^m/(\mathbf{x},\mathbf{y})J^m$, where J is the ideal generated by all antisymmetric polynomials in $\mathbb{C}[\mathbf{x},\mathbf{y}]$, as in Proposition 6.1.5.

CONJECTURE 7.1.3. *Denote by $b^{(m)}(\lambda)$ the number of cells $x \in \lambda$ whose arm and leg satisfy $l(x) \leq a(x) \leq l(x) + m$. For the higher q,t-Catalan polynomials we have the formula*

$$C_n^{(m)}(q,t) = \sum_{\lambda \subseteq m\delta_n} q^{m\binom{n}{2} - |\lambda|} t^{b^{(m)}(\lambda)}.$$

The other conjecture I want to discuss is Haglund and Loehr's conjecture for the Hilbert series of the whole diagonal coinvariant ring, or by Theorem 4.2.5, for the polynomial

$$D_n(q,t) \underset{\text{def}}{=} \langle e_1^n, \nabla e_n \rangle.$$

Recall that $D_n(q,1)$ enumerates parking functions by weight. Haglund and Loehr define an analog for parking functions of the statistic $b(\lambda)$ in Theorem 7.1.1.

We have seen that a parking function can be represented by a standard tableau on the skew shape $(\lambda + (1^n))/\lambda$, where $\lambda \subseteq \delta_n$. Here is example (15) again to illustrate this.

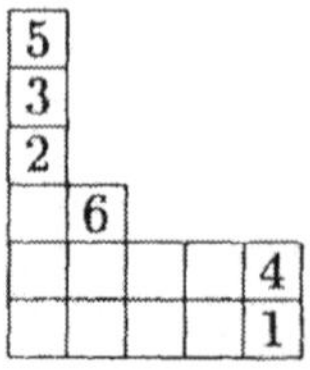

Call the *k-th diagonal* the set of cells (i,j) with $i + j = k$. Now let f be a parking function, T the corresponding tableau, $\lambda \subseteq \delta_n$ the corresponding shape, and define $b(f)$ to be the number of pairs of cells $x, y \in (\lambda + (1^n))/\lambda$ such that the label on x is less than the label on y, and either:

(i) x and y are on the same diagonal and y is in a higher row than x, or
(ii) x and y are on consecutive diagonals k and $k+1$, respectively, and y is in a lower row than x.

In the example above we have $b(f) = 3$, for the pairs $\{1,3\}$, $\{3,4\}$ and $\{4,5\}$. For any given λ the maximum value of $b(f)$ is $b(\lambda)$, attained by the unique tableau T formed by labelling the cells in increasing order of diagonals, and from lowest row to highest within each diagonal.

CONJECTURE 7.1.4 ([**36**]). *The Hilbert series of the diagonal coinvariant ring is given by the sum*

$$D_n(q,t) = \sum_f q^{w(f)} t^{b(f)}$$

over parking functions f on $\{1,\dots,n\}$, with weight $w(f) = \binom{n}{2} - |\lambda|$ and $b(f)$ as defined above.

7.2. Extensions to other groups. It it natural to ask which of the results about the Hilbert scheme, diagonal coinvariants, and $n!$ conjecture might hold and in what form with a more general finite group G in the role of S_n.

The first obvious possibility is to take for G a Coxeter group acting diagonally on two copies $\mathfrak{h} \oplus \mathfrak{h}$ of its defining representation, or more generally a complex reflection group acting on $\mathfrak{h} \oplus \mathfrak{h}^*$. In this setting only a limited generalization of the $n!$ conjecture seems to be possible, but it appears that the phenomena involving diagonal coinvariants generalize beautifully. Very recently, I. Gordon [**34**] has proved a conjecture of mine along these lines.

A second possibility is to take for G the wreath product $\Gamma_n = \Gamma \wr S_n$, where Γ is a finite subroup of $SL_2(\mathbb{C})$. The possible groups Γ are classified by Dynkin diagrams of type ADE, via the *McKay correspondence* [**64**]. Associated to each affine ADE diagram and its fundamental weight at the affine node are *quiver varieties* of Nakajima [**66**], which provide crepant resolutions of $\mathbb{C}^{2n}/\Gamma_n$, with the Hilbert scheme as the special case when Γ is trivial. In this setting there appears to be a good generalization of the $n!$ conjecture.

Most interesting is the intersection of the above two settings, when $\Gamma = \mathbb{Z}/r\mathbb{Z}$ is cyclic, so $G = \Gamma \wr S_n$ is the complex reflection group $G(r,1,n)$ with its doubled reflection representation $\mathfrak{h} \oplus \mathfrak{h}^*$. Here as for S_n we have the $\mathbb{T}^2$ action, with fixed points indexed by r-tuples of partitions. Conjecturally we expect analogs of the bundle P on the Hilbert scheme, with the graded characters of its fibers given by new *wreath Macdonald polynomials*.

Let me now explain some of these developments in more detail.

7.2.1. *Principal nilpotent pairs.* The *commuting variety* of a semisimple Lie algebra $\mathfrak{g}$ is the set

$$C(\mathfrak{g}) = \{(X,Y) \in \mathfrak{g} : [X,Y] = 0\}.$$

It is not known whether the equations $[X,Y] = 0$ generate the ideal of the commuting variety, so in principle we should distinguish the possibly non-reduced *commuting scheme* defined by these equations. It is known that $C(\mathfrak{g})$ is irreducible, of dimension $\dim \mathfrak{g} + \operatorname{rk} \mathfrak{g}$. For $\mathfrak{gl}_n$ this is an old theorem of Motzkin and Taussky [**65**]; it was extended to all $\mathfrak{g}$ by Richardson [**73**]. Equivalently, the commuting regular semisimple pairs are dense in $C(\mathfrak{g})$.

One proves easily that the dimension of the Zariski tangent space to the commuting scheme at (X,Y) is equal to $\dim \mathfrak{g} + \dim \mathfrak{z}(X,Y)$, where $\mathfrak{z}(X,Y)$ is their common centralizer. Set

$$C_{\text{reg}}(\mathfrak{g}) = \{(X,Y) \in C(\mathfrak{g}) : \dim \mathfrak{z} = \operatorname{rk} \mathfrak{g}\},$$

the nonsingular locus of the commuting scheme, which at least on this locus clearly coincides with the commuting variety. A point of $C(\mathfrak{gl}_n)$ is just a pair of commuting $n \times n$ matrices X, Y, and they make $\mathbb{C}^n$ into a $\mathbb{C}[x,y]$-module. Neubauer and Saltman [**69**] showed that $(X,Y) \in C_{\mathrm{reg}}(\mathfrak{gl}_n)$ if and only if $\mathbb{C}^n$ considered as a $\mathbb{C}[x,y]$-module is locally cyclic or cocylic at every point $(x,y) \in \mathbb{C}^2$. Their theorem implies that for every $(X,Y) \in C_{\mathrm{reg}}(\mathfrak{gl}_n)$, the ideal of relations $I = \{f(x,y) \in \mathbb{C}[x,y] : f(X,Y) = 0\}$ has $\dim_{\mathbb{C}} \mathbb{C}[x,y]/I = n$, yielding a morphism

$$C_{\mathrm{reg}}(\mathfrak{gl}_n) \to H_n \tag{103}$$

of $C_{\mathrm{reg}}(\mathfrak{gl}_n)$ onto the Hilbert scheme.

The points of $C_{\mathrm{reg}}(\mathfrak{gl}_n)$ lying over a $\mathbb{T}^2$-fixed point in H_n are exactly the ***principal nilpotent pairs***, as defined more generally for any $\mathfrak{g}$ by Ginzburg.

DEFINITION 7.2.1 ([**33**]). A ***principal nilpotent pair*** is an element $(e_1, e_2) \in C_{\mathrm{reg}}(\mathfrak{g})$ such that the orbit $\mathrm{Ad}(G)(e_1,e_2)$ is a fixed point for the action of $\mathbb{T}^2$ on the set of such orbits.

Here the $\mathbb{T}^2$ action is the obvious one on the two components of $(X,Y) \in C(\mathfrak{g})$.

The principal nilpotent pairs up to $\mathrm{Ad}(G)$ conjugacy have been classified by Elashvili and Panyushev [**16**]. For $\mathfrak{gl}_n$, two pairs dual to each other lie over each fixed point I_μ in the Hilbert scheme (or one self-dual pair if μ is rectangular). For other types they are much more rare, and the elements e_1, e_2 can belong only to very special nilpotent orbits.

Now let $\mathfrak{h}$ be a Cartan subalgebra of $\mathfrak{g}$ and W the Weyl group. To each principal nilpotent pair $\mathbf{e}$ (apart from an exception in type E_7) Ginzburg associates a W-antisymmetric polynomial $\Delta_{\mathbf{e}} \in \mathbb{C}[\mathfrak{h} \oplus \mathfrak{h}]$ generalizing our polynomial Δ_μ in (77), which is the $\mathfrak{gl}_n$ case. Letting $J_{\mathbf{e}} \subseteq \mathbb{C}[\mathfrak{h} \oplus \mathfrak{h}]$ be the ideal of polynomials f such that $f(\partial)\Delta_{\mathbf{e}} = 0$, the analog of our ring R_μ would appear naively to be

$$R_{\mathbf{e}} \underset{\mathrm{def}}{=} \mathbb{C}[\mathfrak{h} \oplus \mathfrak{h}]/J_{\mathbf{e}}.$$

It would be tempting to conjecture that $\dim R_{\mathbf{e}} = |W|$ and $R_{\mathbf{e}}$ affords the regular representation of W, but that is false for $\mathfrak{g} = \mathfrak{sp}_6$.

There seem to be two ways to fix this trouble. The first has been proposed by Kumar and Thomsen [**54**], who would replace the ring $R_{\mathbf{e}}$ by the coordinate ring $D_{\mathbf{e}}$ of the scheme-theoretic intersection of $\mathfrak{h} \oplus \mathfrak{h}$ with the closure $\overline{O}_{\mathbf{e}}$, where $O_{\mathbf{e}}$ is the orbit of $\mathbf{e}$ under a group $G^{\mathbf{e}}$ which is $\mathrm{Ad}(G)$ enlarged by certain outer automorphisms of $\mathfrak{g}$. By definition their ring $D_{\mathbf{e}}$ is a again a quotient of $\mathbb{C}[\mathfrak{h} \oplus \mathfrak{h}]$ by a doubly homogeneous W-invariant ideal, and they conjecture that $D_{\mathbf{e}}$ is Gorenstein and affords the regular representation of W. Unfortunately, it appears their conjecture cannot be true as stated since the properties of $D_{\mathbf{e}}$ would imply that $R_{\mathbf{e}} = D_{\mathbf{e}}$. However, I think a small modification may work.

CONJECTURE 7.2.2. *Let $\tilde{O}_{\mathbf{e}}$ be the normalization of the orbit closure $\overline{O}_{\mathbf{e}}$ defined by Kumar and Thomsen, when it is irreducible, or a suitable seminormalization when it is not. Then the coordinate ring $\tilde{D}_{\mathbf{e}}$ of the scheme-theoretic fiber product of $\tilde{O}_{\mathbf{e}}$ and $\mathfrak{h} \oplus \mathfrak{h}$ over $C(\mathfrak{g})$ is Gorenstein, has dimension $|W|$, and affords the regular representation of $|W|$.*

Observe that the original ring $D_{\mathbf{e}}$ is a subring of $\tilde{D}_{\mathbf{e}}$, the image of the canonical homomorphism $\mathbb{C}[\mathfrak{h} \oplus \mathfrak{h}] \to \tilde{D}_{\mathbf{e}}$. It seems plausible to expect that $D_{\mathbf{e}} = R_{\mathbf{e}}$, so $D_{\mathbf{e}}$ should in fact be Gorenstein, but its dimension will be too small in general. In the case of $\mathfrak{gl}_n$, when $R_{\mathbf{e}} = R_\mu$ has the full dimension $n!$, the normalization step is unnecessary, and the Kumar-Thomsen conjecture should be correct in its original form. Even for $\mathfrak{gl}_n$ it is open.

There is another way to enlarge $R_{\mathbf{e}}$ to what should turn out to be the same ring $\tilde{D}_{\mathbf{e}}$ as in Conjecture 7.2.2. Restriction of functions defines a homomorphism of coordinate rings

$$\mathbb{C}[C(\mathfrak{g})] \to \mathbb{C}[\mathfrak{h} \oplus \mathfrak{h}],$$

and by an important theorem of Wallach [**84**] and Joseph [**46**], it induces an isomorphism of invariants

$$\mathbb{C}[C(\mathfrak{g})]^G \cong \mathbb{C}[\mathfrak{h} \oplus \mathfrak{h}]^W.$$

Thus we get a ring homomorphism $\mathbb{C}[\mathfrak{h} \oplus \mathfrak{h}]^W \to \mathbb{C}[C(\mathfrak{g})]$ and a G-invariant morphism of algebraic varieties $C(\mathfrak{g}) \to (\mathfrak{h} \oplus \mathfrak{h})/W$. Now define the ***isospectral commuting variety*** $X(\mathfrak{g})$ to be the *normalized, reduced* fiber product

$$\begin{array}{ccc} X(\mathfrak{g}) & \longrightarrow & \mathfrak{h} \oplus \mathfrak{h} \\ \downarrow & & \downarrow \\ C(\mathfrak{g}) & \longrightarrow & (\mathfrak{h} \oplus \mathfrak{h})/W. \end{array} \tag{104}$$

The Weyl group W acts on $X(\mathfrak{g})$, and the projection $X(\mathfrak{g}) \to C(\mathfrak{g})$ is finite of degree $|W|$ with W acting by the regular representation on its generic fibers.

CONJECTURE 7.2.3. *The isospectral commuting variety $X(\mathfrak{g})$ is Cohen-Macaulay and Gorenstein. In particular the coordinate ring of its fiber over each point of $C_{\mathrm{reg}}(\mathfrak{g})$ is a Gorenstein ring on which W acts by the regular representation.*

Note that the fiber ring over a point $(X, Y) \in C(\mathfrak{g})$ depends only on its $\mathrm{Ad}(G)$ orbit. The most interesting fibers are those over principal nilpotent pairs $\mathbf{e}$, where the coordinate ring $\hat{D}_{\mathbf{e}}$ of the fiber is doubly graded. Assuming the conjecture holds, the naive ring $R_{\mathbf{e}}$ is the image of the canonical map $\mathbb{C}[\mathfrak{h} \oplus \mathfrak{h}] \to \hat{D}_{\mathbf{e}}$, and it seems reasonable to expect that $\hat{D}_{\mathbf{e}}$ is isomorphic to the ring $\tilde{D}_{\mathbf{e}}$ in Conjecture 7.2.2.

For the case of $\mathfrak{gl}_n$, Conjecture 7.2.3 in its full strength is open. Over the regular locus C_{reg}, however, it is true—this is equivalent to Theorem 5.2.1. The conjecture implies the famous conjecture generally attributed to Hochster that $C(\mathfrak{gl}_n)$ is Cohen-Macaulay. In the $\mathfrak{gl}_n$ case, I expect that the reduced fiber product in (104) is already normal. For other types it is not.

7.2.2. *Diagonal coinvariants and Gordon's theorem.* Let W be a Coxeter group and $\mathfrak{h}$ its (complexified) defining representation. Then $\mathfrak{h}$ is self-dual and it is best to identify the diagonal representation with the W-module $\mathfrak{h} \oplus \mathfrak{h}^*$, which has a natural symplectic structure. For $W = S_n$, we have seen that the coinvariant ring

$$C^W = \mathbb{C}[\mathfrak{h} \oplus \mathfrak{h}^*]/(\mathbb{C}[\mathfrak{h} \oplus \mathfrak{h}^*]^W_+)$$

has dimension $(n+1)^{n-1}$. It is natural for many reasons to predict that the analogous dimension formula in general should be $(h+1)^n$, where h is the Coxeter number of W and $n = \dim \mathfrak{h}$ is the rank, but this turns out not to be quite right. For W of type B_4, for example, the formula gives 9^4, but the actual dimension is $9^4 + 1$. In [**37**] I conjectured that after passing to a suitable quotient of the coinvariant ring, one should have the following result, which has now been proven by Gordon.

THEOREM 7.2.4 (Gordon [**34**]). *There is a canonically defined doubly graded quotient ring R^W of the coinvariant ring C^W with the following properties.*

(i) $\dim R^W = (h+1)^n$, *where h is the Coxeter number and n is the rank;*
(ii) *The Hilbert series of R^W satisfies $H_{R^W}(t^{-1},t) = t^{-hn/2}[h+1]_t^n$;*
(iii) *The image of $\mathbb{C}[\mathfrak{h}]$ in R^W is the classical coinvariant ring;*
(iv) *If W is a Weyl group, then $\varepsilon \otimes R^W$ is isomorphic as a W-module to the permutation representation on $Q/(h+1)Q$, where Q is the root lattice.*

Let us remark that (ii) obviously implies (i). In fact (ii) also implies (iii), because the image of $\mathbb{C}[\mathfrak{h}]$ is in any event a quotient of the classical coinvariant ring. If it were a proper quotient, then its socle would be killed and there would be no $t^{hn/2}$ term in $H_{R^W}(t^{-1},t)$.

Note that $H_{R^W}(t^{-1},t)$ is the Hilbert series of R^W in the grading that assigns degree 1 to the coordinates on $\mathfrak{h}$ and -1 to those on $\mathfrak{h}^*$, which is the natural grading from the symplectic point of view. In fact Gordon fully determines the graded character of R^W in this grading, which implies both (ii) and (iv). It seems to be difficult to understand the second grading using Gordon's approach. It also seems to be difficult to describe the ideal of R^W in C^W. In particular, for $W = S_n$, Gordon's theorem only implies that $(n+1)^{n-1}$ is a lower bound for $\dim R_n$.

Gordon proves Theorem 7.2.4 using representation theory of Cherednik algebras, more specifically their rational degenerations H_c which have been studied by Dunkl [**15**], Opdam and Rouqier [unpublished], Berest, Etingof and Ginzburg [**5**, **6**], and Guay [unpublished] among others. To see how Cherednik algebras enter the picture, we first desribe the coinvariant ring in a different way.

Consider the skew group algebra $\mathbb{C}[\mathfrak{h} \oplus \mathfrak{h}^*]W$, and let $e = \frac{1}{|W|}\sum_{w \in W} w$ be the invariant idempotent. A $\mathbb{C}[\mathfrak{h} \oplus \mathfrak{h}^*]W$-module is the same thing as a W-equivariant $\mathbb{C}[\mathfrak{h} \oplus \mathfrak{h}^*]$-module, and the subalgebra $e\mathbb{C}[\mathfrak{h} \oplus \mathfrak{h}^*]We$ is the ring of invariants $\mathbb{C}[\mathfrak{h} \oplus \mathfrak{h}^*]^W$. Regard $\mathbb{C}$ as a trivial $\mathbb{C}[\mathfrak{h} \oplus \mathfrak{h}^*]^W$-module annihilated by the ideal $\mathbb{C}[\mathfrak{h} \oplus \mathfrak{h}^*]^W_+$. Then the coinvariant ring can be described as

$$C^W = \mathbb{C}[\mathfrak{h} \oplus \mathfrak{h}^*]We \otimes_{e\mathbb{C}[\mathfrak{h}\oplus\mathfrak{h}^*]We} \mathbb{C}.$$

Alternatively, let $e_\varepsilon = \frac{1}{|W|}\sum_{w \in W} \varepsilon(w)w$ be the sign idempotent. Then we also have

$$\varepsilon \otimes C^W = \mathbb{C}[\mathfrak{h} \oplus \mathfrak{h}^*]We_\varepsilon \otimes_{e_\varepsilon\mathbb{C}[\mathfrak{h}\oplus\mathfrak{h}^*]We_\varepsilon} \mathbb{C}. \tag{105}$$

The *rational Cherednik algebra* H_c (depending on a parameter c) is the algebra of operators on $\mathbb{C}[\mathfrak{h}]$ generated by

(i) multiplication operators by coordinates x_i on $\mathfrak{h}$;
(ii) Dunkl-differential operators

$$y_i = \partial/\partial x_i - c\sum_{s \in S}\langle \alpha_s, x_i^* \rangle \frac{1-s}{\alpha_s},$$

where $\{x_i^*\}$ is a dual basis of coordinates on $\mathfrak{h}^*$, S is the set of reflections in W, and α_s is a linear form vanishing on the fixed hyperplane of s;
(iii) the group W.

The operators y_i commute. Thus H_c contains a copy of $\mathbb{C}[\mathfrak{h}]$ generated by the x_i's, a copy of $\mathbb{C}[\mathfrak{h}^*]$ generated by the y_i's, and a copy of $\mathbb{C}W$. It has a Poincaré-Birkhoff-Witt decomposition

$$H_c \cong \mathbb{C}[\mathfrak{h}] \otimes \mathbb{C}W \otimes \mathbb{C}[\mathfrak{h}^*] \tag{106}$$

and commutation relations

$$[y_i, x_j] = \delta_{ij} - c \sum_{s \in S} \langle \alpha_s, x_i^* \rangle \langle x_i, \alpha_s^\vee \rangle s,$$

where $\alpha_s^\vee$ is the coroot vector such that $sv = v - \langle \alpha_s, v \rangle \alpha_s^\vee$. Under the filtration of H_c assigning degree zero to $\mathbb{C}W$ and degree 1 to x_i, y_i, we have

$$\operatorname{gr} H_c \cong \mathbb{C}[\mathfrak{h} \oplus \mathfrak{h}^*]W. \tag{107}$$

The decomposition (106) implies that for each irreducible W-module τ there is a standard "Verma" module $M_c(\tau)$ isomorphic to $\mathbb{C}[\mathfrak{h}] \otimes \tau$ as a $\mathbb{C}[\mathfrak{h}]W$ module and annihilated by $\mathbb{C}[\mathfrak{h}^*]_+$. It has a unique irreducible quotient $L_c(\tau)$. Put $\mathfrak{h}_k = \wedge^k \mathfrak{h}$, which is always an irreducible W-module.

Gordon proves that for the special value $c = (1+h)/h$ of the parameter, the following things happen. First, there is a decomposition

$$L_c(\mathfrak{h}_0) = \sum_k (-1)^k M_c(\mathfrak{h}_k)$$

in the Grothendieck group of H_c-modules. Second, the degree operator $d = \frac{1}{2}\sum_i (x_i y_i + y_i x_i)$ induces a grading on these modules, such that $\deg(x_i) = 1$, $\deg(y_i) = -1$, and $M_c(\mathfrak{h}_k)$ has the same grading as $\mathbb{C}[\mathfrak{h}] \otimes \mathfrak{h}_k$ with the generators in degree $-hn/2 + k(h+1)$. It follows that the Hilbert series of $L_c(\mathfrak{h}_0)$ is equal to

$$t^{-hn/2}[h+1]_t^n,$$

as we want to prove for R^W. The same computation also determines the character of $L_c(\mathfrak{h}_0)$ as a graded W-module. Finally, the subalgebra $e_\varepsilon H_c e_\varepsilon$ has a unique 1-dimensional unital module $\mathbb{C}$ (with degree zero), and $L_c(\mathfrak{h}_0)$ is the induced module

$$L_c(\mathfrak{h}_0) \cong H_c e_\varepsilon \otimes_{e_\varepsilon H_c e_\varepsilon} \mathbb{C}.$$

Passing to the associated graded using (107), and comparing with (105), we get a canonical surjection

$$C^W \to \varepsilon \otimes \operatorname{gr} L_c(\mathfrak{h}_0)$$

of $\mathbb{C}[\mathfrak{h} \oplus \mathfrak{h}^*]W$-modules. Hence $R^W \underset{\text{def}}{=} \varepsilon \otimes \operatorname{gr} L_c(\mathfrak{h}_0)$ has the desired properties.

Shortly before Gordon's work appeared I had conjectured another way to pick out the correct quotient of C^W, as follows.

CONJECTURE 7.2.5. *Let $\tilde{R}^W$ be the quotient of C^W by the unique largest ideal $J \subseteq C^W$ such that $J \cap (C^W)^\epsilon = 0$. Then $\tilde{R}^W$ has the properties in Theorem 7.2.4.*

Conjecture 7.2.5 is true for $W = S_n$, that is, we have $R_n^\epsilon = \operatorname{soc} R_n$. This follows from [**40**, Theorem 4.1]. I can prove it for W of type B_n, and have verified it by computer for type D_4.

PROPOSITION 7.2.6. *If Conjecture 7.2.5 holds for W, then the ring $\tilde{R}^W$ in the conjecture is the same as Gordon's ring R_W.*

PROOF. Suppose Conjecture 7.2.5 holds. Then $\dim(C^W)^\varepsilon = \dim(\tilde{R}^W)^\varepsilon = \dim(R^W)^\varepsilon$. This implies that the defining ideal of Gordon's ring R^W is contained in the ideal J in the conjecture. But $\dim \tilde{R}^W = \dim R^W = (h+1)^n$, so $\tilde{R}^W = R^W$. □

7.2.3. *Quiver varieties.* In this section and the next, the role of S_n will be played by the wreath product $\Gamma_n = \Gamma \wr S_n$, where Γ is a finite subgroup of $SL_2(\mathbb{C})$. Then Γ^n acts on $\mathbb{C}^{2n}$ and is normalized by S_n; the wreath product Γ_n is their semidirect product. The presentation here is a synopsis of a preprint [**41**] that I hope to make available on the servers soon.

The quotient singularities $\mathbb{C}^{2n}/\Gamma_n$ have crepant resolutions by quiver varieties, generalizing the Hilbert scheme in the case $\Gamma = 1$. Our first goal will be to formulate a conjectured analog of the Cohen-Macaulay property of the isospectral Hilbert scheme, Theorem 5.2.1.

The possible groups Γ are classified by the McKay correspondence. Let $\chi_0, \chi_1, \dots, \chi_{r-1}$ be a list of the irreducible characters of Γ, with $\chi_0 = 1$, and let ζ be the character of the defining representation $\mathbb{C}^2$. The *McKay graph* is the graph on vertex set $I = \{0, 1, \dots, r-1\}$ with an edge $\{i, j\}$ if $\langle \chi_i \otimes \zeta, \chi_j \rangle \neq 0$. Since ζ is self-dual, this relation is symmetric in i and j. The McKay graphs turn out to be the affine Dynkin diagrams of type $\hat{A}$, $\hat{D}$ and $\hat{E}$, as follows.

- Type $\hat{A}_{r-1}$: $\Gamma = \mathbb{Z}/r\mathbb{Z}$ is cyclic.
- Type $\hat{D}_{r+2}$: Γ is the *binary dihedral group*, the preimage of the dihedral group D_{2r} by the double cover $SU_2(\mathbb{C}) \to SO_3(\mathbb{R})$.
- Type $\hat{E}_6$, $\hat{E}_7$, $\hat{E}_8$: the binary tetrahedral, octahedral and icosahedral groups.

Fix Γ and its associated affine Dynkin diagram. Let P be the weight lattice, P^+ the dominant weights, Q the root lattice, Q^+ the cone spanned by the simple roots $\alpha_0, \dots, \alpha_{r-1}$, and δ the smallest positive imaginary root. The corresponding root lattice Q_0 of finite type can be identified with $Q/\mathbb{Z}\delta$. To each character $\chi = \sum_i m_i \chi_i$ corresponds the element $\sum_i m_i \alpha_i \in Q^+$, and a key property of the McKay correspondence is that δ corresponds to the character 1_1^Γ of the regular representation.

Denote by $\mathfrak{M}_\theta^\Gamma(\lambda, \mu)$ the *quiver variety* associated by Nakajima [**66**] to the McKay graph I of Γ, elements $\lambda \in P^+$ and $\mu \in Q^+$, and a *stability condition* given by a linear function $\theta\colon Q \to \mathbb{R}$. The quiver varieties $\mathfrak{M}_\theta^\Gamma(\lambda, \mu)$ have many remarkable properties. For generic θ they are nonsingular and have a symplectic (and even a hyper-Kähler) structure. Nakajima has shown that for each λ the top cohomology groups $\bigoplus_\mu H^{\mathrm{top}}(\mathfrak{M}_\theta^\Gamma(\lambda, \mu), \mathbb{C})$ carry the irreducible highest-weight representation $L(\lambda)$ of the Kac-Moody algebra attached to I, with $\mathfrak{M}_\theta^\Gamma(\lambda, \mu)$ contributing the weight space $L(\lambda)_{\lambda-\mu}$.

We will only be concerned with $\mathfrak{M}_\theta^\Gamma(\lambda, \mu)$ for $\lambda = \lambda_0$, the fundamental weight at the affine node of I, and $\theta(\alpha_i) = 1$ for all i, the *standard* stability condition. The highest-weight module $L(\lambda_0)$ is the *basic representation*. The stabilizer of λ_0 in the affine Weyl group W is the finite Weyl group W_0, so every element of Q_0 has a unique representative $\nu_0 \in Q$ such that $\lambda_0 - \nu_0$ is W-conjugate to λ_0. The weight spaces $L(\lambda_0)_{\lambda_0 - \nu_0}$ are one-dimensional. We write every $\nu \in Q$ uniquely as

$$\nu = \nu_0 + n\delta, \quad n \in \mathbb{Z},$$

with ν_0 the distinguished representative of its coset. The weight space $L(\lambda_0)_{\lambda_0-\nu}$ is non-zero, and the quiver variety $\mathfrak{M}_1^\Gamma(\lambda_0, \nu)$ is non-empty, if and only if $n \geq 0$. For such ν set

$$Y_{\Gamma,\nu} = \mathfrak{M}_{\mathbf{1}}^{\Gamma}(\lambda_0, \nu).$$

PROPOSITION 7.2.7. *With $\nu = \nu_0 + n\delta$ as above, the quiver variety $Y_{\Gamma,\nu}$ is a crepant resolution of $\mathbb{C}^{2n}/\Gamma_n$.*

One way to prove this is to use Nakajima's reflection functors [**68**] to identify $Y_{\Gamma,\nu}$ with $\mathfrak{M}_{\theta}^{\Gamma}(\lambda_0, n\delta)$ for a different stability condition θ. Now $\mathfrak{M}_{\theta}^{\Gamma}(\lambda_0, n\delta)$ is projective and birational over the affine quiver variety $\mathfrak{M}_{0}^{\Gamma}(\lambda_0, n\delta)$ and it is not hard to show that $\mathfrak{M}_{0}^{\Gamma}(\lambda_0, n\delta) \cong \mathbb{C}^{2n}/\Gamma_n$.

There is another, more illuminating, way to understand Proposition 7.2.7. The quiver varities $Y_{\Gamma,\nu}$ are exactly the irreducible components of the fixed loci H_m^{Γ} in the Hilbert schemes H_m, for all possible m. This fact, which is well-known to the experts, is a consequence of the McKay correspondence and results of Kronheimer and Nakajima [**53**], Nakajima [**67**], and Crawley-Boevey [**14**]. It develops that these fixed loci also have the following geometrically explicit description.

PROPOSITION 7.2.8. *For each distinguished coset representative ν_0 as above, there is a unique Γ-invariant ideal $I_{\nu_0} \subseteq R = \mathbb{C}[x,y]$ such that $\operatorname{char}(R/I_{\nu_0}) = \chi_{\nu_0}$, the character of Γ corresponding to ν_0. Set $\nu = \nu_0 + n\delta$ and $m = \chi_{\nu_0}(1) + n|\Gamma|$. The quiver variety $Y_{\Gamma,\nu}$ is the closure of the open set in H_m^{Γ} parametrizing subschemes $S \subseteq \mathbb{C}^2$ of the form $V(I_{\nu_0}) \cup T$, where T is a union of n disjoint non-zero Γ-orbits.*

From the proposition we see that the image of $Y_{\Gamma,\nu}$ under the Chow morphism consists of algebraic cycles of the form $\chi_{\nu_0}(1) \cdot 0 + C$, where $C = \sum_{i=1}^{n} \sum_{g \in \Gamma} gP_i$. But $\mathbb{C}^{2n}/\Gamma_n$ can be identified with the set of cycles C, and subtracting the terms $\chi_{\nu_0}(1) \cdot 0$ gives an isomorphism of the image of $Y_{\Gamma,\nu}$ onto $\mathbb{C}^{2n}/\Gamma_n$.

DEFINITION 7.2.9. An ideal I_{ν_0} as in Proposition 7.2.8—*i.e.*, a Γ-invariant ideal in $\mathbb{C}[x,y]$ with no Γ-invariant deformations—is a Γ-*core*.

The motivation for this definition is the remarkable fact that for $\Gamma = \mathbb{Z}/r\mathbb{Z}$, the Γ-cores are exactly the monomial ideals I_μ such that the partition μ is an r-*core* in the classical sense (see §7.2.4, below).

In terms of the picture given by Proposition 7.2.8 we distinguish an open set $U \subseteq Y_{\Gamma,\nu}$ consisting of points with at most one degeneracy: either one orbit collapses to zero or two orbits coincide. For a given n, the quiver varieties $Y_{\Gamma,\nu}$ are all birational to $\mathbb{C}^{2n}/\Gamma_n$ and hence to each other.

PROPOSITION 7.2.10. *For fixed n and $\nu = \nu_0 + n\delta$, $\nu' = \nu_0' + n\delta$, the birational map $Y_{\Gamma,\nu} \sim Y_{\Gamma\nu'}$ is given by an isomorphism of the open set $U \subseteq Y_{\Gamma,\nu}$ described above with the corresonding open set $U' \subseteq Y_{\Gamma,\nu'}$. Moreover these open sets meet every divisor.*

The proposition identifies all the Picard groups $\operatorname{Pic}(Y_{\Gamma,\nu})$ for a given n with the Picard group $\operatorname{Pic}(U_{\Gamma,n})$ of their common open subvariety. We can regard the quiver varieties $Y_{\Gamma,\nu}$ as different projective completions of $U_{\Gamma,n}$ over $\mathbb{C}^{2n}/S_n$, distinguished by different ample cones in $\operatorname{Pic}(U_{\Gamma,n})$.

Each quiver variety $Y_{\Gamma,\nu}$ with $\nu = \nu_0 + n\delta$ carries a *tautological bundle* M whose fibers are $\mathbb{C}[x,y]\Gamma$-modules with character $n \cdot 1_1^{\Gamma}$ (the character corresponding to $n\delta$) and a distinguished Γ-invariant section. To define M, we once again identify $Y_{\Gamma,\nu}$ with $\mathfrak{M}_{\theta}^{\Gamma}(\lambda_0, n\delta)$ for a different stability condition θ; then M is the usual tautological bundle of quiver data.

We are now almost ready to formulate our conjectured extension of Theorem 5.2.1. It would be absurd to conjecture that $Y_{\Gamma,\nu}$ is isomorphic to the Hilbert scheme $\mathbb{C}^{2n}//\Gamma_n$. Not only do we get different crepant resolutions as ν_0 varies, but in general $\mathbb{C}^{2n}//\Gamma_n$ is not a crepant resolution (this already happens for $\Gamma = \mathbb{Z}/2\mathbb{Z}$ and $n = 2$). Our conjecture instead will be that $Y_{\Gamma,\nu}$ is a *moduli space of stable constellations.*

DEFINITION 7.2.11. Let G be a finite subgroup of $GL(V)$. A *G-constellation* is a G-equivariant $\mathbb{C}[V]$ module that affords the regular representation of V. Let $\theta\colon \mathcal{X}(G) \to \mathbb{R}$ be a linear function on characters of G with $\theta(1_1^G) = 0$. A constellation M is *θ-stable* if for every proper G-invariant $\mathbb{C}[V]$-submodule $N \subsetneq M$ we have $\theta(N) < 0$. A *family of G-constellations* on a scheme Y is a locally free sheaf P of G-equivarient $\mathcal{O}_Y \otimes \mathbb{C}[V]$-modules whose fibers as a vector bundle over Y are G-constellations.

When $\theta(1) = 1$ and $\theta(\chi) < 0$ for all $\chi \neq 1$, a constellation is stable if and only if it is a quotient of $\mathbb{C}[V]$, in which case it is called a *G-cluster.* By standard geometric invariant theory techniques one constructs for generic θ a scheme M_θ projective over V/G that parametrizes θ-stable G-constellations, in the technical sense that it represents the relevant functor of families. The scheme M_θ is the *moduli scheme of θ-stable constellations.* The G-Hilbert scheme $V//G$ is an irreducible component of the moduli scheme of clusters. The definition and construction go back to Kronheimer [**52**], although the terminology of constellations and clusters seems to be due to Reid.

Note that if P is a family of Γ_n-constellations, normalized so that the line bundle of invariants P^{Γ_n} is trivial, then its subbundle of Γ_{n-1}-invariants $P^{\Gamma_{n-1}}$ is a family of $\mathbb{C}[x, y]\Gamma$-modules with a distinguished Γ-invariant section and its fibers have character $n \cdot 1_1^\Gamma$.

PROPOSITION 7.2.12. *On the open set $U_{\Gamma,n} \subseteq Y_{\Gamma,\nu}$ there is a unique family of Γ_n-constellations P such that $P^{\Gamma_{n-1}}$ coincides with the tautological bundle M.*

We remark that although the open set $U_{\Gamma,n}$ is common to every $Y_{\Gamma,\nu}$, the restriction of the tautological bundle M to $U_{\Gamma,n}$ depends on which Weyl chamber in Q_0 contains ν_0).

CONJECTURE 7.2.13. *On every $Y_{\Gamma,\nu}$ there is a family of Γ_n-constellations P, unique by Proposition 7.2.12, such that $P^{\Gamma_{n-1}}$ coincides with M. Moreover, P is a family of θ-stable constellations for some θ.*

The proof of the Bridgeland-King-Reid theorem in [**10**] goes through almost verbatim with $V//G$ replaced by the component birational to V/G in a moduli space of stable G-constellations. Hence Conjecture 7.2.13 has the following consequence, which should be important for the study of the quiver varieties $Y_{\Gamma,\nu}$.

COROLLARY 7.2.14. *Assume Conjecture 7.2.13 holds. Then the functor*

$$\Phi = R\Gamma(P \otimes -)\colon D(Y_{\Gamma,\nu}) \to D^{\Gamma_n}(\mathbb{C}^{2n})$$

is an equivalence of categories.

Conjecture 7.2.13 is true for ν_0 sufficiently far from the walls of its Weyl chamber. In this case $Y_{\Gamma,\nu}$ coincides with the Hilbert scheme of points $\mathrm{Hilb}^n(X_\Gamma)$ on the unique minimal resolution X_Γ of $\mathbb{C}^2/\Gamma$, and the conjecture can be deduced from Corollary 5.2.2 applied to X_Γ^n/S_n.

I have checked Conjecture 7.2.13 by computer for $\Gamma = \mathbb{Z}/2\mathbb{Z}$ and $n = 3$. This case implies that Conjecture 7.2.3 holds locally at the points of $X(\mathfrak{sp}_6)$ lying over principal nilpotent pairs in $\mathfrak{sp}_6$, confirming our improved analog of the $n!$ conjecture. Recall that among these pairs are couterexamples to the more naive analog of the $n!$ conjecture.

Further evidence for Conjecture 7.2.13 lies in phenomena it predicts for $\Gamma = \mathbb{Z}/r\mathbb{Z}$, as I will now explain.

7.2.4. *Cores, wreath Macdonald polynomials, and a new positivity conjecture.* For the rest of the discussion, fix r and $\Gamma = \mathbb{Z}/r\mathbb{Z}$, the group of 2×2 matrices

$$\begin{bmatrix} \omega^k & 0 \\ 0 & \omega^{-k} \end{bmatrix}, \quad \omega = e^{2\pi i/r}.$$

In this case, the representation of Γ_n on $\mathbb{C}^{2n}$ splits into two invariant subspaces $\mathfrak{h} \oplus \mathfrak{h}^*$, with Γ_n acting as the complex reflection group $G(r, 1, n)$ on $\mathfrak{h}$.

The torus group $\mathbb{T}^2$ in (82) commutes with Γ and acts on $Y_{\Gamma,\nu}$. By Proposition 7.2.8, each Γ-core I_{ν_0} must be a monomial ideal. Using this to identify ν_0 with a partition, the $\mathbb{T}^2$-fixed points of $Y_{\Gamma,\nu}$ are monomial ideals I_μ for suitable partitions $\mu \supseteq \nu_0$. To see which partitions are involved, we first need to recall the combinatorial theory of r-cores and r-quotients (see James and Kerber [**44**] or Macdonald [**61**]).

A *ribbon* in a partition λ is a connected skew subdiagram λ/ν containing no 2×2 rectangle. To each cell $x \in \lambda$ corresponds a ribbon of size equal to the hook length $h(x)$, running from the end of the leg of x to the end of its arm.

DEFINITION 7.2.15. The partition λ is an *r-core* if it contains no ribbon of size r. The *r-core* $\mathrm{Core}_r(\lambda)$ of any λ is the partition that remains after one removes as many r-ribbons in succession as possible (the result is independent of choices made).

Here is an example for $r = 4$, with the numbered cells showing one possible sequence of 4-ribbon removals. The blank cells form the 4-core.

2	2	2			
4	4	2	1	1	1
	4	4	3	3	1
				3	3

Define the *content* of a cell $x = (i, j) \in \lambda$ to be $c(x) = i - j$. Let y, z be the cells at the end of the arm and leg of x, and define the *row-residue* $\rho(x)$ and *column-residue* $\gamma(x)$ to be the residues of the contents $c(y)$, $c(z)$ (mod r). Observe that $h(x) \equiv 0 \pmod r$ if and only if $\gamma(x) \equiv \rho(x) + 1$. For each residue class i, the cells $x \in \lambda$ with $\gamma(x) \equiv i$ and $\rho(x) \equiv i - 1$ form an "exploded" copy of the diagram of a partition λ^i, with rows and columns not necessarily adjacent, and with a shifted origin.

DEFINITION 7.2.16. The *r-quotient* of a partition λ is the sequence of partitions

$$\mathrm{Quot}_r(\lambda) \underset{\text{def}}{=} (\lambda^0, \dots, \lambda^{r-1})$$

constructed as above.

Here is an example with $r = 3$. In the first picture the ends of rows and columns are labelled with their contents (mod 3). In the second, the cells forming the exploded diagrams are labelled by their $\gamma(x)$ value.

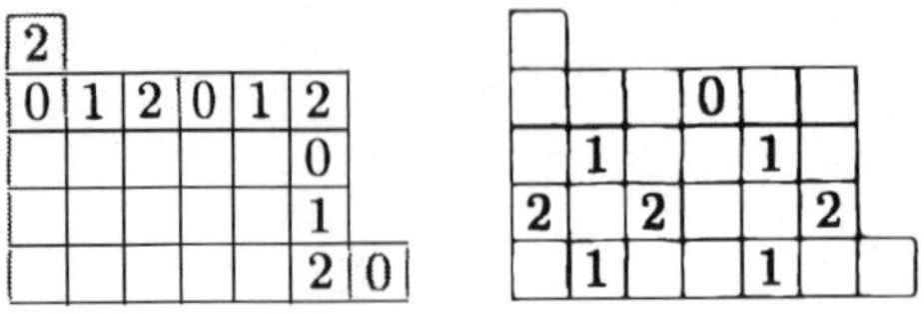

$\mathrm{Quot}_3(\lambda) =$ 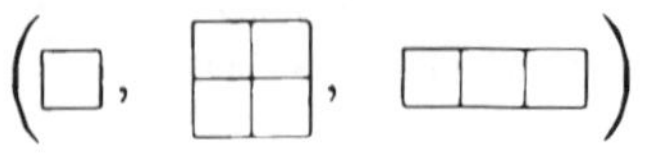

PROPOSITION 7.2.17. *For any partition* λ, *if* $\mathrm{Core}_r(\lambda) = \nu$ *is its* r*-core and* $\mathrm{Quot}_r(\lambda) = (\lambda^0, \dots, \lambda^{r-1})$ *is its* r*-quotient, then*

$$|\lambda| = |\nu| + r\sum_i |\lambda^i|.$$

For any fixed r*-core* ν_0, *the map* $\lambda \mapsto \mathrm{Quot}_r(\lambda)$ *is a bijection from* $\{\lambda : \mathrm{Core}_r(\lambda) = \nu_0\}$ *onto the set of all* r*-tuples of partitions.*

Now we can describe the fixed points of $Y_{\Gamma,\nu}$.

PROPOSITION 7.2.18. *For* $\Gamma = \mathbb{Z}/r\mathbb{Z}$, *the* Γ*-cores are exactly the monomial ideals* I_{ν_0}, *where the partition* ν_0 *is an* r*-core, and the* $\mathbb{T}^2$*-fixed points of* $Y_{\Gamma,\nu_0+n\delta}$ *are the monomial ideals* I_μ *where* $|\mu| = |\nu_0| + nr$ *and* $\mathrm{Core}_r(\mu) = \nu_0$. *In particular, they are in natural bijective correspondence with* r*-tuples of partitions* $(\mu^0, \dots, \mu^{r-1})$ *of total size* $\sum_i |\mu^i| = n$.

We have abused notation by writing ν_0 for both an r-core and a character of Γ. Explicitly, the character corresponding to ν_0 is $\sum_i m_i\chi_i$, where m_i is the number of cells $x \in \nu_0$ with content $c(x) \equiv i \pmod r$, and $\chi_i(\left[\begin{smallmatrix} \omega^k & 0 \\ 0 & \omega^{-k} \end{smallmatrix}\right]) = \omega^{ik}$.

Recall the standard indexing of irreducible characters of Γ_n by r-tuples $(\lambda^0, \dots, \lambda^{r-1})$ of partitions with $\sum_i |\lambda^i| = n$. If V^λ is an irreducible S_n-module and $W_i = \mathbb{C}$ considered as a Γ-module with character χ_i, then $V^\lambda \otimes W_i$ becomes a Γ_n-module via natural homomorphisms $\Gamma_n \to S_n$ and $\Gamma_n \to \Gamma$. Setting $k_i = |\lambda^i|$, the induced module

$$\left(\bigotimes_{i=0}^{r-1} (V^{\lambda^i} \otimes W_i)\right) \uparrow_{\Gamma_{k_0} \times \cdots \times \Gamma_{k_{r-1}}}^{\Gamma_n}$$

is irreducible, and we write $\chi^{(\lambda^0, \dots, \lambda^{r-1})}$ for its character.

Assuming Conjecture 7.2.13 holds, the fiber of the family of constellations P over a fixed point I_μ in $Y_{\Gamma,\nu}$ is a doubly-graded Γ_n-equivariant $\mathbb{C}[\mathbf{x}, \mathbf{y}]$-module affording the regular representation of Γ_n, and we should expect its character to be an analog of the Macdonald polynomial $\tilde{H}_\mu(z; q, t)$. We could define the Frobenius series of a graded Γ_n-module as a symmetric function in r sets of variables, but it is simpler to work in the space $\mathcal{X}_{q,t}(\Gamma_n)$ of virtual characters with coefficients in $\mathbb{Q}(q, t)$. So, for a doubly graded Γ_n-module $A = \bigoplus_{i,j} A_{i,j}$, set

$$F_A(q, t) = \sum_{i,j} t^i q^j \operatorname{char} A_{i,j}.$$

Conjecture 7.2.13 and some plausible extra assumptions about the geometry of $Y_{\Gamma,\nu}$ lead to the following conjecture/definition. The first part of the conjecture is an analog for Γ_n characters of the definition of Macdonald polynomials (Definition 3.5.2) and makes no explicit reference to quiver varieties.

CONJECTURE 7.2.19. *Fix r, $\Gamma = \mathbb{Z}/r\mathbb{Z}$ and an r-core ν_0. Let $\mathfrak{h} = \mathbb{C}^n$ be the defining representation of the complex reflection group $G(r,1,n) \cong \Gamma_n$.*

Existence: *there exists a basis $\{H_\mu(q,t)\}$ of $\mathcal{X}_{q,t}(\Gamma_n)$ indexed by partitions μ of size $|\mu| = |\nu_0| + nr$ with $\mathrm{Core}_r(\mu) = \nu_0$, characterized by the following properties.*

(i) $H_\mu(q,t) \otimes \sum_i (-q)^i \operatorname{char}(\wedge^i \mathfrak{h}) \in \mathbb{Q}(q,t)\{\chi^{\mathrm{Quot}_r(\lambda)} : \lambda \geq \mu,\ \mathrm{Core}_r(\lambda) = \nu_0\}$;
(ii) $H_\mu(q,t) \otimes \sum_i (-t)^{-i} \operatorname{char}(\wedge^i \mathfrak{h}) \in \mathbb{Q}(q,t)\{\chi^{\mathrm{Quot}_r(\lambda)} : \lambda \leq \mu,\ \mathrm{Core}_r(\lambda) = \nu_0\}$;
(iii) $\langle H_\mu(q,t), 1_{\Gamma_n} \rangle = 1$.

Terminology: *The characters $H_\mu(q,t)$ (assuming they exist) are* wreath Macdonald polynomials.

Positivity: *The wreath Macdonald polynomials have coefficients in $\mathbb{N}[q,t,q^{-1},t^{-1}]$.*

Geometry: *The wreath Macdonald polynomials are the graded characters of the fibers $P(I_\mu)$ of the family of Γ_n-constellations in Conjecture 7.2.13 on $Y_{\Gamma,\nu_0+n\delta}$,*

$$H_\mu(q,t) = F_{P(I_\mu)}(q,t).$$

I will close with a few remarks about this conjecture. Like Conjecture 7.2.13 (and for the same reasons), it is true for ν_0 sufficiently far from the walls of its Weyl chamber. The wreath Macdonald polynomials for such ν_0 depend only on the chamber and can be written explicitly in terms of the classical Macdonald polynomials $\tilde{H}_\mu(z;q,t)$. Note that although there are infinitely many r-cores ν_0, they induce for each n only finitely many different orderings on the characters $\chi^{\mathrm{Quot}_r(\lambda)}$, and the wreath Macdonald polynomials depend only on the ordering. Similarly, there are only finitely many non-isomorphic quiver varieties $Y_{\Gamma,\nu_0+n\delta}$ for each Γ and n.

The existence and positivity assertions in the conjecture are easy to check by computer, and I have done so for a fairly large number of cases. The geometry assertion is much harder to verify. The basic trouble is that there is a description analogous to the $n!$ conjecture only for the image of $\mathbb{C}[\mathbf{x},\mathbf{y}]$ in $P(I_\mu)$, which is nearly always a proper subalgebra. However, the existence and positivity can themselves be regarded as evidence for the geometry, especially because the geometry predicts other properties of the characters $H_\mu(q,t)$ that are confirmed by the computations.

Let q^{-k} be the lowest power of q that occurs in $H_\mu(q,t)$. We can regard $H_\mu(t) \underset{\mathrm{def}}{=} q^k H_\mu(q,t)|_{q=0}$ as wreath Hall-Littlewood polynomials. They are characterized (for fixed ν_0) by suitable orthogonality and triangularity properties. The generalized Green functions for $G(r,1,n)$ defined by Shoji [**79**] satisfy the same orthogonality as our $H_\mu(t)$ but a different kind of triangularity. For us, the ordering of characters depends on ν_0 via the r-quotient, whereas Shoji uses an ordering based on generalized Lusztig symbols. Shoji's ordering is incompatible with ours in general, resulting in different polynomials, although some of them are (accidentally?) the same for small r and n.

Finally, when $r = 2$, Γ_n is the Weyl group of type B_n or C_n. It turns out that $\nu_0 = 0$, corresponding to the empty partition, is type C_n, while $\nu_0 = \alpha_0$, corresponding to the partition (1), is type B_n. The reason for this is that there are canonical $\mathrm{Ad}(G)$-invariant morphisms

$$\begin{aligned} C_{\mathrm{reg}}(\mathfrak{sp}_{2n}) &\to Y_{\Gamma,0+n\delta} \\ C_{\mathrm{reg}}(\mathfrak{so}_{2n+1}) &\to Y_{\Gamma,\alpha_0+n\delta}. \end{aligned} \tag{108}$$

The preimages of $\mathbb{T}^2$-fixed points in $Y_{\Gamma,\nu}$ are the principal nilpotent pairs in $C_{\mathrm{reg}}(\mathfrak{g})$. In contrast to (103), the above morphisms are not surjective, and $Y_{\Gamma,\nu}$ has many

$\mathbb{T}^2$-fixed points that are not images of principal nilpotent pairs. One can show that in this situation, the restriction of Conjecture 7.2.13 to the image of the morphism in (108) is equivalent to Conjecture 7.2.3, restricted to the part of $X(\mathfrak{g})$ lying above $C_{\mathrm{reg}}(\mathfrak{g})$.

References

[1] M. F. Atiyah and R. Bott, *A Lefschetz fixed point formula for elliptic complexes. II. Applications*, Ann. of Math. (2) **88** (1968), 451–491.

[2] Victor V. Batyrev, *Non-Archimedean integrals and stringy Euler numbers of log-terminal pairs*, J. Eur. Math. Soc. (JEMS) **1** (1999), no. 1, 5–33, arXiv:math.AG/9803071.

[3] Arnaud Beauville, *Variétés Kähleriennes dont la première classe de Chern est nulle*, J. Differential Geom. **18** (1983), no. 4, 755–782 (1984).

[4] A. A. Beĭlinson, J. Bernstein, and P. Deligne, *Faisceaux pervers*, Analysis and topology on singular spaces, I (Luminy, 1981), Soc. Math. France, Paris, 1982, pp. 5–171.

[5] Y. Berest, P. Etingof, and V. Ginzburg, *Cherednik algebras and differential operators on quasi-invariants*, Electronic preprint, arXiv:math.QA/0111005.

[6] ———, *Finite dimensional representations of rational Cherednik algebras* , Electronic preprint, arXiv:math.RT/0208138.

[7] F. Bergeron, A. M. Garsia, M. Haiman, and G. Tesler, *Identities and positivity conjectures for some remarkable operators in the theory of symmetric functions*, Methods Appl. Anal. **6** (1999), no. 3, 363–420.

[8] N. Bergeron and A. M. Garsia, *On certain spaces of harmonic polynomials*, Hypergeometric functions on domains of positivity, Jack polynomials, and applications (Tampa, FL, 1991), Amer. Math. Soc., Providence, RI, 1992, pp. 51–86.

[9] Joël Briançon, *Description de* $\mathrm{Hilb}^n C\{x,y\}$, Invent. Math. **41** (1977), no. 1, 45–89.

[10] Tom Bridgeland, Alastair King, and Miles Reid, *The McKay correspondence as an equivalence of derived categories*, J. Amer. Math. Soc. **14** (2001), 535–554, arXiv:math.AG/9908027.

[11] Lynne M. Butler, *Combinatorial properties of partially ordered sets associated with partitions and finite abelian groups*, Ph.D. thesis, Massachusetts Institute of Technology, 1986.

[12] L. Carlitz and J. Riordan, *Two element lattice permutation numbers and their q-generalization*, Duke Math. J. **31** (1964), 371–388.

[13] Jan Cheah, *Cellular decompositions for nested Hilbert schemes of points*, Pacific J. Math. **183** (1998), no. 1, 39–90.

[14] William Crawley-Boevey, *Decomposition of Marsden-Weinstein reductions for representations of quivers*, Compositio Math. **130** (2002), no. 2, 225–239.

[15] Charles F. Dunkl, *Differential-difference operators and monodromy representations of Hecke algebras*, Pacific J. Math. **159** (1993), no. 2, 271–298.

[16] Alexander G. Elashvili and Dmitri I. Panyushev, *A classification of the principal nilpotent pairs in simple Lie algebras and related problems*, J. London Math. Soc. (2) **63** (2001), no. 2, 299–318.

[17] Geir Ellingsrud and Stein Arild Strømme, *On a cell decomposition of the Hilbert scheme of points in the plane*, Invent. Math. **91** (1988), no. 2, 365–370.

[18] John Fogarty, *Algebraic families on an algebraic surface*, Amer. J. Math **90** (1968), 511–521.

[19] J. S. Frame, G. de B. Robinson, and R. M. Thrall, *The hook graphs of the symmetric group*, Canadian J. Math. **6** (1954), 316–324.

[20] A. M. Garsia and J. Haglund, *A positivity result in the theory of Macdonald polynomials*, Proc. Natl. Acad. Sci. USA **98** (2001), no. 8, 4313–4316 (electronic).

[21] ———, *A proof of the q,t-Catalan positivity conjecture*, Discrete Mathematics (to appear), Proceedings of the September 2000 Montreal conference on Algebraic Combinatorics.

[22] A. M. Garsia and M. Haiman, *A graded representation model for Macdonald's polynomials*, Proc. Nat. Acad. Sci. U.S.A. **90** (1993), no. 8, 3607–3610.

[23] ———, *A remarkable q,t-Catalan sequence and q-Lagrange inversion*, J. Algebraic Combin. **5** (1996), no. 3, 191–244.

[24] ———, *Some natural bigraded* S_n*-modules and q,t-Kostka coefficients*, Electron. J. Combin. **3** (1996), no. 2, Research Paper 24, approx. 60 pp. (electronic), The Foata Festschrift.

[25] A. M. Garsia, M. Haiman, and G. Tesler, *Explicit plethystic formulas for Macdonald q,t-Kostka coefficients*, Sém. Lothar. Combin. **42** (1999), Art. B42m, 45 pp. (electronic), The Andrews Festschrift (Maratea, 1998).

[26] A. M. Garsia and C. Procesi, *On certain graded S_n-modules and the q-Kostka polynomials*, Adv. Math. **94** (1992), no. 1, 82–138.

[27] A. M. Garsia and J. Remmel, *Plethystic formulas and positivity for q,t-Kostka coefficients*, Mathematical essays in honor of Gian-Carlo Rota (Cambridge, MA, 1996), Birkhäuser Boston, Boston, MA, 1998, pp. 245–262.

[28] A. M. Garsia and G. Tesler, *Plethystic formulas for Macdonald q,t-Kostka coefficients*, Adv. Math. **123** (1996), no. 2, 144–222.

[29] A. M. Garsia and M. Zabrocki, *Polynomiality of the q,t-Kostka revisited*, Algebraic combinatorics and computer science, Springer Italia, Milan, 2001, pp. 473–491.

[30] Adriano M. Garsia, *A q-analogue of the Lagrange inversion formula*, Houston J. Math. **7** (1981), no. 2, 205–237.

[31] Ira Gessel, *A noncommutative generalization and q-analog of the Lagrange inversion formula*, Trans. Amer. Math. Soc. **257** (1980), no. 2, 455–482.

[32] Ira Gessel and Da Lun Wang, *Depth-first search as a combinatorial correspondence*, J. Combin. Theory Ser. A **26** (1979), no. 3, 308–313.

[33] Victor Ginzburg, *Principal nilpotent pairs in a semisimple Lie algebra. I*, Invent. Math. **140** (2000), no. 3, 511–561.

[34] Iain Gordon, *On the quotient ring by diagonal invariants*, Invent. Math. (to appear), arXiv:math.RT/0208126.

[35] Alexander Grothendieck, *Techniques de construction et théorèmes d'existence en géométrie algébrique. IV. Les schémas de Hilbert (Exp. no. 221)*, Séminaire Bourbaki, Vol. 6, Année 1960/61, Soc. Math. France, Paris, 1995, pp. 249–276.

[36] J. Haglund and N. Loehr, *A conjectured combinatorial formula for the Hilbert series for diagonal harmonics*, Proceedings of the 14th International Conference on Formal Power Series and Algebraic Combinatorics, Melbourne 2002.

[37] Mark Haiman, *Conjectures on the quotient ring by diagonal invariants*, J. Algebraic Combin. **3** (1994), no. 1, 17–76.

[38] ———, *t,q-Catalan numbers and the Hilbert scheme*, Discrete Math. **193** (1998), no. 1-3, 201–224, Selected papers in honor of Adriano Garsia (Taormina, 1994).

[39] ———, *Hilbert schemes, polygraphs and the Macdonald positivity conjecture*, J. Amer. Math. Soc. **14** (2001), no. 4, 941–1006, arXiv:math.AG/0010246.

[40] ———, *Vanishing theorems and character formulas for the Hilbert scheme of points in the plane*, Invent. Math. **149** (2002), 371–407, arXiv:math.AG/0201148.

[41] ———, *Cores, quivers and n! conjectures*, Preprint, U.C. Berkeley, 2003.

[42] R. Hotta and T. A. Springer, *A specialization theorem for certain Weyl group representations and an application to the Green polynomials of unitary groups*, Invent. Math. **41** (1977), no. 2, 113–127.

[43] Y. Ito and I. Nakamura, *McKay correspondence and Hilbert schemes*, Proc. Japan Acad. Ser. A Math. Sci. **72** (1996), no. 7, 135–138.

[44] Gordon James and Adalbert Kerber, *The representation theory of the symmetric group*, Encyclopedia of Mathematics and its Applications, vol. 16, Addison-Wesley Publishing Co., Reading, Mass., 1981, With a foreword by P. M. Cohn, With an introduction by Gilbert de B. Robinson.

[45] Nai Huan Jing, *Vertex operators and Hall-Littlewood symmetric functions*, Adv. Math. **87** (1991), no. 2, 226–248.

[46] Anthony Joseph, *On a Harish-Chandra homomorphism*, C. R. Acad. Sci. Paris Sér. I Math. **324** (1997), no. 7, 759–764.

[47] S. V. Kerov, A. N. Kirillov, and N. Yu. Reshetikhin, *Combinatorics, the Bethe ansatz and representations of the symmetric group*, Zap. Nauchn. Sem. Leningrad. Otdel. Mat. Inst. Steklov. (LOMI) **155** (1986), no. Differentsialnaya Geometriya, Gruppy Li i Mekh. VIII, 50–64, 193.

[48] A. N. Kirillov and N. Yu. Reshetikhin, *The Bethe ansatz and the combinatorics of Young tableaux*, Zap. Nauchn. Sem. Leningrad. Otdel. Mat. Inst. Steklov. (LOMI) **155** (1986), 65–115, English translation in J. Soviet Math. **41**, no. 2 (1988), 916–924.

[49] Anatol N. Kirillov and Masatoshi Noumi, *Affine Hecke algebras and raising operators for Macdonald polynomials*, Duke Math. J. **93** (1998), no. 1, 1–39, arXiv:q-alg/9605004.
[50] Friedrich Knop, *Integrality of two variable Kostka functions*, J. Reine Angew. Math. **482** (1997), 177–189, arXiv:q-alg/9603027.
[51] G. Kreweras, *Une famille de polynômes ayant plusieurs propriétés énumeratives*, Period. Math. Hungar. **11** (1980), no. 4, 309–320.
[52] P. B. Kronheimer, *The construction of ALE spaces as hyper-Kähler quotients*, J. Differential Geom. **29** (1989), no. 3, 665–683.
[53] Peter B. Kronheimer and Hiraku Nakajima, *Yang-Mills instantons on ALE gravitational instantons*, Math. Ann. **288** (1990), no. 2, 263–307.
[54] Shrawan Kumar and Jesper Funch Thomsen, *A conjectural generalization of n! result to arbitrary groups*, Transformation Groups (to appear), arXiv:math.AG/0201205.
[55] Luc Lapointe and Luc Vinet, *Rodrigues formulas for the Macdonald polynomials*, Adv. Math. **130** (1997), no. 2, 261–279.
[56] Alain Lascoux and Marcel-P. Schützenberger, *Le monoïde plaxique*, Noncommutative structures in algebra and geometric combinatorics (Naples, 1978), CNR, Rome, 1981, pp. 129–156.
[57] Alain Lascoux and Marcel-Paul Schützenberger, *Sur une conjecture de H. O. Foulkes*, C. R. Acad. Sci. Paris Sér. A-B **286** (1978), no. 7, A323–A324.
[58] G. Lusztig, *Green polynomials and singularities of unipotent classes*, Adv. in Math. **42** (1981), no. 2, 169–178.
[59] George Lusztig, *Character sheaves. V*, Adv. in Math. **61** (1986), no. 2, 103–155.
[60] I. G. Macdonald, *A new class of symmetric functions*, Actes du 20e Séminaire Lotharingien, vol. 372/S-20, Publications I.R.M.A., Strasbourg, 1988, pp. 131–171.
[61] ———, *Symmetric functions and Hall polynomials*, second ed., The Clarendon Press, Oxford University Press, New York, 1995, With contributions by A. Zelevinsky, Oxford Science Publications.
[62] ———, *Orthogonal polynomials associated with root systems*, Sém. Lothar. Combin. **45** (2000/01), Art. B45a, 40 pp. (electronic), arXiv:math.QA/0011046, text of a 1987 preprint.
[63] C. L. Mallows and John Riordan, *The inversion enumerator for labeled trees*, Bull. Amer. Math. Soc. **74** (1968), 92–94.
[64] John McKay, *Cartan matrices, finite groups of quaternions, and Kleinian singularities*, Proc. Amer. Math. Soc. **81** (1981), no. 1, 153–154.
[65] T. S. Motzkin and Olga Taussky, *Pairs of matrices with property L. II*, Trans. Amer. Math. Soc. **80** (1955), 387–401.
[66] Hiraku Nakajima, *Quiver varieties and Kac-Moody algebras*, Duke Math. J. **91** (1998), no. 3, 515–560.
[67] ———, *Lectures on Hilbert schemes of points on surfaces*, American Mathematical Society, Providence, RI, 1999.
[68] ———, *Reflection functors for quiver varieties and Weyl group actions*, Preprint, 2001.
[69] Michael G. Neubauer and David J. Saltman, *Two-generated commutative subalgebras of $M_n(F)$*, J. Algebra **164** (1994), no. 2, 545–562.
[70] Christian Peskine and Lucien Szpiro, *Syzygies et multiplicités*, C. R. Acad. Sci. Paris Sér. A **278** (1974), 1421–1424.
[71] Claudio Procesi, *On the n! conjecture*, Séminaire Bourbaki **54** (2001/02), Société Mathématique de France, Exposé no. 898.
[72] M. Reid, *McKay correspondence*, Electronic preprint, arXiv:alg-geom/9702016, 1997.
[73] R. W. Richardson, *Commuting varieties of semisimple Lie algebras and algebraic groups*, Compositio Math. **38** (1979), no. 3, 311–327.
[74] Paul Roberts, *Le théorème d'intersection*, C. R. Acad. Sci. Paris Sér. I Math. **304** (1987), no. 7, 177–180.
[75] ———, *Intersection theorems*, Commutative algebra (Berkeley, CA, 1987), Springer, New York, 1989, pp. 417–436.
[76] Siddhartha Sahi, *Interpolation, integrality, and a generalization of Macdonald's polynomials*, Internat. Math. Res. Notices **10** (1996), 457–471.
[77] M.-P. Schützenberger, *La correspondance de Robinson*, Combinatoire et représentation du groupe symétrique (Actes Table Ronde CNRS, Univ. Louis-Pasteur Strasbourg, Strasbourg, 1976), Springer, Berlin, 1977, Lecture Notes in Mathematics, Vol. 579, pp. 59–113.

[78] Toshiaki Shoji, *Geometry of orbits and Springer correspondence*, Astérisque (1988), no. 168, 9, 61–140, Orbites unipotentes et représentations, I.

[79] ———, *Green functions associated to complex reflection groups*, J. Algebra **245** (2001), no. 2, 650–694.

[80] Roy Mikael Skjelnes, *On the representability of* $\mathrm{Hilb}^n k[x]_{(x)}$, J. London Math. Soc. (2) **62** (2000), no. 3, 757–770.

[81] T. A. Springer, *A construction of representations of Weyl groups*, Invent. Math. **44** (1978), no. 3, 279–293.

[82] Richard P. Stanley, *Enumerative combinatorics. Vol. 2*, Cambridge University Press, Cambridge, 1999, With a foreword by Gian-Carlo Rota and appendix 1 by Sergey Fomin.

[83] R. W. Thomason, *Une formule de Lefschetz en K-théorie équivariante algébrique*, Duke Math. J. **68** (1992), no. 3, 447–462.

[84] Nolan R. Wallach, *Invariant differential operators on a reductive Lie algebra and Weyl group representations*, J. Amer. Math. Soc. **6** (1993), no. 4, 779–816.

[85] Hermann Weyl, *The Classical Groups. Their Invariants and Representations*, Princeton University Press, Princeton, N.J., 1939.

[86] Andrey V. Zelevinsky, *Representations of finite classical groups*, Lecture Notes in Mathematics, vol. 869, Springer-Verlag, Berlin, 1981, A Hopf algebra approach.

DEPT. OF MATHEMATICS, UNIVERSTITY OF CALIFORNIA, BERKELEY, CA, 94720
E-mail address: `mhaiman@math.berkeley.edu`

Current Developments in Mathematics, 2002
Pages 113 – 133

Periods of Limit Mixed Hodge Structures

Richard Hain

To Wilfred Schmid on the occasion of this 60th birthday

1. Introduction

The first goal of this paper is to explain some important results of Wilfred Schmid from his fundamental paper [**30**] in which he proves very general results which govern the behaviour of the *periods* of a of smooth projective variety X_t as it degenerates to a singular variety. As has been known since classical times, the periods of a smooth projective variety sometimes contain significant information about the geometry of the variety, such as in the case of curves where the periods determine the curve. Likewise, information about the asymptotic behaviour of the periods of a variety as it degenerates sometimes contain significant information about the degeneration and the singular fiber. For example, the *Hodge norm* estimates, which are established in [**30**] and [**3**], describe the asymptotics of the Hodge norm of a cohomology class as the variety degenerates in terms of its monodromy. They are an essential ingredient in the study of the L_2 cohomology of smooth varieties with coefficients in a variation of Hodge structure [**37, 4**].

A second goal is to give some idea of how geometric and arithmetic information can be extracted from the limit periods, both in the geometric case and in the case of the limits of the mixed Hodge structures on fundamental groups of curves.

To get oriented, recall that if X is a compact Riemann surface of genus g, then for each choice of a symplectic (w.r.t. the intersection form) basis $a_1, \dots, a_g, b_1, \dots, b_g$ of $H_1(X, \mathbb{Z})$, there is a basis $w_1, \dots, w_g$ of the holomorphic differentials $H^0(X, \Omega^1)$ such that

$$\int_{a_j} w_k = \delta_{jk}.$$

The $g \times g$ matrix

$$\Omega := \left(\int_{b_j} w_k \right)$$

is called the *period matrix* of X. The classical Riemann bilinear relations assert that Ω is symmetric and that it has positive definite imaginary part. By the classical Torelli Theorem [**17**], the period matrix determines the Riemann surface up to isomorphism.

A more conceptual way to view the period matrix is to note that the augmented period matrix $(I|\Omega)$ is the set of coordinates of the g-plane $H^0(X,\Omega^1)$ in the grassmannian of g planes in the $2g$-dimensional vector space $H^1(X,\mathbb{C})$ with respect to the integral basis $a_1,\dots,b_g$ of $H_1(X,\mathbb{Z})$. The g-plane $H^0(C,\Omega^1)$ lies in the closed submanifold of the grassmannian consisting of g-planes isotropic with respect to the cup product.

2. Quick Review of Hodge Theory

2.1. The Hodge Theorem. For any complex manifold, we can define

$$H^{p,q}(X) := \frac{\text{closed } (p,q)\text{-forms}}{(p,q)\text{-forms that are exact}}.$$

This is simply the subspace of $H^{p+q}(X,\mathbb{C})$ consisting of classes that can be represented by a closed (p,q)-form.

For a general complex manifold, these do not give a decomposition of the cohomology of X. However, if X is compact Kähler (for example, a smooth projective variety), then the natural mapping

$$\bigoplus_{p+q=k} H^{p,q}(X) \to H^k(X,\mathbb{C})$$

is an isomorphism. Denote the class in $H^{1,1}(X)$ of the Kähler form by w. The Hard Lefschetz Theorem states that if $\dim X = n$ and $k \geq 0$, then

$$\_\wedge w^k : H^{n-k}(X,\mathbb{C}) \to H^{n+k}(X,\mathbb{C})$$

is an isomorphism. One can then define the ***primitive*** cohomology in degree k by

$$PH^k(X,\mathbb{C}) = \ker\{\_\wedge w^{k+1} : H^{n-k}(X,\mathbb{C}) \to H^{n+k+2}(X,\mathbb{C})\}.$$

The Hodge decomposition of $H^k(X)$ restrictions to give one of the primitive cohomology:

$$PH^k(X,\mathbb{C}) \cong \bigoplus_{p+q=k} PH^{p,q}(X)$$

where $PH^{p,q}(X) = H^{p,q}(X) \cap PH^{p+q}(X,\mathbb{C})$.

Define a form $Q : H^{n-k}(X) \otimes H^{n-k}(X) \to \mathbb{C}$ by

$$Q(\xi,\eta) = \int_X \xi \wedge \eta \wedge w^k.$$

This form is always defined over $\mathbb{R}$ as w is a real cohomology class, and over $\mathbb{Z}$ when X is smooth projective and w is the class of a hyperplane section.

The Riemann-Hodge bilinear relations generalize the classical Riemann bilinear relations for a compact Riemann surface. They state that

i. Q vanishes on $PH^{p,q}(X) \otimes PH^{r,s}(X)$ unless $p = s$ and $q = r$;
ii. if $\xi \in PH^{p,q}(X)$ is non-zero, then

$$i^{p-q}(-1)^{k(k-1)/2} Q(\xi,\bar{\xi}) > 0,$$

where $k = p + q$.

In degree 2, the Riemann-Hodge bilinear relations imply the Hodge Index Theorem.

The primitive cohomology $P^k H^{p,q}(X)$ together with the form $S := (-1)^{k(k-1)/2} Q$ is the prototypical example of a ***polarized Hodge structure of weight k***.

DEFINITION 1. A *Hodge structure V of weight k* consists of

i. a finitely generated abelian group $V_{\mathbb{Z}}$ and
ii. a bigrading

$$V_{\mathbb{C}} = \bigoplus_{p+q=k} V^{p,q}$$

of its complexification, which satisfies $\overline{V^{p,q}} = V^{q,p}$.

A *polarized Hodge structure of weight* $k \in \mathbb{Z}$ is a Hodge structure V of weight k together with a $(-1)^k$ symmetric bilinear form

$$S : V_{\mathbb{Z}} \otimes V_{\mathbb{Z}} \to \mathbb{Z}$$

that satisfies:

i. $S(V^{p,q}, V^{r,s}) = 0$ unless $p = s$ and $q = r$,
ii. if $v \in V^{p,q}$ is non-zero, then $i^{p-q} S(v, \overline{v}) > 0$.

The Hodge norm of $v \in V$ is defined by

$$\|v\|^2 = S(Cv, \overline{v})$$

where C is the linear operator whose restriction to $V^{p,q}$ is i^{p-q}.

3. Periods

The *Hodge filtration* of a Hodge structure V of weight k is the decreasing filtration

$$\cdots \supseteq F^p V \supseteq F^{p+1} V \supseteq F^{p+2} V \supseteq \cdots$$

defined by

$$F^p V := \bigoplus_{s \geq p} V^{s,k-s}.$$

One can recover the bigrading from the Hodge filtration and the real structure by

$$V^{p,k-p} = F^p V \cap \overline{F}^{k-p},$$

where $\overline{F}^{\bullet}$ is the complex conjugate of the Hodge filtration.

The reason for working with the Hodge filtration rather than the bigrading is the observation of Griffiths [**16**] that the Hodge filtration varies holomorphically[1] in families, whereas the (p,q) pieces generally do not. Indeed, if $H^{p,q}$ varies holomorphically, then $H^{q,p}$ varies anti-holomorphically. So if both $H^{p,q}$ and $H^{q,p}$ vary holomorphically, then both are locally constant.

To be more precise, suppose that $f : X \to B$ is a family of compact Kähler manifolds. This means that f is a proper holomorphic mapping, that X is Kähler and that each fiber X_b is smooth and has the induced Kähler structure. One has the local system (i.e., locally constant sheaf)

$$\mathbb{H}^k := R^k f_* \mathbb{Z} = \{H^k(X_b, \mathbb{Z})\}_{b \in B}$$

on B. We can complexify this to obtain a flat holomorphic vector bundle

$$\mathcal{H}^k := \mathbb{H}^k \otimes_{\mathbb{Z}} \mathcal{O}_B$$

over B. Denote the natural flat connection on $\mathcal{H}^k$ by ∇. Let

$$\mathcal{F}^p := \left\{F^p H^k(X_b)\right\}_{b \in B} \subseteq \mathcal{H}^k.$$

[1] With respect to the locally constant structure coming from the lattice of integral cohomology.

Fundamental results of Griffiths [**16**] assert that the $\mathcal{F}^p$ are holomorphic sub-bundles of $\mathcal{H}^k$ and that if one differentiates a section of $\mathcal{F}^p$ along a holomorphic vector field, then the result lies in $\mathcal{F}^{p-1}$:

$$(1) \qquad \nabla : \mathcal{F}^p \to \mathcal{F}^{p-1} \otimes \Omega^1_B.$$

This is the prototypical example of a *variation of Hodge structure*. Replacing $\mathbb{H}^k$ by the bundle of primitive cohomology

$$P\mathbb{H}^k := \{PH^k(X_b)\}_{b\in B}$$

yields the prototypical example of a *polarized variation of Hodge structure* — a variation of Hodge structure with an inner product S which is parallel with respect to the connection and which polarizes each fiber.

DEFINITION 2. A *variation of Hodge structure of weight k* over a complex manifold B is a $\mathbb{Z}$-local system $\mathbb{V}$ together with a flag

$$\cdots \supseteq \mathcal{F}^p \supseteq \mathcal{F}^{p+1} \supseteq \cdots$$

of holomorphic sub-bundles of the flat bundle $\mathcal{V} := \mathbb{V} \otimes_{\mathbb{Z}} \mathcal{O}_B$ which satisfy:

i. $\nabla : \mathcal{F}^p \to \mathcal{F}^{p-1} \otimes \Omega^1_B$,
ii. Each fiber, endowed with the induced Hodge filtration, is a Hodge structure of weight k.

The variation is *polarized* if it has a $(-1)^k$-symmetric inner product which is parallel with respect to the flat structure and polarizes the Hodge structure on each fiber.

Given a polarized variation of Hodge structure $\mathbb{V}$ of weight k over a base B, one has a period mapping

$$\widetilde{B} \to \{\text{set of flags } \{F^p\} \text{ that satisfy } S(F^p, F^{k-p+1}) = 0\}$$

defined on the universal covering of the base. The flags should be considered to live in some fixed fiber of $\mathbb{V} \to B$; other fibers being compared with this one using the flat structure. The flags should have the "same shape" as the Hodge filtration on this reference fiber.

Griffiths' results assert that this mapping is holomorphic and lies in the open subset D of flags that satisfy

$$i^{p-q} S(v, \overline{v}) > 0$$

whenever v is a non-zero element of $F^p \cap \overline{F}^{k-p}$ and $p + q = k$.

When the variation is $P\mathbb{H}^k$ this mapping really is a period mapping in the sense that the coordinates in the flag manifold of the image of $b \in \widetilde{B}$ will be given by periods over integral cycles of a basis of $PH^k(X_b)$ adapted to the Hodge filtration.

At this point it is useful to revisit the period mapping for curves.

EXAMPLE 3. Fix a genus $g \geq 1$. To know the Hodge filtration on a compact Riemann surface of genus g, one simply has to know the location of the holomorphic differentials $F := H^{1,0}(X_b)$ in $H^1(X, \mathbb{C})$ with respect to an integral symplectic basis. The polarization S on $H^1(X_b)$ is the one given by cup product. The first Riemann bilinear relation says that $S(F, F) = 0$.

Choose a base point $b_o \in B$. Let $H = H^1(X_{b_o})$. The image of the period mapping is contained in the closed submanifold Y of the grassmannian of g planes in $H_{\mathbb{C}}$ consisting of those g-planes F that satisfy $S(F, F) = 0$. It is a complex manifold of dimension $g(g+1)/2$. The period mapping $\widetilde{B} \to Y$ is holomorphic, and has image contained in the open subset

$$U = \{F \in Y : iS(v, \overline{v}) > 0 \text{ if } v \neq 0, v \in F\}$$

of Y; it is a complex manifold. One can check that

$$\begin{aligned} U &= \text{ the symmetric space of } \mathrm{Sp}_g(\mathbb{R}) \\ &= \mathrm{Sp}_g(\mathbb{R})/U(g) \\ &= \text{ Siegel's upper half plane } \mathfrak{h}_g \text{ of rank } g \\ &= \{g \times g \text{ symmetric complex matrices } \Omega \text{ with } \operatorname{Im} \Omega > 0\}. \end{aligned}$$

The period map takes X_b to its period matrix $\Omega(b)$, described in the Introduction. That it is holomorphic is equivalent to the statement that $\Omega(b)$ depends holomorphically on $b \in B$.

The monodromy representation of the variation is a homomorphism

$$\pi_1(B, b_o) \to \mathrm{Sp}_g(\mathbb{Z}).$$

The period mapping descends to the mapping

$$B \to \mathrm{Sp}_g(\mathbb{Z})\backslash\mathfrak{h}_g = \mathcal{A}_g$$

to the moduli space of principally polarized abelian varieties that takes $b \in B$ to the jacobian of X_b.

4. The Limit Hodge Filtration

Schmid's work concerns the behaviour of the period mapping of a polarized variation of Hodge structure as the the Hodge structures degenerate. In the geometric case, this happens when the varieties X_b degenerate to a singular variety. To keep things simple I'll only discuss the local case in one variable, which he considers in [**30**]. The several variable case is more subtle and is worked out in [**30**] and in [**3**].

Suppose that $\mathbb{V} \to \Delta^*$ is a polarized variation of Hodge structure of weight k over the punctured disk. The geometrically inclined may prefer to think of this as the local system of primitive cohomology of degree k associated to a proper family $X \to \Delta$ over the disk, where X is smooth and Kähler and where each fiber X_t is smooth when $t \neq 0$.

Let

$$T : V_{t_o} \to V_{t_o}$$

be the monodromy operator associated to the positive generator of $\pi_1(\Delta^*, t_o)$. Landman in the geometric case, and Borel in general, (reproved by Schmid [**30**, p. 245]) proved that the eigenvalues of T are roots of unity, so that T is quasi unipotent; that is, there exist integers e and m such that

$$(T^e - I)^m = 0.$$

By pulling $\mathbb{V}$ back along the e-fold covering $t \mapsto t^e$ of $\Delta^* \to \Delta^*$ we may assume that T is unipotent. The logarithm of T is then defined by the usual power series. Set

$$N = \log T/2\pi i.$$

This is nilpotent and preserves the polarization in the sense that

$$S(Nx, y) + S(x, Ny) = 0. \tag{2}$$

We shall view it as a flat section of the bundle End $\mathbb{V}$, so that it acts on all fibers, not just the one over t_o.

Denote the flat holomorphic vector bundle $\mathbb{V} \otimes_{\mathbb{Z}} \mathcal{O}_{\Delta^*}$ by $\mathcal{V}$ and its connection by ∇.

Note that, unless T has finite order, neither T nor N is a morphism of Hodge structures $V_{t_o} \to V_{t_o}$. One of Schmid's results, which is explained below, is that N is compatible with the limit Hodge filtration and acts as an endomorphism of the "limit mixed Hodge structure", which allows one to relate Hodge theory and monodromy.

4.1. Deligne's Canonical Extension. In order to discuss the limit of the Hodge structure V_t as $t \to 0$, we need to extend the flat bundle $\mathcal{V} \to \Delta^*$ to a holomorphic vector bundle $\widetilde{\mathcal{V}} \to \Delta$ over the entire disk. The way to do this is to use Deligne's *canonical extension* [**6**].

First note that any trivialization of $\mathcal{V}$ over Δ^* gives an extension of $\mathcal{V}$ to Δ; the holomorphic sections of the extended bundle are those whose restriction to Δ^* are of the form

$$\sum_j f_j \phi_j$$

where each f_j is holomorphic on Δ, and $\phi_1, \dots, \phi_m$ is a holomorphic frame that gives the trivialization over Δ^*. Two framings of $\mathcal{V}$ over Δ^* give the same extension if and only if they differ by a holomorphic mapping $g : \Delta \to GL_m(\mathbb{C})$. It follows that the set of extensions of $\mathcal{V}$ to Δ can be identified with the homogeneous space

$$GL_m(\mathcal{O}(\Delta^*))/GL_m(\mathcal{O}(\Delta)),$$

which is large.

To construct Deligne's extension, we choose multivalued flat sections $\phi_1(t), \dots, \phi_m(t)$ defined on Δ^*. The extension is constructed by regularizing these. Specifically, set

$$e_j(t) = \phi_j(t) t^{-N},$$

where for a square matrix A, t^A is defined to be $\exp(A \log t)$. The $e_j(t)$ comprise a single valued framing of $\mathcal{V}$ over Δ^* as each $\phi_j(t)$ changes to

$$\phi_j(t) \cdot T$$

when analytically continued around the circle, whereas t^{-N} changes to

$$\exp\big((-\log t - 2\pi i)N\big) = e^{-2\pi i N} t^{-N} = T^{-1} t^{-N}$$

when continued around the origin. The framing e_j determines Deligne's extension $\widetilde{\mathcal{V}}$ of $\mathcal{V}$ to all of Δ. With respect to Deligne's framing $e_j(t)$, the connection is

$$\nabla = d - N\frac{dt}{t}.$$

In other words, the connection on $\widetilde{\mathcal{V}}$ is meromorphic with a simple pole at the origin and the residue of the connection at $t = 0$ is $-N$. In particular, the extended connection has a regular singular point at the origin with nilpotent residue. This property characterizes Deligne's extension up to isomorphism. This fact, proved by Deligne in more generality in [**6**], is a consequence of the classical result about differential equations stated below, a proof of which can be found in [**35**, Chapt. **II**].

LEMMA 4. *Suppose that $A : \Delta \to \operatorname{End} V$ is holomorphic. If no two eigenvalues of $A_0 := A(0)$ differ by a non-zero integer (e.g., if A_0 is nilpotent), then there is a unique holomorphic function $P : \Delta \to \operatorname{Aut} V$ with $P(0) = \mathrm{id}_V$ such that each (local) solution $v : \Delta^* \to V$ of the differential equation*

$$tv'(t) = v(t)A(t)$$

is of the form

$$v(t) = v_0\, t^{A_0} P(t).$$

Note that the monodromy operator T is:

$$v(t) = v_0\, t^{A_0} P(t) \mapsto v_0\, t^{A_0} e^{2\pi i A_0} P(t) = v(t) \exp\big(2\pi i P(t)^{-1} A_0\, P(t)\big).$$

The operator $\log T/2\pi i$ on the fiber over t is thus right multiplication by

$$N(t) = P(t)^{-1} A_0\, P(t) \ \in \operatorname{Aut} V.$$

The Deligne trivializing sections are therefore of the form

$$v(t)t^{-N(t)} = v(t)P(t)^{-1}t^{-A_0}P(t) = v_0\, P(t),$$

which converges to v_0 as $t \to 0$. This also implies that the function $P(t)$ is the difference between the Deligne framing and the given trivialization.

COROLLARY 5. *If A_0 is nilpotent, then*

$$\lim_{t\to 0} v(t)t^{-A_0} = \lim_{t \to 0} v(t)t^{-N(t)} = v_0$$

where the limit is taken along any angular ray.

PROOF. The second equality was proved above. As for the first, we have

$$v(t)t^{-A_0} = v_0\, t^{A_0} P(t) t^{-A_0}.$$

If $A_0^{k+1} = 0$, then there is a constant C such that

$$\|t^{A_0}\| \text{ and } \|t^{-A_0}\| \leq C\big(\log 1/|t|\big)^k$$

when $0 < |t| \leq R$, for some $R > 0$. Writing $P(t) = I + \sum_{n\geq 1} P_n t^n$, we have

$$\|t^{A_0}P(t)t^{-A_0} - I\| \leq 2C|t|\big(\log 1/|t|\big)^k \sum_{n=1}^{\infty} \|P_n\| |t|^{n-1}$$

which goes to zero along each angular ray. □

This formula will prove useful in Section 6.2 when computing the period of the path $[0,1]$ in the limit MHS on the unipotent completion in the set of paths in $\mathbb{P}^1 - \{0, 1, \infty\}$ from 0 to 1.

REMARK 6. Note that $v(t)t^{-A_0}$ is not, in general, single-valued on the punctured disk even though its limit as $t \to 0$ along each radial ray exists.

4.2. The Nilpotent Orbit Theorem. Schmid's first main result is the Nilpotent Orbit Theorem. We shall state it as two separate results.

THEOREM 7 (Schmid). *The Hodge sub-bundles $\mathcal{F}^p$ of $\mathcal{V}$ extend to holomorphic sub-bundles $\widetilde{\mathcal{F}}^p$ of Deligne's canonical extension $\widetilde{\mathcal{V}} \to \Delta$. That is, both $\widetilde{\mathcal{F}}^p$ and $\widetilde{\mathcal{V}}/\widetilde{\mathcal{F}}^p$ are vector bundles.*

This shows that, when taken appropriately, the limit of the Hodge filtration exists. The limit Hodge filtration $\widetilde{\mathcal{F}}_0^\bullet$ is a filtration of the fiber V_0 of the canonical extension $\widetilde{\mathcal{V}}$ over 0.

In the geometric case, the extended Hodge bundles can be constructed explicitly. This was done by Steenbrink in [**31**].

So far we have constructed a complex vector space V_0 and a Hodge filtration $F^\bullet V_0 := \widetilde{\mathcal{F}}_0^p$ on it. The polarization also extends to the central fiber as at t^N preserves the inner product since N preserves it infinitesimally by (2).

We can also construct an integral form $V_\mathbb{Z}$ of V_0. If the ϕ_j are integral, the integral lattice $V_\mathbb{Z}$ is simply the $\mathbb{Z}$-linear span of the $e_j(0)$. It is important to note, however, that except when the monodromy is trivial, $(V_\mathbb{Z}, F_0^\bullet)$ is not a Hodge structure even though it is a limit of Hodge structures.

In general, the integral structure depends on the holomorphic parameter t chosen for the disk. This dependence is quite transparent, and is a straightforward consequence of Corollary 5.

LEMMA 8. *The lattice $V_\mathbb{Z}$ in V_0 depends only on the parameter t to first order, and hence only on the tangent vector $\partial/\partial t$. The lattice corresponding to the tangent vector $\lambda\partial/\partial t$, where λ is a sufficiently small non-zero complex number, is $\lambda^N \cdot V_\mathbb{Z}$.* □

This extends to a "nilpotent orbit" over $\mathbb{C}^*$. Its fiber over $t \in \mathbb{C}^*$ consists of:

i. the complex vector space V_0,
ii. the limit Hodge filtration $F_0^\bullet$ on V_0,
iii. the lattice $t^N V_\mathbb{Z}$ in V_0.

It has a flat connection given by the limit connection

$$\nabla_0 = d - N\frac{dt}{t}.$$

with respect to the constant trivialization given by V_0. It satisfies the Griffiths infinitesimal period relation (1).

The Deligne frame gives a connection preserving isomorphism

$$\begin{array}{ccc} V_0 \times \Delta & \longrightarrow & \widetilde{\mathcal{V}} \\ \downarrow & & \downarrow \\ \Delta & = & \Delta \end{array}$$

between $\widetilde{\mathcal{V}}$ and the restriction of the nilpotent orbit to a neighbourhood of the origin which is the identity on the fiber over the origin. The second part of Schmid's Nilpotent Orbit Theorem can be stated as follows.

THEOREM 9 (Schmid). *The nilpotent orbit satisfies Griffiths infinitesimal period relation (1). There is an $\epsilon > 0$ such that if $0 < |t| < \epsilon$, then the fiber of the nilpotent orbit over t is a Hodge structure with the same Hodge numbers* *as those of the original variation. Moreover, there is a holomorphic mapping* $g : \Delta \to \mathrm{Aut}(V_0, S)$ *such that* $g(0)$ *is the identity and* g *carries the Hodge bundle of the nilpotent orbit onto the Hodge bundle of* $\widetilde{\mathcal{V}}$.

So this gives us our first precise information about the asymptotics of periods of polarized variations of Hodge structure. Basically, the crudest model of a degeneration of Hodge structure is a nilpotent orbit. General degenerations are obtained from these by perturbing the Hodge filtration.

It is useful to think of the nilpotent orbit as being defined on the punctured tangent space $T_0\Delta - \{0\}$ of Δ at $t = 0$. Then in a sufficiently small neighbourhood of the origin, the nilpotent orbit is a variation of Hodge structure that approximates the original variation. The nilpotent orbit can be thought of as the restriction of the canonical extension of the variation $\mathbb{V}$ to $\operatorname{Spec} \mathbb{C}[t]/(t^2)$.

5. The SL_2-orbit Theorem

Schmid's second main theorem, and the deeper of the two, is the SL_2-orbit Theorem. It gives more precise information about how nilpotent orbits approximate degenerations of polarized variations of Hodge structure and is analogous to the Hard Lefschetz Theorem for the cohomology of a smooth projective variety. In fact, the SL_2-orbit Theorem corresponds to the Hard Lefschetz Theorem for Calabi-Yau manifolds under mirror symmetry.[2] It also leads naturally to mixed Hodge structures, which, like Hodge structures, have a lattice and Hodge filtration, but also have a ***weight filtration***. In the case of a limit of Hodge structures, the weight filtration is constructed from the monodromy logarithm.

5.1. The Monodromy Weight Filtration. Suppose that V is a finite dimensional vector space over a field of characteristic zero, and suppose that N is a nilpotent endomorphism of V. Then there is a unique filtration $W(N)_\bullet$

$$\cdots \subseteq W_n(N) \subseteq W_{n+1}(N) \subseteq \cdots$$

of V such that

i. $N(W_n(N)) \subseteq W_{n-2}(N)$ and
ii. $N^k : W_k/W_{k-1} \to W_{-k}/W_{-k-1}$ is an isomorphism.

This is an easy exercise using the Jordan canonical form of N. It suffices to prove the result for a single Jordan block, where the result is evident.

In our case, the monodromy logarithm acts on V_0. We therefore have the monodromy weight filtration $W_\bullet(N)$ of V_0. In the present situation, we shift the filtration by the weight k of the variation — define:

$$W_n V_0 = W_{k-n}(N).$$

Our central fiber now has the Hodge filtration $F^\bullet V_0$, the nilpotent orbit $t^N V_\mathbb{Z}$ of integral structures, and the monodromy weight filtration $W_\bullet$. One of the main consequences of the SL_2-Orbit Theorem is:

---

[2]As Eduardo Cattani pointed out to me (cf. [**2**]), a precursor of this already appears in the papers [**3, 4**] of Cattani-Kaplan-Schmid. They prove a general result, a special case of which constructs out of $(H^\bullet(X), F^\bullet)$, the cohomology and Hodge filtration of a compact Kähler manifold X, a polarized variation of Hodge structure over the complexified Kähler cone of X. This can be restricted to the the (p,p) part of $H^\bullet(X)$, and this is what should correspond to the polarized variation of Hodge structure of the degeneration of the mirror, as was pointed out to David Morrison by Cattani in 1994. Deligne [**10**] has also studied certain nilpotent orbits arising in mirror symmetry.

THEOREM 10 (Schmid). *For each $t \in \mathbb{C}^*$, the collection*

$$(t^N V_{\mathbb{Z}}, F_0^\bullet, W_\bullet)$$

is a mixed Hodge structure. Moreover, $N : V_0 \to V_0$ is defined over $\mathbb{Q}$ and is is a morphism of $\mathbb{Q}$-mixed Hodge structure of type $(-1,-1)$.

So, even if one is interested only in compact Kähler manifolds, one is inexorably lead to mixed Hodge structures when one studies their degeneration.

DEFINITION 11 (Deligne). A *mixed Hodge structure* V consists of a finitely generated abelian group $V_{\mathbb{Z}}$, an increasing filtration

$$\cdots \subseteq W_{m-1}V_{\mathbb{Q}} \subseteq W_m V_{\mathbb{Q}} \subseteq W_{m+1}V_{\mathbb{Q}} \subseteq \cdots$$

(called the *weight filtration*) of $V_{\mathbb{Q}} := V_{\mathbb{Z}} \otimes_{\mathbb{Z}} \mathbb{Q}$, and a decreasing filtration

$$\cdots \supseteq F^{p-1}V_{\mathbb{C}} \supseteq F^p V_{\mathbb{C}} \supseteq F^{p+1}V_{\mathbb{C}} \supseteq \cdots$$

(called the *Hodge filtration*) of $V_{\mathbb{C}} := V_{\mathbb{Z}} \otimes_{\mathbb{Z}} \mathbb{C}$. These are required to satisfy the condition that, for each m,

$$\operatorname{Gr}_m^W V := W_m V / W_{m-1} V$$

with the induced Hodge filtration

$$F^p \operatorname{Gr}_m^W V := \operatorname{im}\{F^p V \cap W_m V \to \operatorname{Gr}_m^W V\}$$

is a Hodge structure of weight m.

Deligne [**7, 8**] proved that the cohomology groups of every complex algebraic variety (not necessarily smooth or compact) have a mixed Hodge structure (MHS) that is natural with respect to morphisms.

This definition takes a while to digest, and it may be helpful to note that, up to some subtle issues of torsion in $\mathbb{V}_{\mathbb{Z}}$, mixed Hodge structures can be constructed as follows.

Let $(H^m)_{m\in\mathbb{Z}}$ be a collection of Hodge structures where H^m has weight m and all but a finite number of the H^m are zero. Then their direct sum

$$H = \bigoplus_m H^m$$

is a mixed Hodge structure with the direct sum Hodge and weight filtrations:

$$F^p H := \bigoplus_k F^p H^m, \quad W_m H = \bigoplus_{k \le m} H^k.$$

Such a MHS is said to be ***split*** as it is the direct sum of Hodge structures. Very few of the MHSs encountered in life are split. Up to the torsion issues alluded to above, all MHSs V with $\operatorname{Gr}_m^W V \cong H^m$ can be constructed as follows.

Let

$$G = \{\phi \in \operatorname{Aut} H_{\mathbb{C}} \text{ that preserve } W_\bullet \text{ and act trivially on all } \operatorname{Gr}_m^W H\}.$$

This is a unipotent group. For each element of g of G, we can define $V(g)$ to be the MHS with $V_{\mathbb{C}} = H_{\mathbb{C}}$, the same Hodge and weight filtrations (over $\mathbb{C}$), and lattice $g \cdot H_{\mathbb{Z}} \hookrightarrow H_{\mathbb{C}}$. The MHS obtained in this way, depends only on the double coset of g^{-1} in

$$G_{\mathbb{Z}} \backslash G / F^0 G$$

where

$$G_{\mathbb{Z}} = G \cap \operatorname{Aut} V_{\mathbb{Z}} \text{ and } F^0 G = \{\phi \in G : \phi \text{ preserves } F^\bullet\}.$$

In fact, this double coset space is the moduli space of MHSs V together with an isomorphism $\mathrm{Gr}_m^W V \cong H^m$.

EXAMPLE 12. Let A and B be Hodge structures of weights a and b, respectively, where $b < a$. The set of all mixed Hodge structures with weight graded quotients isomorphic to A and B, together with these *framings* is

$$\mathrm{Hom}(A_{\mathbb{C}}, B_{\mathbb{C}}) / \big(\mathrm{Hom}(A_{\mathbb{Z}}, B_{\mathbb{Z}}) + F^0 \mathrm{Hom}(A, B)\big)$$

as the group G in this case is $\mathrm{Hom}(A_{\mathbb{C}}, B_{\mathbb{C}})$. This gives the computation of $\mathrm{Ext}^1_{\mathrm{Hodge}}(A, B)$ in the category of MHSs given by Carlson in [**1**].

5.2. The Hodge Norm Estimates. Another important consequence of the SL_2-orbit Theorem is the Hodge norm estimates, which are important in applications to L_2-cohomology of algebraic varieties with coefficients in a polarized variation of Hodge structure — cf. [**37**] in the one-variable case, and [**4**] and [**28**] in general.

First note that the monodromy weight filtration is defined on the cohomology of each fiber and is preserved by the connection.

THEOREM 13 (Schmid). *If $\mathbb{V}$ is a polarized variation of Hodge structure of weight k over the punctured disk Δ^*, then a flat section $v(t)$ lies in W_m if and only if the square of its Hodge norm satisfies*

$$S(Cv(t), \bar{v}(t)) = O\big(\big(\log(1/|t|)\big)^{m-k}\big)$$

when t goes to 0 along a radial ray. Here $C : V_t \to V_t$ is the operator that is multiplication by i^{p-q} on $V_t^{p,q}$.

5.3. Topology and Geometry of the Limit MHS. Suppose that $f : X \to \Delta$ is a proper holomorphic mapping from a Kähler manifold to the disk. Suppose that the fiber X_t over $t \in \Delta$ is smooth whenever $t \neq 0$ and that X_0 is a reduced divisor with normal crossings. This implies that the monodromy operator $T \in \mathrm{Aut}\, H^m(X_t)$ is unipotent.

The local system

$$\mathbb{V} := R^m (f|_{\Delta^*})_* \mathbb{Z}$$

underlies a polarized variation of Hodge structure of weight m over Δ^*. By Schmid's Theorems, for each choice of a non-zero tangent vector $\vec{v}$ of Δ at 0, there is a limit mixed Hodge structure of $\mathbb{V}$, which one can think of as a MHS on the cohomology group $H^m(X_{\vec{v}})$ of the first order deformation $X_{\vec{v}}$ of X_0 given by $\vec{v}$.

Basic results about the compatibility of this MHS with others associated to the degeneration follow from the geometric constructions of the limit MHS given by Clemens [**5**] and Steenbrink [**31**]. These compatibilities aid in extracting geometric and topological information from $H^\bullet(X_{\vec{v}})$.

To begin to understand the limit MHS, $H^m(X_{\vec{v}})$, it is helpful to think of X_v as being built out of X_0. In order to explain this, we denote the normalization of X_0 by $\nu : \widetilde{X}_0 \to X_0$. Denote the inverse image under $\nu : \widetilde{X}_0 \to X_0$ of the singular locus of X_0 by D. It is a normal crossings divisor. Set

$$X_0' = \text{ real oriented blowup of } \widetilde{X}_0 \text{ along } D$$

which is a manifold with corners. One can think of each $X_{\vec{v}}$ as being obtained from X_0' by gluing, the data for which is given by the combinatorics of D, its local defining equation of X_0, and by $\vec{v}/|\vec{v}|$. There are therefore mappings

$$\widetilde{X}_0 - D \hookrightarrow X_0' \to X_{\vec{v}} \to X_0$$

which induce homomorphisms

$$H^m(X_0) \xrightarrow{\beta} H^m(X_v) \xrightarrow{\alpha} H^m(\widetilde{X}_0 - D).$$

THEOREM 14. *The mappings α and β are both morphisms of MHS.*

The statement that β is a morphism is proved by both Clemens and Steenbrink. It is closely related to the *Local Invariant Cycle Theorem*, which is stated below. That α is a morphism follows from Steenbrink's construction of $H^\bullet(X_{\vec{v}})$ in [**31**].

For each $t \in \Delta$, there is a restriction mapping

$$j_t^* : H^m(X, \mathbb{Q}) \to H^m(X_t, \mathbb{Q})$$

which corresponds, via duality, to intersection with the fiber

$$H_n(X, X|_{\partial\Delta}, \mathbb{Q}) \to H_{n-2}(X_t, \mathbb{Q})$$

and to the mapping β above after taking into account the fact that the inclusion $X_0 \hookrightarrow X$ is a deformation retraction.

The following result was first established in the ℓ-adic case by Deligne and then in the complex analytic case by Clemens [**5**] and Steenbrink [**31**]. Although not expressed in terms of limit Hodge theory, it is a non-trivial application of its existence and construction.

THEOREM 15 (Local Invariant Cycle Theorem). *If $t \neq 0$, then the image of j_t^* is the space*

$$\ker\{(T - I) : H^m(X_t, \mathbb{Q}) \to H^m(X_t, \mathbb{Q})\}$$

of invariant cohomology classes.

5.4. Degenerations of Curves. In this section, we give a relatively simple example to illustrate how geometric information can be extracted from the limit mixed Hodge structure. To get the full power from this theory, one needs to combine Schmid's results with the complementary results of Clemens [**5**] and the explicit constructions of Steenbrink [**31**] surveyed in the previous paragraph.

Suppose that $C \to \Delta$ is a stable degeneration of compact Riemann surfaces of genus g. The total space C is assumed to be smooth and the fiber C_t over t is assumed to be smooth when $t \neq 0$. The central fiber is assumed to be reduced and stable (i.e., its automorphism group is finite.)

Let B be the set of the homology classes of the vanishing cycles. Then, using the Picard-Lefschetz formula, one sees that the monodromy is unipotent and satisfies $(T-I)^2 = 0$. This implies that $N = T - I$ and that the monodromy weight filtration has length 3:

$$0 \subseteq W_0H^1(C_t) \subseteq W_1H^1(C_t) \subseteq W_2H^1(C_t) = H^1(C_t).$$

The defining properties of the monodromy weight filtration imply that

$$W_0H^1(C_t) = \operatorname{im} N \text{ and } W_1H^1(C_t) = \ker N.$$

The Picard-Lefschetz formula implies that $W_0H^1(C_t)$ is spanned by the Poincaré duals of the vanishing cycles.

Fix a tangent vector $\vec{v}$ of 0 in Δ and denote the corresponding limit MHS by $H^1(C_{\vec{v}})$.

Denote the normalization of C_0 by $\widetilde{C}_0$. Let D be the inverse image in $\widetilde{C}_0$ of the double points of C_0. Results of the previous paragraph imply that the natural morphisms

$$H^1(C_0) \xrightarrow{\beta} H^1(C_{\vec{v}}) \xrightarrow{\alpha} H^1(C_0 - D)$$

are morphisms of MHS.

The fact that $F^0H^1(C_{\vec{v}}) = H^1(C_{\vec{v}}, \mathbb{C})$ and $F^2H^1(C_{\vec{v}}) = 0$ imply that

$$\mathrm{Gr}^W_m H^1(C_{\vec{v}})$$

is of type $(0,0)$ when $m = 0$, of type $(1,1)$ when $m = 2$, and is a polarized Hodge structure of with Hodge numbers $(1,0)$ and $(0,1)$ when $m = 1$. Basic topology implies that β injective, from which it follows that β induces a natural isomorphism

$$\mathrm{Gr}^W_1 H^1(C_{\vec{v}}) \cong H^1(\widetilde{C}_0) \cong \bigoplus_{\substack{\text{components}\\ E \text{ of } C_0}} H^1(E)$$

of polarized Hodge structures. It also implies that $W_0H^1(C_{\vec{v}})$ is subspace of $H^1(C_{\vec{v}})$ generated by the Poincaré duals of the vanishing cycles and that β induces an isomorphism

(3) $$W_1H^1(C_{\vec{v}}) \cong H^1(C_0)$$

The classical Torelli Theorem, the fact that the theta divisor of the jacobian of a smooth curve is irreducible and the semi-simplicity of polarized Hodge structures of weight 1 imply that the polarized limit MHS, $H^1(C_{\vec{v}})$, determines the normalization of C_0. Using this, and by approximating the period mapping by the nilpotent orbit, one can give proofs of weak versions (i.e., mod t) of the two results [**13**, Cor. 3.2, Cor 3.8] in Fay's book.

If the normalization of C_0 is connected, then one can determine the divisor classes in $\mathrm{Jac}\,\widetilde{C}_0$ of the pairs of points that are identified to obtain C_0. This information is extracted from the extension

$$0 \to W_0H^1(C_{\vec{v}}) \to W_1H^1(C_{\vec{v}}) \to \mathrm{Gr}^W_1 H^1(C_{\vec{v}}) \to 0.$$

using the work of Carlson [**1**] and the isomorphism (3). This implies that when C_0 is irreducible, the limit MHS $H^1(C_{\vec{v}})$ determines C_0 up to isomorphism.

On the other hand, if the dual graph of C_0 is a tree, then all vanishing cycles are trivial in homology, and the Picard-Lefschetz formula implies that $N = 0$. In this case, the monodromy weight filtration is trivial, and the limit MHS $H^1(C_{\vec{v}})$ is a polarized Hodge structure of weight 1. By the argument above, this determines the non-rational components of $\widetilde{C}_0$ up to isomorphism. But since there is no extension data, it gives no information about how to reassemble C_0 from its irreducible components. In this case, one can use the limit MHS on the truncation of the group ring of the fundamental group of the smooth fiber by the 4th power of its augmentation ideal[3] to determine C_0 up to finite ambiguity using the limit mixed Hodge structure on the fundamental group which is constructed in [**20**]. A proof of this assertion is not published, but most of the technical constructions needed to prove it are worked out in detail in the doctoral thesis of Rainer Kaenders [**24**]. More generally, I believe that for all stable degenerations of curves, the limit mixed Hodge structure on the truncation of the integral group ring of the fundamental group of

[3] The mixed Hodge structure on the fundamental groups is discussed in the next section. Also, to apply it, one needs to choose a section $\sigma : \Delta \to C$ of base points.

the smooth fiber by the 5th power of its augmentation ideal should determine C_0 up to finite ambiguity.

6. Periods of $\pi_1(\mathbb{P}^1 - \{0,1,\infty\}, \vec{01})$

In this section, we sketch a fundamental example of computing the periods of a limit MHS on the fundamental group of a the thrice punctured line. The periods turn out to be Euler's the mixed zeta numbers (cf. [**36**], [**15**]).

6.1. Hodge Theory of Homotopy Groups. This is a very brief introduction to the Hodge theory of homotopy groups of complex algebraic varieties.

First, suppose that π is a discrete group. For a commutative ring R, denote the group algebra of π over R by $R\pi$. There is a natural augmentation

$$\epsilon : R\pi \to R$$

defined by taking $\sum r_j g_j$ to $\sum r_j$, where each $r_j \in R$ and $g_j \in \pi$. The augmentation ideal J_R is the kernel of the augmentation. The mapping

$$J_R/J_R^2 \to H_1(\pi, R) \quad (g-1) + J_R^2 \to [g]$$

is a group isomorphism, which can be thought of as a kind of Hurewicz isomorphism.

THEOREM 16 (Morgan [**29**], Hain [**19**]). *If X is a complex algebraic variety and $x \in X$, then for each $s \geq 0$, there is a natural MHS on the truncated group ring $\mathbb{Z}\pi_1(X,x)/J^{s+1}$. These form an inverse system in the category of MHS. The mixed Hodge structure on J/J^2 is dual to Deligne's MHS on $H^1(X)$.*

These mixed Hodge structures are constructed using Chen's iterated integrals.

Suppose that M is a manifold and that $w_1, \dots, w_r$ are smooth $\mathbb{C}$-valued 1-forms on M. For each piecewise smooth path $\gamma : [0,1] \to M$, we can define

$$\int_\gamma w_1 \dots w_r = \int \cdots \int_{0 \leq t_1 \leq \cdots \leq t_r \leq 1} f_1(t_1) \cdots f_r(t_r)\, dt_1 \dots dt_r,$$

where $\gamma^* w_j = f_j(t)\, dt$ for each j. This is viewed as a $\mathbb{C}$-valued function

$$\int_\gamma w_1 \dots w_r : PM \to \mathbb{C}$$

on the the space of piecewise smooth paths in M. When $r = 1$, $\int_\gamma w$ is just the usual line integral. An *iterated integral* is any function $PM \to \mathbb{C}$ which is a linear combination of a constant function and basic iterated integrals

$$\int_\gamma w_1 \dots w_r.$$

An exposition of the construction of the MHS in J/J^{s+1} can be found in [**18**].

6.2. The MHS on $\pi_1(\mathbb{P}^1 - \{0,1,\infty\}, t)$. One can describe the MHS on the J-adic completion

$$\mathbb{Q}\pi_1(\mathbb{P}^1 - \{0,1,\infty\}, t)^\wedge := \varprojlim_s \mathbb{Q}\pi_1(\mathbb{P}^1 - \{0,1,\infty\}, t)/J^s.$$

of $\pi_1(\mathbb{P}^1 - \{0,1,\infty\}, t)$ quite directly.

Consider the non-commutative power series ring

$$A := \mathbb{C}\langle\langle X_0, X_1 \rangle\rangle$$

freely generated by the indeterminates X_0 and X_1. Set

$$w_0 = \frac{dz}{z} \text{ and } w_1 = \frac{dz}{1-z}.$$

Now consider the A-valued iterated integral

$$T = 1 + \int w_0 X_0 + \int w_1 X_1 + \cdots + \int w_{j_1} \ldots w_{j_r} X_{j_1} \ldots X_{j_r} + \cdots$$

It is not difficult to use the definition of iterated integrals to show that the value $T(\gamma)$ of this on a path γ depends only on its homotopy class relative to its endpoints. In addition, one can show that if α and β are composable paths, then[4]

(4) $$T(\alpha\beta) = T(\alpha)T(\beta).$$

For each $t \in \mathbb{P}^1 - \{0,1,\infty\}$, we can thus define a homomorphism

$$\pi_1(\mathbb{P}^1 - \{0,1,\infty\}, t) \to \text{ the group of units of } A$$

by taking the class of the loop γ based at t to $T(\gamma)$. This induces a homomorphism

$$\Theta_t : \mathbb{C}\pi_1(\mathbb{P}^1 - \{0,1,\infty\}, t)^\wedge \to A$$

which can easily be shown to be an isomorphism by using universal mapping properties of free groups and free algebras.

We can use this to construct a MHS on $\mathbb{Q}\pi_1(\mathbb{P}^1 - \{0,1,\infty\}, t)^\wedge$. Give each generator X_j type $(-1,-1)$. Extend this bigrading to all monomials in A in the standard way — the monomial X_I will have Hodge type $(-|I|, -|I|)$ where $I = (i_1, \ldots, i_r)$ is a multi index and $|I|$ is its length r. Thus X_I has weight $-2|I|$. It is natural to define

$$W_m A = \text{ the closure of the span of } X_I \text{ where } -2|I| \leq m$$

(which is the ℓth power of the maximal ideal when $m = -2\ell$) and

$$F^p A = \text{ the span of the } X_I \text{ where } -|I| \geq p$$

which is finite dimensional for all p. Pulling back these filtrations along Θ_t defines the standard MHS on $\mathbb{Q}\pi_1(\mathbb{P}^1 - \{0,1,\infty\}, t)^\wedge$.

The periods of the MHS on $\mathbb{Q}\pi_1(\mathbb{P}^1 - \{0,1,\infty\}, t)^\wedge$ are the coefficients of the monomials X_I in the power series $T(\gamma)$, where $\gamma \in \pi_1(\mathbb{P}^1 - \{0,1,\infty\}, t)$.

REMARK 17. In the Hodge theory of complex varieties, the periods are typically (local) coordinates in an appropriate moduli space of MHS, such as the one

$$G_{\mathbb{Z}} \backslash G / F^0 G$$

discussed near the end of Section 5.1. However, in more arithmetic situations, where the MHS arises from a variety defined (say) over $\mathbb{Q}$, the de Rham invariant together with its Hodge and weight filtrations has a natural $\mathbb{Q}$-form. In this case, one usually takes the periods to be the integrals over rational (betti) cycles of a basis of the $\mathbb{Q}$-de Rham version of the invariant which is adapted to the Hodge and weight filtrations.

This is the case above as $(\mathbb{P}^1 - \{0,1,\infty\}, t)$ is defined over $\mathbb{Q}$ when t is $\mathbb{Q}$-rational. There is an algebraic de Rham theorem for the unipotent fundamental group in this case (cf. [**21**]), which is particularly easy to describe in the case above.

[4]Note that I use the topologists convention that if $\alpha, \beta : [0,1] \to X$ are two paths with $\alpha(1) = \beta(0)$, then $\alpha\beta$ is the path obtained by first traversing α, then β. This is the opposite of the convention used by many papers in this field.

The dual of $\mathbb{C}\pi_1(\mathbb{P}^1 - \{0,1,\infty\},t)/J^{N+1}$ is, by Chen's π_1 de Rham Theorem (or elementary arguments in this case),

$$V_{\mathbb{C}}^{\mathrm{DR}} := \left\{ \sum_{r \le N} a_I \int w_{i_1} \dots w_{i_r} : I \subset \{0,1\}^r,\ |I| = r,\ a_I \in \mathbb{C} \right\}.$$

This has the natural $\mathbb{Q}$-form

$$V_{\mathbb{Q}}^{\mathrm{DR}} := \left\{ \sum_{r \le N} a_I \int w_{i_1} \dots w_{i_r} : I \subset \{0,1\}^r,\ |I| = r,\ a_I \in \mathbb{Q} \right\}.$$

The basis of $\mathbb{C}\pi_1(\mathbb{P}^1 - \{0,1,\infty\},t)/J^{N+1}$ dual to basis $\{\int w_{i_1} \dots w_{i_r}\}$ of $V_{\mathbb{Q}}^{\mathrm{DR}}$ is the set of monomials $X_I := X_{i_1} \dots X_{i_r}$. The periods of the MHS on $\mathbb{C}\pi_1(\mathbb{P}^1 - \{0,1,\infty\},t)/J^{N+1}$ are thus the integrals of the $\int w_{i_1} \dots w_{i_r}$ over elements of $\mathbb{Q}\pi_1(\mathbb{P}^1 - \{0,1,\infty\},t)/J^{N+1}$.

6.3. The MHS on $\pi_1(\mathbb{P}^1 - \{0,1,\infty\},\vec{01})$. In general, the periods of the MHS on $\pi_1(\mathbb{P}^1 - \{0,1,\infty\},t)$ are difficult to compute as the values of the coefficients of T on paths based at t do not appear to be readily recognizable numbers when t is general. However, if we let t approach 0, the periods (or their asymptotics) become more recognizable as we shall explain.

When t goes to 0, the MHS on $\mathbb{Q}\pi_1(\mathbb{P}^1 - \{0,1,\infty\},t)^\wedge$ degenerates as all of the integrals of length ≥ 2 that begin or end with w_0, such as $\int w_0 w_1$, will diverge. (The MHS on the fundamental group can degenerate when the variety becomes singular, and also when the base point runs off the edge of the space as it is here.) Just as there is Schmid's theory of limits of Hodge structures, there is a theory of limits of MHSs, at least in the geometric case [**32**]. (There is a fledgling theory in the abstract case too, being developed by Kaplan and Pearlstein [**25**].) We will compute the periods of the limit mixed Hodge structure as $t \to 0$ with respect to the tangent vector $\vec{01} := \partial/\partial t$, where t is the standard parameter on $\mathbb{P}^1$.

Deligne [**9**] introduced the idea of the fundamental group of (say) an affine curve C with base point a non-zero tangent vector $\vec{v}$ at one of the cusps P. It is simple and elegant. The fundamental group $\pi_1(C,\vec{v})$ is the set of homotopy classes of loops γ in $C \cup \{P\}$ based at P, that leave P with velocity vector $\vec{v}$ and return to P with velocity vector $-\vec{v}$. It is also required that γ does not return to P when $0 < t < 1$. It is naturally isomorphic to the standard fundamental group.[5] This will carry a limit MHS associated to the tangent vector $\vec{v}$.

DEFINITION 18. For integers $n_1,\dots,n_r$, where $n_r > 1$, define

$$\zeta(n_1,\dots,n_r) = \sum_{0<k_1<\dots<k_r} \frac{1}{k_1^{n_1} k_2^{n_2} \dots k_r^{n_r}}.$$

These numbers generalize the classical values of the Riemann zeta function at integers > 1 and were first considered by Euler. They have recently resurfaced in the works of Zagier [**36**] and Goncharov [**15**]. Their $\mathbb{Q}$-linear span in $\mathbb{R}$ is a subalgebra MZN due to combinatorial identities such as

$$\zeta(n)\zeta(m) = \zeta(n,m) + \zeta(m,n) + \zeta(m+n).$$

[5] An elegant way to prove this is to note that $\pi_1(C,\vec{v})$ is naturally isomorphic to $\pi_1(\widetilde{C},[\vec{v}])$ where $\widetilde{C}$ is the real oriented blowup of C at P and $[\vec{v}]$ is the point on the exceptional circle $\partial\widetilde{C}$ corresponding to the oriented ray determined by $\vec{v}$.

Since $\zeta(2) = \pi^2/6$,

$$\mathrm{MZN}_{\mathbb{C}} := \mathrm{MZN} \oplus i\pi\mathrm{MZN}$$

is a $\mathbb{Q}$-subalgebra of $\mathbb{C}$.

Mixed zeta numbers can be expressed as iterated integrals (cf. [**36**]):

$$\zeta(n_1, \dots, n_r) = \int_0^1 w_1 \overbrace{w_0 \dots w_0}^{n_1-1} w_1 \overbrace{w_0 \dots w_0}^{n_2-1} w_1 \dots w_1 \overbrace{w_0 \dots w_0}^{n_r-1}.$$

Here the path of integration is along the unit interval, and the integral converges if and only if $n_r > 1$. This identity follows directly from the definition of iterated integrals using the following slightly more general form:

$$\int_0^x w_1 \overbrace{w_0 \dots w_0}^{n_1-1} w_1 \overbrace{w_0 \dots w_0}^{n_2-1} w_1 \dots w_1 \overbrace{w_0 \dots w_0}^{n_r-1} = \sum_{0<k_1<\dots<k_r} \frac{x^{k_r}}{k_1^{n_1} k_2^{n_2} \dots k_r^{n_r}}.$$

The basic result we wish to explain is the following "folk theorem." Goncharov in [**15**] has recently written down a proof of a considerably more refined statement.

THEOREM 19. *The space of periods of the limit mixed Hodge structure on* $\mathbb{Q}\pi_1(\mathbb{P}^1 - \{0, 1, \infty\}, \vec{01})^\wedge$ *is precisely* $\mathrm{MZN}_{\mathbb{C}}$.

It is convenient to give a proof of this theorem by appealing to the MHS on spaces of paths. Denote the space of piecewise smooth paths from a to b in $\mathbb{P}^1 - \{0, 1, \infty\}$ by $P_{a,b}(\mathbb{P}^1 - \{0, 1, \infty\})$. Endow it with the compact open topology. Then

$$H_0(P_{a,b}(\mathbb{P}^1 - \{0, 1, \infty\}), \mathbb{Q})$$

is the $\mathbb{Q}$-vector space generated by the homotopy classes of paths in $\mathbb{P}^1 - \{0, 1, \infty\}$ that go from a to b. It is a module over $\mathbb{Q}\pi_1(\mathbb{P}^1 - \{0, 1, \infty\}, a)$ and can be completed in the J-adic topology. Denote this completion by

$$H_0(P_{a,b}(\mathbb{P}^1 - \{0, 1, \infty\}), \mathbb{Q})^\wedge.$$

(It is also a $\mathbb{Q}\pi_1(\mathbb{P}^1 - \{0, 1, \infty\}, b)$ module. The completion with respect to this action is identical with the completion above.) As in the case where $a = b$, we can define a MHS on it by pulling back the Hodge and weight filtrations of A along the mapping

$$\Theta_{a,b} : H_0(P_{a,b}(\mathbb{P}^1 - \{0, 1, \infty\}), \mathbb{Q})^\wedge \to A$$

induced by taking γ to $T(\gamma)$, which is an isomorphism. The natural mappings

$$H_0(P_{a,b}(\mathbb{P}^1 - \{0, 1, \infty\}), \mathbb{Q})^\wedge \otimes H_0(P_{b,c}(\mathbb{P}^1 - \{0, 1, \infty\}), \mathbb{Q})^\wedge \to H_0(P_{a,c}(\mathbb{P}^1 - \{0, 1, \infty\}), \mathbb{Q})^\wedge$$

are easily seen to be morphisms of MHS because of the formula (4).

By taking limit MHS, we can replace a and b by non-zero tangent vectors at the cusps. In particular, we have a mixed Hodge structure on

$$H_0(P_{\vec{01},\vec{10}}(\mathbb{P}^1 - \{0, 1, \infty\}), \mathbb{Q})^\wedge$$

where $\vec{10}$ is the tangent vector $\partial/\partial(1-t)$ at $t = 1$.

The following result is essentially a restatement of a result of Le and Murakami [**26**, Thm. A.9]. (See also, [**15**, Lemma 5.4].)

LEMMA 20. *The periods of the path* $[0,1] \in P_{\vec{01},\vec{10}}(\mathbb{P}^1 - \{0,1,\infty\})$ *under the mapping*

$$\Theta_{\vec{01},\vec{10}} : H_0(P_{\vec{01},\vec{10}}(\mathbb{P}^1 - \{0,1,\infty\})) \to A$$

are mixed zeta numbers. All elements of MZN *occur.*

PROOF. Let $\Phi \in A$ be the image of $[0,1]$ under the regularized period mapping $\Theta_{\vec{01},\vec{10}} : H_0(P_{\vec{01},\vec{10}}\mathbb{P}^1 - \{0,1,\infty\}) \to A$. I claim that this is Drinfeld's associator [**12**]. The result will then follow as Le and Murakami have shown that the coefficients of the Drinfeld associator are mixed zeta numbers and all mixed zeta numbers occur.

To see that Φ is the Drinfeld associator, we use the prescription given in Corollary 5 for computing the regularized limit periods of a flat section. Applying this directly, we see that the renormalized value of $T([0,1])$ is

$$\lim_{t\to 0} t^{X_0} T([t, 1-t]) t^{X_1}$$

as the residue at $t = 0$ of the connection for which $T[t, 1-t]$ is a flat section is

$$\text{left multiplication by } X_0 + \text{ right multiplication by } X_1$$

The limit is a well-known expression for Drinfeld's associator. (Cf. [**12**], [**26**].) □

Consider the paths illustrated in figure 1

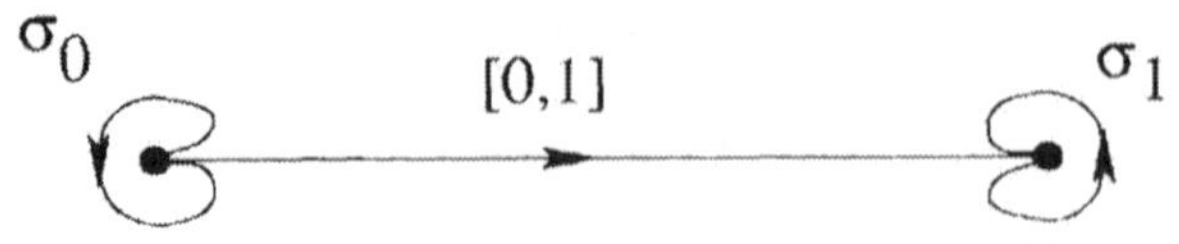

FIGURE 1

where $\sigma_0 \in P_{\vec{01},\vec{01}}\mathbb{P}^1 - \{0,1,\infty\}$ and $\sigma_1 \in P_{\vec{10},\vec{10}}\mathbb{P}^1 - \{0,1,\infty\}$.

Theorem 19 now follows as one can check, as in the proof of the previous lemma, that under the renormalized homomorphism

$$\Theta_{\vec{01}} : \pi_1(\mathbb{P}^1 - \{0,1,\infty\}, \vec{01}) \to A$$

the image of the loop σ_0 is $\exp(2\pi i X_0)$ and that the image of σ_1 under

$$\Theta_{\vec{10}} : \pi_1(\mathbb{P}^1 - \{0,1,\infty\}, \vec{10}) \to A$$

is $\exp(-2\pi i X_1)$. Since $\pi_1(\mathbb{P}^1 - \{0,1,\infty\}, \vec{01})$ is generated by σ_0 and $[0,1]\sigma_1[0,1]^{-1}$, and since

$$T([0,1]\sigma_1[0,1]^{-1}) = \Phi e^{2\pi i X_1} \Phi^{-1}$$

we see that all periods of $\mathbb{Q}\pi_1(\mathbb{P}^1 - \{0,1,\infty\}, \vec{01})^\wedge$ lie in $\mathrm{MZN}_\mathbb{C}$.

7. Mixed Tate Motives

The question arises as to why one should expect the periods of $\pi_1(\mathbb{P}^1 - \{0,1,\infty\}, \vec{01})$ to be so special and to have such a nice description in terms of mixed zeta numbers. Deligne and Goncharov [**11**], building on previous work of both authors, especially [**9**] and several of Goncharov's papers in arXiv.org, have developed a satisfying and deep explanation in terms of the theory of mixed Tate motives.

Thanks to the work of Voevodsky [**34**], Levine [**27**], and Deligne-Goncharov [**11**], we now know that there is a tannakian category of mixed Tate motives over

Spec of the ring of S-integers in a number field with the expected ext groups. So, in particular, there is a tannakian category of mixed Tate motives over Spec $\mathbb{Z}$. Such mixed Tate motives have a Hodge realization, which is a mixed Hodge structure, all of whose weight graded quotients are of type (p,p). These mixed Hodge structures will have two $\mathbb{Q}$-structures — one coming from a $\mathbb{Q}$-de Rham structure, and another from topology (the "Betti realization"). The entries of the matrices relating them are, by definition, the periods of the motive.

Deligne and Goncharov [**11**] show that $\mathbb{Q}\pi_1(\mathbb{P}^1 - \{0,1,\infty\}, \vec{01})^\wedge$, equivalently the Lie algebra of the unipotent fundamental group of $(\mathbb{P}^1 - \{0,1,\infty\}, \vec{01})$, is a pro-object in the category of mixed Tate motives over Spec $\mathbb{Z}$. One reason one should suspect this is that

$$\mathbb{P}^1_{\mathbb{Z}} - \{0,1,\infty\} := \operatorname{Spec} \mathbb{Z}[t, t^{-1}, (1-t)^{-1}]$$

has everywhere good reduction. However, the pointed variety $(\mathbb{P}^1_{\mathbb{Z}} - \{0,1,\infty\}, N)$, where $N \in \mathbb{Z} - \{0,1\}$, will not have good reduction at primes dividing N as then the base point will move to a cusp. In order for $(\mathbb{P}^1 - \{0,1,\infty\}, t)$ to have everywhere good reduction, we are forced to take t to be $\vec{01}$, or one of its 5 other images under $\operatorname{Aut}(\mathbb{P}^1, \{0,1,\infty\})$.

The fact that $\mathbb{Q}\pi_1(\mathbb{P}^1 - \{0,1,\infty\}, \vec{01})^\wedge$ is a mixed Tate motive over Spec $\mathbb{Z}$ has a remarkable consequence for the transcendence properties of mixed zeta numbers [**33**] and for the Galois action on its $\mathbb{Q}_\ell$-form [**22**]. In the first case, Zagier conjectured (unpublished), and Terasoma [**33**] has proved, that the dimension of the $\mathbb{Q}$-linear span of the mixed zeta numbers MZN_m of weight[6] m is bounded by the dimension of the mth weight graded quotient of the graded algebra

$$\mathbb{Q}[Z_2] \otimes \mathbb{Q}\langle Z_3, Z_5, Z_7, \dots \rangle$$

where Z_m has weight m and $\mathbb{Q}\langle Z_3, Z_5, Z_7, \dots \rangle$ denotes the free associative algebra generated by the Z_{odd}.

One currently unresolved question is whether the periods of all mixed Tate motives over Spec $\mathbb{Z}$ lie in $\mathrm{MZN}_{\mathbb{C}}$. Goncharov [**14**, p. 385] has conjectured that this is the case. To see why this might be, we need to discuss the motivic Lie algebra of Spec $\mathbb{Z}$ and its connection with the absolute Galois group $G_{\mathbb{Q}} := \operatorname{Gal} \overline{\mathbb{Q}}/\mathbb{Q}$ and its action on $\mathbb{Q}_\ell\pi_1(\mathbb{P}^1 - \{0,1,\infty\}, \vec{01})^\wedge$.

Since the category of mixed Tate motives over Spec $\mathbb{Z}$ is tannakian, it is the category of representations of a pro-algebraic group $\mathcal{A}$ over $\mathbb{Q}$. Since the tannakian fundamental group of the category of pure Tate motives over Spec $\mathbb{Z}$ is $\mathbb{G}_m$, this group is an extension

$$1 \to \mathcal{U} \to \mathcal{A} \to \mathbb{G}_m \to 1$$

where $\mathcal{U}$ is a prounipotent group. That the ext groups in the category of mixed Tate motives over $\mathbb{Z}$ have the desired relationship to $K_\bullet(\mathbb{Z})$ implies that the Lie algebra of $\mathcal{U}$ is a free pronilpotent Lie algebra on generators $z_3, z_5, z_7, \dots$, where z_n has weight $-2n$, cf. [**11**] and [**23**]. The Lie algebra $\mathfrak{a}$ of $\mathcal{A}$ is itself a pro-object of the category of mixed Tate motives, and formal arguments show that all periods of mixed Tate motives over Spec $\mathbb{Z}$ occur as periods of $\mathfrak{a}$. Thus one should try to prove that all periods of the MHS of $\mathfrak{a}$ lie in $\mathrm{MZN}_{\mathbb{C}}$. One way to approach this is by studying the action of the absolute Galois group on the algebraic fundamental group of the thrice punctured line over $\overline{\mathbb{Q}}$.

[6]The weight of $\zeta(n_1, \dots, n_r)$ is, by definition, $n_1 + \cdots + n_r$.

Every mixed Tate motive V has an ℓ-adic realization for each rational prime ℓ. This is a representation $\rho_\ell : G_{\mathbb{Q}} \to \operatorname{Aut} V_{\mathbb{Q}_\ell}$, where $V_{\mathbb{Q}_\ell}$ denotes $V \otimes_{\mathbb{Q}} \mathbb{Q}_\ell$. These actions induce a Zariski dense homomorphism

$$G_{\mathbb{Q}} \to \mathcal{A}(\mathbb{Q}_\ell)$$

through which $\mathcal{A} \otimes \mathbb{Q}_\ell \to \operatorname{Aut} V$ factors. The homomorphism ρ_ℓ is the composite of this with the canonical homomorphism $\mathcal{A} \to \operatorname{Aut} V$:

$$G_{\mathbb{Q}} \to \mathcal{A}(\mathbb{Q}_\ell) \to \operatorname{Aut} V_{\mathbb{Q}_\ell}.$$

The image of $\mathcal{A} \otimes \mathbb{Q}_\ell$ in $\operatorname{Aut} V_{\mathbb{Q}_\ell}$ is thus the Zariski closure of the image of $G_{\mathbb{Q}}$ (cf. [**23**]).

In the case of the unipotent fundamental group of $(\mathbb{P}^1 - \{0,1,\infty\}, \vec{01})$, ρ_ℓ is the natural Galois action

$$G_{\mathbb{Q}} \to \mathcal{A}(\mathbb{Q}_\ell) \to \operatorname{Aut} \mathbb{Q}_\ell \pi_1(\mathbb{P}^1 - \{0,1,\infty\}, \vec{01})^\wedge$$

induced by the action of $G_{\mathbb{Q}}$ on its algebraic fundamental group. The Zariski closure of the image of this action is the image of the homomorphism

$$\mathcal{A} \otimes \mathbb{Q}_\ell \to \operatorname{Aut} \mathbb{Q}_\ell \pi_1(\mathbb{P}^1 - \{0,1,\infty\}, \vec{01})^\wedge.$$

This homomorphism is injective if and only if its derivative

$$\mathfrak{a} \to \operatorname{Der} \mathbb{Q}_\ell \pi_1(\mathbb{P}^1 - \{0,1,\infty\}, \vec{01})^\wedge$$

is. Since the periods of $\operatorname{Der} \mathbb{Q}_\ell \pi_1(\mathbb{P}^1 - \{0,1,\infty\}, \vec{01})^\wedge$ lie in $\mathrm{MZN}_{\mathbb{C}}$, the periods of $\mathfrak{a}$ will too if the derivative is injective. At present it is not known whether or not this is the case, although there are computer results that show that in "small weights" the derivative is injective.

References

[1] J. Carlson: *Extensions of mixed Hodge structures*, in Journées de Géometrie Algébrique d'Angers, Juillet 1979, Sijthoff & Noordhoff, Alphen aan den Rijn, 1980, 107–127.

[2] E. Cattani: *Personal communication*, April 17,2003.

[3] E. Cattani, A. Kaplan, W. Schmid: *Degeneration of Hodge structures*, Ann. of Math. (2) 123 (1986), 457–535.

[4] E. Cattani, A. Kaplan, W. Schmid: L^2 and intersection cohomologies for a polarizable variation of Hodge structure, Invent. Math. 87 (1987), 217–252.

[5] C. H. Clemens: *Degeneration of Kähler manifolds*, Duke Math. J. 44 (1977), 215–290.

[6] P. Deligne: *Équations différentielles à points singuliers réguliers*, Lecture Notes in Mathematics, Vol. 163, Springer-Verlag, Berlin-New York, 1970.

[7] P. Deligne: *Théorie de Hodge. II*, Inst. Hautes Études Sci. Publ. Math. No. 40 (1971), 5–57.

[8] P. Deligne: *Théorie de Hodge. III*, Inst. Hautes Études Sci. Publ. Math. No. 44 (1974), 5–77.

[9] *Le groupe fondamental de la droite projective moins trois points*, Galois groups over $\mathbb{Q}$ (Berkeley, CA, 1987), 79–297, Math. Sci. Res. Inst. Publ., 16, Springer, New York, 1989.

[10] P. Deligne: *Local behavior of Hodge structures at infinity*, Mirror symmetry, II, 683–699, AMS/IP Stud. Adv. Math., 1, Amer. Math. Soc., Providence, RI, 1997.

[11] P. Deligne, A. Goncharov: *Groupes fondamentaux motiviques de Tate mixte*, preprint 2003, math.NT/0302267.

[12] V. Drinfeld: *On quasitriangular quasi-Hopf algebras and on a group that is closely connected with* $\operatorname{Gal}(\overline{\mathbb{Q}}/\mathbb{Q})$, (Russian) Algebra i Analiz 2 (1990), 149–181; translation in Leningrad Math. J. 2 (1991), 829–860

[13] J. Fay: *Theta functions on Riemann surfaces*, Lecture Notes in Mathematics, Vol. 352. Springer-Verlag, Berlin-New York, 1973.

[14] A. Goncharov: *Polylogarithms in arithmetic and geometry*, Proceedings of the International Congress of Mathematicians, Vol. 1, 2 (Zürich, 1994), 374–387, Birkhäuser, Basel, 1995.

[15] A. Goncharov: *Multiple polylogarithms and mixed Tate motives*, preprint math.AG/0103059.

[16] P. Griffiths: *Periods of integrals on algebraic manifolds III, Some global differential-geometric properties of the period mapping*, Inst. Hautes Études Sci. Publ. Math. No. 38 (1970), 125–180.

[17] P. Griffiths, J. Harris: *Principles of algebraic geometry*, Wiley-Interscience, New York, 1978.

[18] R. Hain: *The geometry of the mixed Hodge structure on the fundamental group*, Algebraic geometry, Bowdoin, 1985 (Brunswick, Maine, 1985), 247–282, Proc. Sympos. Pure Math., 46, Part 2, Amer. Math. Soc., Providence, RI, 1987.

[19] R. Hain: *Mixed Hodge structures on homotopy groups*, Bull. Amer. Math. Soc. 14 (1986), 111–114.

[20] R. Hain: *The de Rham homotopy theory of complex algebraic varieties II*, K-Theory 1 (1987), 481–497.

[21] R. Hain: *Iterated integrals and algebraic cycles: examples and prospects*, Contemporary trends in algebraic geometry and algebraic topology (Tianjin, 2000), Nankai Tracts Math., 5, World Sci. Publishing, 2002, 55–118.

[22] R. Hain, M. Matsumoto: *Weighted Completion of Galois Groups and Galois Actions on the Fundamental Group of* $\mathbb{P}^1 - \{0, 1, \infty\}$, Compositio Math. to appear, math.AG/0006158.

[23] R. Hain, M. Matsumoto: *Tannakian Fundamental Groups Associated to Galois Groups*, preprint, math.AG/0010210.

[24] R. Kaenders: *New Period Mappings for Plane Curve Singularities*, preprint, math.AG/9802011.

[25] A. Kaplan, G. Pearlstein: *Singularities of variations of mixed Hodge structure*, preprint 2000, math.AG/0007040.

[26] T. Le, J. Murakami: *Kontsevich's integral for the Kauffman polynomial*, Nagoya Math. J. 142 (1996), 39–65.

[27] M. Levine: *Mixed motives*, Mathematical Surveys and Monographs 57, American Mathematical Society, 1998.

[28] E. Looijenga: *L^2-cohomology of locally symmetric varieties*, Compositio Math. 67 (1988), 3–20.

[29] J. Morgan: *The algebraic topology of smooth algebraic varieties*, Inst. Hautes Études Sci. Publ. Math. No. 48 (1978), 137–204; Correction: Inst. Hautes Études Sci. Publ. Math. No. 64 (1986), 185.

[30] W. Schmid: *Variation of Hodge structure: the singularities of the period mapping*, Invent. Math. 22 (1973), 211–319.

[31] J. Steenbrink: *Limits of Hodge structures*, Invent. Math. 31 (1975/76), 229–257.

[32] J. Steenbrink, S. Zucker: *Variation of mixed Hodge structure I*, Invent. Math. 80 (1985), 489–542.

[33] T. Terasoma: *Mixed Tate motives and multiple zeta values*, Invent. Math. 149 (2002), 339–369.

[34] V. Voevodsky, A. Suslin, E. Friedlander: *Cycles, transfers, and motivic homology theories*, Annals of Mathematics Studies 143, Princeton University Press, 2000.

[35] W. Wasow: *Asymptotic expansions for ordinary differential equations*, Reprint of the 1976 edition. Dover Publications, Inc., New York, 1987.

[36] D. Zagier: *Values of zeta functions and their applications*, First European Congress of Mathematics, Vol. II (Paris, 1992), 497–512, Progr. Math., 120, Birkha"user, Basel, 1994.

[37] S. Zucker: *Hodge theory with degenerating coefficients, L_2 cohomology in the Poincaré metric*, Ann. of Math. (2) 109 (1979), 415–476.

Current Developments in Mathematics, 2002
Pages 135 – 179

Modular forms and arithmetic geometry

Stephen S. Kudla

The aim of these notes is to describe some examples of modular forms whose Fourier coefficients involve quantities from arithmetical algebraic geometry. At the moment, no general theory of such forms exists, but the examples suggest that they should be viewed as a kind of arithmetic analogue of theta series and that there should be an arithmetic Siegel–Weil formula relating suitable averages of them to special values of derivatives of Eisenstein series. We will concentrate on the case for which the most complete picture is available, the case of generating series for cycles on the arithmetic surfaces associated to Shimura curves over $\mathbb{Q}$, expanding on the treatment in [**40**]. A more speculative overview can be found in [**41**].

In section 1, we review the basic facts about the arithmetic surface $\mathcal{M}$ associated to a Shimura curve over $\mathbb{Q}$. These arithmetic surfaces are moduli stacks over $\mathrm{Spec}(\mathbb{Z})$ of pairs (A,ι) over a base S, where A is an abelian scheme of relative dimension 2 and ι is an action on A of a maximal order O_B in an indefinite quaternion algebra B over $\mathbb{Q}$. In section 2, we recall the definition of the arithmetic Chow group $\widehat{\mathrm{CH}}^1(\mathcal{M})$, following Bost, [**7**], and we discuss the metrized Hodge line bundle $\hat{\omega}$ and the conjectural value of $\langle\hat{\omega},\hat{\omega}\rangle$, where $\langle\ ,\ \rangle$ is the height pairing on $\widehat{\mathrm{CH}}^1(\mathcal{M})$. In the next two sections, we describe divisors $\mathcal{Z}(t)$, $t\in\mathbb{Z}_{>0}$, on $\mathcal{M}$. These are defined as the locus of (A,ι,x)'s where x is a special endomorphism (Definition 3.1) of (A,ι) with $x^2=-t$. Since such an x gives an action on (A,ι) of the order $\mathbb{Z}[\sqrt{-t}]$ in the imaginary quadratic field $k_t=\mathbb{Q}(\sqrt{-t})$, the cycles $\mathcal{Z}(t)$ can be viewed as analogues of the familiar CM points on modular curves. In section 3, the complex points and hence the horizontal components of $\mathcal{Z}(t)$ are determined. In section 4, the vertical components of $\mathcal{Z}(t)$ are determined using the p-adic uniformization of the fibers $\mathcal{M}_p$ of bad reduction of $\mathcal{M}$. In section 5, we construct Green functions $\Xi(t,v)$ for the divisors $\mathcal{Z}(t)$, depending on a parameter $v\in\mathbb{R}^\times_{>0}$. When $t<0$, the series defining $\Xi(t,v)$ becomes a smooth function on $\mathcal{M}(\mathbb{C})$. These Green functions are used in section 6 to define classes $\widehat{\mathcal{Z}}(t,v)\in\widehat{\mathrm{CH}}^1(\mathcal{M})$, for $t\in\mathbb{Z}$, $t\neq 0$, and an additional class $\widehat{\mathcal{Z}}(0,v)$ is defined using $\hat{\omega}$. The main result of section 6 (Theorem 6.3) says that generating series

$$\widehat{\theta}(\tau)=\sum_{t\in\mathbb{Z}}\widehat{\mathcal{Z}}(t,v)\,q^t,\qquad \tau=u+iv,\qquad q=e(\tau),$$

Partially supported by NSF grant DMS-0200292 and by a Max-Planck Research Prize from the Max-Planck Society and Alexander von Humboldt Stiftung.

is the q-expansion of a (nonholomorphic) modular form of weight $\frac{3}{2}$, which we call an arithmetic theta function. The proof of this result is sketched in section 7. The main ingredients are (i) the fact that the height pairing of $\widehat{\theta}(\tau)$ with various classes in $\widehat{\mathrm{CH}}^1(\mathcal{M})$, e.g., $\hat{\omega}$, can be shown to be modular, and (ii) the result of Borcherds, [**5**], which says that a similar generating series with coefficients in the usual Chow group of the generic fiber $\mathrm{CH}^1(\mathcal{M}_{\mathbb{Q}})$ is a modular form of weight $\frac{3}{2}$. In section 8, we use the arithmetic theta function to define an arithmetic theta lift

$$\widehat{\theta}: S_{\frac{3}{2}} \longrightarrow \widehat{\mathrm{CH}}^1(\mathcal{M}), \qquad f \mapsto \widehat{\theta}(f) = \langle\, f, \widehat{\theta}\, \rangle_{\mathrm{Pet}},$$

from a certain space of modular forms of weight $\frac{3}{2}$ to the arithmetic Chow group. This lift is an arithmetic analogue of the classical theta lift from modular forms of weight $\frac{3}{2}$ to automorphic forms of weight 2 for $\Gamma = O_B^\times$. According to the results of Waldspurger, reviewed in section 9, the nonvanishing of this classical lift is controlled by a combination of local obstructions and, most importantly, the central value $L(1, F)$ of the standard Hecke L-function[1] of the cusp form F of weight 2 coming from f via the Shimura lift. In section 10, we describe a doubling integral representation (Theorem 10.1) of the Hecke L-function, involving f and an Eisenstein series $\mathcal{E}(\tau, s, B)$ of weight $\frac{3}{2}$ and genus 2. At the central point $s = 0$, $\mathcal{E}(\tau, 0, B) = 0$. In the case in which the root number of the L-function is -1, we obtain a formula (Corollary 10.2)

$$\langle\, \mathcal{E}_2'(\begin{pmatrix} \tau_1 & \\ & -\bar{\tau}_2 \end{pmatrix}, 0; B),\, \overline{f(\tau_2)}\, \rangle_{\mathrm{Pet},\, \tau_2} = f(\tau_1) \cdot C(0) \cdot L'(\frac{1}{2}, \pi),$$

for an explicit constant $C(0)$, whose vanishing is controlled by local obstructions. Finally, in section 11, we state a conjectural identity (Conjecture 11.1)

$$\langle\, \widehat{\theta}(\tau_1), \widehat{\theta}(\tau_2)\, \rangle \overset{??}{\to} = \mathcal{E}_2'(\begin{pmatrix} \tau_1 & \\ & -\bar{\tau}_2 \end{pmatrix}, 0; B)$$

relating the height pairing of the arithmetic theta function and the restriction to the diagonal of the derivative at $s = 0$ of the weight $\frac{3}{2}$ Eisenstein series. This identity is equivalent to a series of identities of Fourier coefficients, (11.1),

$$\langle\, \widehat{\mathcal{Z}}(t_1, v_1), \widehat{\mathcal{Z}}(t_2, v_2)\, \rangle \cdot q_1^{t_1} q_2^{t_2} \quad = \sum_{\substack{T \in \mathrm{Sym}_2(\mathbb{Z})^\vee \\ \mathrm{diag}(T) = (t_1, t_2)}} \mathcal{E}_{2,T}'(\begin{pmatrix} \tau_1 & \\ & \tau_2 \end{pmatrix}, 0; B).$$

Here

$$\mathrm{Sym}_2(\mathbb{Z})^\vee = \{\, T = \begin{pmatrix} t_1 & m \\ m & t_2 \end{pmatrix} \mid t_1, t_2 \in \mathbb{Z},\ m \in \frac{1}{2}\mathbb{Z}\, \}$$

is the dual lattice of $\mathrm{Sym}_2(\mathbb{Z})$ with respect to the trace pairing. We sketch the proof of these identities in the case where $t_1 t_2$ is not a square (Theorem 11.2). As a consequence, we prove Conjecture 11.1 up to a linear combination of theta series for quadratic forms in one variable (Corollary 11.3). Assuming that f is orthogonal to such theta series, we can substitute the height pairing $\langle\, \widehat{\mathcal{Z}}(t_1, v_1), \widehat{\mathcal{Z}}(t_2, v_2)\, \rangle$ for the derivative of the Eisenstein series in the doubling identity and obtain,

$$\langle\, \widehat{\theta}(\tau_1), \widehat{\theta}(f)\, \rangle = f(\tau_1) \cdot C(0) \cdot L'(\frac{1}{2}, \pi),$$

---

[1] Here we assume that F is a newform, so that $L(1, F) = L(\frac{1}{2}, \pi)$, where π is the corresponding cuspidal automorphic representation.

in the case of root number -1. This yields the arithmetic inner product formula

$$\langle \widehat{\theta}(f), \widehat{\theta}(f) \rangle = \langle f, f \rangle \cdot C(0) \cdot L'(\frac{1}{2}, \pi),$$

analogous to the Rallis inner product formula for the classical theta lift. Some discussion of the relation of this result to the Gross-Kohnen-Zagier formula, [**24**], is given at the end of section 11.

Most of the results described here are part of a long term collaboration with Michael Rapoport and Tonghai Yang. I would also like to thank J.-B. Bost, B. Gross, M. Harris, J. Kramer, and U. Kühn for their comments and suggestions.

Contents

1. Shimura curves and arithmetic surfaces

Let B be an indefinite quaternion algebra over $\mathbb{Q}$ and let $D(B)$ be the product of the primes p for which $B_p = B \otimes_{\mathbb{Q}} \mathbb{Q}_p$ is a division algebra. For the moment, we allow the case $B = M_2(\mathbb{Q})$, where $D(B) = 1$. The three dimensional $\mathbb{Q}$-vector space

$$V = \{x \in B \mid \text{tr}(x) = 0\} \tag{1.1}$$

is equipped with the quadratic form $Q(x) = \nu(x) = -x^2$. Here ν (resp. tr) is the reduced norm (resp. trace) on B; in the case $B = M_2(\mathbb{Q})$, this is the usual determinant (resp. trace). The bilinear form associated to Q is given by $(x, y) = \text{tr}(xy^\iota)$, where $x \mapsto x^\iota$ is the involution on B given by $x^\iota = \text{tr}(x) - x$. The action of $H = B^\times$ on V by conjugation, $h : x \to hxh^{-1}$, preserves the quadratic form and induces an isomorphism

$$H \xrightarrow{\sim} \text{GSpin(V)}, \tag{1.2}$$

where GSpin(V) is the spinor similitude group of V. Since B is indefinite, i.e., since $B_{\mathbb{R}} = B \otimes_{\mathbb{Q}} \mathbb{R} \simeq M_2(\mathbb{R})$, V has signature $(1, 2)$. Let

$$D = \{\, w \in V(\mathbb{C}) \mid (w, w) = 0, \ (w, \bar{w}) < 0 \,\}/\mathbb{C}^\times \subset \mathbb{P}(V(\mathbb{C})), \tag{1.3}$$

so that D is an open subset of a quadric in $\mathbb{P}(V(\mathbb{C}))$. Then the group $H(\mathbb{R})$ acts naturally on D, and, if we fix an isomorphism $B_{\mathbb{R}} \simeq M_2(\mathbb{R})$, then there is an identification

$$\mathbb{C} \setminus \mathbb{R} \xrightarrow{\sim} D, \qquad z \mapsto w(z) := \begin{pmatrix} z & -z^2 \\ 1 & -z \end{pmatrix} \mod \mathbb{C}^\times, \tag{1.4}$$

which is equivariant for the action of $H(\mathbb{R}) \simeq \mathrm{GL}_2(\mathbb{R})$ on $\mathbb{C} \setminus \mathbb{R}$ by fractional linear transformations.

Let O_B be a maximal order in B and let $\Gamma = O_B^\times$. In the case $B = M_2(\mathbb{Q})$, one may take $O_B = M_2(\mathbb{Z})$, so that $\Gamma = \mathrm{GL}_2(\mathbb{Z})$. Also let

$$K = (\hat{O}_B)^\times \subset H(\mathbb{A}_f), \tag{1.5)/}$$

where $\hat{O}_B = O_B \otimes_{\mathbb{Z}} \hat{\mathbb{Z}}$, for $\hat{\mathbb{Z}} = \varprojlim_N \mathbb{Z}/N\mathbb{Z}$. Then the quotient

$$M(\mathbb{C}) = H(\mathbb{Q}) \backslash \Big(D \times H(\mathbb{A}_f)/K \Big) \simeq \Gamma \backslash D, \tag{1.6}$$

which should be viewed as an orbifold, is the set of complex points of a Shimura curve M, if $D(B) > 1$, or of the modular curve (without its cusp), if $D(B) = 1$. From now on, we assume that $D(B) > 1$, although much of what follows can be carried over for $D(B) = 1$ with only slight modifications. The key point is to interpret M as a moduli space.

Let $\mathcal{M}$ be the moduli stack over $\mathrm{Spec}(\mathbb{Z})$ for pairs (A, ι) where A is an abelian scheme over a base S with an action $\iota : O_B \to \mathrm{End}_S(A)$ satisfying the determinant condition, [**49**], [**36**], [**9**],

$$\det(\iota(b); \mathrm{Lie(A)}) = \nu(\mathrm{b}). \tag{1.7}$$

Over $\mathbb{C}$, such an (A, ι) is an abelian surface with O_B action. For example, for $z \in D \simeq \mathbb{C} \setminus \mathbb{R}$, the isomorphism

$$\lambda_z : B_{\mathbb{R}} \simeq M_2(\mathbb{R}) \xrightarrow{\sim} \mathbb{C}^2, \qquad b \mapsto b \cdot \begin{pmatrix} z \\ 1 \end{pmatrix} = \begin{pmatrix} w_1 \\ w_2 \end{pmatrix} \tag{1.8}$$

determines a lattice $L_z = \lambda_z(O_B) \subset \mathbb{C}^2$. The complex torus $A_z = \mathbb{C}^2/L_z$ is an abelian variety with a natural O_B action given by left multiplication, and hence defines an object $(A_z, \iota) \in \mathcal{M}(\mathbb{C})$. Two points in D give the same lattice if and only if they are in the same $O_B^\times$-orbit, and, up to isomorphism, every (A, ι) over $\mathbb{C}$ arises in this way. Thus, the construction just described gives an isomorphism

$$[\Gamma \backslash D] \xrightarrow{\sim} \mathcal{M}(\mathbb{C}) \tag{1.9}$$

of orbifolds, and $\mathcal{M}$ gives a model of $M(\mathbb{C})$ over $\mathrm{Spec}(\mathbb{Z})$ with generic fiber

$$M = \mathcal{M} \times_{\mathrm{Spec}(\mathbb{Z})} \mathrm{Spec}(\mathbb{Q}), \tag{1.10}$$

the Shimura curve over $\mathbb{Q}$.

Since we are assuming that $D(B) > 1$, $\mathcal{M}$ is proper of relative dimension 1 over $\mathrm{Spec}(\mathbb{Z})$ and smooth over $\mathrm{Spec}(\mathbb{Z}[D(B)^{-1}])$. We will ignore the stack aspect from now on and simply view $\mathcal{M}$ as an arithmetic surface over $\mathrm{Spec}(\mathbb{Z})$.

The surface $\mathcal{M}$ has bad reduction at primes $p \mid D(B)$ and this reduction can be described via p-adic uniformization [**9**], [**46**]. Let $\hat{\Omega}_p$ be Drinfeld's p-adic upper half plane. It is a formal scheme over $\mathbb{Z}_p$ with a natural action of $\mathrm{PGL}_2(\mathbb{Q}_p)$. Let $W = W(\bar{\mathbb{F}}_p)$ be the Witt vectors of $\bar{\mathbb{F}}_p$ and let

$$\hat{\Omega}_W = \hat{\Omega}_p \times_{\mathrm{Spf}(\mathbb{Z}_p)} \mathrm{Spf(W)} \tag{1.11}$$

be the base change of $\hat{\Omega}_p$ to W. Also, let $\hat{\Omega}_W^\bullet = \hat{\Omega}_W \times \mathbb{Z}$, and let $g \in \mathrm{GL}_2(\mathbb{Q}_p)$ act on $\hat{\Omega}_W^\bullet$ by

$$g : (z, i) \to (\, g(z), i + \mathrm{ord}_p(\det(g))\,). \tag{1.12}$$

Let $B^{(p)}$ be the definite quaternion algebra over $\mathbb{Q}$ with invariants

$$\mathrm{inv}_\ell(\mathrm{B}^{(p)}) = \begin{cases} -\mathrm{inv}_\ell(\mathrm{B}) & \text{if } \ell = p,\ \infty, \\ \ \ \mathrm{inv}_\ell(\mathrm{B}) & \text{otherwise.} \end{cases} \tag{1.13}$$

Let $H^{(p)} = (B^{(p)})^\times$ and

$$V^{(p)} = \{\ x \in B^{(p)} \mid \mathrm{tr}(x) = 0\ \}. \tag{1.14}$$

For convenience, we will often write $B' = B^{(p)}$, $H' = H^{(p)}$ and $V' = V^{(p)}$ when p has been fixed. Fix isomorphisms

$$H'(\mathbb{Q}_p) \simeq \mathrm{GL}_2(\mathbb{Q}_p), \qquad \text{and} \qquad H'(\mathbb{A}_f^p) \simeq H(\mathbb{A}_f^p). \tag{1.15}$$

Let $\widehat{\mathcal{M}}_p$ be the base change to W of the formal completion of $\mathcal{M}$ along its fiber at p. Then, the Drinfeld–Cherednik Theorem gives an isomorphism of formal schemes over W

$$\widehat{\mathcal{M}}_p \ \overset{\sim}{\longrightarrow}\ H'(\mathbb{Q})\backslash\Big(\ \hat{\Omega}^\bullet_W \times H(\mathbb{A}_f^p)/K^p\ \Big) \simeq \Gamma'\backslash\hat{\Omega}^\bullet_W, \tag{1.16}$$

where $K = K_pK^p$ and $\Gamma' = H'(\mathbb{Q}) \cap H'(\mathbb{Q}_p)K^p$. The special fiber $\hat{\Omega}_p \times_{\mathbb{Z}_p} \mathbb{F}_p$ of $\hat{\Omega}_p$ is a union of projective lines $\mathbb{P}_{[\Lambda]}$ indexed by the vertices $[\Lambda]$ of the building $\mathcal{B}$ of $\mathrm{PGL}_2(\mathbb{Q}_p)$. Here $[\Lambda]$ is the homothety class of the $\mathbb{Z}_p$–lattice Λ in $\mathbb{Q}_p^2$. The crossing points of these lines are double points indexed by the edges of $\mathcal{B}$, and the action of $\mathrm{PGL}_2(\mathbb{Q}_p)$ on components is compatible with its action on $\mathcal{B}$. Thus, the dual graph of the special fiber of $\widehat{\mathcal{M}}_p$ is isomorphic to $\Gamma'\backslash\mathcal{B}^\bullet$, where $\mathcal{B}^\bullet = \mathcal{B} \times \mathbb{Z}$.

2. Arithmetic Chow groups

The modular forms of interest in these notes will take values in the arithmetic Chow groups of $\mathcal{M}$. We will use the version of these groups with real coefficients defined by Bost, [7], section 5.5. Let $\widehat{Z}^1(\mathcal{M})$ be the real vector space spanned by pairs (Z, g), where Z is a real linear combination of Weil divisors on $\mathcal{M}$ and g is a Green function for Z. In particular, if Z is a Weil divisor, g is a C^∞ function on $\mathcal{M}(\mathbb{C}) \setminus Z(\mathbb{C})$, with a logarithmic singularity along the $Z(\mathbb{C})$, and satisfies the Green equation

$$dd^c g + \delta_Z = [\omega_Z], \tag{2.1}$$

where ω_Z is a smooth $(1,1)$-form on $\mathcal{M}(\mathbb{C})$, and $[\omega_Z]$ is the corresponding current. If $Z = \sum_i c_i Z_i$ is a real linear combination of Weil divisors, then $g = \sum_i c_i g_i$ is a real linear combination of such Green functions. By construction, $\alpha \cdot (Z, g) = (\alpha Z, \alpha g)$ for $\alpha \in \mathbb{R}$. The first arithmetic Chow group, with real coefficients, $\widehat{\mathrm{CH}}^1_{\mathbb{R}}(\mathcal{M})$, is then the quotient of $\widehat{Z}^1(\mathcal{M})$ by the subspace spanned by pairs $\widehat{\mathrm{div}}(f) = (\mathrm{div}(f), -\log|f|^2)$ where f is a rational function on $\mathcal{M}$, and $\mathrm{div}(f)$ is its divisor. Finally, we let

$$\widehat{\mathrm{CH}}^1(\mathcal{M}) = \widehat{\mathrm{CH}}^1_{\mathbb{R}}(\mathcal{M}) \otimes_{\mathbb{R}} \mathbb{C}. \tag{2.2}$$

Note that restriction to the generic fiber yields a degree map

$$\deg_{\mathbb{Q}} : \widehat{\mathrm{CH}}^1(\mathcal{M}) \longrightarrow \mathrm{CH}^1(\mathrm{M}) \otimes \mathbb{C} \overset{\deg}{\longrightarrow} \mathbb{C}. \tag{2.3}$$

The group $\widehat{\mathrm{CH}}^2(\mathcal{M})$ is defined analogously, and the arithmetic degree map yields an isomorphism

$$\widehat{\deg} : \widehat{\mathrm{CH}}^2(\mathcal{M}) \xrightarrow{\sim} \mathbb{C}. \tag{2.4}$$

Moreover, there is a symmetric $\mathbb{R}$–bilinear height pairing[2]

$$\langle\ ,\ \rangle : \widehat{\mathrm{CH}}^1_{\mathbb{R}}(\mathcal{M}) \times \widehat{\mathrm{CH}}^1_{\mathbb{R}}(\mathcal{M}) \longrightarrow \mathbb{R}. \tag{2.5}$$

According to the index Theorem, cf. [**7**], Theorem 5.5, this pairing is nondegenerate and has signature $(+,-,-,\dots)$. We extend it to an Hermitian pairing on $\widehat{\mathrm{CH}}^1(\mathcal{M})$, conjugate linear in the second argument.

Let $\mathcal{A}$ be the universal abelian scheme over $\mathcal{M}$ with zero section ϵ, and let

$$\omega = \epsilon^*\Omega^2_{\mathcal{A}/\mathcal{M}} \tag{2.6}$$

be the Hodge line bundle on $\mathcal{M}$. We define the natural metric on ω by letting

$$||s_z||^2_{\mathrm{nat}} = \left| \left(\frac{i}{2\pi}\right)^2 \int_{A_z} s_z \wedge \bar{s}_z \right|. \tag{2.7}$$

for any section $s : z \mapsto s_z$, where, for $z \in \mathcal{M}(\mathbb{C})$, A_z is the associated abelian variety. As in section 3 of [**49**], we set

$$||\ ||^2 = e^{-2C}\,||\ ||^2_{\mathrm{nat}} \tag{2.8}$$

where $2C = \log(4\pi) + \gamma$, where γ is Euler's constant. The reason for this choice of normalization is explained in the introduction to [**49**]. The pair $\widehat{\omega} = (\omega, ||\ ||)$ defines an element of $\widehat{\mathrm{Pic}}(\mathcal{M})$, the group of metrized line bundles on $\mathcal{M}$. We write $\widehat{\omega}$ for the image of this class in $\widehat{\mathrm{CH}}^1_{\mathbb{R}}(\mathcal{M})$ under the natural map $\widehat{\mathrm{Pic}}(\mathcal{M}) \to \widehat{\mathrm{CH}}^1_{\mathbb{R}}(\mathcal{M})$.

The pullback to D of the restriction of ω to $\mathcal{M}(\mathbb{C})$ is trivialized by the section α defined as follows. For $z \in D$, let α_z be the holomorphic 2–form on $A_z = \mathbb{C}^2/L_z$ given by

$$\alpha_z = D(B)^{-1}(2\pi i)^2\, dw_1 \wedge dw_2 \tag{2.9}$$

where w_1 and w_2 are the coordinates on the right side of (1.6). Then

$$\omega_{\mathbb{C}} = \big[\, \Gamma\backslash(D \times \mathbb{C}) \,\big], \tag{2.10}$$

where the action of $\gamma \in \Gamma$ is given by

$$\gamma : (z, \zeta) \longmapsto (\gamma(z), (cz+d)^2\,\zeta). \tag{2.11}$$

[2] Here the symmetry must still be checked in the case of a stack, cf. section 4 of [**49**].

Thus, on $\mathcal{M}(\mathbb{C})$, ω is isomorphic to the canonical bundle $\Omega^1_{\mathcal{M}(\mathbb{C})}$, under the map which sends α_z to dz. The resulting metric on $\Omega^1_{\mathcal{M}(\mathbb{C})}$ is

$$\begin{aligned} ||dz||^2 &= ||\alpha_z||^2 = e^{-2C}\left|\left(\frac{i}{2\pi}\right)^2\int_{A_z}\alpha_z\wedge\bar{\alpha}_z\right| \\ &= e^{-2C}\,(2\pi)^{-2}\,(2\pi)^4\,D(B)^{-2}\,\mathrm{vol.}\,(\mathrm{M}_2(\mathbb{R})/\mathrm{O}_\mathrm{B})\cdot\mathrm{Im(z)}^2 \\ &= e^{-2C}\,(2\pi)^2\cdot\mathrm{Im}(z)^2. \qquad (2.12)\end{aligned}$$

In [**49**], it was conjectured that

$$(2.13)\qquad \langle\hat{\omega},\hat{\omega}\rangle \overset{??}{\to} = \zeta_{D(B)}(-1)\left[\,2\frac{\zeta'(-1)}{\zeta(-1)}+1-2C-\sum_{p|D(B)}\frac{p\log(p)}{p-1}\,\right],$$

where

$$(2.14)\qquad \zeta_{D(B)}(s)=\zeta(s)\cdot\prod_{p|D(B)}(1-p^{-s}),$$

and $2C=\log(4\pi)+\gamma$, as before. In the case $D(B)=1$, i.e., for a modular curve, the analogous value for $\langle\hat{\omega},\hat{\omega}\rangle$ was established, independently, by Bost, [**8**], and Kühn, [**53**], cf. the introduction to [**49**] for a further discussion of normalizations. It will be convenient to define a constant $\mathbf{c}$ by

$$(2.15)\qquad \frac{1}{2}\,\deg_{\mathbb{Q}}(\hat{\omega})\cdot\mathbf{c}:=\langle\hat{\omega},\hat{\omega}\rangle-\text{RHS of (2.13)}.$$

In particular, $\langle\hat{\omega},\hat{\omega}\rangle$ has the conjectured value if and only if $\mathbf{c}=0$. It seems likely that this conjecture can be proved using recent work of Bruinier, Burgos and Kühn, [**12**], on heights of curves on Hilbert modular surfaces. Their work uses an extended theory of arithmetic Chow groups, developed by Burgos, Kramer and Kühn, [**13**], which allows metrics with singularities of the type which arise on compactified Shimura varieties. In addition, they utilize results of Bruinier, [**10**], [**11**], concerning Borcherds forms. Very general conjectures about such arithmetic degrees and their connections with the equivariant arithmetic Riemann–Roch formula have been given by Maillot and Roessler, [**55**]. Some connections of arithmetic degrees with Fourier coefficients of derivatives of Eisenstein series are discussed in [**39**].

3. Special cycles, horizontal components

Just as the CM points on the modular curves are constructed as the points where the corresponding elliptic curves have additional endomorphisms, cycles on $\mathcal{M}$ can be defined by imposing additional endomorphisms as follows.

Definition 3.1 For a given (A,ι), the space of **special endomorphisms** of (A,ι) is

$$(3.1)\qquad V(A,\iota)=\{\,x\in\mathrm{End}_S(A)\;\mid\;\mathrm{tr}(x)=0\text{ and }x\circ\iota(b)=\iota(b)\circ x,\ \forall b\in O_B\,\}.$$

Over a connected base, this free $\mathbb{Z}$–module of finite rank has a $\mathbb{Z}$–valued quadratic form Q given by

$$(3.2)\qquad -x^2=Q(x)\cdot\mathrm{id_A}.$$

Definition 3.2 For a positive integer t, let $\mathcal{Z}(t)$ be the moduli stack over $\text{Spec}(\mathbb{Z})$ of triples (A, ι, x), where (A, ι) is as before and $x \in V(A, \iota)$ is a special endomorphism with $Q(x) = t$.

There is a natural morphism

$$(3.3) \qquad \mathcal{Z}(t) \longrightarrow \mathcal{M}, \qquad (A, \iota, x) \mapsto (A, \iota),$$

which is unramified, and, by a slight abuse of notation, we write $\mathcal{Z}(t)$ for the divisor on $\mathcal{M}$ determined by this morphism.

Over $\mathbb{C}$, such a triple (A, ι, x) is an abelian surface with an action of $O_B \otimes_{\mathbb{Z}} \mathbb{Z}[\sqrt{-t}]$, i.e., with additional 'complex multiplication' by the order $\mathbb{Z}[\sqrt{-t}]$ in the imaginary quadratic field $\boldsymbol{k}_t = \mathbb{Q}(\sqrt{-t})$. Suppose that $A \simeq A_z = \mathbb{C}^2/L_z$, so that, by (1.8), the tangent space $T_e(A_z)$ is given as $B_{\mathbb{R}} \simeq \mathbb{C}^2 = T_e(A_z)$. Since the lift $\tilde{x}_{\mathbb{R}}$ of x in the diagram

$$(3.4) \qquad \begin{array}{ccccc} B_{\mathbb{R}} & \simeq & \mathbb{C}^2 & \longrightarrow & A_z \\ \tilde{x}_{\mathbb{R}} = r(j_x) \downarrow & & \tilde{x} \downarrow & & x \downarrow \\ B_{\mathbb{R}} & \simeq & \mathbb{C}^2 & \longrightarrow & A_z \end{array}$$

commutes with the left action of O_B and carries O_B into itself, it is given by right multiplication $r(j_x)$ by an element $j_x \in O_B \cap V$ with $\nu(j_x) = t$. Since the map $\tilde{x}$ is holomorphic, it follows that $z \in D_x$, the fixed point set of j_x on D. To simplify notation, we will write x in place of j_x. Then, we find that

$$(3.5) \qquad \mathcal{Z}(t)(\mathbb{C}) = [\,\Gamma \backslash D_t\,],$$

where

$$(3.6) \qquad D_t = \coprod_{\substack{x \in O_B \cap V \\ Q(x) = t}} D_x.$$

In particular, $\mathcal{Z}(t)(\mathbb{C})$, which can be viewed as a set of CM points on the Shimura curve $M(\mathbb{C})$, is nonempty if and only if the imaginary quadratic field $\boldsymbol{k}_t$ embeds in B.

The horizontal part $\mathcal{Z}^{\mathrm{hor}}(t)$ of $\mathcal{Z}(t)$ is obtained by taking the closure in $\mathcal{M}$ of these CM–points, i.e.,

$$\mathcal{Z}^{\mathrm{hor}}(t) := \overline{\mathcal{Z}(t)_{\mathbb{Q}}}. \tag{3.7}$$

Finally, if we write $4t = n^2 d$ where $-d$ is the discriminant of the imaginary quadratic field $\mathbb{k}_t = \mathbb{Q}(\sqrt{-t})$, then the degree of the divisor $\mathcal{Z}(t)_{\mathbb{Q}}$ is given by

$$\deg_{\mathbb{Q}} \mathcal{Z}(t) = 2\delta(d, D(B))\, H_0(t, D(B)), \tag{3.8}$$

where

$$\delta(d, D(B)) = \prod_{p|D(B)} (1 - \chi_d(p)), \tag{3.9}$$

and

$$\begin{aligned} H_0(t, D(B)) &= \sum_{c|n} \frac{h(c^2 d)}{w(c^2 d)} \\ &= \frac{h(d)}{w(d)} \cdot \Bigg(\sum_{\substack{c|n \\ (c,D(B))=1}} c \prod_{\ell|c} (1 - \chi_d(\ell)\ell^{-1}) \Bigg). \end{aligned}$$

Here $h(c^2d)$ is the class number of the order O_{c^2d} in $\mathbb{k}_t$ of conductor c, $w(c^2d)$ is the number of roots of unity in O_{c^2d}, and χ_d is the Dirichlet character for the field $\mathbb{k}_t$. Note that, in this formula, we are counting points on the orbifold $[\Gamma\backslash D]$, so that each point $\mathrm{pr}(z)$, $z \in D$, is counted with multiplicity e_z^{-1} where $e_z = |\Gamma_z|$ is the order of the stabilizer of z in Γ. For example, suppose that $z \in D_x$ for $x \in V \cap O_B$ with $Q(x) = t$. Then, since $\mathbb{Z}[x] \simeq \mathbb{Z}[\sqrt{-t}]$ is an order of conductor n, $\mathbb{Q}[x] \cap O_B \supset \mathbb{Z}[x]$ is an order of conductor c for some $c \mid n$, and $e_z = w(c^2d) = |\, (\mathbb{Q}[x] \cap O_B)^\times|$.

4. Special cycles, vertical components

In this section, we describe the vertical components of the special cycles $\mathcal{Z}(t)$ in some detail, following [**46**]. In the end, we obtain a 'p-adic uniformization', (4.27), quite analogous to the expression (3.5) for $\mathcal{Z}(t)(\mathbb{C})$. We first review the construction of the p-adic uniformization isomorphism (1.16).

Let $\mathbb{B}$ be the division quaternion algebra over $\mathbb{Q}_p$ and let O_B be its maximal order. Let $\mathbb{Z}_{p^2}$ be the ring of integers in the unramified quadratic extension $\mathbb{Q}_{p^2}$ of $\mathbb{Q}_p$. Fixing an embedding of $\mathbb{Q}_{p^2}$ into $\mathbb{B}$, we have $\mathbb{Z}_{p^2} \hookrightarrow O_B$, and we can choose an element $\Pi \in O_B$ with $\Pi^2 = p$ such that $\Pi a = a^\sigma \Pi$, for all $a \in \mathbb{Q}_{p^2}$, where σ is the generator of the Galois group of $\mathbb{Q}_{p^2}$ over $\mathbb{Q}_p$. Then $O_B = \mathbb{Z}_{p^2}[\Pi]$.

Recall that $W = W(\bar{\mathbb{F}}_p)$. Let Nilp be the category of W–schemes S such that p is locally nilpotent in O_S, and for $S \in$ Nilp, let $\bar{S} = S \times_W \bar{\mathbb{F}}_p$.

A ***special formal*** (s.f.) O_B–module over a W scheme S is a p-divisible formal group X over S of dimension 2 and height 4 with an action $\iota : O_B \hookrightarrow \mathrm{End}_S(X)$. The Lie algebra Lie(X), which is a $\mathbb{Z}_{p^2} \otimes O_S$–module, is required to be free of rank 1 locally on S.

Fix a s.f. O_B–module $\mathbb{X}$ over $\mathrm{Spec}(\bar{\mathbb{F}}_p)$. Such a module is unique up to O_B–linear isogeny, and

$$\mathrm{End}^0_{O_B}(\mathbb{X}) \simeq M_2(\mathbb{Q}_p). \tag{4.1}$$

We fix such an isomorphism. Consider the functor

$$\mathcal{D}^\bullet : \mathrm{Nilp} \to \mathrm{Sets} \tag{4.2}$$

which associates to each $S \in \mathrm{Nilp}$ the set of isomorphism classes of pairs (X, ρ), where X is a s.f. O_B–module over S and

$$\rho : \mathbb{X} \times_{\mathrm{Spec}(\bar{\mathbb{F}}_p)} \bar{S} \longrightarrow X \times_S \bar{S} \tag{4.3}$$

is a quasi-isogeny[3]. The group $\mathrm{GL}_2(\mathbb{Q}_p)$ acts on $\mathcal{D}^\bullet$ by

$$g : (X, \rho) \mapsto (X, \rho \circ g^{-1}). \tag{4.4}$$

There is a decomposition

$$\mathcal{D}^\bullet = \coprod_i \mathcal{D}^i, \tag{4.5}$$

where the isomorphism class of (X, ρ) lies in $\mathcal{D}^i(S)$ if ρ has height i. The action of $g \in \mathrm{GL}_2(\mathbb{Q}_p)$ carries $\mathcal{D}^i$ to $\mathcal{D}^{i-\mathrm{ord}\det(g)}$. Drinfeld showed that $\mathcal{D}^\bullet$ is representable by a formal scheme, which we also denote by $\mathcal{D}^\bullet$, and that there is an isomorphism of formal schemes

$$\mathcal{D}^\bullet \xrightarrow{\sim} \hat{\Omega}^\bullet_W \tag{4.6}$$

which is equivariant for the action of $\mathrm{GL}_2(\mathbb{Q}_p)$.

Similarly, using the notation, B, O_B, H, etc., of section 1, the formal scheme $\widehat{\mathcal{M}}_p$ represents the functor on Nilp which associates to S the set of isomorphism classes of triples $(A, \iota, \bar{\eta})$ where A is an abelian scheme of relative dimension 2 over S, up to prime to p–isogeny, with an action $O_B \hookrightarrow \mathrm{End}_S(A)$ satisfying the determinant condition (1.7), and $\bar{\eta}$ is K^p–equivalence class of O_B–equivariant isomorphisms

$$\eta : \hat{V}^p(A) \xrightarrow{\sim} B(\mathbb{A}^p_f), \tag{4.7}$$

where

$$\hat{V}^p(A) \xrightarrow{\sim} \prod_{\ell \neq p} T_\ell(A) \otimes \mathbb{Q} \tag{4.8}$$

is the rational Tate module of A. Two isomorphisms η and η' are equivalent iff there exists an element $k \in K^p$ such that $\eta' = r(k) \circ \eta$.

Fix a base point $(A_0, \iota_0, \bar{\eta}_0)$ in $\widehat{\mathcal{M}}_p(\bar{\mathbb{F}}_p)$, and let

$$\widehat{\mathcal{M}}_p^\sim : \mathrm{Nilp} \longrightarrow \mathrm{Sets} \tag{4.9}$$

be the functor which associates to S the set of isomorphism classes of tuples $(A, \iota, \bar{\eta}, \psi)$, where $(A, \iota, \bar{\eta})$ is as before, and

$$\psi : A_0 \times_{\mathrm{Spec}(\bar{\mathbb{F}}_p)} \bar{S} \longrightarrow A \times_S \bar{S} \tag{4.10}$$

is an O_B–equivariant p-primary isogeny.

---

[3] This means that, locally on S, there is an integer r such that $p^r\rho$ is an isogeny.

To relate the functors just defined, we let

$$(4.11)\qquad B' = \mathrm{End}^0(A_0, \iota_0), \qquad \text{and} \qquad H' = (B')^\times,$$

and fix $\eta_0 \in \bar{\eta}_0$. Since the endomorphisms of $\hat{V}^p(A_0)$ coming from $B'(\mathbb{A}_f^p) = B' \otimes_{\mathbb{Q}} \hat{\mathbb{Z}}^p$ commute with O_B, the corresponding endomorphisms of $B(\mathbb{A}_f^p)$, obtaine via η_0, are given by right multiplications by elements of $B(\mathbb{A}_f^p)$. Thus we obtain identifications, as in (1.15),

$$(4.12)\quad B'(\mathbb{A}_f^p) \xrightarrow{\sim} B(\mathbb{A}_f^p)^{\mathrm{op}}, \qquad \text{and} \qquad H'(\mathbb{A}_f^p) \xrightarrow{\sim} H(\mathbb{A}_f^p)^{\mathrm{op}},$$

where the order of multiplication is reversed in B^{op} and H^{op}. We also identify $\mathbb{B}$ with B_p and take $\mathbb{X} = A_0(p)$, the p-divisible group of A_0, with the action of $\mathcal{O}_B = O_B \otimes_{\mathbb{Z}} \mathbb{Z}_p$ coming from ι_0. Note that we also obtain an identification $B'_p \simeq \mathrm{GL}_2(\mathbb{Q}_p)$, via (4.1).

Once these identifications have been made, there is a natural isomorphism

$$(4.13)\qquad \widehat{\mathcal{M}}_p^{\sim} \xrightarrow{\sim} \mathcal{D}^\bullet \times H(\mathbb{A}_f^p)/K^p$$

defined as follows. To a given $(A, \iota, \bar{\eta}, \psi)$ over S, we associate:

$$\begin{aligned} X &= A(p) = \text{ the p-divisible group of } A, \\ \iota &= \text{ the action of } \mathcal{O}_B = O_B \otimes_{\mathbb{Z}} \mathbb{Z}_p \text{ on } A(p), \\ \rho &= \rho(\psi) = \text{the quasi-isogeny} \\ (4.14)\qquad \rho(\psi) : \mathbb{X} \times_{\mathrm{Spec}(\bar{\mathbb{F}}_p)} \bar{S} &\longrightarrow A(p) \times_S \bar{S} \end{aligned}$$

determined by ψ,

so that (X, ρ) defines an element of $\mathcal{D}^\bullet$. For $\eta \in \bar{\eta}$, there is also a diagram

$$(4.15)\qquad \begin{array}{ccc} \hat{V}^p(A_0) & \overset{\eta_0}{\underset{\sim}{\longrightarrow}} & B(\mathbb{A}_f^p) \\ \psi_* \downarrow & & \downarrow r(g) \\ \hat{V}^p(A) & \overset{\eta}{\underset{\sim}{\longrightarrow}} & B(\mathbb{A}_f^p) \end{array}$$

where $r(g)$ denotes right multiplication[4] by an element $g \in H(\mathbb{A}_f^p)$. The coset gK^p is then determined by the equivalence class $\bar{\eta}$, and the isomorphism (4.13) sends $(A, \iota, \bar{\eta}, \psi)$ to $((X, \rho), gK^p)$.

The Drinfeld-Cherednik Theorem says that, by passing to the quotient under the action of $H'(\mathbb{Q})$, we have

$$(4.16)\qquad \begin{array}{ccc} \widehat{\mathcal{M}}_p^{\sim} & \xrightarrow{\sim} & \mathcal{D}^\bullet \times H(\mathbb{A}_f^p)/K^p \\ \downarrow & & \downarrow \\ \widehat{\mathcal{M}}_p & \xrightarrow{\sim} & H'(\mathbb{Q})\backslash\left(\mathcal{D}^\bullet \times H(\mathbb{A}_f^p)/K^p\right). \end{array}$$

Via the isomorphism $\mathcal{D}^\bullet \overset{\sim}{\to} \to \hat{\Omega}_W^\bullet$ of (4.6), this yields (1.16).

---

[4]This is the reason for the "op" in the isomorphism (4.12), since we ultimately want a left action of $H'(\mathbb{A}_f^p)$.

We can now describe the formal scheme determined by the cycle $\mathcal{Z}(t)$, following section 8 of [**46**]. Let[5] $\widehat{\mathcal{C}}_p(t)$ be the base change to W of the formal completion of $\mathcal{Z}(t)$ along its fiber at p, and let $\widehat{\mathcal{C}}_p^{\sim}(t)$ be the fiber product:

$$\begin{array}{ccc} \widehat{\mathcal{C}}_p^{\sim}(t) & \longrightarrow & \widehat{\mathcal{M}}_p^{\sim} \\ \downarrow & & \downarrow \\ \widehat{\mathcal{C}}_p(t) & \longrightarrow & \widehat{\mathcal{M}}_p. \end{array} \tag{4.17}$$

A point of $\widehat{\mathcal{C}}_p(t)$ corresponds to a collection $(A, \iota, \bar{\eta}, x)$, where $x \in V(A, \iota)$ is a special endomorphism with $Q(x) = t$. In addition, since we are now working with A's up to prime to p isogeny, we also require that the endomorphism $\eta_*(x)$ of $B(\mathbb{A}_f^p)$, obtained by transfering, via η, the endomorphism of $\hat{V}^p(A)$ induced by x, is given by right multiplication by an element $j_f^p(x) \in V(\mathbb{A}_f^p) \cap \hat{O}_B$. This condition does not depend on the choice of η in the K^p–equivalence class $\bar{\eta}$.

Next, we would like to determine the image of $\widehat{\mathcal{C}}_p^{\sim}(t)$ in $\mathcal{D}^\bullet \times H(\mathbb{A}_f^p)/K^p$ under the isomorphism in the top line of (4.16). Let

$$V' = \{\, x \in B' \mid \operatorname{tr}(x) = 0 \,\} = V(A_0, \iota_0) \otimes_{\mathbb{Z}} \mathbb{Q}, \tag{4.18}$$

where $V(A_0, \iota_0)$ is the space of special endomorphisms of (A_0, ι_0). We again write Q for the quadratic form on this space. For a point of $\widehat{\mathcal{C}}_p^{\sim}(t)$ associated to a collection $(A, \iota, \bar{\eta}, x, \psi)$, let $((X, \rho), gK^p)$ be the corresponding point in $\mathcal{D}^\bullet \times H(\mathbb{A}_f^p)/K^p$. The special endomorphism $x \in V(A, \iota)$ induces an endomorphism of $A \times_S \bar{S}$ and, thus, via the isogeny ψ, there is an associated endomorphism $\psi^*(x) \in V'(\mathbb{Q})$. This element satisfies two compatibility conditions with the other data:

(i) $\psi^*(x)$ determines an element $j = j(x) \in V'(\mathbb{Q}_p) = \operatorname{End}^0_{O_B}(\mathbb{X})$. By construction, this element has the property that the corresponding element

$$\rho \circ j \circ \rho^{-1} \in \operatorname{End}^0_{O_B}(X \times_S \bar{S}) \tag{4.19}$$

is, in fact, the restriction of an element of $\operatorname{End}_{O_B}(X)$, viz. the endomorphism induced by original x. Said another way, j defines an endomorphism of the reduction $X \times_S \bar{S}$ which lifts to an endomorphism of X.

(ii) Via the diagram (4.15),

$$g^{-1}\, \psi^*(x)\, g \in V(\mathbb{A}_f^p) \cap \hat{O}_B^p. \tag{4.20}$$

Here we are slightly abusing notation and, in effect, are identifying $\psi^*(x) \in V'(\mathbb{Q})$ with an element of $V(\mathbb{A}_f^p)$ obtained via the identification

$$V'(\mathbb{A}_f^p) \overset{\sim}{\longrightarrow} V(\mathbb{A}_f^p) \tag{4.21}$$

coming from (4.12).

Condition (i) motivates the following basic definition, [**46**], Definition 2.1.

Definition 4.1 *For a special endomorphism $j \in V'(\mathbb{Q}_p)$ of $\mathbb{X}$, let $Z^\bullet(j)$ be the closed formal subscheme of $\mathcal{D}^\bullet$ consisting of the points (X, ρ) such that $\rho \circ j \circ \rho^{-1}$ lifts to an endomorphism of X.*

We will also write $Z(j) \subset \mathcal{D}$ for the subschemes where the height of the quasi-isogeny ρ is 0. We will give a more detailed description of $Z(j)$ in a moment.

---

[5] We change to $\widehat{\mathcal{C}}$ to avoid confusion with the notation $\widehat{\mathcal{Z}}(t, v)$ used for classes in the arithmetic Chow group.

As explained above and in more detail in [**46**], section 8, there is a map

$$\widehat{\mathcal{C}}_p^{\sim}(t) \hookrightarrow V'(\mathbb{Q}) \times \mathcal{D}^{\bullet} \times H(\mathbb{A}_f^p)/K^p \tag{4.22}$$

whose image is the set:

$$(\star) := \left\{ (y,(X,\rho),gK^p) \;\middle|\; \begin{array}{c} (i) \quad Q(y)=t \\ (ii) \quad (X,\rho) \in Z^{\bullet}(j(y)) \\ (iii) \quad y \in g\,(V(\mathbb{A}_f^p) \cap \hat{O}_B^p)\,g^{-1} \end{array} \right\}. \tag{4.23}$$

Taking the quotient by the group $H'(\mathbb{Q})$, we obtain the following p-adic uniformization of the special cycle:

Proposition 4.2 The construction aboves yields isomorphisms:

$$\begin{array}{ccc} \widehat{\mathcal{C}}_p(t) & \xrightarrow{\sim} & H'(\mathbb{Q})\backslash(\star) \\ \downarrow & & \downarrow \\ \widehat{\mathcal{M}}_p & \xrightarrow{\sim} & H'(\mathbb{Q})\backslash\Big(\mathcal{D}^{\bullet} \times H(\mathbb{A}_f^p)/K^p\Big) \end{array}$$

of formal schemes over W.

Remark 4.3. In fact, in this discussion, the compact open subgroup K^p giving the level structure away from p can be arbitrary. In the case of interest, where $K^p = (\hat{O}_B^p)^{\times}$, the last diagram can be simplified as follows. Let

$$L' = V'(\mathbb{Q}) \cap \big(B'(\mathbb{Q}_p) \times \hat{O}_B^p\big), \tag{4.24}$$

so that L' is a $\mathbb{Z}[p^{-1}]$-lattice in $V'(\mathbb{Q})$. Also, as in section 1, let

$$\Gamma' = H'(\mathbb{Q}) \cap \big(H'(\mathbb{Q}_p) \times K^p\big), \tag{4.25}$$

so that (the projection of) Γ' is an arithmetic subgroup of $H'(\mathbb{Q}_p) \simeq \mathrm{GL}_2(\mathbb{Q}_p)$. Finally, let

$$\mathcal{D}_t^{\bullet} = \coprod_{\substack{y \in L' \\ Q(y)=t}} Z^{\bullet}(j(y)). \tag{4.26}$$

Then,

$$\begin{array}{ccc} \widehat{\mathcal{C}}_p(t) & \xrightarrow{\sim} & [\Gamma'\backslash\mathcal{D}_t^{\bullet}] \\ \downarrow & & \downarrow \\ \widehat{\mathcal{M}}_p & \xrightarrow{\sim} & [\Gamma'\backslash\mathcal{D}^{\bullet}]. \end{array} \tag{4.27}$$

Of course, we should now view these quotients as orbifolds, and, in fact, should formulate the discussion above in terms of stacks.

To complete the picture of the vertical components of our cycle $\mathcal{Z}(t)$, we need a more precise description of the formal schemes $Z^\bullet(j)$ for $j \in V'(\mathbb{Q}_p)$, as obtained in the first four sections of [**46**]. Note that we are using the quadratic form $Q(j) = \det(j)$, whereas, in [**46**], the quadratic form $q(j) = j^2 = -Q(j)$ was used. It is most convenient to give this description in the space $\hat{\Omega}^\bullet_W = \hat{\Omega}_W \times \mathbb{Z}$. In fact, we will just consider $Z(j) \subset \hat{\Omega}_W$, and we will assume that $p \neq 2$. The results for $p = 2$, which are very similar, are described in section 11 of [**49**]. Recall that the special fiber of $\hat{\Omega}_W$ is a union of projective lines $\mathbb{P}_{[\Lambda]}$, indexed by the vertices $[\Lambda]$ of the building $\mathcal{B}$ of $\mathrm{PGL}_2(\mathbb{Q}_p)$.

The first result, proved in section 2 of [**46**], describes the support of $Z(j)$ in terms of the building.

Proposition 4.4
(i)

$$\mathbb{P}_{[\Lambda]} \cap Z(j) \neq \emptyset \iff j(\Lambda) \subset \Lambda.$$

In particular, if $Z(j) \neq \emptyset$*, then* $\mathrm{ord}_p(Q(j)) \geq 0$.
(ii)

$$j(\Lambda) \subset \Lambda \quad \iff \quad d([\Lambda], \mathcal{B}^j) \leq \frac{1}{2} \cdot \mathrm{ord}_p(Q(j)).$$

Here $\mathcal{B}^j$ *is the fixed point set of* j *on* $\mathcal{B}$*, and* $d(x,y)$ *is the distance between the points* x *and* $y \in \mathcal{B}$.

Recall that the distance function on the building $\mathcal{B}$ is $\mathrm{PGL}_2(\mathbb{Q}_p)$–invariant and gives each edge length 1.

In effect, if we write $Q(j) = \epsilon p^\alpha$, for $\epsilon \in \mathbb{Z}_p^\times$, then the support of $Z(j)$ lies in the set of $\mathbb{P}_{[\Lambda]}$'s indexed by vertices of $\mathcal{B}$ in the 'tube' $\mathcal{T}(j)$ of radius $\frac{\alpha}{2}$ around the fixed point set $\mathcal{B}^j$ of j.

Next, the following observation of Genestier is essential, [**46**], Theorem 3.1: Write $Q(j) = \epsilon p^\alpha$. Then

$$Z(j) = \begin{cases} (\hat{\Omega}_W)^j & \text{if } \alpha = 0, \\ (\hat{\Omega}_W)^{1+j} & \text{if } \alpha > 0. \end{cases} \tag{4.28}$$

where $(\hat{\Omega}_W)^x$ denotes the fixed point set of the element $x \in \mathrm{GL}_2(\mathbb{Q}_p)$ acting on $\hat{\Omega}_W$, via its projection to $\mathrm{PGL}_2(\mathbb{Q}_p)$. Using this fact, one can obtain local equations for $Z(j)$ in terms of the local coordinates on $\hat{\Omega}_W$, cf. [**46**], section 3. Recall that there are standard coordinate neighborhoods associated to each vertex $[\Lambda]$ and each edge $[\Lambda_0, \Lambda_1]$ of $\mathcal{B}$, cf. [**46**], section 1. By Proposition 4.4, it suffices to compute in the neighborhoods of those $\mathbb{P}_{[\Lambda]}$'s which meet the support of $Z(j)$. For the full result, see Propositions 3.2 and 3.3 in [**46**]. In particular, embedded components can occur. But since these turn out to be negligible, e.g., for intersection theory, cf. section 4 of [**46**], we can omit them from $Z(j)$ to obtain the divisor $Z(j)^{\mathrm{pure}}$ which has the following description.

Proposition 4.5
Write $Q(j) = \epsilon p^{\alpha}$, *and let*

$$\mu_{[\Lambda]}(j) = \max\{\, 0,\ \frac{\alpha}{2} - d([\Lambda], \mathcal{B}^j)\,\}.$$

(i) If α *is even and* $-\epsilon \in \mathbb{Z}_p^{\times,2}$, *then*

$$Z(j)^{\mathrm{pure}} = \sum_{[\Lambda]} \mu_{[\Lambda]}(j) \cdot \mathbb{P}_{[\Lambda]}.$$

(ii) If α *is even and* $-\epsilon \notin \mathbb{Z}_p^{\times,2}$, *then*

$$Z(j)^{\mathrm{pure}} = Z(j)^h + \sum_{[\Lambda]} \mu_{[\Lambda]}(j) \cdot \mathbb{P}_{[\Lambda]},$$

where the horizontal part $Z(j)^h$ *is the disjoint union of two divisors projecting isomorphically to* Spf(W) *and meeting the special fiber in 'ordinary special' points of* $\mathbb{P}_{[\Lambda(j)]}$, *where* $[\Lambda(j)]$ *is the unique vertex* $\mathcal{B}^j$ *of* $\mathcal{B}$ *fixed by* j.

(iii) If α *is odd, then*

$$Z(j)^{\mathrm{pure}} = Z(j)^h + \sum_{[\Lambda]} \mu_{[\Lambda]}(j) \cdot \mathbb{P}_{[\Lambda]},$$

where the horizontal divisor $Z(j)^h$ *is* Spf(W′) *where* W' *is the ring of integers in a ramified quadratic extension of* $W \otimes_{\mathbb{Z}} \mathbb{Q}$ *and* $Z(j)^h$ *meets the special fiber in the double point* $\mathrm{pt}_{\Delta(j)}$ *where* $\Delta(j)$ *is the edge of* $\mathcal{B}$ *containing the unique fixed point* $\mathcal{B}^j$ *of* j.
Thus $Z(j)^{\mathrm{pure}}$ *is a sum with multiplicities of regular one dimensional formal schemes.*

In case (i), the split case, $\mathbb{Q}_p(j) \simeq \mathbb{Q}_p \oplus \mathbb{Q}_p$, the element j lies in a split torus A in $\mathrm{GL}_2(\mathbb{Q}_p)$, and $\mathcal{B}^j$ is the corresponding apartment in $\mathcal{B}$. More concretely, if e_0 and e_1 are eigenvectors of j giving a basis of $\mathbb{Q}_p^2$, then $\mathcal{B}^j$ is the geodesic arc connecting the vertices $[\mathbb{Z}_p e_0 \oplus p^r \mathbb{Z}_p e_1]$, for $r \in \mathbb{Z}$. The $\mathbb{P}_{[\Lambda]}$'s for $[\Lambda] \in \mathcal{B}^j$ have multiplicity $\frac{\alpha}{2}$ in $Z(j)$, and the multiplicity decreases linearly with the distance from $\mathcal{B}^j$. The cycle $Z(j)$ is infinite and there are no horizontal components.

In case (ii), the inert case, $\mathbb{Q}_p(j)$ is an unramified quadratic extension of $\mathbb{Q}_p$, the element j lies in the corresponding nonsplit Cartan subgroup of $\mathrm{GL}_2(\mathbb{Q}_p)$ and $\mathcal{B}^j = [\Lambda(j)]$ is a single vertex. The corresponding $\mathbb{P}_{[\Lambda(j)]}$ occurs with multiplicity $\frac{\alpha}{2}$ in $Z(j)$, and the multiplicity of the vertical components $\mathbb{P}_{[\Lambda]}$ decreases linearly with the distance $d([\Lambda], [\Lambda(j)])$.

Finally, in case (iii), the ramified case, $\mathbb{Q}_p(j)$ is a ramified quadratic extension of $\mathbb{Q}_p$, the element j lies in the corresponding nonsplit Cartan subgroup of $\mathrm{GL}_2(\mathbb{Q}_p)$ and $\mathcal{B}^j$ is the midpoint of a unique edge $\Delta(j) = [\Lambda_0, \Lambda_1]$. The vertical components $\mathbb{P}_{[\Lambda_0]}$ and $\mathbb{P}_{[\Lambda_1]}$ occur with multiplicity $\frac{\alpha-1}{2}$ in $Z(j)$, and, again, the multiplicity of the vertical components $\mathbb{P}_{[\Lambda]}$ decreases linearly with the distance $d([\Lambda], \mathcal{B}^j)$, which is now a half-integer.

This description of the $Z(j)$'s, together with the p-adic unformization of Proposition 4.2, gives a fairly complete picture of the vertical components of the cycles $\mathcal{Z}(t)$ in the fibers $\mathcal{M}_p$, $p \mid D(B)$, of bad reduction. Several interesting features are evident.

For example, the following result gives a criterion for the occurrence of such components.

Proposition 4.6

For $p \mid D(B)$, the cycle $\mathcal{Z}(t)$ contains components of the fiber $\mathcal{M}_p$ of bad reduction if and only if $\text{ord}_p(t) \geq 2$, and no prime $\ell \neq p$ with $\ell \mid D(B)$ is split in $\boldsymbol{k}_t$.

Note that the condition amounts to (i) $\text{ord}_p(t) \geq 2$, and (ii) the field $\boldsymbol{k}_t$ embeds into $B^{(p)}$

For example, if more than one prime $p \mid D(B)$ splits in the quadratic field $\boldsymbol{k}_t = \mathbb{Q}(\sqrt{-t})$, then $\mathcal{Z}(t)$ is empty. If $p \mid D(B)$ splits in $\boldsymbol{k}_t$, and all other primes $\ell \mid D(B)$ are not split in $\boldsymbol{k}_t$, then the generic fiber $\mathcal{Z}(t)_{\mathbb{Q}}$ is empty, and $\mathcal{Z}(t)$ is a vertical cycle in the fiber at p, provided $\text{ord}_p(t) \geq 2$. In general, if $p \mid D(B)$, and $\boldsymbol{k}_t$ embeds in $B^{(p)}$, then the vertical component in $\mathcal{M}_p$ of the cycle $\mathcal{Z}(p^{2r}t)$ grows as r goes to infinity, while the horizontal part does not change. Indeed, if we change t to $p^{2r}t$, then, by (3.8), $\deg_{\mathbb{Q}} \mathcal{Z}(p^{2r}t) = \deg_{\mathbb{Q}} \mathcal{Z}(t)$, while both the radius of the tube $\mathcal{T}(p^r j)$ and the multiplicity function $\mu_{[\Lambda]}(p^r j)$ increase.

Analogues of these results about cycles defined by special endomorphisms are obtained in [**45**] and [**47**] for Hilbert–Blumenthal varieties and Siegel modular varieties of genus 2 respectively.

5. Green functions

To obtain classes in the arithmetic Chow group $\widehat{\text{CH}}^1_{\mathbb{R}}(\mathcal{M})$ from the $\mathcal{Z}(t)$'s, it is necessary to equip them with Green functions. These are defined as follows; see [**37**] for more details. For $x \in V(\mathbb{R})$, with $Q(x) \neq 0$, let

$$D_x = \{\, z \in D \mid (x, w(z)) = 0 \,\}. \tag{5.1}$$

Here $w(z) \in V(\mathbb{C})$ is any vector with image z in $\mathbb{P}(V(\mathbb{C}))$. The set D_x consists of two points if $Q(x) > 0$, and is empty if $Q(x) < 0$. By (3.5) and (3.6), we have

$$\mathcal{Z}(t)(\mathbb{C}) = \sum_{\substack{x \in O_B \cap V \\ Q(x)=t \\ \text{mod } \Gamma}} \text{pr}(D_x) \tag{5.2}$$

where $\text{pr} : D \to \Gamma\backslash D$. For $x \in V(\mathbb{R})$, with $Q(x) \neq 0$, and $z \in D$, let

$$R(x, z) = |(x, w(z))|^2 |(w(z), \overline{w(z)})|^{-1}. \tag{5.3}$$

This function on D vanishes precisely on D_x. Let

$$\beta_1(r) = \int_1^{\infty} e^{-ru} u^{-1}\, du = -\text{Ei}(-r) \tag{5.4}$$

be the exponential integral. Note that

$$\beta_1(r) = \begin{cases} -\log(r) - \gamma + O(r), & \text{as } r \to 0, \\ O(e^{-r}), & \text{as } r \to \infty. \end{cases} \tag{5.5}$$

Thus, the function

$$\xi(x, z) = \beta_1(2\pi R(x, z)) \tag{5.6}$$

has a logarithmic singularity on D_x and decays exponentially as z goes to the boundary of D. A straightforward calculation, [**37**], section 11, shows that $\xi(x, \cdot)$ is a Green function for D_x.

Proposition 5.1 As currents on D,

$$dd^c\xi(x) + \delta_{D_x} = [\varphi^0_\infty(x)\,\mu],$$

where, for $z \in D \simeq \mathbb{C} \setminus \mathbb{R}$ with $y = \mathrm{Im}(z)$,

$$\mu = \frac{1}{2\pi}\,\frac{i}{2}\,\frac{dz \wedge d\bar{z}}{y^2}$$

is the hyperbolic volume form and

$$\varphi^0_\infty(x, z) = \big[\, 4\pi(\, R(x, z) + 2Q(x)\,) - 1\,\big] \cdot e^{-2\pi R(x,z)}.$$

Recall that $dd^c = -\frac{1}{2\pi i}\partial\bar{\partial}$.

Remark. For fixed $z \in D$,

$$(x, x)_z = (x, x) + 2\,R(x, z) \tag{5.7}$$

is the majorant attached to z, [**63**]. Thus, the function

$$\varphi_\infty(x, z) \cdot \mu = \varphi^0_\infty(x, z) \cdot e^{-2\pi Q(x)} \cdot \mu \tag{5.8}$$

is (a very special case of) the Schwartz function valued in smooth $(1, 1)$–forms on D defined in [**43**]. In fact, the function $\xi(x, z)$ was first obtained by solving the Green equation of Proposition 5.1 with this right hand side.

Because of the rapid decay of $\xi(x, \cdot)$, we can average over lattice points.

Corollary 5.2 For $v \in \mathbb{R}^\times_{>0}$, let

$$\Xi(t, v)(z) := \sum_{\substack{x \in O_B \cap V \\ Q(x) = t}} \xi(v^{\frac{1}{2}}x, z).$$

(i) For $t > 0$, $\Xi(t, v)$ defines a Green function for $\mathcal{Z}(t)$.
(ii) For $t < 0$, $\Xi(t, v)$ defines a smooth function on $M(\mathbb{C})$.

Note, for example, that for $t > 0$,

$$(5.9) \qquad \begin{matrix} \mathcal{Z}(t)(\mathbb{C}) = \emptyset \\ \text{and } \Xi(t,v) = 0 \end{matrix} \iff \{x \in O_B \cap V,\ Q(x) = t\} = \emptyset \iff \begin{matrix} \boldsymbol{k}_t \text{ does not} \\ \text{embed in } B. \end{matrix}$$

An explicit construction of Green functions for divisors in general locally symmetric varieties is given by Oda and Tsuzuki, [**58**], by a different method.

6. The arithmetic theta series

At this point, we can define a family of classes in $\widehat{\mathrm{CH}}^1(\mathcal{M})$. These can be viewed as an analogue for the arithmetic surface $\mathcal{M}$ of the Hirzebruch-Zagier classes T_N in the middle cohomology of a Hilbert modular surface, [**29**].

Definition 6.1 For $t \in \mathbb{Z}$, with $t \neq 0$, and for a parameter $v \in \mathbb{R}_+^\times$, define classes in $\widehat{\mathrm{CH}}^1_{\mathbb{R}}(\mathcal{M})$ by

$$\widehat{\mathcal{Z}}(t,v) = \begin{cases} (\,\mathcal{Z}(t), \Xi(t,v)\,) & \text{if } t > 0, \\ (\,0, \Xi(t,v)\,) & \text{if } t < 0. \end{cases}$$

For $t = 0$, define

$$\widehat{\mathcal{Z}}(0,v) = -\hat{\omega} - (0, \log(v)) + (0, \mathbf{c}),$$

where $\mathbf{c}$ is the constant defined by (2.15).

We next construct a generating series for these classes; again, this can be viewed as an arithmetic analogue of the Hirzebruch–Zagier generating series for the T_N's. For $\tau = u + iv \in \mathfrak{H}$, the upper half plane, let $q = e(\tau) = e^{2\pi i \tau}$.

Definition 6.2 The arithmetic theta series is the generating series

$$\widehat{\theta}(\tau) = \sum_{t \in \mathbb{Z}} \widehat{\mathcal{Z}}(t,v)\, q^t \ \in \widehat{\mathrm{CH}}^1(\mathcal{M})[[q]].$$

Note that, since the imaginary part v of τ appears as a parameter in the coefficient $\widehat{\mathcal{Z}}(t,v)$, this series is not a holomorphic function of τ.

The arithmetic theta function $\widehat{\theta}(\tau)$ is closely connected with the generating series for quadratic divisors considered by Borcherds, [**5**], and the one for Heegner points introduced by Zagier, [**78**]. The following result, [**42**], justifies the terminology. Its proof, which will be sketched in section 7, depends on the results of [**49**] and of Borcherds, [**5**].

Theorem 6.3 The arithmetic theta series $\widehat{\theta}(\tau)$ is a (nonholomorphic) modular form of weight $\frac{3}{2}$ valued in $\widehat{\mathrm{CH}}^1(\mathcal{M})$.

As explained in Proposition 7.1 below, the arithmetic Chow group $\widehat{\mathrm{CH}}^1(\mathcal{M})$ can be written as direct sum

$$\widehat{\mathrm{CH}}^1(\mathcal{M}) = \widehat{\mathrm{CH}}^1(\mathcal{M},\mu)_{\mathbb{C}} \oplus C_0^\infty(\mathcal{M}(\mathbb{C}))$$

where $\widehat{\mathrm{CH}}^1(\mathcal{M},\mu)_{\mathbb{C}}$ a finite dimensional complex vector space, the Arakelov Chow group of $\mathcal{M}$ for the hyperbolic metric μ, (7.7), and $C_0^\infty(\mathcal{M}(\mathbb{C}))$ is the space of smooth functions on $\mathcal{M}(\mathbb{C})$ with integral 0 with respect to μ. Theorem 6.3 then means that there is a smooth function of ϕ_{Ar} of τ valued in $\widehat{\mathrm{CH}}^1(\mathcal{M},\mu)_{\mathbb{C}}$ and a smooth function $\phi(\tau,z)$ on $\mathfrak{H}\times\mathcal{M}(\mathbb{C})$, with

$$\int_{\mathcal{M}(\mathbb{C})} \phi(\tau,z)\,d\mu(z) = 0,$$

and such that the sum $\phi(\tau) = \phi_{\mathrm{Ar}}(\tau) + \phi(\tau,z)$ satisfies the usual transformation law for a modular form of weight $\frac{3}{2}$, for $\Gamma_0(4D(B))$, and such that the q-expansion of $\phi(\tau)$ is the formal generating series $\widehat{\theta}(\tau)$ of Definition 6.2. By abuse of notation, we write $\widehat{\theta}(\tau)$ both for $\phi(\tau)$ and for its q-expansion.

7. Modularity of the arithmetic theta series

Although the arithmetic theta series $\widehat{\theta}(\tau)$ can be viewed as a kind of generating series for lattice vectors in the spaces $V(A,\iota)$ of special endomorphisms, there is no evident analogue of the Poisson summation formula, which is the key ingredient of the proof of modularity of classical theta series. Instead, the modularity of $\widehat{\theta}(\tau)$ is proved by computing its height pairing with generators of the group $\widehat{\mathrm{CH}}^1(\mathcal{M})$ and identifying the resulting functions of τ with known modular forms.

First we recall the structure of the arithmetic Chow group $\widehat{\mathrm{CH}}^1_{\mathbb{R}}(\mathcal{M})$, [20], [66], [7]. There is a map

$$a : C^\infty(M(\mathbb{C})) \longrightarrow \widehat{\mathrm{CH}}^1_{\mathbb{R}}(\mathcal{M}), \qquad \phi \longmapsto (0,\phi) \tag{7.1}$$

from the space of smooth functions on the curve $M(\mathbb{C})$. Let $\mathbb{1} = a(1)$ be the image of the constant function. Let $\mathrm{Vert} \subset \widehat{\mathrm{CH}}^1_{\mathbb{R}}(\mathcal{M})$ be the subspace generated by classes of the form $(Y_p,0)$, where Y_p is a component of a fiber $\mathcal{M}_p$. The relation $\widehat{\mathrm{div}}(p) = (\mathcal{M}_p, -\log(p)^2) \equiv 0$ implies that $(\mathcal{M}_p,0) \equiv 2\log(p)\cdot\mathbb{1}$, so that $\mathbb{1} \in \mathrm{Vert}$. The spaces $a(C^\infty(M(\mathbb{C}))$ and Vert span the kernel of the restriction map

$$\mathrm{res}_{\mathbb{Q}} : \widehat{\mathrm{CH}}^1_{\mathbb{R}}(\mathcal{M}) \longrightarrow \mathrm{CH}^1(\mathcal{M}_{\mathbb{Q}})_{\mathbb{R}} \tag{7.2}$$

to the generic fiber. Here $\mathrm{CH}^1(\mathcal{M}_{\mathbb{Q}})_{\mathbb{R}} = \mathrm{CH}^1(\mathcal{M}_{\mathbb{Q}}) \otimes_{\mathbb{Z}} \mathbb{R}$.

Let

$$\mathrm{MW}_{\mathbb{R}} = \mathrm{MW}(M)_{\mathbb{R}} = \mathrm{Jac(M)}(\mathbb{Q}) \otimes_{\mathbb{Z}} \mathbb{R} \tag{7.3}$$

be the Mordell–Weil space of the Shimura curve $M = \mathcal{M}_{\mathbb{Q}}$, and recall that this space is the kernel of the degree map

$$\mathrm{MW}_{\mathbb{R}} \longrightarrow \mathrm{CH}^1(\mathrm{M})_{\mathbb{R}} \xrightarrow{\deg} \mathbb{R}. \tag{7.4}$$

Note that the degree of the restriction of the class $\widehat{\omega}$ to the generic fiber is positive, since it is given by the integral of the hyperbolic volume form over $[\Gamma\backslash D]$.

Finally, let $C_0^\infty = C^\infty(M(\mathbb{C}))_0$ be the space of smooth functions which are orthogonal to the constants with respect to the hyperbolic volume form.

The following is a standard result in the Arakelov theory of arithmetic surfaces, [**32**], [**15**], [**7**].

Proposition 7.1 Let

$$\widetilde{\mathrm{MW}} := \Big(\mathbb{R}\widehat{\omega} \oplus \mathrm{Vert} \oplus \mathrm{a}(\mathrm{C}_0^\infty) \Big)^{\perp}$$

be the orthogonal complement of $\mathbb{R}\widehat{\omega} \oplus \mathrm{Vert} \oplus \mathrm{a}(\mathrm{C}_0^\infty)$ with respect to the height pairing. Then

$$\widehat{\mathrm{CH}}^1_{\mathbb{R}}(\mathcal{M}) = \widetilde{\mathrm{MW}} \oplus \Big(\mathbb{R}\widehat{\omega} \oplus \mathrm{Vert} \Big) \oplus \mathrm{a}(\mathrm{C}_0^\infty),$$

where the three summands are orthogonal with respect to the height pairing. Moreover, the restriction map $\mathrm{res}_{\mathbb{Q}}$ induces an isometry

$$\mathrm{res}_{\mathbb{Q}} : \widetilde{\mathrm{MW}} \quad \overset{\sim}{\longrightarrow} \quad \mathrm{MW}$$

with respect to the Gillet–Soulé–Arakelov height pairing on $\widetilde{\mathrm{MW}}$ and the negative of the Neron–Tate height pairing on MW.

In addition, there are some useful formulas for the height pairings of certain classes. For example, for any $\widehat{\mathcal{Z}} \in \widehat{\mathrm{CH}}^1(\mathcal{M})$,

$$\langle \widehat{\mathcal{Z}}, \mathbb{1} \rangle = \frac{1}{2} \deg_{\mathbb{Q}}(\widehat{\mathcal{Z}}). \tag{7.5}$$

In particular, $\mathbb{1}$ is in the radical of the restiction of the height pairing to Vert, and $\langle \widehat{\omega}, \mathbb{1} \rangle = \frac{1}{2} \deg_{\mathbb{Q}}(\widehat{\omega}) > 0$. Also, for ϕ_1 and $\phi_2 \in C_0^\infty$,

$$\langle a(\phi_1), a(\phi_2) \rangle = \frac{1}{2} \int_{M(\mathbb{C})} dd^c\phi_1 \cdot \phi_2. \tag{7.6}$$

Note that the subgroup

$$\widehat{\mathrm{CH}}^1(\mathcal{M}, \mu)_{\mathbb{R}} = \widetilde{\mathrm{MW}} \oplus \mathbb{R}\widehat{\omega} \oplus \mathrm{Vert} \tag{7.7}$$

is the Arakelov Chow group (with real coefficients) for the hyperbolic metric μ.

Returning to the arithmetic theta series $\widehat{\theta}(\tau)$, we consider its pairing with various classes in $\widehat{\mathrm{CH}}^1_{\mathbb{R}}(\mathcal{M})$. To describe these, we first introduce an Eisenstein series of weight $\frac{3}{2}$ associated to the quaternion algebra B, [**49**]. Let $\Gamma' = \mathrm{SL}_2(\mathbb{Z})$ and let Γ'_∞ be the stabilizer of the cusp at infinity. For $s \in \mathbb{C}$, let

$$\mathcal{E}(\tau, s; B) = v^{\frac{1}{2}(s-\frac{1}{2})} \sum_{\gamma \in \Gamma'_\infty \backslash \Gamma'} (c\tau + d)^{-\frac{3}{2}} \, |c\tau + d|^{-(s-\frac{1}{2})} \, \Phi^B(\gamma, s), \tag{7.8}$$

where $\gamma = \begin{pmatrix} a & b \\ c & d \end{pmatrix}$ and $\Phi^B(\gamma, s)$ is a function of γ and s depending on B. This series converges absolutely for $\mathrm{Re}(s) > 1$ and has an analytic continuation to the whole s–plane. It satisfies the functional equation, [**49**], section 16,

$$\mathcal{E}(\tau, s; B) = \mathcal{E}(\tau, -s; B), \tag{7.9}$$

normalized as in Langlands theory.

The following result is proved in [**49**].

Theorem 7.2

(i)

$$2 \cdot \langle \widehat{\theta}(\tau), 1\!\!1 \rangle \quad = -\mathrm{vol.}\ (\mathrm{M}(\mathbb{C})) + \textstyle\sum_{\mathrm{t}>0} \deg(\mathcal{Z}(\mathrm{t})_{\mathbb{Q}})\, \mathrm{q}^{\mathrm{t}}$$

$$= \mathcal{E}(\tau, \tfrac{1}{2}; B).$$

(ii)

$$\langle \widehat{\theta}(\tau), \widehat{\omega} \rangle \quad = \textstyle\sum_t \langle \widehat{\mathcal{Z}}(t, v), \widehat{\omega} \rangle\, q^t$$

$$= \mathcal{E}'(\tau, \tfrac{1}{2}; B).$$

This result is proved by a direct computation of the Fourier coefficients on the two sides of (i) and (ii). For (i), the identity amounts to the formula

$$\text{(7.10)} \qquad \deg(\mathcal{Z}(t)_{\mathbb{Q}})\, q^t = 2\,\delta(d; D(B))\, H_0(t; D(B))\, q^t = \mathcal{E}'_t(\tau, \frac{1}{2}; D(B)),$$

with the notation as in (3.8), (3.9) and (3.10) above. Recall that we write $4t = n^2 d$ where $-d$ is the discriminant of the field $\boldsymbol{k}_t$.

For example, if $D(B) = 1$, i.e., in the case of $B = M_2(\mathbb{Q})$,

$$\text{7.11} \qquad H_0(t; 1) = \sum_{c|n} \frac{h(c^2 d)}{w(c^2 d)}.$$

Thus, $2\, H_0(t; 1) = H(4t)$, where $H(t)$ is the 'class number' which appears in the Fourier expansion of Zagier's nonholomorphic Eisenstein series of weight $\frac{3}{2}$, [**14**], [**78**]:

$$\text{(7.12)} \qquad \mathcal{F}(\tau) = -\frac{1}{12} + \sum_{t>0} H(t)\, q^t + \sum_{m \in \mathbf{Z}} \frac{1}{16\pi}\, v^{-\frac{1}{2}} \int_1^\infty e^{-4\pi m^2 v r}\, r^{-\frac{3}{2}}\, dr\, q^{-m^2}.$$

In fact, when $D(B) = 1$, the value of our Eisenstein series is

(7.13)

$$\mathcal{E}(\tau, \frac{1}{2}; 1) = -\frac{1}{12} + \sum_{t>0} 2\, H_0(t; 1)\, q^t + \sum_{m \in \mathbf{Z}} \frac{1}{8\pi}\, v^{-\frac{1}{2}} \int_1^\infty e^{-4\pi m^2 v r}\, r^{-\frac{3}{2}}\, dr \cdot q^{-m^2}.$$

As in (i) of Theorem 7.2, both of these series have an interpretation in term of degrees of 0–cycles of CM points on the modular curve, [**76**]. Their relation to a regularized integral of a theta series is proved by J. Funke in [**16**], [**17**].

The computations involved in (ii) are significantly more difficult. For example, if $t > 0$ and $\mathcal{Z}(t)_{\mathbb{Q}}(\mathbb{C})$ is nonempty, then, [**49**], Theorem 8.8, the t-th Fourier coefficient of $\mathcal{E}'(\tau, \frac{1}{2}; B)$ is

$$\begin{aligned}\mathcal{E}'_t(\tau, \frac{1}{2}; B) \\ &= 2\delta(d; D)\, H_0(t; D)\cdot q^t \cdot \Big[F\frac{1}{2}\log(d) + \frac{L'(1,\chi_d)}{L(1,\chi_d)} - \frac{1}{2}\log(\pi) - \frac{1}{2}\gamma \\ &+ \frac{1}{2}J(4\pi t v) + \sum_{\substack{p \\ p\nmid D}} \Big(\log|n|_p - \frac{b'_p(n,0;D)}{b_p(n,0;D)} \Big) + \sum_{\substack{p \\ p\mid D}} K_p \log(p) \Big].\end{aligned} \tag{7.14}$$

Here, we write D for $D(B)$, $4t = n^2 d$, as before, and

$$K_p = \begin{cases} -k + \frac{(p+1)(p^k-1)}{2(p-1)} & \text{if } \chi_d(p) = -1, \text{ and} \\ -1 - k + \frac{p^{k+1}-1}{p-1} & \text{if } \chi_d(p) = 0, \end{cases} \tag{7.15}$$

with $k = k_p = \mathrm{ord}_p(n)$. Also,

$$J(x) = \int_0^\infty e^{-xr}\big[\,(1+r)^{\frac{1}{2}} - 1\,\big]\, r^{-1}\, dr. \tag{7.16}$$

Finally, for a prime $p \nmid D$,

$$\frac{1}{\log p}\cdot\frac{b'_p(n,0;D)}{b_p(n,0;D)} = \frac{\chi_d(p) - \chi_d(p)\,(2k+1)p^k + (2k+2)p^{k+1})}{1 - \chi_d(p) + \chi_d(p)\,p^k - p^{k+1}} - \frac{2p}{1-p} \tag{7.17}$$

The connection of this rather complicated quantity with arithmetic geometry is not at all evident. Nonetheless, the identity of part (ii) of Theorem 7.2 asserts that

$$\mathcal{E}'_t(\tau, \frac{1}{2}; B) = \langle\, \widehat{\mathcal{Z}}(t,v), \widehat{\omega}\,\rangle\, q^t. \tag{7.18}$$

Recall that the points of the generic fiber $\mathcal{Z}(t)_{\mathbb{Q}}$ correspond to abelian surfaces A with an action of $O_B \otimes_{\mathbb{Z}} \mathbb{Z}[\sqrt{-t}]$, an order in $M_2(\boldsymbol{k}_t)$. The contribution to $\langle\, \widehat{\mathcal{Z}}(t,v), \widehat{\omega}\,\rangle$ of the associated horizontal component of $\mathcal{Z}(t)$ is the Faltings height of A. Due to the action of $O_B \otimes_{\mathbb{Z}} \mathbb{Z}[\sqrt{-t}\,]$, A is isogenous to a product $E \times E$ where E is an elliptic curve with CM by the order O_d, and so the Faltings heights are related by

$$h_{\mathrm{Fal}}(A) = 2\, h_{\mathrm{Fal}}(E) + \text{an isogeny correction.} \tag{7.19}$$

Up to some combinatorics involving counting of the points $\mathcal{Z}(t)_{\mathbb{Q}}$, the terms on the first line of the right side of (7.14) come from $h_{\mathrm{Fal}}(E)$, while sum on $p \nmid D(B)$ on the second line arises from the isogeny correction. The $J(4\pi t v)$ term comes from the contribution of the Green function $\Xi(t,v)$ with parameter v. Finally, the cycle $\mathcal{Z}(t)$ can have vertical components, and the sum on $p \mid D(B)$ comes from their pairing with $\widehat{\omega}$.

The analogue of (ii) also holds for modular curves, i.e., when $D(B) = 1$. In this case, there are some additional terms, as in the nonholomorphic parts of (7.12) and (7.13). These terms are contributions from a class in $\widehat{\mathrm{CH}}^1_{\mathbb{R}}(\mathcal{M})$ supported in the cusp, [**76**], [**51**].

To continue the proof of modularity of $\widehat{\theta}(\tau)$, we next consider the height pairing of $\widehat{\theta}(\tau)$ with classes of the form $(Y_p, 0)$ in Vert, and with classes of the form $(0, \phi)$, for $\phi \in C_0^\infty$, which might be thought of as 'vertical at infinity'.

First we consider $\langle \hat{\theta}(\tau), (Y_p, 0) \rangle$. For $p \neq 2$, the intersection number of a component Y_p indexed by $[\Lambda]$ with a cycle $\mathcal{Z}(t)$ is calculated explicitly in [**46**]. For $p = 2$, the computation is very similar. Using this result, we obtain, [**42**]:

Theorem 7.3 Assume that $p \mid D(B)$ and that $p \neq 2$. Then, for a component Y_p of $\mathcal{M}_p$ associated to a homothety class of lattices $[\Lambda]$, there is a Schwartz function $\varphi_{[\Lambda]} \in S(V^{(p)}(\mathbb{A}_f))$ and associated theta function of weight $\frac{3}{2}$

$$\theta(\tau, \varphi_{[\Lambda]}) = \sum_{x \in V^{(p)}(\mathbb{Q})} \varphi_{[\Lambda]}(x)\, q^{Q(x)},$$

such that

$$\langle \hat{\theta}(\tau), (Y_p, 0) \rangle = \theta(\tau, \varphi_{[\Lambda]}).$$

Next consider the height pairings with classes of the form $a(\phi) = (0, \phi)$ for $\phi \in C_0^\infty$. Here we note that, for a class $(Z, g_Z) \in \widehat{\mathrm{CH}}^1_R(\mathcal{M})$,

$$\langle (Z, g_Z), a(\phi) \rangle = \frac{1}{2} \int_{\mathcal{M}(\mathbb{C})} \omega_Z\, \phi, \tag{7.20}$$

where $\omega_Z = dd^c g_Z + \delta_Z$ is the smooth $(1,1)$–form on the right side of the Green equation. The map $(Z, g_Z) \mapsto \omega_Z$ defines a map, [**20**], [**66**]

$$\omega : \widehat{\mathrm{CH}}^1_{\mathbb{R}}(\mathcal{M}) \longrightarrow A^{(1,1)}(\mathcal{M}(\mathbb{C})). \tag{7.21}$$

By the basic construction of the Green function $\Xi(t, v)$ and Proposition 5.1, we have

$$\begin{aligned} \omega(\hat{\theta}(\tau)) &= \textstyle\sum_{x \in O_B \cap V} \varphi^0_\infty(x v^{\frac{1}{2}}, z)\, q^{Q(x)} \cdot \mu \\ &= \theta(\tau, \varphi^0_\infty)(z) \cdot \mu, \end{aligned} \tag{7.22}$$

where $\theta(\tau, \varphi^0_\infty)(z)$ is the theta series of weight $\frac{3}{2}$ for the rational quadratic space V of signature $(1,2)$. Thus,

$$\langle \hat{\theta}(\tau), a(\phi) \rangle = \frac{1}{2} \int_{\mathcal{M}(\mathbb{C})} \theta(\tau, \varphi^0_\infty)(z)\, \phi(z)\, d\mu(z), \tag{7.23}$$

is the classical theta lift of ϕ. To describe this more precisely, recall that

$$dd^c \phi = \frac{1}{2} \Delta\{\phi\} \cdot \mu \tag{7.24}$$

where Δ is the hyperbolic Laplacian, and consider functions $\phi_\lambda \in C_0^\infty(\mathcal{M}(\mathbb{C}))$ satisfying $\Delta\phi_\lambda + \lambda\phi_\lambda = 0$ for $\lambda > 0$. Let ϕ_{λ_i}, for $0 < \lambda_1 \leq \lambda_2 \leq \ldots$ be a basis of such eigenfunctions, orthonormal with respect to μ. These are just the Maass forms of weight 0 for the cocompact Fuchsian group $\Gamma = O_B^\times$. Recalling (7.6), we see that the classes $a(\phi_\lambda) \in \widehat{\mathrm{CH}}^1_{\mathbb{R}}(\mathcal{M})$ are orthogonal with respect to the height pairing and span the subspace $a(C_0^\infty)$. By (7.23), we have the following, [**42**].

Theorem 7.4 For a Maass form ϕ_λ of weight 0 for Γ, let

$$\theta(\tau;\phi_\lambda) := \int_{\mathcal{M}(\mathbb{C})} \theta(\tau,\varphi^0_\infty)(z)\,\phi_\lambda(z)\,d\mu(z)$$

be its classical theta lift, a Maass form of weight $\frac{3}{2}$ and level $4D(B)$. Then

$$\langle\,\widehat{\theta}(\tau), a(\phi_\lambda)\,\rangle = \frac{1}{2}\,\theta(\tau;\phi_\lambda).$$

Finally, we must consider the component $\widehat{\theta}_{\mathrm{MW}}(\tau)$ of $\widehat{\theta}(\tau)$ in the space $\widetilde{\mathrm{MW}}$. Recall the isomorphism

$$\text{(7.25)} \qquad \mathrm{res}_{\mathbb{Q}} : \widetilde{\mathrm{MW}} \overset{\sim}{\longrightarrow} \mathrm{MW} \subset \mathrm{CH}^1(\mathrm{M}).$$

Write

$$\text{(7.26)} \qquad \theta_B(\tau) := \mathrm{res}_{\mathbb{Q}}(\widehat{\theta}(\tau)) = -\omega + \sum_{t>0} \mathrm{Z(t)}\,\mathrm{q}^{\mathrm{t}} \quad \in \mathrm{CH}^1(\mathrm{M}),$$

where $Z(t)$ (resp. ω) is the class of the 0–cycle $\mathcal{Z}(t)_{\mathbb{Q}}$ (resp. the line bundle $\omega = \mathrm{res}_{\mathbb{Q}}(\widehat{\omega})$) in $\mathrm{CH}^1(\mathrm{M})$. This series is essentially a special case of the generating function for divisors considered by Borcherds, [**5**], [**6**]. Assuming that certain spaces of vector valued modular forms have bases with rational Fourier coefficients, Borcherds proved that his generating series are modular forms, and McGraw, [**56**], verified Borcherds assumption. The main point in Borcherds' proof is the existence of enough relations among the divisors in question, and such relations can be explicitly given via Borcherds construction of meromorphic modular forms with product expansions, [**3**], [**4**]. Thus, the series $\theta_B(\tau)$ is a modular form of weight $\frac{3}{2}$ valued in $\mathrm{CH}^1(\mathrm{M})$.

By (i) of Theorem 7.2, we have

$$\text{(7.27)} \qquad \deg(\theta_B(\tau)) = \mathcal{E}(\tau,\frac{1}{2};B),$$

and so, the function

$$\text{(7.28)} \qquad \theta_{\mathrm{MW}}(\tau) := \theta_B(\tau) - \mathcal{E}(\tau,\frac{1}{2};B)\cdot\frac{\omega}{\deg(\omega)} \quad \in \mathrm{MW}$$

is also modular of weight $\frac{3}{2}$. Thus we obtain, [**42**],

Theorem 7.5 *The image of $\widehat{\theta}_{\mathrm{MW}}(\tau)$ under the isomorphism (7.25), is given by*

$$\text{(1)} \qquad \mathrm{res}_{\mathbb{Q}}(\,\widehat{\theta}_{\mathrm{MW}}(\tau)\,) = \mathrm{res}_{\mathbb{Q}}(\,\widehat{\theta}(\tau) - \mathcal{E}(\tau,\tfrac{1}{2};\mathrm{B})\,\deg_{\mathbb{Q}}(\widehat{\omega})^{-1}\cdot\widehat{\omega}\,)$$

$$\text{(2)} \qquad = \theta_{\mathrm{MW}}(\tau) \in \mathrm{MW}.$$

Thus, $\widehat{\theta}_{\mathrm{MW}}(\tau)$ is a modular form of weight $\frac{3}{2}$.

This completes the proof of the modularity of the arithmetic theta function $\widehat{\theta}(\tau)$.

8. The arithmetic theta lift

The arithmetic theta function $\widehat{\theta}(\tau)$ can be used to define an arithmetic theta lift

$$\widehat{\theta}: S_{\frac{3}{2}} \longrightarrow \widehat{\mathrm{CH}}^1(\mathcal{M}), \qquad f \mapsto \widehat{\theta}(f),$$

where $S_{\frac{3}{2}}$ is the space of cusp forms of weight $\frac{3}{2}$ for $\Gamma' = \Gamma_0(4D(B))$, as follows. Given a cusp form f of weight $\frac{3}{2}$ for $\Gamma' = \Gamma_0(4D(B))$, let

$$\widehat{\theta}(f) = \langle f, \widehat{\theta} \rangle_{\mathrm{Pet}} = \int_{\Gamma'\backslash\mathfrak{H}} f(\tau)\, \overline{\widehat{\theta}(\tau)}\, v^{\frac{3}{2}} \frac{du\, dv}{v^2} \quad \in \widehat{\mathrm{CH}}^1(\mathcal{M}). \tag{8.1}$$

This construction is analogous to the construction of Niwa, [**57**], of the classical Shimura lift, [**64**], from $S_{\frac{3}{2}}$ to modular forms of weight 2 for $O_B^\times$. Recall that we extend the height pairing $\langle\ ,\ \rangle$ on $\widehat{\mathrm{CH}}^1_{\mathbb{R}}(\mathcal{M})$ to an Hermitian pairing on $\widehat{\mathrm{CH}}^1(\mathcal{M})$, conjugate linear in the second argument. By adjointness, and Theorem 7.2,

$$\langle \widehat{\theta}(f), \mathbb{1} \rangle = \langle f, \langle \widehat{\theta}, \mathbb{1} \rangle \rangle_{\mathrm{Pet}} = \frac{1}{2} \langle f, \mathcal{E}(\frac{1}{2}; B) \rangle_{\mathrm{Pet}} = 0, \tag{8.2}$$

and

$$\langle \widehat{\theta}(f), \widehat{\omega} \rangle = \langle f, \langle \widehat{\theta}, \widehat{\omega} \rangle \rangle_{\mathrm{Pet}} = \langle f, \mathcal{E}'(\frac{1}{2}; B) \rangle_{\mathrm{Pet}} = 0. \tag{8.3}$$

If f is holomorphic, then, by Theorem 7.4, for any $\phi \in C_0^\infty(\mathcal{M}(\mathbb{C}))$,

$$\langle \widehat{\theta}(f), a(\phi) \rangle = \langle f, \langle \widehat{\theta}, a(\phi) \rangle \rangle = \langle f, \theta(\phi) \rangle_{\mathrm{Pet}} = 0, \tag{8.4}$$

since $\theta(\tau; \phi)$ is a combination of Maass forms of weight $\frac{3}{2}$. Thus, for f holomorphic

$$\widehat{\theta}(f) \in \widetilde{\mathrm{MW}} \oplus \mathrm{Vert}^0 \subset \widehat{\mathrm{CH}}^1(\mathcal{M}) \tag{8.5}$$

where

$$\mathrm{Vert}^0 = \mathrm{Vert} \cap \ker\langle \cdot, \widehat{\omega} \rangle. \tag{8.6}$$

It remains to describe the components of $\hat{\theta}(f)$ in the spaces Vert and $\widetilde{\mathrm{MW}}$. This is best expressed in terms of automorphic representations.

9. Theta dichotomy: Waldspurger's theory

We begin with a brief review of Waldspurger's theory of the correspondence between cuspidal automorphic representations of the metaplectic cover of SL_2 and automorphic representations of PGL_2 and its inner forms. For a more detailed survey, the reader can consult [**59**], [**70**], [**31**], as well as the original papers [**68**], [**69**], and especially, [**73**].

We fix the additive character ψ of $\mathbb{A}/\mathbb{Q}$ which has trivial conductor, i.e., is trivial on $\widehat{\mathbb{Z}} = \prod_{p<\infty} \mathbb{Z}_p$ and has archimedean component $\psi_\infty(x) = e(x) = e^{2\pi i x}$. Since ψ is fixed, we suppress it from the notation.

Let $G = \mathrm{SL}_2$ and let $G'_{\mathbb{A}}$ be the 2–fold cover of $G(\mathbb{A})$ which splits over $G(\mathbb{Q})$. Let

$$\begin{aligned} \mathcal{A}_0(G') &= \text{the space of genuine cusp forms for } G'_{\mathbb{A}}. \\ \mathcal{A}_{00}(G') &= \text{the space of genuine cusp forms for } G'_{\mathbb{A}}. \\ &\qquad \text{orthogonal to all } O(1) \text{ theta series} \end{aligned} \tag{9.1}$$

For an irreducible cuspidal automorphic representation $\sigma \simeq \otimes_{p\le\infty}\sigma_p$ in $\mathcal{A}_{00}(G')$, Waldspurger constructs an irreducible cuspidal automorphic representation

$$\pi = \pi(\sigma) = \mathrm{Wald}(\sigma) = \mathrm{Wald}(\sigma, \psi) \tag{9.2}$$

of $\mathrm{PGL}_2(\mathbb{A})$, which serves as a kind of reference point for the description of the global theta lifts for various ternary quadratic spaces. For each $p \le \infty$, there is a corresponding local construction $\sigma_p \mapsto \mathrm{Wald}(\sigma_p, \psi_p)$, and the local and global constructions are compatible, i.e.,

$$\mathrm{Wald}(\sigma, \psi) \simeq \otimes_{p\le\infty}\mathrm{Wald}(\sigma_p, \psi_p).$$

Since we have fixed the additive character $\psi = \otimes_p\psi_p$, we will often omit it from the notation.

For a quaternion algebra B over $\mathbb{Q}$, let

$$V^B = \{\, x \in B \mid \mathrm{tr}(x) = 0 \,\}, \tag{9.3}$$

with quadratic form[6] $Q(x) = -x^2 = \nu(x)$, and let

$$H^B = B^\times \simeq \mathrm{GSpin}(V). \tag{9.4}$$

For a Schwartz function $\varphi \in S(V^B(\mathbb{A}))$, $g' \in G'_{\mathbb{A}}$ and $h \in H^B(\mathbb{A})$, define the theta kernel by

$$\theta(g', h; \varphi) = \sum_{x\in V^B(\mathbb{Q})} \omega(g')\varphi(h^{-1}x). \tag{9.5}$$

Here $\omega = \omega_\psi$ is the Weil representation of $G'_{\mathbb{A}}$ on $S(V(\mathbb{A}))$. For $f \in \sigma$ and $\varphi \in S(V^B(\mathbb{A}))$, the classical theta lift is

$$\theta(f;\varphi) = \langle\, f, \theta(\varphi)\,\rangle_{\mathrm{Pet}} = \int_{G'_{\mathbb{Q}}\backslash G'_{\mathbb{A}}} f(g')\,\overline{\theta(g', h; \varphi)}\,dg' \quad \in \mathcal{A}_0(H^B). \tag{9.6}$$

The global theta lift of σ to H^B is the space

$$\theta(\sigma; V^B) \subset \mathcal{A}_0(H^B), \tag{9.7}$$

spanned by the $\theta(f;\varphi)$'s for $f \in \sigma$ and $\varphi \in S(V(\mathbb{A}))$. Here $\mathcal{A}_0(H^B)$ is the space of cusp forms on $H^B(\mathbb{Q})\backslash H^B(\mathbb{A})$. Then $\theta(\sigma; V^B)$ is either zero or is an irreducible cuspidal representation of $H^B(\mathbb{A})$, [**73**], Prop. 20, p. 290.

[6] Note that Waldspurger uses the opposite sign.

For any irreducible admissible genuine representation σ_p of G'_p, the metaplectic cover of $G(\mathbb{Q}_p)$, there are analogous local theta lifts $\theta(\sigma_p, V_p^B)$. Each of them is either zero or is an irreducible admissible representation of H_p^B. The local and global theta lifts are compatible in the sense that

$$\theta(\sigma; V^B) \simeq \begin{cases} \otimes_p \theta(\sigma_p; V_p^B) & \text{or} \\ 0. \end{cases} \tag{9.8}$$

In particular, the global theta lift is zero if any local $\theta(\sigma_p; V_p^B) = 0$, but the global theta lift can also vanish even when there is no such local obstruction, i.e., even if $\otimes_{p\leq\infty}\theta(\sigma_p; V_p^B) \neq 0$.

For each p, let $B_p^{\pm}$ be the quaternion algebra over $\mathbb{Q}_p$ with invariant

$$\mathrm{inv_p}(\mathrm{B_p^{\pm}}) = \pm 1. \tag{9.9}$$

Then the two ternary quadratic spaces

$$V_p^{\pm} = \{\ x \in B_p^{\pm} \mid \mathrm{tr}(x) = 0\ \} \tag{9.10}$$

have the same discriminant and opposite Hasse invariants. A key local fact established by Waldspurger is:

Theorem 9.1 **(Local theta dichotomy)** For an irreducible admissible genuine representaton σ_p of G'_p, precisely one of the spaces $\theta(\sigma_p; V_p^+)$ and $\theta(\sigma_p; V_p^-)$ is nonzero.

Definition 9.2. Let $\epsilon_p(\sigma_p) = \pm 1$ be the unique sign such that with

$$\theta(\sigma_p, V_p^{\epsilon_p(\sigma_p)}) \neq 0.$$

Examples: (i) If p is a finite prime and σ_p is an unramified principal series representation, then $\epsilon_p(\sigma_p) = +1$ and $\theta(\sigma_p; V_p^+)$ is a principal series representation of $\mathrm{GL}_2(\mathbb{Q}_p)$. It is unramified if σ_p is unramified.

(ii) For a finite prime p and a character μ_p of $\mathbb{Q}_p^\times$, with $\mu_p^2 = |\ |$, there is a special representation $\sigma_p(\mu_p)$ of G'_p. For $\sigma_p = \sigma_p(\mu_p)$, with $\mu_p = |\ |^{\frac{1}{2}}$,

$$\begin{aligned} \theta(\sigma_p, V_p^-) &= 1\!\!1 \neq 0, \qquad B_p^- = \mathbb{B}_p \\ \theta(\sigma_p, V_p^+) &= 0, \qquad B_p^+ = M_2(\mathbb{Q}_p) \\ \mathrm{Wald}(\sigma_p, \psi_p) &= \text{unramified special } \sigma(|\ |^{\frac{1}{2}}, |\ |^{-\frac{1}{2}}) \text{ of } GL_2(\mathbb{Q}_p). \end{aligned}$$

If $\mu_p \neq |\ |^{\frac{1}{2}}$, then

$$\theta(\sigma_p, V^+) = \mathrm{Wald}(\sigma_p, \psi_p) = \sigma(\mu, \mu^{-1})$$

is a special representation of $\mathrm{GL}_2(\mathbb{Q}_p)$ and $\theta(\sigma_p, V_p^-) = 0$.

(iii) If $\sigma_\infty = \mathrm{HDS}_{\frac{3}{2}}$, the holomorphic discrete series representation of $G'_\mathbb{R}$ of weight $\frac{3}{2}$, then

$$\begin{aligned} \theta(\sigma_\infty, V_\infty^-) &= 1\!\!1 \neq 0, \qquad B_\infty^- = \mathbb{H} \\ \theta(\sigma_\infty, V_\infty^+) &= 0, \qquad B_\infty^+ = M_2(\mathbb{R}) \\ \mathrm{Wald}(\sigma_\infty, \psi_\infty) &= \mathrm{DS}_2 = \text{weight 2 disc. series of } GL_2(\mathbb{R}). \end{aligned}$$

Since we will be considering only holomorphic cusp forms of weight $\frac{3}{2}$ and level $4N$ with N odd and square free, these examples give all of the relevant local information.

The local root number $\epsilon_p(\frac{1}{2}, \mathrm{Wald}(\sigma_p)) = \pm 1$ and the invariant $\epsilon_p(\sigma_p)$ are related as follows. There is an element $-\mathbf{1}$ in the center of G'_p which maps to $-1 \in G_p$, and Waldspurger defines a sign $\epsilon(\sigma_p, \psi_p)$, the central sign of σ_p, by

$$\sigma_p(-\mathbf{1}) = \epsilon(\sigma_p, \psi_p)\, \chi_\psi(-1) \cdot I_{\sigma_p}, \tag{9.11}$$

[**73**], p.225. Then,

$$\epsilon_p(\sigma_p) = \epsilon_p(\frac{1}{2}, \mathrm{Wald}(\sigma_p, \psi_p))\, \epsilon(\sigma_p, \psi_p). \tag{9.12}$$

Note that

$$\prod_p \epsilon(\sigma_p, \psi_p) = 1. \tag{9.13}$$

Thus, for a given global $\sigma \simeq \otimes_p \sigma_p$,

$$\begin{aligned} \epsilon(\frac{1}{2}, \mathrm{Wald}(\sigma, \psi)) = +1 \quad &\Longleftrightarrow \text{ there is a } B/\mathbb{Q} \text{ with } \mathrm{inv_p(B)} = \epsilon_{\mathrm{p}}(\sigma_{\mathrm{p}}) \text{ for all } p \\ &\Longleftrightarrow \text{ there is a } B/\mathbb{Q} \text{ with } \otimes_p \theta(\sigma_p, V_p^B) \neq 0. \end{aligned}$$

10. The doubling integral

The key to linking Waldspurger's theory to the arithmetic theta lift is the doubling integral representation of the Hecke L-function, [60], [54], [52], and in classical language, [18], [2]. Let $\widetilde{G} = \mathrm{Sp}_2$ be the symplectic group of rank 2 over $\mathbb{Q}$, and let $\widetilde{G}'_{\mathbb{A}}$ be the 2–fold metaplectic cover of $\widetilde{G}(\mathbb{A})$. Recall that $G = \mathrm{SL}_2 = \mathrm{Sp}_1$. Let $i_0 : G \times G \to \widetilde{G}$ be the standard embedding:

$$(10.1) \qquad i_0 : \begin{pmatrix} a_1 & b_1 \\ c_1 & d_1 \end{pmatrix} \times \begin{pmatrix} a_2 & b_2 \\ c_2 & d_2 \end{pmatrix} \longmapsto \begin{pmatrix} a_1 & & b_1 & \\ & a_2 & & b_2 \\ c_1 & & d_1 & \\ & c_2 & & d_2 \end{pmatrix}.$$

For $g \in G$, let

$$(10.2) \qquad g^\vee = \mathrm{Ad}\begin{pmatrix} 1 & \\ & -1 \end{pmatrix} \cdot g,$$

and let

$$(10.3) \qquad i(g_1, g_2) = i_0(g_1, g_2^\vee).$$

Let $\widetilde{P} \subset \widetilde{G}$ be the standard Siegel parabolic, and, for $s \in \mathbb{C}$, let $\widetilde{I}(s)$ be the induced representation

$$(10.4) \qquad \widetilde{I}(s) = \mathrm{Ind}_{\widetilde{P}'_{\mathbb{A}}}^{\widetilde{G}'_{\mathbb{A}}}(\delta^{s+\frac{3}{2}}) \simeq \otimes_p \widetilde{I}_p(s),$$

where δ is a certain lift to the cover $\widetilde{P}'_{\mathbb{A}}$ of the modulus character of $\widetilde{P}(\mathbb{A})$. Here we are using unnormalized induction. For a section $\Phi(s) \in \widetilde{I}(s)$, there is a Siegel–Eisenstein series

$$(10.5) \qquad E(g', s, \Phi) = \sum_{\gamma \in \widetilde{P}'_{\mathbb{Q}} \backslash \widetilde{G}'_{\mathbb{Q}}} \Phi(\gamma g', s),$$

convergent for $\mathrm{Re}(s) > \frac{3}{2}$, and with an analytic continuation in s satisfying a functional equation relating s and $-s$. For $\sigma \simeq \otimes_p \sigma_p$ a cuspidal representation in $\mathcal{A}_0(G')$, as above, the doubling integral is defined as follows:
For vectors f_1, $f_2 \in \sigma$ and $\Phi(s) \in \widetilde{I}(s)$,

$$(10.6) \qquad Z(s, f_1, f_2, \Phi) = \int_{G'_{\mathbb{Q}} \backslash G'_{\mathbb{A}} \times G'_{\mathbb{Q}} \backslash G'_{\mathbb{A}}} \overline{f_1(g'_1)}\, f_2(g'_2)\, E(i(g'_1, g'_2), s, \Phi)\, dg'_1\, dg'_2.$$

If f_1, f_2 and $\Phi(s)$ are factorizable and unramified outside of a set of places S, including ∞ and 2, then, [60], [54], [44][7],

$$(10.7) \qquad Z(s, f_1, f_2, \Phi) = \frac{1}{\zeta^S(2s+2)}\, L^S(s + \frac{1}{2}, \mathrm{Wald}(\sigma)) \cdot \prod_{p \in S} Z_p(s, f_{1,p}, f_{2,p}, \Phi_p),$$

where $Z_p(s, f_{1,p}, f_{2,p}, \Phi_p)$ is a local zeta integral depending on the local components at p.

[7] Note that [60] and [44] deal with the symplectic case, while the metaplectic case needed here is covered in [54].

Now suppose that $\sigma_\infty = \mathrm{HDS}_{\frac{3}{2}}$, and take the archimedean local components $f_{1,\infty} = f_{2,\infty}$ to be the weight $\frac{3}{2}$ vectors. We can identify f_1 and f_2 with classical cusp forms of weight $\frac{3}{2}$. Taking $\Phi_\infty(s) \in \tilde{I}_\infty(s)$ to be the standard weight $\frac{3}{2}$ section $\Phi_\infty^{\frac{3}{2}}(s)$, we can write the Eisenstein series as a classical Siegel Eisenstein series $E(\tau, s, \Phi_f)$ of weight $\frac{3}{2}$, where $\tau \in \mathfrak{H}_2$, the Siegel space of genus 2, and $\Phi_f(s)$ is the finite component of $\Phi(s)$.

For a given indefinite quaternion algebra B, there is a section $\Phi_f(s)$ defined as follows. For a finite prime p, the group $\tilde{G}'_p$ acts on the Schwartz space $S((V_p^{\pm})^2)$ via the Weil representation ω determined by ψ_p, and there is a map

$$\lambda_p : S((V_p^{\pm})^2) \longrightarrow \tilde{I}_p(0), \qquad \varphi_p \mapsto \lambda_p(\varphi_p)(g') = \omega(g')\varphi_p(0). \tag{10.8}$$

Here $V_p^{\pm}$ is the ternary quadratic space defined in (5.10). Note that a section $\Phi_p(s)$ is determined by its restriction to the compact open subgroup K'_p, the inverse image of $\tilde{G}(\mathbb{Z}_p)$ in $\tilde{G}'_p$. A section is said to be standard if this restriction is independent of s. The function $\lambda_p(\varphi_p) \in \tilde{I}_p(0)$ has a unique extension to a standard section of $\tilde{I}_p(s)$. Fix a maximal order $R_p^{\pm}$ in $B_p^{\pm}$, and let $\varphi_p^{\pm} \in S((V_p^{\pm})^2)$ be the characteristic function of $(R_p^{\pm} \cap V_p^{\pm})^2$. Also let $R_p^e \subset R_p^+$ be the Eichler order[8] of index p, and let φ_p^e be the characteristic function of $(R_p^e \cap V_p^+)^2$. We then have standard sections $\Phi_p^0(s)$, $\Phi_p^-(s)$ and $\Phi_p^e(s)$ whose restrictions to K'_p are $\lambda_p(\varphi_p^+)$, $\lambda_p(\varphi_p^-)$, and $\lambda_p(\varphi_p^e)$ respectively. Following [**46**], let

$$\tilde{\Phi}_p(s) = \Phi_p^-(s) + A_p(s)\Phi_p^0(s) + B_p(s)\,\Phi_p^e(s), \tag{10.9}$$

where $A_p(s)$ and $B_p(s)$ are entire functions of s such that

$$A_p(0) = B_p(0) = 0, \qquad \text{and} \qquad A'_p(0) = -\frac{2}{p^2-1}\log(p), \quad B'_p(0) = \frac{1}{2}\frac{p+1}{p-1}\log(p). \tag{10.10}$$

Then let

$$\tilde{\Phi}^B(s) = \Big(\otimes_{p|D(B)}\tilde{\Phi}_p(s)\Big) \otimes \Big(\otimes_{p\nmid D(B)}\Phi_p^0(s)\Big). \tag{10.11}$$

Using this section, we can define a normalized Siegel Eisenstein series of weight $\frac{3}{2}$ and genus 2 attached to B by

$$\mathcal{E}_2(\tau, s; B) = \eta(s, B)\,\zeta_{D(B)}(2s+2)\,E(\tau, s; \tilde{\Phi}^B), \tag{10.12}$$

where $\eta(s, B)$ is a certain normalizing factor and the partial zeta function is as in (2.14). Then we have a precise version of the doubling identity, [**52**]:

[8] So $R_p^+ = M_2(\mathbb{Z}_p)$ and $x \in R_p^e$ iff $c \equiv 0 \mod p$.

Theorem 10.1[9] For every indefinite quaternion algebra B over $\mathbb{Q}$, the associated Siegel–Eisenstein series $\mathcal{E}_2(\tau, s; B)$ of genus 2 and weight $\frac{3}{2}$ has the following property. For each holomorphic 'newform' f of weight $\frac{3}{2}$ and level $4D(B)$ associated to an irreducible cuspidal representation $\sigma = \otimes_p \sigma_p$:

$$\langle\, \mathcal{E}_2(\begin{pmatrix} \tau_1 & \\ & -\bar{\tau}_2 \end{pmatrix}, s; B)\,,\, \overline{f(\tau_2)}\,\rangle_{\mathrm{Pet},\,\tau_2}$$
$$= C(s)\, C(s;\sigma;B)\, L(s+\tfrac{1}{2}, \mathrm{Wald}(\sigma)) \cdot f(\tau_1),$$

where[10] $C(0) \neq 0$, and

$$C(s;\sigma;B) = \prod_{p|D(B)} C_p(s;\sigma_p;B_p),$$

with

$$C_p(0;\sigma_p;B_p) = \begin{cases} 1 & \text{if } \epsilon_p(\sigma_p) = -1, \\ 0 & \text{if } \epsilon_p(\sigma_p) = +1, \end{cases}$$

and, if $\epsilon_p(\sigma_p) = +1$,

$$C_p'(0;\sigma_p;B_p) = \log(p).$$

Note that $\epsilon_p(\sigma_p) = 1$ for $p \nmid 4D(B)$, and that

$$\epsilon(\frac{1}{2}, \mathrm{Wald}(\sigma)) = - \prod_{p|D(B)} \epsilon_p(\sigma_p). \tag{10.13}$$

Thus, for example, for $\epsilon(\frac{1}{2}, \mathrm{Wald}(\sigma)) = +1$, an odd number of $p \mid D(B)$ have $\epsilon_p(\sigma_p) = +1$.

Corollary 10.2 With the notation and assumptions of Theorem 10.1,

$$\langle\, \mathcal{E}_2'(\begin{pmatrix} \tau_1 & \\ & -\bar{\tau}_2 \end{pmatrix}, 0; B)\,,\, \overline{f(\tau_2)}\,\rangle_{\mathrm{Pet},\,\tau_2}$$
$$= f(\tau_1) \cdot C(0) \cdot \begin{cases} L'(\frac{1}{2}, \mathrm{Wald}(\sigma)), & \text{if } \epsilon_p(\sigma_p) = -1 \\ & \text{for all } p \mid D(B), \\ L(\frac{1}{2}, \mathrm{Wald}(\sigma)) & \text{if } \epsilon_p(\sigma_p) = +1 \\ \quad \times \log(p), & \text{for a unique } p \mid D(B), \\ 0 & \text{otherwise.} \end{cases}$$

11. The arithmetic inner product formula

Returning to the arithmetic theta lift and arithmetic theta function $\widehat{\theta}(\tau)$, there should be a second relation between derivatives of Eisenstein series and arithmetic geometry, [**37**]:

[9] At the time of this writing, we must assume that $2 \nmid D(B)$ and that $\epsilon_2(\sigma_2) = +1$. It should not be difficult to remove these restrictions.

[10] These are explicit elementary factors.

Conjecture 11.1

$$\langle \widehat{\theta}(\tau_1), \widehat{\theta}(\tau_2) \rangle = \mathcal{E}_2'(\begin{pmatrix} \tau_1 & \\ & -\bar{\tau}_2 \end{pmatrix}, 0; B).$$

Here recall that $\langle\ ,\ \rangle$ has been extended to be conjugate linear in the second factor. Additional discussion can be found in [**38**], [**40**], [**41**], [**51**]. This conjecture amounts to identities on Fourier coefficients:

$$\langle \widehat{\mathcal{Z}}(t_1, v_1), \widehat{\mathcal{Z}}(t_2, v_2) \rangle \cdot q_1^{t_1} q_2^{t_2}$$

$$(11.1) \qquad = \sum_{\substack{T \in \mathrm{Sym}_2(\mathbb{Z})^\vee \\ \mathrm{diag}(T)=(t_1,t_2)}} \mathcal{E}_{2,T}'(\begin{pmatrix} \tau_1 & \\ & \tau_2 \end{pmatrix}, 0; B).$$

If $t_1 t_2$ is not a square, then any $T \in \mathrm{Sym}_2(\mathbb{Z})^\vee$ with $\mathrm{diag}(T) = (t_1, t_2)$ has $\det(T) \neq 0$. Thus, only the nonsingular Fourier coefficients of $\mathcal{E}_2(\tau, s, B)$ contribute to the right hand side of (11.1) in this case. Under the same condition, the cycles $\mathcal{Z}(t_1)$ and $\mathcal{Z}(t_2)$ do not meet on the generic fiber, although they may have common vertical components in the fibers of bad reduction.

On the other hand, if $t_1 t_2 = m^2$, then the singular matrices $T = \begin{pmatrix} t_1 & \pm m \\ \pm m & t_2 \end{pmatrix}$ occur on the right side of (11.1), and the cycles $\mathcal{Z}(t_1)$ and $\mathcal{Z}(t_2)$ meet in the generic fiber and have common horizontal components.

The results of [**37**] and [**46**] yield the following.

Theorem 11.2 Suppose that $t_1 t_2$ is not a square. In addition, assume that (11.11) (*resp.* (11.26)) below holds for $p = 2$ if $2 \nmid D(B)$ (resp. $2 \mid D(B)$). Then the Fourier coefficient identity (11.1) holds.

We now briefly sketch the proof of Theorem 11.2. The basic idea is that there is a decomposition of the height pairing on the left hand side of (11.1) into terms indexed by $T \in \mathrm{Sym}_2(\mathbb{Z})^\vee$ with $\mathrm{diag}(T) = (t_1, t_2)$. One can prove identities between terms on the two sides corresponding to a given T.

Recalling the modular definition of the cycles given in section 3, the intersection $\mathcal{Z}(t_1) \cap \mathcal{Z}(t_2)$ can be viewed as the locus of triples $(A, \iota, \mathbf{x})$, where $\mathbf{x} = [x_1, x_2]$ is a pair of special endomorphisms $x_i \in V(A, \iota)$ with $Q(x_i) = t_i$. Associated to $\mathbf{x}$ is the 'fundamental matrix' $Q(\mathbf{x}) = \frac{1}{2}((x_i, x_j)) \in \mathrm{Sym}_2(\mathbb{Z})^\vee$, where $(\ ,\ )$ is the bilinear form on the quadratic lattice $V(A, \iota)$. Thus we may write

$$(11.2) \qquad \mathcal{Z}(t_1) \cap \mathcal{Z}(t_2) = \coprod_{\substack{T \in \mathrm{Sym}_2(\mathbb{Z})^\vee \\ \mathrm{diag}(T)=(t_1,t_2)}} \mathcal{Z}(T),$$

where $\mathcal{Z}(T)$ is the locus of triples with $Q(\mathbf{x}) = T$. Note that the fundamental matrix is always positive semidefinite, since the quadratic form on $V(A, \iota)$ is positive definite. On the other hand, if $\det(T) \neq 0$, then $\mathcal{Z}(T)_{\mathbb{Q}}$ is empty, since the space of special endomorphisms $V(A, \iota)$ has rank 0 or 1 in characteristic 0.

For a nonsingular $T \in \mathrm{Sym}_2(\mathbb{Q})$, there is a unique global ternary quadratic space V_T with discriminant -1 which represents T; the matrix of the quadratic form on this space is

$$Q_T = \begin{pmatrix} T & \\ & \det(T)^{-1} \end{pmatrix}. \tag{11.3}$$

The space V_T is isometric to the space of trace zero elements for some quaternion algebra B_T over $\mathbb{Q}$, and the local invariants of this algebra must differ from those of the given indefinite B at a finite set of places

$$\mathrm{Diff(T,B)} := \{\ \mathrm{p} \le \infty \mid \mathrm{inv_p(B_T)} = -\mathrm{inv_p(B)}\ \}, \tag{11.4}$$

with $|\mathrm{Diff(T,B)}|$ even[11].

The nonsingular Fourier coefficients of $\mathcal{E}_2(\tau, s; B)$ have a product formula

$$\mathcal{E}_{2,T}(\tau, s; B) = \eta(s; B)\,\zeta_{D(B)}(2s+2) \cdot W_{T,\infty}(\tau, s; \frac{3}{2}) \cdot \prod_p W_{T,p}(s; \tilde{\Phi}_p^B), \tag{11.5}$$

where $\tilde{\Phi}^B(s) = \otimes_p \tilde{\Phi}_p^B(s)$ is given by (10.11). Moreover, for a finite prime p,

$$p \in \mathrm{Diff(T,B)} \iff \mathrm{W_{T,p}(0, \tilde{\Phi}_p^B) = 0}, \tag{11.6}$$

while

$$\mathrm{ord}_{s=0}\, W_{T,\infty}(\tau, s; \frac{3}{2}) = \begin{cases} 0 & \text{if } T > 0, \text{ and} \\ 1 & \text{if } \mathrm{sig}(T) = (1,1) \text{ or } (0,2). \end{cases} \tag{11.7}$$

First suppose that $T > 0$, so that $\infty \in \mathrm{Diff(T,B)}$, since V^B has signature $(1,2)$ and hence cannot represent T. If $|\mathrm{Diff(T,B)}| \ge 4$, then $\mathcal{Z}(T)$ is empty and $\mathcal{E}'_{2,T}(\tau, 0; B) = 0$, so there is no contribution of such T's on either side of (11.1). If $T > 0$ and $|\mathrm{Diff(T,B)}| = 2$, then $\mathrm{Diff(T,B)} = \{\infty, \mathrm{p}\}$ for a unique finite prime p. In this situation, it turns out that $\mathcal{Z}(T)$ is supported in the fiber $\mathcal{M}_p$ at p. There are two distinct cases:

1. "(i)" If $p \nmid D(B)$, then $\mathcal{Z}(T)$ is a finite set of points in $\mathcal{M}_p$.
2. "(ii)" If $p \mid D(B)$, then $\mathcal{Z}(T)$ can be a union of components Y_p of the fiber at p, with multiplicities.

In case (i), the contribution to the height pairing on the left side of (11.1) is $\log(p)$ times the sum of the local multiplicities of points in $\mathcal{Z}(T)$. It turns out that all points have the same multiplicity, $e_p(T)$, so that the contribution to (11.1) has the form

$$e_p(T) \cdot |\mathcal{Z}(T)(\bar{\mathbb{F}}_p)|. \tag{11.8}$$

[11] This is a slightly different definition than that used in [37], where ∞ is taken to be in the Diff set for T indefinite. Hence the difference in parity.

The computation of $e_p(T)$ can be reduced to a special case of a problem in the deformation theory of p-divisible groups which was solved by Gross and Keating, [**23**]. As explained in section 14 of [**37**], their result yields the formula

$$e_p(T) = \begin{cases} \sum_{j=0}^{\frac{\alpha-1}{2}} (\alpha+\beta-4j)\,p^j & \text{if } \alpha \text{ is odd,} \\ \sum_{j=0}^{\frac{\alpha}{2}-1} (\alpha+\beta-4j)\,p^j + \frac{1}{2}(\beta-\alpha+1)\,p^{\frac{\alpha}{2}} & \text{if } \alpha \text{ is even,} \end{cases} \tag{11.9}$$

where, for $p \neq 2$, T is equivalent, via the action of $\mathrm{GL}_2(\mathbb{Z}_p)$, to $\mathrm{diag}(\epsilon_1 p^\alpha, \epsilon_2 p^\beta)$, with $\epsilon_1, \epsilon_2 \in \mathbb{Z}_p^\times$ and $0 \leq \alpha \leq \beta$. The same result holds for $p = 2$, but with a slightly different definition, [**23**], of the invariants α and β of T.

On the other hand, if $\mathrm{Diff}(T, B) = \{\infty, p\}$ with $p \nmid D(B)$, then

$$\begin{aligned} \mathcal{E}'_{2,T}(\tau, 0; B) &= \eta(0; B)\,\zeta_{D(B)}(2) \cdot W_{T,\infty}(\tau, 0; \tfrac{3}{2}) \\ &\quad \times \frac{W'_{T,p}(0,\Phi^0_p)}{W_{T,p}(0,\Phi^-_p)} \cdot \left(W_{T,p}(0, \Phi^-_p) \cdot \textstyle\prod_{\ell\neq p} W_{T,\ell}(0, \widetilde{\Phi}^B_\ell) \right). \end{aligned} \tag{11.10}$$

Here recall that the local sections $\widetilde{\Phi}^B_\ell(s)$ are as defined in (10.9) and (10.11). For $p \neq 2$, formulas of Kitaoka, [**35**], for representation densities $\alpha_p(S,T)$ of binary forms T by unimodular quadratic forms S can be used to compute the derivatives of the local Whittaker functions, and yield, [**37**], Proposition 8.1 and Proposition 14.6,

$$\frac{W'_{T,p}(0, \Phi^0_p)}{W_{T,p}(0, \Phi^-_p)} = \frac{1}{2}(p-1)\log(p) \cdot e_p(T), \tag{11.11}$$

where $e_p(T)$ is precisely the multiplicity (11.9)! On the other hand, up to simple constants[12],

$$W_{T,\infty}\left(\begin{pmatrix} \tau_1 & \\ & \tau_2 \end{pmatrix}, 0; \frac{3}{2}\right) \doteq q_1^{t_1} q_2^{t_2}, \tag{11.12}$$

and

$$W_{T,p}(0, \Phi^-_p) \cdot \prod_{\ell \neq p} W_{T,\ell}(0, \widetilde{\Phi}^B_\ell) \doteq |\mathcal{Z}(T)(\bar{\mathbb{F}}_p)|. \tag{11.13}$$

Note that to obtain (11.11) in the case $p = 2$, one needs to extend Kitaoka's representation density formula to this case; work on this is in progress.

[12] Hence the notation $\doteq$.

Next, we turn to case (ii), where the component $\mathcal{Z}(T)$ is attached to T with $\text{diag}(T) = (t_1, t_2)$ and $\text{Diff}(T, B) = \{\infty, p\}$ with $p \mid D(B)$. This case is studied in detail in [**46**] under the assumption that $p \neq 2$, using the p-adic uniformization described in section 4 above. First, we can base change to $\mathbb{Z}_{(p)}$ and use the intersection theory explained in section 4 of [**46**]. The contribution of $\mathcal{Z}(T)$ to the height pairing is then

$$(11.14) \qquad \chi(\,\mathcal{Z}(T), \mathcal{O}_{\mathcal{Z}(t_1)} \overset{\mathbb{L}}{\to} \otimes \mathcal{O}_{\mathcal{Z}(t_2)}\,) \cdot \log(p).$$

Here χ is the Euler–Poincaré characteristic and, for quasicoherent sheaves $\mathcal{F}$ and $\mathcal{G}$ on $\mathcal{M} \times_{\text{Spec}(\mathbb{Z})} \text{Spec}(\mathbb{Z}_{(p)})$, with $\text{supp}(\mathcal{F}) \cap \text{supp}(\mathcal{G})$ contained in the special fiber and proper over $\text{Spec}(\mathbb{Z}_{(p)})$,

$$(11.15) \qquad \chi(\mathcal{F} \overset{\mathbb{L}}{\to} \otimes \mathcal{G}) \;=\; \chi(\mathcal{F} \otimes \mathcal{G}) - \chi(\mathcal{T}or_1(\mathcal{F}, \mathcal{G})) + \chi(\mathcal{T}or_2(\mathcal{F}, \mathcal{G})).$$

Recall that, for $i = 1, 2$, $\widehat{\mathcal{C}}_p(t_i)$ is the base change to W of the formal completion of $\mathcal{Z}(t_i)$ along its fiber at p. Similarly, we write $\widehat{\mathcal{C}}_p(T)$ for the analogous formal scheme over W determined by $\mathcal{Z}(T)$. By Lemma 8.4 of [**46**],

$$(11.16) \qquad \chi(\,\mathcal{Z}(T), \mathcal{O}_{\mathcal{Z}(t_1)} \overset{\mathbb{L}}{\to} \otimes \mathcal{O}_{\mathcal{Z}(t_2)}\,) \;=\; \chi(\,\widehat{\mathcal{C}}_p(T), \mathcal{O}_{\widehat{\mathcal{C}}_p(t_1)} \overset{\mathbb{L}}{\to} \otimes \mathcal{O}_{\widehat{\mathcal{C}}_p(t_2)}\,),$$

so that we can calculate after passing to the formal situation. The same arguments which yield the p-adic uniformization, Proposition 4.2, of $\widehat{\mathcal{C}}_p(t)$ yields a diagram

$$(11.17) \qquad \begin{array}{ccc} \widehat{\mathcal{C}}_p(T) & \overset{\sim}{\longrightarrow} & H'(\mathbb{Q})\backslash(\star\star) \\ \downarrow & & \downarrow \\ \widehat{\mathcal{M}}_p & \overset{\sim}{\longrightarrow} & H'(\mathbb{Q})\backslash\Big(\,\mathcal{D}^{\bullet} \times H(\mathbb{A}_f^p)/K^p\,\Big) \end{array}$$

where

$$(11.18) \qquad (\star\star) := \left\{ (\mathbf{y}, (X, \rho), gK^p) \;\middle|\; \begin{array}{c} (i) \quad Q(\mathbf{y}) = T \\ (ii) \quad (X, \rho) \in Z^{\bullet}(j(\mathbf{y})) \\ (iii) \quad \mathbf{y} \in \big(g\,(V(\mathbb{A}_f^p) \cap \hat{O}_B^p)\,g^{-1}\big)^2 \end{array} \right\}.$$

Proceeding as in Remark 4.3, we obtain

$$(11.19) \qquad \widehat{\mathcal{C}}_p(T) \;\overset{\sim}{\longrightarrow}\; [\Gamma'\backslash\mathcal{D}_T^{\bullet}],$$

where

$$11.20 \qquad \mathcal{D}_T^{\bullet} = \coprod_{\substack{\mathbf{y} \in (L')^2 \\ Q(\mathbf{y}) = T}} Z^{\bullet}(j(\mathbf{y})).$$

Here, recall from (4.25) that

$$\Gamma' = H'(\mathbb{Q}) \cap \big(H'(\mathbb{Q}_p) \times K^p \big) = \big(O'_B[\tfrac{1}{p}] \big)^\times. \tag{11.21}$$

Since any $\mathbf{y} \in L'$ with $Q(\mathbf{y}) = T$ spans a nondegenerate 2–plane in V',

$$\Gamma'_{\mathbf{y}} = \Gamma' \cap Z(\mathbb{Q}) \simeq \big(\mathbb{Z}[\tfrac{1}{p}]\big)^\times = \{\pm 1\} \times p^{\mathbb{Z}}, \tag{11.22}$$

and the central element p acts on $\mathcal{D}^\bullet$ by translation by 2, i.e. carries $\mathcal{D}^i$ to $\mathcal{D}^{i+2}$. Thus, unfolding, as in [**46**], p.216, we have

$$\chi(\hat{\mathcal{C}}_p(T), \mathcal{O}_{\hat{\mathcal{C}}_p(t_1)} \overset{\mathbb{L}}{\to} \otimes \mathcal{O}_{\hat{\mathcal{C}}_p(t_2)}) = \sum_{\substack{\mathbf{y} \in (L')^2 \\ Q(\mathbf{y})=T \\ \mathrm{mod}\ \Gamma'}} \chi(\, Z(\mathbf{j}), \mathcal{O}_{Z(j_1)} \overset{\mathbb{L}}{\to} \otimes \mathcal{O}_{Z(j_2)}\,). \tag{11.23}$$

Here we have used the orbifold convention, which introduces a factor of $\frac{1}{2}$ from the ± 1 in $\Gamma'_{\mathbf{y}}$ which acts trivially, and the fact that the two 'sheets' $Z(\mathbf{j}) = Z^0(\mathbf{j})$ and $Z^1(\mathbf{j})$ make the same contribution.

Two of the main results of [**46**], Theorem 5.1 and 6.1, give the following:

Theorem 11.3 (i) The quantity

$$e_p(T) := \chi(\, Z(\mathbf{j}), \mathcal{O}_{Z(j_1)} \overset{\mathbb{L}}{\to} \otimes \mathcal{O}_{Z(j_2)}\,)$$

is the intersection number, [**46**], section 4, of the cycles $Z(j_1)$ and $Z(j_2)$ in the formal scheme $\mathcal{D}_p$. It depends only on the $\mathrm{GL}_2(\mathbb{Z}_p)$–equivalence class of T.
(ii) For $p \neq 2$, and for $T \in \mathrm{Sym}_2(\mathbb{Z}_p)$ which is $\mathrm{GL}_2(\mathbb{Z}_p)$–equivalent to $\mathrm{diag}(\epsilon_1 p^\alpha, \epsilon_2 p^\beta)$, with $0 \le \alpha \le \beta$ and $\epsilon_1, \epsilon_2 \in \mathbb{Z}_p^\times$,

$$e_p(T) = \alpha+\beta+1-\begin{cases} p^{\alpha/2} + 2\frac{p^{\alpha/2}-1}{p-1} & \text{if } \alpha \text{ is even and } (-\epsilon_1, p)_p = -1, \\ (\beta-\alpha+1)\,p^{\alpha/2} + 2\frac{p^{\alpha/2}-1}{p-1} & \text{if } \alpha \text{ is even and } (-\epsilon_1, p)_p = 1, \\ 2\frac{p^{(\alpha+1)/2}-1}{p-1} & \text{if } \alpha \text{ is odd.} \end{cases}$$

Part (i) of Theorem 11.3, together with (11.16) and (11.23), yields

$$\chi(\, \mathcal{Z}(T), \mathcal{O}_{\mathcal{Z}(t_1)} \overset{\mathbb{L}}{\to} \otimes \mathcal{O}_{\mathcal{Z}(t_2)}\,) \;=\; e_p(T) \cdot \Big(\sum_{\substack{\mathbf{y} \in (L')^2 \\ Q(\mathbf{y})=T \\ \mathrm{mod}\ \Gamma'}} 1 \Big), \tag{11.24}$$

which is the analogue of (11.8) in the present case.

On the other hand, if $\mathrm{Diff}(T,B)=\{\infty,p\}$ with $p\mid D(B)$, the term on the right side of (11.1) is

$$\mathcal{E}'_{2,T}(\tau,0;B)=c\cdot W_{T,\infty}(\tau,0;\frac{3}{2})\cdot W'_{T,p}(0,\tilde{\Phi}_p)\cdot\prod_{\ell\neq p}W_{T,\ell}(0,\tilde{\Phi}^B_\ell). \tag{11.25}$$

where $c=\eta(0;B)\zeta_{D(B)}(2)$. By [**46**], Corollary 11.4, the section $\tilde{\Phi}_p(s)$ in (10.9) satisfies,

$$W'_{T,p}(0,\tilde{\Phi}_p)=p^{-2}(p+1)\log(p)\cdot e_p(T) \tag{11.26}$$

while

$$\prod_{\ell\neq p}W_{T,\ell}(0,\Phi_\ell)\ \stackrel{\circ}{=}\ \Big(\sum_{\substack{y\in(L')^2\\ Q(y)=T\\ \mathrm{mod}\ \Gamma'}}1\Big). \tag{11.27}$$

can be thought of as the number of 'connected components' of $\mathcal{Z}(T)$. Note that, the particular choice (10.10) of the coefficients $A_p(s)$ and $B_p(s)$ in the definition of $\tilde{\Phi}_p(s)$ was dictated by the identity (11.16), the proof of which is based on results of Tonghai Yang, [**74**], on representation densities $\alpha_p(S,T)$ of binary forms T by nonunimodular forms S. To obtain (11.16) in the case $p=2$, one needs to extend both the intersection calculations of [**46**] and the density formulas of [**74**] to this case. The first of these tasks, begun in the appendix to section 11 of [**50**], is now complete. The second is in progress.

Finally, the contribution to the right side of (11.1) of the terms for T of signature $(1,1)$ or $(0,2)$, which can be calculated using the formulas of [**65**], coincides with the contribution of the star product of the Green functions $\Xi(t_1,v_1)$ and $\Xi(t_2,v_2)$ to the height pairing. This is a main result of [**37**]; for a sketch of the ideas involved, cf. [**38**].

This completes the sketch of the proof of Theorem 11.2. □ □

Before turning to consequences, we briefly sketch how part (ii) of Theorem 11.3 is obtained. By part (i) of that Theorem, it will suffice to compute the intersection number $(Z(j_1),Z(j_2))$ for j_1 and $j_2\in V'(\mathbb{Q}_p)$ with $j_1^2=-Q(j_1)=-\epsilon_1p^\alpha$, $j_2^2=-Q(j_2)=-\epsilon_2p^\beta$ and $(j_1,j_2)=j_1j_2+j_2j_1=0$. By Theorem 5.1 of [**46**], $(Z(j_1),Z(j_2))=(Z(j_1)^{\mathrm{pure}},Z(j_2)^{\mathrm{pure}})$, so that we may use the description of $Z(j)^{\mathrm{pure}}$ given in Proposition 4.5, above.

The following result, which combines Lemmas 4.7, 4.8, and 4.9 of [**46**], describes the intersection numbers of individual components. Recall that the vertical components $\mathbb{P}_{[\Lambda]}$ of $\mathcal{D}$ are indexed by vertices $[\Lambda]$ in the building $\mathcal{B}$ of $\mathrm{PGL}_2(\mathbb{Q}_p)$.

Lemma 11.4
(i) For a pair of vertices $[\Lambda]$ and $[\Lambda']$,

$$(\mathbb{P}_{[\Lambda]}, \mathbb{P}_{[\Lambda']}) = \begin{cases} 1 & \text{if } ([\Lambda],[\Lambda']) \text{ is an edge,} \\ -(p+1) & \text{if } [\Lambda] = [\Lambda'], \text{ and} \\ 0 & \text{otherwise.} \end{cases}$$

(ii)

$$(Z(j_1)^h, Z(j_2)^h) = \begin{cases} 1 & \text{if } \alpha \text{ and } \beta \text{ are odd,} \\ 0 & \text{otherwise.} \end{cases}$$

(iii)

$$(Z(j_1)^h, \mathbb{P}_{[\Lambda]}) = \begin{cases} 2 & \text{if } \alpha \text{ is even, } (-\epsilon_1, p)_p = -1, \text{ and } \mathcal{B}^{j_1} = [\Lambda], \\ 1 & \text{if } \alpha \text{ is odd and } d([\Lambda], \mathcal{B}^{j_1}) = \frac{1}{2}, \text{ and} \\ 0 & \text{otherwise.} \end{cases}$$

and similarly for $Z(j_2)^h$.

The computation of the intersection number $(Z(j_1)^{\text{pure}}, Z(j_2)^{\text{pure}})$ is thus reduced to a combinatorial problem. Recall from Proposition 4.5 above that the multiplicity in $Z(j_1)^{\text{pure}}$ of a vertical component $\mathbb{P}_{[\Lambda]}$ indexed by a vertex $[\Lambda]$ is determined by the distance of $[\Lambda]$ from the fixed point set $\mathcal{B}^{j_1}$ of j_1 on $\mathcal{B}$ by the formula

$$\mu_{[\Lambda]}(j_1) = \max\{\, 0, \, \frac{\alpha}{2} - d([\Lambda], \mathcal{B}^{j_1})\,\}.$$

In particular, this multiplicity is zero outside of the tube $\mathcal{T}(j_1)$ of radius $\frac{\alpha}{2}$ around $\mathcal{B}^{j_1}$. Of course, the analogous description holds for $Z(j_2)^{\text{pure}}$. Our assumption that the matrix of inner products of j_1 and j_2 is diagonal implies that j_1 and j_2 anticommute, and hence the relative position of the fixed point sets and tubes $\mathcal{T}(j_1)$ and $\mathcal{T}(j_2)$ is particularly convenient.

For example, consider the case in which α and β are both even, with $(-\epsilon_1,p)_p = 1$ and $(-\epsilon_2,p)_p = -1$. These conditions mean that $\mathbb{Q}_p(j_1)^\times$ is a split Cartan in $GL_2(\mathbb{Q}_p)$ and $\mathbb{Q}_p(j_2)^\times$ is a nonsplit, unramified, Cartan. The fixed point set $\mathcal{A} = \mathcal{B}^{j_1}$ is an apartment, and $\mathcal{T}(j_1)$ is a tube of radius $\frac{\alpha}{2}$ around it. Moreover, $Z(j_1)^h = \emptyset$, so that $Z(j_1)$ consists entirely of vertical components. The fixed point set $\mathcal{B}^{j_2}$ is a vertex $[\Lambda_0]$ and $\mathcal{T}(j_2)$ is a ball of radius $\frac{\beta}{2}$ around it. Since j_1 and j_2 anticommute, the vertex $[\Lambda_0]$ lies in the apartment $\mathcal{A} = \mathcal{B}^{j_1}$. For any vertex $[\Lambda]$, the geodesic from $[\Lambda_0]$ to $[\Lambda]$ runs a distance ℓ inside the apartment $\mathcal{A}$ and then a distance r outside of it. By (i) of Lemma 11.4, the contribution to the intersection number of the vertices with $\ell = 0$ is

$$(11.28)\qquad -(p+1)\frac{\alpha}{2} + (1-p)\sum_{r=1}^{\alpha/2-1}\left(\frac{\alpha}{2} - r\right)(p-1)p^{r-1} = 1 - \alpha - p^{\alpha/2}.$$

Here the first term is the contribution of $[\Lambda_0]$, since

$$(11.29)\qquad (\mathbb{P}_{[\Lambda_0]}, Z(j_2)^v) = -(p+1)\frac{\beta}{2} + (p+1)\left(\frac{\beta}{2} - 1\right) = -(p+1),$$

where $Z(j_2)^v$ is the vertical part of $Z(j_2)^{\text{pure}}$. Similarly, for $r > 0$, there are $(p-1)p^{r-1}$ vertices $[\Lambda]$ at distance r (with $\ell = 0$), and each contributes

$$(11.30)\quad (\mathbb{P}_{[\Lambda]}, Z(j_2)^v) = \left(\frac{\beta}{2} - r + 1\right) - (p+1)\left(\frac{\beta}{2} - r\right) + p\left(\frac{\beta}{2} - r - 1\right) = 1 - p.$$

Here we have used the fact that all such vertices lie inside the ball $\mathcal{T}(j_2)$. Next, consider vertices with $1 \le \ell \le (\beta-\alpha)/2$. Note that, if such a vertex $[\Lambda]$ has $r = \frac{\alpha}{2}$, then $[\Lambda] \in \mathcal{T}(j_2)$. The contribution of vertices with ℓ values in this range is

$$(11.31)\ 2(1-p)\sum_{\ell=1}^{(\beta-\alpha)/2}\left(\frac{\alpha}{2} + \sum_{r=1}^{\alpha/2-1}\left(\frac{\alpha}{2} - r\right)(p-1)p^{r-1}\right) = (\alpha-\beta)(p^{\alpha/2-1} - 1).$$

Next, in the range $(\beta-\alpha)/2 < \ell < \beta/2$ and with $r < \beta/2 - \ell - 1$, $[\Lambda] \in \mathcal{T}(j_2)$, and such $[\Lambda]$'s contribute

$$(11.32)\ 2(1-p)\sum_{\ell=(\beta-\alpha)/2+1}^{\beta/2-1}\left(\frac{\alpha}{2} + \sum_{r=1}^{\beta/2-\ell-1}\left(\frac{\alpha}{2} - r\right)(p-1)p^{r-1}\right) = 2\alpha - 4\frac{p^{\alpha/2-1}-1}{p-1}.$$

Finally, the vertices on the boundary of the ball $\mathcal{T}(j_2)$ contribute

$$(11.33)\qquad \alpha + 2\sum_{\ell=(\beta-\alpha)/2+1}^{\beta/2-1}\left(\frac{\alpha}{2} - \left(\frac{\beta}{2} - \ell\right)\right)(p-1)p^{\beta/2-\ell-1} = 2\frac{p^{\alpha/2-1}-1}{p-1}.$$

Summing these contributions and adding $(\mathbb{P}_{[\Lambda_0]}, Z(j_2)^h)\cdot\alpha/2 = \alpha$, we obtain the quantity claimed in (ii) of Theorem 11.3 in this case. The other cases are similar. □

We now describe the consequences of Theorem 11.2.

Corollary 11.5 Assume that the p-adic density identity (11.11) (resp. (11.26)) holds for $p = 2$ when $2 \mid D(B)$ (resp. $2 \nmid D(B)$). Then

$$\langle \widehat{\theta}(\tau_1), \widehat{\theta}(\tau_2) \rangle = \mathcal{E}_2'(\begin{pmatrix} \tau_1 & \\ & -\bar{\tau}_2 \end{pmatrix}, 0; B) + \sum_{t_1} \Big(\sum_{\substack{t_2 \\ t_1 t_2 = \text{square}}} c(t_1, t_2, v_1, v_2) \cdot \bar{q}_2^{t_2} \Big) q_1^{t_1}.$$

for some coefficients $c(t_1, t_2, v_1, v_2)$.

The extra term on the right hand side should vanish, according to Conjecture 11.1, but, in any case, the coefficient of each $q_1^{t_1}$ is a modular form of weight $\frac{3}{2}$ in τ_2 with only one square class of Fourier coefficients, i.e., is a distinguished form. But such forms all come from $O(1)$'s, and so are orthogonal to cusp forms in $\mathcal{A}_{00}(G')$, [**19**].

Corollary 11.6 With the notation and assumptions of Theorem 10.1 and assuming that the p-adic density identity (11.11) holds for $p = 2$,

$$\langle \widehat{\theta}(\tau_1), \widehat{\theta}(f) \rangle = f(\tau_1) \cdot C(0) \cdot \begin{cases} L'(\frac{1}{2}, \mathrm{Wald}(\sigma)), \\ L(\frac{1}{2}, \mathrm{Wald}(\sigma)) \cdot \log(p), \\ 0 \end{cases}.$$

Proof

$$\begin{aligned} \langle \widehat{\theta}(\tau_1), \widehat{\theta}(f) \rangle &= \langle \widehat{\theta}(\tau_1), \langle \widehat{\theta}, f \rangle_{\mathrm{Pet}} \rangle \\ &= \langle \langle \widehat{\theta}(\tau_1), \widehat{\theta}(\tau_2) \rangle, \overline{f(\tau_2)} \rangle_{\mathrm{Pet}} \\ &= \langle \mathcal{E}_2'(\begin{pmatrix} \tau_1 & \\ & -\bar{\tau}_2 \end{pmatrix}, 0; B), \overline{f(\tau_2)} \rangle_{\mathrm{Pet}} \end{aligned}$$

$$= f(\tau_1) \cdot C(0) \cdot \begin{cases} L'(\frac{1}{2}, \mathrm{Wald}(\sigma)), \\ L(\frac{1}{2}, \mathrm{Wald}(\sigma)) \cdot \log(p), \\ 0 \end{cases} \tag{11.34}$$

□

Corollary 11.7 With the notation and assumptions of Corollary 11.6,
(i) If $\epsilon_p(\sigma_p) = +1$ for more than one $p \mid D(B)$, then $\widehat{\theta}(f) = 0$.
Otherwise
(ii) If $\epsilon(\frac{1}{2}, \mathrm{Wald}(\sigma)) = +1$, then there is a unique prime $p \mid D(B)$ such that $\epsilon_p(\sigma_p) = +1$, $\widehat{\theta}(f) \in \mathrm{Vert}_p \subset \mathrm{Vert}$, and

$$\langle \widehat{\theta}(f), \widehat{\theta}(f) \rangle = C(0) \cdot \langle f, f \rangle \cdot L(\frac{1}{2}, \mathrm{Wald}(\sigma)) \cdot \log(p).$$

(iii) If $\epsilon(\frac{1}{2}, \mathrm{Wald}(\sigma)) = -1$, then $\widehat{\theta}(f) \in \widetilde{\mathrm{MW}}$ and

$$\langle \widehat{\theta}(f), \widehat{\theta}(f) \rangle = C(0) \cdot \langle f, f \rangle \cdot L'(\frac{1}{2}, \mathrm{Wald}(\sigma)).$$

This result gives an analogue of Waldspurger's theory, described in section 9 above, and of the Rallis inner product formula, [**61**], for the arithmetic theta lift (8.1). Of course, the result only applies to forms of weight $\frac{3}{2}$, and we have made quite strong restrictions on the level of f. The restrictions on the level can most likely be removed with some additional work, but the restriction on the weight is essential to our setup.

In the case $\epsilon(\frac{1}{2}, \mathrm{Wald}(\sigma)) = +1$, part (ii) gives a geometric interpretation of the value $L(\frac{1}{2}, \mathrm{Wald}(\sigma))$ in terms of vertical components of $\mathcal{M}$, analogous to the geometric interpretation of the value of the base change L-function given by Gross in his Montreal paper, [**21**].

Finally, if $\widehat{\theta}(f) \neq 0$, let

$$\widehat{\mathcal{Z}}(t)(f) = \widehat{\mathcal{Z}}(t,v)(f) = \frac{\langle \widehat{\mathcal{Z}}(t,v), \widehat{\theta}(f) \rangle}{\langle \widehat{\theta}(f), \widehat{\theta}(f) \rangle} \cdot \widehat{\theta}(f) \tag{11.35}$$

be the component of the cycle $\widehat{\mathcal{Z}}(t,v)$ along the line spanned by $\widehat{\theta}(f)$. Note that, by Corollary 11.6, this projection does not depend on v. The following is an analogue of the Gross-Kohnen-Zagier relation, [**24**], [**78**]. Note that it holds in both cases $\epsilon(\frac{1}{2}, \mathrm{Wald}(\sigma)) = +1$ or $\epsilon(\frac{1}{2}, \mathrm{Wald}(\sigma)) = -1$.

Corollary 11.8

$$\sum_t \widehat{\mathcal{Z}}(t)(f) \cdot q^t = \frac{f(\tau) \cdot \widehat{\theta}(f)}{\langle f, f \rangle}.$$

Thus, the Fourier coefficients of f encode the position of the cycles $\widehat{\mathcal{Z}}(t)(f)$ on the $\widehat{\theta}(f)$ line. Both sides are invariant under scaling of f by a constant factor.

For the related results of Gross–Kohnen–Zagier in the case $\epsilon(\frac{1}{2}, \mathrm{Wald}(\sigma)) = -1$, cf. [**24**], especially Theorem C, p. 503, the discussion on pp. 556–561, and the examples in [**78**], where the analogies with the work of Hirzebruch–Zagier are also explained and used in the proof! Our class $\widehat{\theta}(f)/ < f, f >$, which arises as the image of f under the arithmetic theta lift, is the analogue of the class y_f in [**24**]. To proved the main part of Theorem C of [**24**] in our case, namely that the π_f–isotypic components, where $\pi = \mathrm{Wald}(\sigma)$, of the classes $\widehat{\mathcal{Z}}(t,v)$ either vanish or lie on a line, we can use the Howe duality Theorem [**30**], [**72**] for the local theta correspondence, cf. [**42**].

In [**22**], Gross gives a beautiful representation theoretic framework in which the Gross–Zagier formula, [**26**], and a result of Waldspurger, [**71**], can be viewed together. He works with unitary similitude groups $G = G^B = GU(2)$, constructed from the choice of a quaternion algebra B over a global field k and a quadratic extension E/k which splits B. The torus T over k with $T(k) = E^\times$ embeds in

$$G^B(k) = (B^\times \times E^\times)/\Delta k^\times.$$

For a place v of k, the results of Tunnel, [**67**], and Saito, [**62**], show that the existence of local $T(k_v)$–invariant functionals on an irreducible admissible representation Π_v of G_v is controlled by the local root number $\epsilon_v(\Pi_v)$ of the Langlands L-function defined by a 4 dimensional symplectic representation of ${}^LG$, [**22**], section 10. There is a dichotomy phenomenon. If v is a place which is not split in E/k, then the local quaternion algebras B_v^+ and B_v^- are both split by E_v, so that the torus T_v embeds in both similitude groups G_v^+ and G_v^-. If Π_v is a discrete series representation of G_v^+, then, by the local Jacquet–Langlands correspondence, there is an associated representation Π_v' of G_v^-, and

$$\dim \mathrm{Hom}_{T_v}(\Pi_v, \mathbb{C}) + \dim \mathrm{Hom}_{T_v}(\Pi_v', \mathbb{C}) = 1.$$

For a global cuspidal automorphic representation Π of $G(\mathbb{A}) = (\mathrm{GL}_2(\mathbb{A}) \times E_{\mathbb{A}}^\times)/\mathbb{A}^\times$, there is a finite collection of quaternion algebras B over k, split by E, with automorphic cuspidal representations Π^B associated to Π by the global Jacquet–Langlands correspondence. When the global root number $\epsilon(\Pi)$ for the degree 4 Langlands L-function $L(s,\Pi)$ is $+1$, Waldspurger's theorem, [**71**], says that the nonvanishing of the automorphic $T(\mathbb{A})$–invariant linear functional, defined on Π^B by integration over $\mathbb{A}^\times T(k)\backslash T(\mathbb{A})$, is equivalent to (i) the nonvanishing of the local linear functionals[13] and (ii) the nonvanishing of the central value $L(\frac{1}{2},\Pi)$ of the L-function. Suppose that k is a totally real number field, E is a totally imaginary quadratic extension, and Π_v is a discrete series of weight 2 at every archimedean place of k. Then, in the case $\epsilon(\Pi) = -1$, the local root number conditions determine a Shimura curve M^B over k, and Gross defines a $T(\mathbb{A}_f)$–invariant linear functional on the associated Mordell–Weil space MW^{B} using the 0–cycle attached to T. Gross conjectures that the nonvanishing of this functional on the Π_f^B component of MW^{B} is equivalent to the nonvanishing of the central derivative $L'(\frac{1}{2},\Pi)$, and that there is an explicit expression for this quantity in terms of the height pairing of a suitable 'test vector'. In the case $k = \mathbb{Q}$, this is the classical Gross–Zagier formula, [**26**], and in there is work of Zhang in the general case [**79**].

Thus, there is a close parallel between Gross's 'arithmetic' version of Waldspurger's central value result [**71**] and our 'arithmetic' version of Waldspurger's results on the Shimura lift. It should be possible to formulate our conjectures above for Shimura curves over an arbitrary totally real field k. It would be interesting to find a direct connection between our constructions and the methods used by Shou-Wu Zhang, [**79**].

Finally, it is possible to formulate a similar theory for central value/derivative of the triple product L-function. This is discussed in [**25**], cf. also [**27**], [**28**]. It may be that all of these examples can be covered by some sort of arithmetic version of Jacquet's relative trace formula, [**33**], [**34**], [**1**]. Further speculations about 'arithmetic theta functions' and derivatives of Eisenstein series can be found in [**40**] and [**41**].

[13] By local root number conditions, this uniquely determines the quaternion algebra B for the given E and Π.

References

[1] E. M. Baruch and Z. Mao, *Central value of automorphic L-functions,* preprint (2002) .

[2] S. Böcherer, *Über die Funktionalgleichung automorpher L-Funktionen zur Siegelschen Modul-gruppe,* J. reine angew. Math. vol. 362 (1985), pp. 146–168 .

[3] R. Borcherds, *Automorphic forms on* $O_{s+2,2}(\mathbf{R})$ *and infinite products,* Invent. math. vol. 120 (1995), pp. 161–213 .

[4] ———, *Automorphic forms with singularities on Grassmannians,* Invent. math. vol. 132 (1998), pp. 491–562 .

[5] ———, *The Gross-Kohnen-Zagier theorem in higher dimensions,* Duke Math. J. vol. 97 (1999), pp. 219–233 .

[6] ———, *Correction to: "The Gross-Kohnen-Zagier theorem in higher dimensions",* Duke Math. J. (2000), vol. 105 pp. 183–184 .

[7] J.-B. Bost, *Potential theory and Lefschetz theorems for arithmetic surfaces,* Ann. Sci. École Norm. Sup. (1999), vol. 32, pp. 241–312 .

[8] ———, Lecture at Univ. of Maryland, Nov. 11, 1998, .

[9] J.-F. Boutot and H. Carayol, *Uniformisation p-adique des courbes de Shimura, Courbes Modulaires et Courbes de Shimura,* Astérisque, vol. **196–197**, 1991, pp. 45–158 .

[10] J. H. Bruinier, *Borcherds products and Chern classes of Hirzebruch-Zagier divisors,* Invent. Math. **138** (1999), 51–83 .

[11] ———, *Borcherds products on O(2,l) and Chern classes of Heegner divisors, Lecture Notes in Math. 1780,* Springer, New York, 2002 .

[12] J. H. Bruinier, J. I. Burgos Gil, and U. Kühn, *Borcherds products in the arithmetic intersection theory of Hilbert modular surfaces,* in preparation .

[13] J. I. Burgos Gil, J. Kramer and U. Kühn, *Cohomological arithmetic Chow rings,* preprint, (2003) .

[14] H. Cohen, *Sums involving the values at negative integers of L-functions of quadratic characters,* Math. Ann. **217** (1975), pp. 271–285 .

[15] G. Faltings, *Calculus on arithmetic surfaces,* Annals of Math. **119** (1984), pp. 387–424.

[16] J. Funke, *Rational quadratic divisors and automorphic forms,* Thesis, University of Maryland (1999) .

[17] ———, *Heegner Divisors and non-holomorphic modular forms,* Compositio Math. (to appear) .

[18] P. Garrett, *Pullbacks of Eisenstein series; applications,* Automorphic Forms of Several Variables, Tanaguchi Symposium, Katata, 1983, Birkhäuser, Boston, 1984 .

[19] S. Gelbart and I. I. Piatetski-Shapiro, *Distinguished representations and modular forms of half-integral weight,* Invent. Math. **59** (1980), pp. 145–188 .

[20] H. Gillet and C. Soulé, *Arithmetic intersection theory,* Publ. Math. IHES **72** (1990), pp. 93–174 .

[21] B. H. Gross, *Heights and special values of L-series,* in Number Theory (Montreal, 1985), (H. Kisilevsky and J. Labute, ed.) CMS Conf, Proc. 7 AMS, Providence: 1987, pp. 115–187.

[22] ———, *Heegner points and representation theory,* Proc. of Conference on Special Values of Rankin L-Series, MSRI, Dec. 2001, to appear.

[23] B. H. Gross and K. Keating, *On the intersection of modular correspondences,* Invent. Math. **112** (1993), pp. 225–245 .

[24] B. H. Gross, W. Kohnen and D. Zagier, *Heegner points and derivatives of L-functions. II,* Math. Annalen, **278** (1987), pp. 497–562 .

[25] B. H. Gross and S. Kudla, *Heights and the central critical values of triple product L-functions,* Compositio Math. **81** (1992), pp. 143–209 .

[26] B. H. Gross and D. Zagier, *Heegner points and the derivatives of L-series,* Inventiones math. **84** (1986), pp. 225–320 .

[27] M. Harris and S. Kudla, *The central critical value of a triple product L-function,* Annals of Math. **133** (1991), pp. 605–672 .

[28] ———, *On a conjecture of Jacquet* , preprint, 2001, arXiv:math.NT/0111238 .

[29] F. Hirzebruch and D. Zagier, *Intersection numbers of curves on Hilbert modular surfaces and modular forms of Nebentypus,* Invent. Math. **36** (1976), pp. 57–113 .

[30] R. Howe, *θ-series and invariant theory,* in P roc. Symp. Pure Math. vol. 33, 1979, pp. 275–285 .

[31] R. Howe and I. I. Piatetski-Shapiro, *Some examples of automorphic forms on* Sp_4, Duke Math. J. **50** (1983), 55–106 .

[32] P. Hriljac, *Heights and arithmetic intersection theory,* Amer. J. Math. **107** (1985), 23–38 .

[33] H. Jacquet *On the nonvanishing of some L-functions* Proc. Indian Acad. Sci. **97** (1987) pp. 117–155 .

[34] ——— *Sur un result de Waldspurger,* Ann. Scinet. Éc. Norm. Sup. **19** (1986), 185–229 .

[35] Y. Kitaoka, *A note on local densities of quadratic forms,* Nagoya Math. J. **92** (1983), pp. 145–152 .

[36] R. Kottwitz, *Points on some Shimura varieties over finite fields,* J. AMS **5** (1992), pp. 373–444 .

[37] S. Kudla, *Central derivatives of Eisenstein series and height pairings,* Ann. of Math. **146** (1997), pp. 545-646 .

[38] ———, *Derivatives of Eisenstein series and generating functions for arithmetic cycles,* Sém. Bourbaki n° 876, Astérisque, vol. 276 2002 pp. 341–368 .

[39] ———, *Integrals of Borcherds forms,* to appear in Compositio Math .

[40] ——— *Eisenstein series and arithmetic geometry,* Proc. of the ICM, Beijing, August 2002.

[41] ———, *Special cycles and derivatives of Eisenstein series,* Proc. of Conference on Special Values of Rankin L-Series, MSRI, Dec. 2001, to appear.

[42] ———, *An arithmetic theta function,* in preparation .

[43] S. Kudla and J. Millson, *The theta correspondence and harmonic forms I,* Math. Annalen, **274** (1986) pp. 353–378 .

[44] S. Kudla and S. Rallis, *A regularized Siegel-Weil formula: the first term identity,* Annals of Math., **140** (1994) 1–80 .

[45] S. Kudla and M. Rapoport, *Arithmetic Hirzebruch–Zagier cycles,* J. reine angew. Math., **515** (1999) 155–244 .

[46] ———, *Height pairings on Shimura curves and p-adic uniformization,* Invent. math., **142** (2000) 153–223 .

[47] ———, *Cycles on Siegel threefolds and derivatives of Eisenstein series,* Ann. Scient. Éc. Norm. Sup., **33** (2000) pp. 695–756 .

[48] S. Kudla, M. Rapoport and T. Yang, *On the derivative of an Eisenstein series of weight 1,* Int. Math. Res. Notices, No.7 (1999), 347–385 .

[49] ———, *Derivatives of Eisenstein series and Faltings heights,* preprint (2001) .

[50] ———, *Derivatives of Eisenstein series and Faltings heights II: modular curves,* in preparation .

[51] ———, *On a Siegel–Eisenstein series,* in preparation .

[52] ———, *An arithmetic inner product formula,* in preparation .

[53] U. Kühn, *Generalized arithmetic intersection numbers,* J. reine angew. Math., **534** (2001), pp. 209–236 .

[54] J.-S. Li, *Non-vanishing theorems for the cohomology of certain arithmetic quotients,* J. reine angew. Math., **428** (1992), pp. 177–217 .

[55] V. Maillot and D. Roessler, *Conjectures sur les dérivées logarithmiques des fonctions L d'Artin aux entiers négatifs,* preprint (2001) .

[56] W. J. McGraw, *On the rationality of vector-valued modular forms,* Math. Ann. (2003) DOI: 10.1007/s00208-003-0413-1 .

[57] S. Niwa, *Modular forms of half integral weight and the integral of certain theta functions,* Nagoya Math. J. **56** (1975), 147–161 .

[58] T. Oda and M. Tsuzuki, *Automorphic Green functions associated to secondary spherical harmonics,* preprint (2000) .

[59] I. I. Piatetski-Shapiro, *Work of Waldspurger,* Lie Group Representations II, Lecture Notes in Math. 1041, (1984), pp. 280–302 .

[60] I. I. Piatetski-Shapiro and S. Rallis, *L-functions for classical groups,* Lecture Notes in Math., vol. 1254, Springer-Verlag, New York, (1987), pp. 1–52 .

[61] S. Rallis, *Injectivity properties of liftings associated to Weil representations,* Compositio Math., **52** (1984), 139–169 .

[62] H. Saito, *On Tunnell's formula for characters of* GL_2, Compositio Math, **85** (1993), 99–108 .

[63] C. L. Siegel, *Lectures on quadratic forms,* Tata Institute, Bombay, 1957. .

[64] G. Shimura, *On modular forms of half integral weight,* Ann. of Math. **97** (1973), 440–481.

[65] ———, *Confluent hypergeometric functions on tube domains,* Math. Ann. **260** (1982) 269-302 .

[66] C. Soulé, D. Abramovich, J.-F. Burnol, and J.Kramer, *Lectures on Arakelov Geometry,* **Cambridge Stud. Adv. Math., vol 33** Cambridge U. Press. 1992.

[67] J. Tunnell, *Local* ϵ*-factors and characters of* GL_2, Amer. J. of Math. **105** (1983), pp. 1277–1308 .

[68] J.-L. Waldspurger, *Correspondance de Shimura,* J. Math. Pures Appl., **59** (1980), pp. 1–132 .

[69] ———, *Sur les coefficients de Fourier des formes modulaires de poids demi-entier,* J. Math. Pures Appl., **60** (1981), 375–484 .

[70] ———, *Correspondance de Shimura,* Séminare de Théorie des Nombres, Paris 1979–80, Progr. Math., 12, Birkhäuser, Boston, 1981, 357–369 .

[71] ———, *Sur les valeurs de certaines fonctions L automorphes en leur centre de symétrie,* Compositio Math **54** (1985), 173–242 .

[72] ———, *Demonstration d'une conjecture de duality de Howe dans le case p-adique,* $p \neq 2$, Festschrift in honor of Piatetski-Shapiro, vol 2, Israel Math. Conf. Proc., 1990, pp. 267–234.

[73] ——— *Correspondances de Shimura et quaternions* Forum Math. **3** (1991), 219–307 .

[74] T. Yang An explicit formula for local densities of quadratic forms, *J. Number Theory,* **72** (1998), pp. 309–356 .

[75] ———, *The second term of an Eisenstein series,* Proc. of the ICCM, (to appear) .

[76] ———, *Faltings heights and the derivative of Zagier's Eisenstein series,* Proc. of MSRI workshop on Heegner points, preprint (2002) .

[77] D. Zagier, Nombres de classes et formes modulaires de poids 3/2, *C. R. Acad. Sc. Paris* **281** (1975), 883–886 .

[78] ———, Modular points, modular curves, modular surfaces and modular forms, Lecture Notes in Math. 1111, Springer, Berlin, (1985), pp. 225–248 .

[79] Shou-Wu Zhang, *Gross–Zagier formula for* GL_2, Asian J. of Math. **5** (2001), pp. 183–290

Current Developments in Mathematics, 2002
Pages 181 – 217

End Invariants and the Classification of Hyperbolic 3-Manifolds

Yair N. Minsky

CONTENTS

1. Introduction

These notes are a biased guide to some recent developments in the deformation theory of hyperbolic 3-manifolds and Kleinian groups. This field has its roots in the work of Poincaré and Klein, and connects to topology via Thurston's geometrization program, to analysis via the Ahlfors-Bers quasiconformal theory, and to complex dynamics via the work of Thurston, Sullivan and others. It encompasses many techniques and ideas and may be too big a subject for a single account. We will focus on the geometric study of ends of hyperbolic 3-manifolds and boundaries of deformation spaces, and in particular on the techniques that led to the recent solution by Brock, Canary and the author [**82, 23**] of the incompressible-boundary case of Thurston's "Ending Lamination Conjecture".

The space of hyperbolic structures on a fixed 3-manifold M is studied by considering representations of $\pi_1(M)$ into the isometry group of hyperbolic 3-space $\mathbb{H}^3$, up to a natural equivalence by conjugation. In this space, called the character variety, the injective representations with discrete image correspond to hyperbolic 3-manifolds that are homotopy-equivalent to M (see §2 for more details).

Mostow/Prasad rigidity tells us that when M is closed or admits a hyperbolic structure of finite volume, the hyperbolic structure is unique, and hence the deformation space is a single point. On the other hand, in the infinite volume case the deformation space has nontrivial interior in the character variety.

This interior has been fairly completely described, in a complex-analytic sense, by the work of Ahlfors, Bers, Kra, Marden, Maskit and Sullivan (see e.g. [**3, 8, 9, 69, 61, 67, 100**]. The boundary continues to pose challenging questions, although considerable progress has been made in recent years. In many ways the general picture we have of this deformation space is analogous to that of the Mandelbrot set in complex dynamics, or more generally to the bifurcation locus in a family of rational maps of the Riemann sphere.

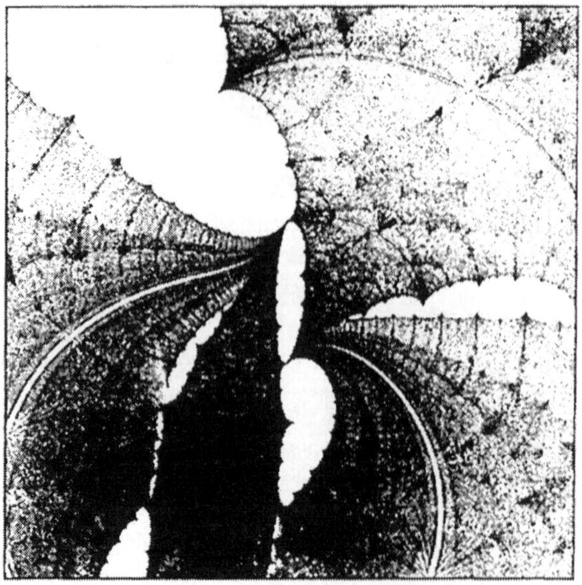

FIGURE 1. A 2-dimensional slice of the 4-dimensional deformation space associated to a once-punctured torus. The interior is the solid white region. (Courtesy C. McMullen and D. Wright)

We can summarize the current state of knowledge in terms of the following basic problems and questions:

- **Classification:** Describe elements in the deformation space in terms of their ***asymptotic geometry***. For elements in the interior this amounts to the Ahlfors-Bers parametrization in terms of Teichmüller parameters. For elements on the boundary this corresponds to Thurston's Ending Lamination Conjecture.
- **Density of the interior:** Bers, Sullivan and Thurston conjectured that the deformation space is the closure of its interior. This is analogous in dynamics to the density of hyperbolic (or Axiom A) systems – false in general but still conjectured and open for rational maps. Density was established, for the incompressible boundary case without cusps, by Brock-Bromberg, and is also a consequence of the Ending Lamination Conjecture.

- **Bumping:** Analogous to bifurcation in dynamics, this is the phenomenon, discovered by Anderson-Canary, that different components of the interior of the deformation space can have intersecting closures. This in particular implies that the topological type of a hyperbolic 3-manifold can change discontinuously in the representation space. Bumping and associated phenomena are the chief remaining obstructions to having a complete topological picture of the deformation space.
- **Tameness:** Marden conjectured that a hyperbolic 3-manifold with finitely-generated fundamental group is *tame*, i.e. homeomorphic to the interior of a compact 3-manifold. This is known for manifolds whose cores have incompressible boundaries but for the compressible boundary case is open, and is the chief obstruction to completing the classification problem.

There are other questions about the *geometry* of the deformation space, e.g. Miyachi's work on the shape of cusps [**84**], Kapovich's work on hyperbolic fixed points of 'renormalization' maps [**54**], or Keen-Series' work on pleating coordinates [**56**], which we will not address at all in this account. For a lovely semi-popular and well-illustrated treatment of many aspects of the deformation theory, see the book *Indra's Pearls* [**34**] by Mumford-Series-Wright.

In §§2–4 we will discuss, in a general way, the structure of the deformation theory and the current state of the problems listed above. In §§5–8 we will give a tutorial-style discussion of some central geometric tools that are used in the analysis of hyperbolic 3-manifolds: Margulis tubes, geometric limits, pleated surfaces, laminations and the complex of curves. In §6, in particular, we will discuss a collection of examples of Kleinian surface groups that illustrate some of the richness of the geometric structures that can occur. In §9 we will indicate in a broad outline how these ideas are put together to give a proof of the Ending Lamination Conjecture. We will conclude in §10 with a discussion of some corollaries of this work, and of some remaining open questions.

2. Hyperbolic geometry and deformations

Hyperbolic geometry. Hyperbolic n-space $\mathbb{H}^n$ is the unique complete simply connected Riemannian n-manifold with all sectional curvatures $K \equiv -1$. A standard model of $\mathbb{H}^n$ is given by the upper half-space $U = \{(x_1, \dots, x_n) : x_n > 0\}$, with the metric

$$\frac{dx_1^2 + \cdots dx_n^2}{x_n^2}.$$

The isometries of $\mathbb{H}^n$ are the Möbius transformations that preserve U. For $n = 2$, $\mathbb{H}^2$ can be identified with $\{z \in \mathbb{C} : \operatorname{Im} z > 0\}$, and the orientation-preserving isometries are just $\mathrm{PSL}_2(\mathbb{R})$, acting by $z \mapsto \frac{az+b}{cz+d}$. For $n = 3$ the orientation preserving isometries are $\mathrm{PSL}_2(\mathbb{C})$, acting on the upper half-space $\{(z,t) : z \in \mathbb{C}, t > 0\}$ by an extension of the natural action on $\mathbb{C}$. In fact the Riemann sphere $\widehat{\mathbb{C}} = \mathbb{C} \cup \{\infty\}$ is a natural boundary for $\mathbb{H}^3$, and the group acts there by conformal homeomorphisms.

A hyperbolic structure on an n-manifold N is a complete metric of constant sectional curvature $K = -1$ (usually considered up to diffeomorphism isotopic to the identity). Every such structure is obtained as the quotient $N \cong \mathbb{H}^n/\Gamma$ where Γ acts discretely on hyperbolic n-space $\mathbb{H}^n$ by isometries. When $n = 2$ Γ is called a ***Fuchsian group***, and when $n = 3$ it is called a ***Kleinian group***.

Geometrization. Thurston conjectured in the 1970's that all compact 3-manifolds admit canonical decompositions into "geometric pieces," where the geometry is one of eight possible types coming from three-dimensional homogeneous spaces: Euclidean space $\mathbb{E}^3$, the sphere S^3, hyperbolic space $\mathbb{H}^3$, and the "fibred geometries", $S^2 \times \mathbb{R}$, $\mathbb{H}^2 \times \mathbb{R}$, $\widetilde{\mathrm{PSL}_2}(\mathbb{R})$, the 3-dimensional solvable group *Sol*, and the 3-dimensional (nilpotent) Heisenberg group *Nil*. By far the most interesting and diverse family is the hyperbolic manifolds. See Scott [**95**] for a thorough discussion.

In particular, let M be the interior of a compact oriented 3-manifold $\overline{M}$. The hyperbolic part of Thurston's conjecture states:

Hyperbolization Conjecture. *M admits a complete hyperbolic structure if and only if*

1. *$\pi_1(M)$ is infinite or M is the 3-ball,*
2. *M is irreducible, and*
3. *M is atoroidal.*

M is irreducible if every embedded 2-sphere bounds a ball, and is atoroidal if no immersed torus is π_1-injective unless it is homotopic into $\partial\overline{M}$. It is easy to see that these conditions are necessary. Thurston proved this conjecture in many cases, and in particular in the case that $\partial\overline{M} \neq \emptyset$, which is the case that concerns us the most in these lectures. (The work of G. Perelman, announced not long after these lectures were given, promises to prove the Geometrization Conjecture in its full generality.)

Rigidity. If M admits a hyperbolic structure, one naturally asks about the set of all such structures, and this will be our concern in what follows. The theorems of Mostow [**86**] and Prasad [**93**] tell us that

Mostow/Prasad Rigidity. *If $n \geq 3$ and a hyperbolic n-manifold M is closed or has finite volume, then its hyperbolic structure is unique.*

In particular for oriented 3-manifolds, $vol(M) < \infty$ is equivalent to the condition that $\partial\overline{M}$ is a union of tori. In this case the ends of M are *cusps*: a neighborhood of a component T^2 of $\partial\overline{M}$ can be parameterized as $T^2 \times (0, \infty)$, where the second coordinate denotes distance and the cross-sectional tori $T^2 \times \{t\}$ contract exponentially with t.

This uniqueness of the hyperbolic structure is of course of great value in topology: any geometric property of the hyperbolic metric is automatically a topological invariant. Since, for example, many knot complements admit hyperbolic structures, this places hyperbolic geometry squarely in the center of 3 dimensional topology.

Deformations. When $\partial\overline{M}$ contains surfaces of genus 2 or greater, there is a very rich deformation theory for hyperbolic structures on M.

Let us first return for a moment to the 2-dimensional case. If S is the interior of a compact surface it is well-known that S admits a hyperbolic structure if and only if $\chi(S) < 0$. Assume for simplicity that S is oriented. Since a hyperbolic structure gives an identification of $\pi_1(S)$ with a discrete subgroup of $\mathrm{PSL}_2(\mathbb{R})$, the space of all such structures can be studied by first considering the quotient (in the sense of algebraic geometry)

$$X(S) = Hom(\pi_1(S), \mathrm{PSL}_2(\mathbb{R}))//\mathrm{PSL}_2(\mathbb{R}).$$

Here $\mathrm{PSL}_2(\mathbb{R})$ acts on $Hom(\pi_1(S), \mathrm{PSL}_2(\mathbb{R}))$ by conjugation of the image, since conjugate representations certainly give the same geometric structure. $X(S)$ is known as the ***character variety*** of $\pi_1(S)$, since it can be parametrized by traces of the images of generators (see e.g. Shalen [**96**]). Within $X(S)$ we consider those (equivalence classes of) representations that are injective with discrete images, and so that the quotient of $\mathbb{H}^2$ by the resulting group action has finite area, and is homeomorphic to S (when S is closed the last two conditions are automatic). This is called the ***Teichmüller space*** of S, denoted $\mathcal{T}(S)$. It is a well-known space that has been studied from various points of view since Riemann, and is in particular homeomorphic to $\mathbb{R}^{6g-6+2b}$, where g is the genus of S and b is the number of boundary components.

$\mathcal{T}(S)$ can also be identified with the space of Riemann surface structures on S modulo diffeomorphism homotopic to the identity. The equivalence with the previous definition is via the Uniformization Theorem.

For our 3-manifold M we can consider a similar definition:

$$X(M) = Hom(\pi_1(M), \mathrm{PSL}_2(\mathbb{C}))//\mathrm{PSL}_2(\mathbb{C}).$$

Within this variety lies $AH(M)$, those equivalence classes $[\rho]$ where $\rho : \pi_1(M) \to \mathrm{PSL}_2(\mathbb{C})$ is injective and has discrete image. (The terminology "AH", due to Thurston, indicates the Algebraic topology on the set of Hyperbolic structures.) The quotient manifold $N_\rho = \mathbb{H}^3/\rho(\pi_1(M))$ is equipped with a homotopy class of homotopy-equivalences $M \to N_\rho$, determined by $[\rho]$. We can think of $AH(M)$ as the set of hyperbolic 3-manifolds equipped with such homotopy-equivalences, called markings.

$AH(M)$ is a closed subset (excluding the elementary cases where $\pi_1(M)$ is abelian): this is a theorem of Jørgensen, using his criterion for discreteness [**53**] (or alternatively the "Margulis lemma" applied to $\mathrm{PSL}_2(\mathbb{C})$, see Thurston [**103**]). It is a consequence of the Ahlfors-Bers quasiconformal deformation theory that $AH(M)$ has a non-empty interior if $\partial\overline{M}$ has components of genus 2 or more.

To gain some appreciation for this fact, consider first a natural decomposition of the Riemann sphere imposed by the dynamics of the action of a discrete group $\Gamma \subset \mathrm{PSL}_2(\mathbb{C})$. The ***limit set*** Λ is the smallest closed invariant subset of $\widehat{\mathbb{C}}$, and the action of Γ on Λ is chaotic in various senses. The complement $\Omega = \widehat{\mathbb{C}} \setminus \Lambda$ is known as the ***domain of discontinuity***, and the quotient Ω/Λ is a Riemann surface, sometimes denoted $\partial_\infty N$.

Let us consider the special (but central) case that $\overline{M} = S \times [-1, 1]$ where S is a closed surface of genus $g \geq 2$. A point in $\mathcal{T}(S)$ is represented by $\rho : \pi_1(S) \to \mathrm{PSL}_2(\mathbb{R})$, which can equally well be considered as a representation of $\pi_1(M)$ into $\mathrm{PSL}_2(\mathbb{C})$. The quotient $\mathbb{H}^3/\rho(\pi_1(M))$ is homeomorphic to M. The limit set of $\rho(\pi_1(M))$ is the circle $\widehat{\mathbb{R}} = \mathbb{R} \cup \{\infty\}$, and Ω is a pair of disks $\Omega_\pm$. The quotient $\Omega/\rho(\pi_1(S))$ is a pair of Riemann surfaces conformally equivalent to the original point in $\mathcal{T}(S)$.

Poincaré and Klein were aware that such a group could be perturbed slightly in $\mathrm{PSL}_2(\mathbb{C})$ to yield a group that is no longer conjugate to a subgroup of $\mathrm{PSL}_2(\mathbb{R})$. Instead, such a small perturbation ρ' will have a limit set that is a Jordan curve but not a circle. The quotients $\Omega'_\pm/\rho'(\pi_1(M))$ are a pair of Riemann surfaces equipped with an identification with S, hence give rise to a new pair of points in $\mathcal{T}(S)$. Bers [**8**] proved the astonishing result that

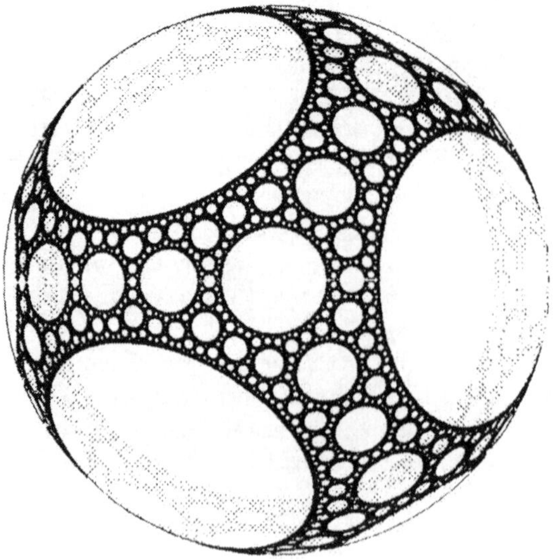

FIGURE 2. The limit set of a Kleinian group (courtesy C. McMullen)

Simultaneous Uniformization Theorem. *Every pair of points in $\mathcal{T}(S)$ are realized as $\Omega_\pm/\rho(\pi_1(S))$ for some $\rho \in AH(S \times [-1,1])$. This correspondence yields a homeomorphism*

$$QF(S) \cong \mathcal{T}(S) \times \mathcal{T}(S).$$

Here $QF(S) \subset AH(M)$ is the subset consisting of representations whose action on $\widehat{\mathbb{C}}$ is topologically (in fact quasiconformally) conjugate to the Fuchsian action. These are known as ***quasifuchsian groups***. Bers used his work with Ahlfors [**2**] on solutions to the Beltrami equation to show that *any* deformation of the conformal structures at infinity, i.e. of the surfaces $\Omega_\pm/\rho(\pi_1(S))$, can be extended to a deformation of the entire group action.

The work of Marden [**66**] and Sullivan [**100**] shows that $QF(S)$ is in fact the interior of $AH(S)$. This picture generalizes to any compact $\overline{M}$: Each component of $int(AH(M))$ is a region of ***structural stability***: two points in the same component correspond to marked hyperbolic manifolds that are in the same marked homeomorphism class, and whose associated actions on $\widehat{\mathbb{C}}$ are quasiconformally conjugate. The Simultaneous Uniformization Theorem generalizes to give a parameterization of each component in terms of a boundary Teichmüller space (see Kra [**61**] and Maskit [**68**]).

In general, quasiconformally conjugate actions on $\widehat{\mathbb{C}}$ correspond to quotient manifolds that are *quasi-isometric* (via e.g. the extension theorem of Douady-Earle [**35**]). Mostow rigidity depends on the fact that, when the manifolds are compact, no quasiconformal deformation of their group actions is possible. Sullivan proved a strong generalization of Mostow rigidity to finitely generated groups with infinite volume quotients, showing in essence that any quasiconformal deformation of a group must be supported on its domain of discontinuity:

FIGURE 3. This Jordan curve is the limit set of a quasifuchsian group (courtesy D. Wright).

Sullivan Rigidity. [**99**] *Let $f : N_1 \to N_2$ be a quasi-isometry between hyperbolic 3-manifolds with finitely-generated fundamental groups, that extends to a conformal homeomorphism from $\partial_\infty N_1$ to $\partial_\infty N_2$. Then f is homotopic to an isometry.*

Equivalently, a quasiconformal conjugacy in $\widehat{\mathbb{C}}$ between two finitely generated Kleinian groups which is conformal on the domain of discontinuity is a Möbius transformation.

3. Cores and ends of hyperbolic 3-manifolds

Let us consider a particular element of $AH(M)$, with quotient manifold N. We can understand, in a coarse sense, the hyperbolic structure of N by considering its decomposition into a *compact core* and *ends*. Scott [**94**] showed that any 3-manifold N with finitely-generated fundamental group contains a compact 3-dimensional submanifold K whose inclusion into N is a homotopy-equivalence. The components of $N \setminus K$ are then in one-to-one correspondence with the ends of N as a topological space, where each component is a neighborhood of an end. Let us assume, as we may by changing our choice of M, that K is homeomorphic to $\overline{M}$.

N also has a *convex core* C_N, which is the smallest closed *convex* submanifold of N whose inclusion is a homotopy-equivalence. The structure of the ends of N is classified by how they intersect with the convex core.

To simplify the discussion, assume from now on that the hyperbolic structure on N has *no cusps*: that is, no element of the associated group $\Gamma = \rho(\pi_1(N))$ is parabolic. In this case we have a dichotomy for the structure of each end of N:

- An end E is *geometrically finite* if it has a neighborhood that is outside the convex core. Such a neighborhood can then be foliated by a family S_t of surfaces, isotopic to a component S of ∂K, so that each S_t is convex, and the size of S_t grows exponentially with distance from K. The metrics on S_t, after conformal rescaling to constant size, converge to a unique hyperbolic metric on S, which therefore gives a point ν_E in the Teichmüller space $\mathcal{T}(S)$.

Equivalently, such an end corresponds to a component of $\partial_\infty N = \Omega/\rho(\pi_1(M))$, which is homeomorphic to S, and ν_E is the natural conformal structure inherited from Ω.

- An end E is *geometrically infinite* if it has a neighborhood that is completely contained in the convex core. (In this case there is no component of $\partial_\infty N$ corresponding to E).

That these are the only two possibilities is a consequence of Ahlfors' Finiteness Theorem [**3**], which implies that the hyperbolic area of ∂C_N is finite, from which it follows that ∂C_N cannot partition a neighborhood of an end into two unbounded pieces.

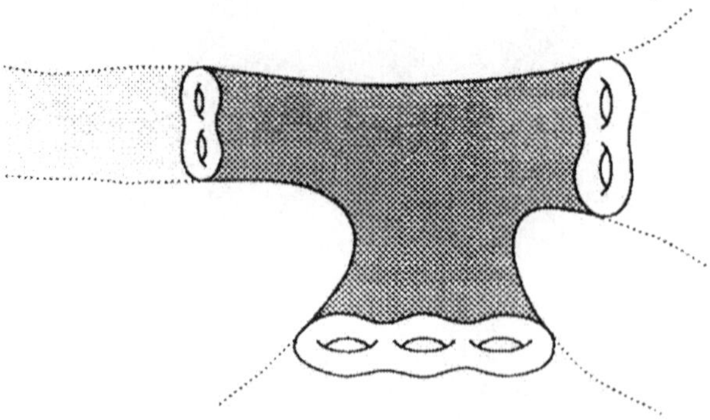

FIGURE 4. Schematic of a compact core (dark grey). The convex core includes the additional geometrically infinite end on the left (light grey).

The geometry of a geometrically finite end is determined, up to uniform bilipschitz equivalence, from any of the surfaces S_t (see Epstein-Marden [**36**]), or equivalently from the limiting structure $\nu_E \in \mathcal{T}(S)$. Furthermore, the general form of the Ahlfors-Bers theory implies that, if N has only geometrically finite ends, then it is uniquely determined by the corresponding points in the Teichmüller space.

Note that, in this description, we did not need to know (because of Scott's theorem) that N was actually homeomorphic to M – the discussion goes through just as well only with the assumption that $\pi_1(N)$ is finitely generated. This is a useful observation for considering arbitrary points in $AH(M)$, which may not yield manifolds homeomorphic to the interior of any compact manifold. The property of being the interior of a compact manifold is known as *topological tameness,* and it is worth recording here Marden's conjecture from [**66**]:

Tameness Conjecture. *A hyperbolic 3-manifold with finitely generated fundamental group is topologically tame.*

Our description of geometrically finite ends implies in particular that any manifold all of whose ends are geometrically finite is topologically tame – as originally shown by Marden in [**66**].

Geometrically infinite ends. What can we say about the structure of geometrically infinite ends? Groups with such ends were constructed by Bers-Maskit [**11**] (and shown to be geometrically infinite by Greenberg [**40**]), but very little was understood about their geometry.

It turns out that the state of our knowledge depends on whether the boundary of $\overline{M}$ is compressible or not – incompressibility means that the map $\pi_1(S) \to \pi_1(M)$ induced by the inclusion of any component S of $\partial\overline{M}$ is injective.

If $\partial\overline{M}$ is incompressible, then the work of Thurston and Bonahon [**102, 13**] gives a preliminary description of the geometry and topology of the geometrically infinite ends of N. Thurston defined the notion of a *simply degenerate* end, and showed that such an end determines a unique *lamination* on the corresponding boundary component of $\overline{M}$, which is called an ending lamination. Thurston also showed that if every geometrically infinite end of M is simply degenerate, then in fact M is topologically tame. Bonahon showed that indeed every geometrically infinite end of M *is* simply degenerate (provided $\partial\overline{M}$ is incompressible). Thus Marden's conjecture is established in the setting of incompressible boundary. We will discuss ending laminations in more detail in the next section.

When $\partial\overline{M}$ is compressible, much less is known. Canary showed [**29**] that, if M is topologically tame, then a suitable notion of simply degenerate ends and ending laminations can be defined, so that the same theory carries through with suitable modifications. There has also been a string of results [**27, 90, 5, 38, 22**] that establish the tameness of certain limits of geometrically finite manifolds. However the general tameness conjecture remains open, and is at this point the biggest remaining problem in the deformation theory of hyperbolic 3-manifolds.

4. Laminations and degenerate ends

4.1. Geodesic laminations on surfaces

Consider, by way of motivation, the case of a torus T^2. The set of simple closed curves in T^2, up to homotopy, can be described via the extended rational numbers $\widehat{\mathbb{Q}} = \mathbb{Q} \cup \{\infty\}$, by associating each curve to its slope, in a chosen identification of $H_1(T^2)$ with $\mathbb{Z}^2$. The natural completion of this countable set to $\widehat{\mathbb{R}} = \mathbb{R} \cup \{\infty\}$ can be given a geometric interpretation in terms of *foliations* on T^2, whose slopes can be irrational.

A surface of higher genus does not admit non-singular foliations because of the Poincaré index theorem, but the related notions of singular measured foliations and measured geodesic laminations serve as good generalizations of the situation for the torus. We will describe this using the language of laminations, but see e.g. [**39, 63**] for the foliation point of view.

Fix a closed hyperbolic surface S. A *geodesic lamination* in S is a compact set, foliated by complete geodesics. A simple closed geodesic is the simplest example, but one can obtain more complicated examples by taking Hausdorff limits of sequences of simple closed geodesics whose lengths go to ∞. In fact the space $\mathcal{GL}(S)$ of all geodesic laminations in S is compact in the topology of Hausdorff convergence.

A *transverse measure* on a geodesic lamination is a family of Borel measures on arcs transverse to the lamination, invariant by holonomy; that is, by sliding along the leaves. (See Figure 5).

Thurston introduced this notion and defined the space $\mathcal{ML}(S)$ of all such *measured laminations*. This space has a natural topology, coming from weak-* convergence of the measures, which makes it homeomorphic to $\mathbb{R}^{6g-6}$ where g is the genus of S. The hyperbolic metric we chose for S turns out to be irrelevant: different choices yield canonically homeomorphic spaces. The *projective lamination space*,

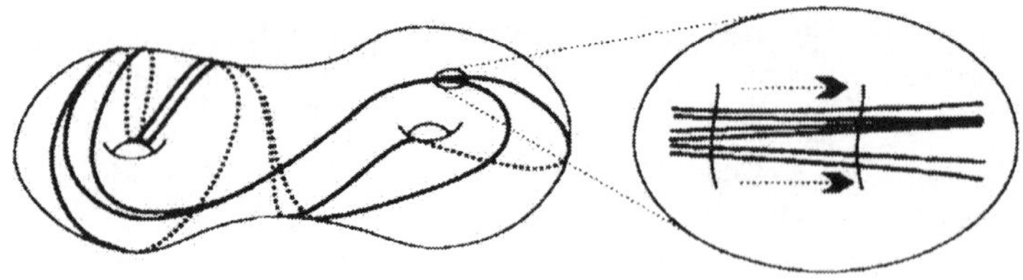

FIGURE 5. A geodesic lamination with closeup showing a transversal sliding along the leaves.

denoted $\mathcal{PML}(S)$, is the quotient of $\mathcal{ML}(S) \setminus \{0\}$ by the $\mathbb{R}_+$ action that multiplies measures by constants (0 denotes the empty lamination). $\mathcal{PML}(S)$ is homeomorphic to the $6g-7$-sphere, and contains the simple closed geodesics as a dense set of "rational" points, in analogy with the torus. See Bonahon [**14**], Casson-Bleiler [**33**] and Penner-Harer [**92**] for more details.

Let us also define the space $\mathcal{UML}(S)$ of "unmeasured laminations", which is just the quotient of $\mathcal{ML}(S) \setminus \{0\}$ under the equivalence relation that forgets the measures.

This space is different from $\mathcal{GL}(S)$ in two ways. First, not all laminations admit transverse measures of full support; $\mathcal{UML}(S)$ contains only laminations that are the supports of measures. Second, the topology of Hausdorff convergence in $\mathcal{GL}(S)$ is different from the quotient topology on $\mathcal{UML}(S)$.

$\mathcal{UML}(S)$ is also different from $\mathcal{PML}(S)$, and the difference is due to the fact that there exist laminations which support a continuum of projectively inequivalent measures. The natural map $\mathcal{PML}(S) \to \mathcal{UML}(S)$ collapses all these measures to a point.

Although $\mathcal{UML}(S)$ is not a Hausdorff space, it contains a natural subset that will be useful to us, and which is Hausdorff. This is the set of ***filling laminations***: A measured lamination λ is called filling if it has non-trivial intersection with every essential closed curve in S. Equivalently λ is filling if it cuts S into a union of ideal hyperbolic polygons (if S is not closed then these can be once-punctured polygons).

The filling laminations are in fact generic in $\mathcal{PML}(S)$, just as the irrational numbers are generic among real numbers. Their image in $\mathcal{UML}(S)$, which we denote by $\mathcal{EL}(S)$, is a Hausdorff space (see Klarreich [**58**]) and we will revisit it in the next section.

4.2. Simply degenerate ends

If an end of a hyperbolic 3-manifold is not geometrically finite, both its geometry and topology are a priori rather mysterious. Thurston discovered a way to "tame" the study of ends by considering the placement of closed geodesics within them (and more usefully, of "pleated surfaces", which we will discuss in §7).

Thurston defined an end E of N (in the incompressible-boundary case) to be ***simply degenerate*** if there is a sequence of simple closed curves α_i on the associated surface S whose geodesic representatives α_i^* in N are eventually contained in any neighborhood of E – we say that α_i^* *exit the end.* Note the contrast between this and the property of being geometrically finite: The existence of the convex surfaces S_t means that, outside S_t, there cannot be *any* closed geodesics. This is because the exterior of a convex surface admits a distance-decreasing projection back to the surface, and hence the shortest representative of any homotopy class of closed curves must lie inside the convex core C_N.

If such a sequence α_i exists, it turns out that its asymptotic behavior as a subset of $\mathcal{UML}(S)$ is quite simple. Thurston proved that a simply degenerate end always has a unique lamination which is the limit point of any such sequence. Furthermore, he showed:

THEOREM 4.1. [**102**] *If $\partial\overline{M}$ is incompressible and N is the quotient manifold of a point in $AH(M)$ without cusps, then, whenever E is a simply degenerate end of N facing a component S of the compact core boundary,*

1. *There is a unique lamination $\nu_E \in \mathcal{UML}(S)$ such that a sequence α_i of simple closed curves in S converges to ν_E in $\mathcal{UML}(S)$ if and only if α_i^* exit the end E.*
2. *There exists a sequence α_i^* exiting E such that $\ell_N(\alpha_i^*) \le L_0$ where L_0 depends only on the topological type of S.*
3. *ν_E is* filling*: it has the property that any homotopically non-trivial simple closed curve in S intersects ν_E nontrivially.*
4. *E is* topologically tame*: it has a neighborhood homeomorphic to $S \times (0, \infty)$.*

(For a few remarks on the case with cusps see §5).

Thurston called ν_E the *ending lamination* of E. He also proved that every limit of geometrically finite elements of $AH(M)$ has ends which are either geometrically finite or simply degenerate. Ends which are either one or the other are called *geometrically tame*. A few years later, Bonahon proved the foundational theorem:

THEOREM 4.2. [**13**] *If $\partial\overline{M}$ is incompressible and $N \in AH(M)$, then every end of N is geometrically tame.*

In other words the ending laminations are well-defined in the setting that will concern us in these notes. To unify notation, the *end invariant* of a geometrically tame end facing a component S of $\partial\overline{M}$ is the associated $\nu_S \in \mathcal{T}(S)$ if the end is geometrically finite, and the associated ending lamination in $\mathcal{EL}(S)$ if it is simply degenerate.

Thurston's Double Limit Theorem (see [**101**] and Ohshika [**89**]), together with the quasiconformal deformation theory, implies that every filling lamination actually occurs as an ending lamination (this justifies our name $\mathcal{EL}(S)$ for the set of filling laminations). We state this a bit more carefully, and in the case without cusps:

THEOREM 4.3. *Let $\overline{M}$ be compact with incompressible boundary and let*

$$(\nu_S \in \mathcal{EL}(S) \cup \mathcal{T}(S))_S$$

be a set of choices of end invariants for components S of ∂M, with the restriction that, when $\overline{M} = S \times [-1,1]$, $\nu_{S\times\{-1\}} \neq \nu_{S\times\{1\}}$ if they lie in $\mathcal{EL}(S)$. Then there exists $N \in AH(M)$ homeomorphic to M, whose associated end invariants are the (ν_S).

In other words, all possible combinations of end invariants are obtained (the restriction when N is a product is necessary: Ending laminations for opposite ends can never be the same).

Thus a complete classification of $AH(M)$ would follow from the following conjecture of Thurston:

Ending Lamination Conjecture. *An element of $AH(M)$ is uniquely determined by its list of end invariants.*

The proof of this conjecture, in the case that ∂M is incompressible (in particular the case $M = S \times [0,1]$) will appear in **[82]** and **[23]**.

5. Tubes, cusps and geometric limits

A central feature of the study of hyperbolic 3-manifolds is the notion of a *thick-thin decomposition*, and we will take a little time now to discuss this decomposition and the structure of the components of the thin part.

Hyperbolic tubes. A *hyperbolic tube* is the quotient of an r-neighborhood of a geodesic in $\mathbb{H}^3$ by a translation or a screw motion. Our goal here is to explain how the geometry of a hyperbolic tube is controlled by the geometry of its *marked boundary*.

More explicitly, given $\lambda \in \mathbb{C}$ with $\operatorname{Re}\lambda > 0$, and $r > 0$, we define $\mathbb{T}(\lambda, r)$ to be the quotient of the r-neighborhood of the vertical line L above $0 \in \mathbb{C}$ in the upper half-space model of $\mathbb{H}^3$ by the loxodromic $\gamma : z \mapsto e^\lambda z$. The action of γ in the third coordinate is just multiplication by $|e^\lambda|$, and in particular γ translates the axis L by a hyperbolic distance $\operatorname{Re}\lambda$, and rotates around it by angle $\operatorname{Im}\lambda$. Any hyperbolic tube is isometric to some $\mathbb{T}(\lambda, r)$, but we note that the imaginary part of λ is, so far, only determined modulo 2π.

Marked boundaries and Teichmüller parameters. If T is an oriented Euclidean torus, a *marking* of it is an ordered pair (α, β) of homotopy classes of unoriented simple closed curves with intersection number 1. There is a unique $t > 0$ and $\omega \in \mathbb{H}^2 = \{z : \operatorname{Im} z > 0\}$ such that T can be identified with $\mathbb{C}/t(\mathbb{Z} + \omega\mathbb{Z})$ by an orientation-preserving isometry, so that the images of $\mathbb{R}$ and $\omega\mathbb{R}$ are in the classes α and β, respectively. The parameter ω describes the conformal structure of T as a point in the Teichmüller space $\mathcal{T}(T) \equiv \mathbb{H}^2$.

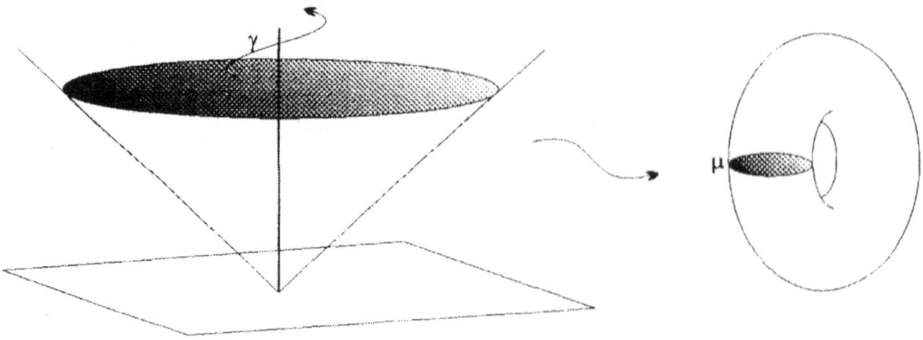

FIGURE 6. A hyperbolic tube has a Euclidean torus boundary. The meridian disk is shaded.

The boundary torus of a hyperbolic tube $\mathbb{T}$ inherits a Euclidean metric and an orientation from $\mathbb{T}$, and it admits an almost uniquely defined marking: Let μ denote the homotopy class of a meridian of the torus (i.e. the boundary of an essential disk in $\mathbb{T}$) and let α denote a homotopy class in $\partial\mathbb{T}$ of simple curves homotopic to the core curve of $\mathbb{T}$. While μ is unique, α is only defined up to multiples of μ. Fixing such a choice (α, μ), we obtain ***boundary parameters*** (ω, t) as above.

The freedom of choice of λ (mod 2π) is related to the freedom of choice of α mod μ, and indeed there is a natural way to choose α given λ. We then obtain (see [**82**, §3]) a bijective correspondence between the parameters (λ, r) and the boundary data (ω, t).

Parabolic tubes. A rank-1 parabolic tube is the quotient of a horoball in $\mathbb{H}^3$ by a cyclic group of parabolic transformations. Explicitly, consider the parabolic transformation $z \mapsto z + t$, acting on the region of height ≥ 1 in the upper half space model of $\mathbb{H}^3$. The quotient is homeomorphic to $S^1 \times \mathbb{R} \times [0, \infty)$, and we note that any rank-1 parabolic tube is obtained in this way for some t.

A rank-2 parabolic tube is the quotient of a horoball by a parabolic group isomorphic to $\mathbb{Z}^2$. It is homeomorphic to $T^2 \times [0, \infty)$ and can be obtained as the quotient of the region of height ≥ 1 by the group generated by $z \mapsto z + t$ and $z \mapsto z + t\omega$ with $\omega \in \mathbb{H}^2$. As for hyperbolic tubes, once a marking (α, μ) for the torus boundary is chosen the parameters (ω, t) are uniquely determined; but note that there is no natural choice of either μ or α in this case.

Thick-thin decomposition. By Margulis' Lemma, or Jørgensen's inequality (see e.g. Thurston [**103**] or Kapovich [**55**]), there is a universal $\epsilon_0 > 0$ such that the ***ϵ-thin part*** of a hyperbolic 3-manifold N, i.e. the set $N_{(0,\epsilon]} = \{x : inj(x) \leq \epsilon\}$, is a disjoint union of hyperbolic and parabolic tubes, whenever $\epsilon \leq \epsilon_0$. The parabolic tubes are usually called ***cusps***.

While the structure of each individual tube is relatively simple, the real challenge in deformation theory is understanding how these tubes are organized within a given hyperbolic 3-manifold. We will see some indication of the complexity of this in the examples of Section 6.

Geometric limits. The ***geometric limit*** of a sequence of Kleinian groups Γ_i is their Gromov-Hausorff limit as subsets of $\mathrm{PSL}_2(\mathbb{C})$. Fixing an origin $0 \in \mathbb{H}^3$, if the translation lengths $d(0, \gamma(0))$ are bounded away from 0 and ∞ for all $1 \neq \gamma \in \Gamma_i$, for all i, then such a limit always exists (after taking a subsequence) and is a nontrivial and discrete group Γ_∞. The equivalent statement for the quotient manifolds $N_i = \mathbb{H}^3/\Gamma_i$ is this: let the ***basepoint*** x_i of N_i be the image of 0 under the quotient map. Then geometric convergence implies that, for any $R > 0$, the R-neighborhood of x_i in N_i is eventually diffeomorphic to the R-neighborhood of x_∞ in N_∞, by a map that is locally converging to an isometry as $i \to \infty$. The opposite implication also holds, up to possibly conjugating the groups by rotations fixing 0. (See e.g. [**30**]).

The notion of geometric limit is extremely useful in the study of hyperbolic 3-manifolds – a lot of our available geometric control comes from compactness or contradiction arguments involving geometric limits.

The simplest types of Kleinian groups are the cyclic groups giving rise to hyperbolic tubes as above. It is instructive to consider geometric limits in this setting. Consider a sequence of hyperbolic tubes $\mathbb{T}_i$ with boundary parameters $(\omega_i, 1)$. If $\omega_i \to \omega$ with $\mathrm{Im}\,\omega > 0$ then it is fairly clear that the tubes (and the corresponding groups) converge geometrically to the tube with parameters $(\omega, 1)$. The case $\mathrm{Im}\,\omega = 0$ involves a non-discrete limit and will not interest us here. More interesting is the case that $|\omega_i| \to \infty$. Here we find that the radii $r_i \to \infty$, and (after restricting to a subsequence and choosing appropriate basepoints) the tubes converge geometrically to a parabolic tube. Which kind of parabolic tube depends on the sequence: If $\operatorname{Im}\omega_i \to \infty$, then the limit is a rank-1 tube. But if $\operatorname{Re}\omega_i \to \infty$ and $\operatorname{Im}\omega_i$ is bounded, the limit is a rank-2 tube. This is because in this case the tori defined by $(\omega_i, 1)$ remain in a bounded set of shapes *up to homeomorphism*, or equivalently the projections of ω_i to the moduli space of the torus remain bounded.

This phenomenon was first discovered by Jørgensen [**52**], and also plays a very important role in Thurston's "Dehn-filling theorem" and its generalizations, which show that hyperbolic 3-manifolds with rank-2 cusps can be geometrically approximated by hyperbolic manifolds where the cusp tubes are replaced by solid tori.

End invariants in the presence of cusps. If N is allowed to have cusps the description of ends from §3 becomes a bit more complicated, but is in principle the same. The main idea is to consider N_0, the complement in N of the (interiors of) parabolic Margulis tubes. This manifold has annuli (and/or tori) in its boundary, and a *relative compact core* K (see McCullough [**72**] and Kulkarni-Shalen [**62**]) which meets ∂N_0 in compact annuli and tori. The components of $N_0 \setminus K$ determine the ends of N_0. Assuming that the components of $\partial K \setminus \partial N_0$ are incompressible, each one will support an end invariant as before, which is either a filling lamination or an element of a Teichmüller space.

The ending lamination conjecture now becomes the statement that all this data – parabolic elements, ending laminations and Teichmüller data – determine N uniquely.

An important example comes from surface groups, if we consider a compact surface $\overline{S}$ with boundary and an element $[\rho] \in AH(\overline{S} \times [-1,1])$ such that ρ takes $\partial\overline{S}$ to parabolic elements. Then the cusps associated to these elements cut N into $\overline{S} \times \mathbb{R}$. Note that the *relative* boundary components of N_0 will be incompressible in this case, in spite of the fact that $\overline{S} \times [0,1]$ is a handlebody.

6. Examples of surface groups

The following collection of examples gives a taste of the huge variety of possibilities for the geometry of Kleinian surface groups.

6.1. Fuchsian and Quasifuchsian groups

If Γ preserves a totally geodesic copy of $\mathbb{H}^2$ inside $\mathbb{H}^3$ as well as its transverse orientation, then $N = \mathbb{H}^3/\Gamma$ contains a totally geodesic copy of the surface $S = \mathbb{H}^2/\Gamma$, and the metric on N can be explicitly described in the coordinates $S \times \mathbb{R}$, as

$$(\cosh^2 t)ds^2 + dt^2 \tag{6.1}$$

where t is the $\mathbb{R}$ parameter, measuring distance from S, and ds is the hyperbolic metric on S. Γ is conjugate to a subgroup of $\mathrm{PSL}_2(\mathbb{R})$, and is called a Fuchsian group, as described in §2.

For a general quasi-Fuchsian group, the geometry outside the convex core can be modeled, up to uniform bilipschitz distortion, by half of the Fuchsian example, namely $S \times [0, \infty)$ with the metric (6.1) (see Epstein-Marden [**36**]). From now on we will concentrate on describing the metric inside the convex core.

6.2. Periodic manifolds

Our next example is a well-known type of manifold with geometrically infinite ends (the convex core is the whole manifold.)

Let $\varphi : S \to S$ be a pseudo-Anosov homeomorphism of the closed surface S (this means that φ leaves no finite set of non-boundary curves invariant up to isotopy). The mapping torus of φ is

$$M_\varphi = S \times \mathbb{R}/\langle (x,t) \mapsto (\varphi(x), t+1)\rangle,$$

a surface bundle over S^1 with fibre S and monodromy φ. Thurston [**101**] showed, as part of his hyperbolization theorem, that M_φ admits a hyperbolic structure which we'll call N_φ (see also Otal [**91**] and McMullen [**74**]). Let $N \cong S \times \mathbb{R}$ be the infinite cyclic cover of N_φ, "unwrapping" the circle direction (Figure 7). After identifying S with some lift of the fibre, we obtain an isomorphism $\rho : \pi_1(S) \to \pi_1(N) \subset \mathrm{PSL}_2(\mathbb{C})$, which gives us an element of $AH(S \times [-1,1])$.

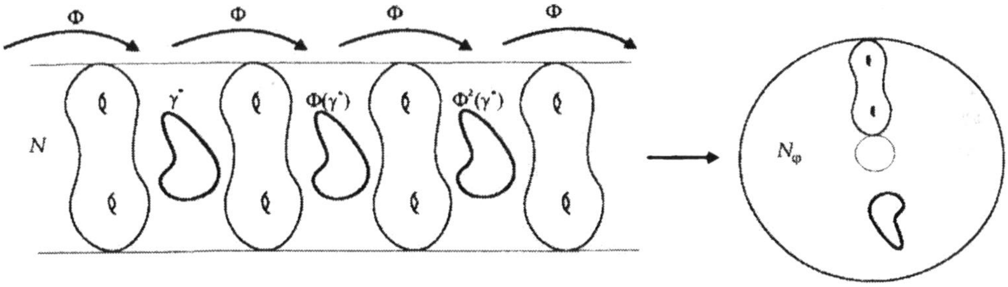

FIGURE 7. N covers the surface bundle N_φ.

The deck translation $\Phi : N \to N$ of the covering, acting as an isometry of the lifted metric, induces $\Phi_* = \varphi_* : \pi_1(S) \to \pi_1(S)$. Note that N has two ends, neither of which can be geometrically finite: for if one of the ends were foliated by convex surfaces S_t, which are expanding exponentially, the isometry Φ could not exist.

We next consider the action of φ on S, which, after isotopy, can be arranged (by Thurston's classification theorem for surface automorphisms, see e.g. [**39, 33**]) to have the following structure: There is a transverse pair (ν_+, ν_-) of laminations so that φ preserves the leaves of both ν_+ and ν_-, stretching the former and contracting the latter. Every simple closed curve γ in S intersects this system of leaves nontrivially, and it follows that repeated application of φ tends to make γ more and more parallel to ν_+ (and application of φ^{-1} brings it more in the direction of ν_-). Indeed in $\mathcal{PML}(S)$ (and in $\mathcal{UML}(S)$) the sequences $\{\varphi^n(\gamma)\}$ and $\{\varphi^{-n}(\gamma)\}$ converge to ν_+ and ν_-, respectively.

We can see $\nu_\pm$ directly in the asymptotic geometry of N: For a curve γ in S, let γ^* be its geodesic representative in N. Now consider $\Phi^n(\gamma^*)$ – these are all geodesics of the same length, marching off to infinity in both directions as $n \to \pm\infty$, and note that $\Phi^n(\gamma^*) = \varphi^n(\gamma)^*$. So, we have a sequence of simple curves in S, converging to ν_+ as $n \to \infty$, whose geodesic representatives "exit the + end" of N (similarly as $n \to \infty$ they converge to ν_- and the geodesics exit the other end). It follows that $\nu_\pm$ are the ending laminations of ρ.

The way that Thurston actually proved the existence of this example was to obtain it as a limit of quasi-Fuchsian groups. For $X, Y \in \mathcal{T}(S)$, let $qf(X,Y)$ denote the quasi-Fuchsian group (up to conjugation) whose surfaces at infinity are X and Y. That is, qf is the Ahlfors-Bers simultaneous uniformization map from §2.

Now if we fix X and consider the sequence

$$\rho_n = qf(\varphi^{-n}(X), \varphi^n(X)),$$

Thurston showed that this converges in $AH(S \times [-1,1])$ to the representation ρ. At the nth stage, the convex core of ρ_n is roughly (up to uniform bilipschitz diffeomorphism) a concatenation of $2n$ fundamental domains of Φ's action on the limiting manifold (see McMullen [**74**]).

6.3. Drilling out a curve

Kerckhoff-Thurston [**57**] constructed the following sequence of quasi-Fuchsian representations ρ_n, which is related to Jørgenson's original observation that the geometric limit of a sequence of hyperbolic tubes can be a rank-2 parabolic tube.

Let γ be an essential curve in S, and let D_γ be a Dehn twist on γ. Then consider the sequence

$$\rho_n = qf(X, D_\gamma^n(X)).$$

The bottom surface of the convex core of N_{ρ_n} is (by Sullivan's theorem [**36**]) uniformly bilipschitz equivalent to X, and so is the top, but by a map which is in the homotopy class of D_γ^n. Kerckhoff and Thurston showed that a uniform bilipschitz model for the geometry of the convex core of N_{ρ_n} can be described as follows:

Let $M = S \times [-1,1]$ and let A be a closed annular collar of γ in S. Let U be the solid torus $A \times [-1/2, 1/2]$ (see figure 8). Place a fixed metric on M for which the torus ∂U is Euclidean. We can describe the meridian of U by choosing an arc a in A connecting the two components of ∂A, and letting

$$\mu = \partial(a \times [-1/2, 1/2]).$$

Now remove U from M, and replace it by a new solid torus U_n, whose meridian μ_n is obtained from μ by twisting its top arc, $a \times \{1/2\}$, n times around the core of $A \times \{1/2\}$. The new manifold is homeomorphic to the old, but the geometry will be different.

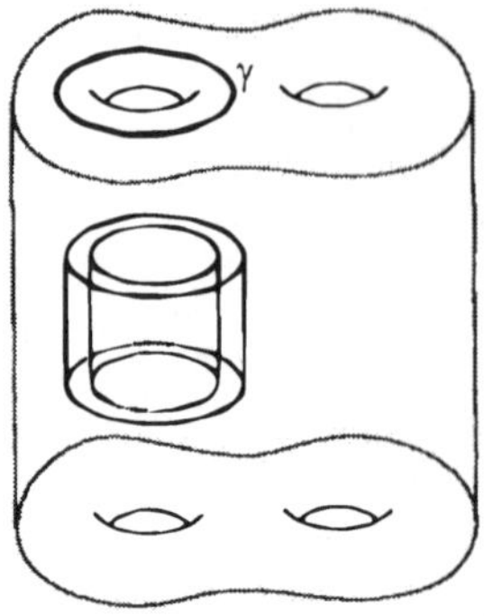

FIGURE 8. The model for the Kerckhoff-Thurston "drilling" construction.

Letting α be the core of $A \times \{1/2\}$, we have a marking (α, μ_n) for ∂U. As discussed in §5, there is a unique hyperbolic tube whose marked boundary has the same geometry as ∂U with this marking, and we can identify U_n with this tube.

The resulting manifold thus has a very short representative γ_n of γ, surrounded by a deep Margulis tube.

As $n \to \infty$, this sequence of models M_n (and hence also the convex cores of the hyperbolic manifolds) converges in the sense of Gromov-Hausdorff to a manifold in which U has been replaced by a ***parabolic tube*** with torus boundary. (This is just the same as the limit discussed in §5, where the boundary parameters ω_n satisfy $\mathrm{Im}\,\omega_n$ bounded and $\mathrm{Re}\,\omega_n \to \infty$.) Equivalently, the geometric limit is homeomorphic to $S \times [-1, 1]$ minus the level curve $\gamma \times \{0\}$.

This example begins to show the role that Margulis tubes play in the structure of surface groups. It also indicates that geometric limits of surface groups do not themselves have to be surface groups.

6.4. Drilling out infinitely many curves

Thurston [**101**] and Bonahon-Otal [**15**] produced iterations of the Kerckhoff-Thurston construction which illustrate that the geometric limit of a sequence of quasi-Fuchsian groups can in fact be infinitely generated.

In this example, consider an infinite sequence $\{\gamma_1, \gamma_2, \ldots\}$ of curves in S, and let D_i be Dehn twists about γ_i. Suppose that the sequence has the following two properties:

- ***Eventual filling property:*** For each i there is some $j > i$ such that the curves $\gamma_i, \ldots, \gamma_j$ ***fill*** the surface, meaning that every essential curve in S must intersect one of these curves.
- ***Non-annular property:*** If $\gamma_i = \gamma_j$ for $i < j$ then there must be some $i < k < j$ such that γ_k intersects γ_i essentially.

Then the sequence

$$\rho_n = qf(X, D_1^{e(n)} \circ \cdots \circ D_n^{e(n)}(X))$$

(where $e(n) \to \infty$ sufficiently fast) has a geometric limit which is homeomorphic to

$$S \times \mathbb{R} \setminus \bigcup_{i=1}^{\infty} \gamma_i \times \{i\}.$$

The missing curves $\gamma_i \times \{i\}$ correspond to rank 2 cusps, and the non-annularity condition is necessary because of the property of hyperbolic manifolds that curves in distinct cusps cannot be homotopic. The "eventual filling" condition is not necessary for the construction, but in its absence the structure of the limit can be even more complicated.

A model for the nth manifold in this sequence can be constructed, similarly to the previous example, by starting with $S \times [0, n+1]$ and performing appropriate integer Dehn surgeries on solid tori centered on $\gamma_i \times \{i\}$, giving the new solid tori the geometry of the appropriate hyperbolic tubes.

6.5. Brock's example

Brock [**20**] gave an example that illustrates that exotic geometric limits do not just involve appearance of rank-2 cusps, as in the previous two cases.

For this example, let $R \subset S$ be a (closed) proper essential subsurface, not an annulus or a 3-holed sphere. Let $\phi_R : R \to R$ be a pseudo-Anosov and let $\phi : S \to S$ be a map which is the identity outside R and equal to ϕ_R in R. This is called a ***partial pseudo-Anosov.***

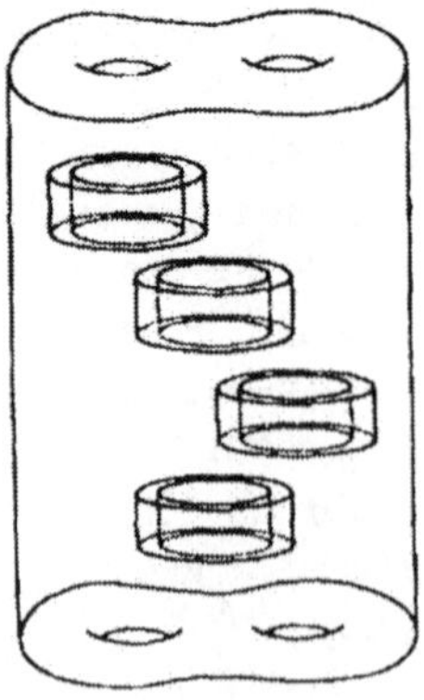

FIGURE 9. The convex hull can have many Margulis tubes, which in the geometric limit become cusps.

If we now let

$$\rho_n = qf(X, \phi^n(X)),$$

Brock showed that the geometric limit of the manifolds N_{ρ_n} is homeomorphic to

$$S \times \mathbb{R} \setminus R \times \{0\}.$$

Moreover, the following is a uniform bilipschitz model for the convex core of the nth manifold: Let Q denote the periodic hyperbolic 3-manifold homeomorphic to $int(R) \times \mathbb{R}$ which covers the mapping torus of ϕ_R, as in §6.2. Note that because R has boundary, Q should have a parabolic tube of rank 1 for each component of ∂R. Let Q_n denote the union of n successive fundamental domains of the deck translation of Q, minus the cusp tubes associated to ∂R. Thus Q_n is homeomorphic to $R \times [0, n]$, endowed with a metric such that each block $R \times [i, i+1]$ is isometric to $R \times [0, 1]$ by a map homotopic to ϕ_R^i.

Now remove from $S \times [-1, 1]$ a tubular neighborhood of $\partial R \times \{0\}$, and then cut out a small collar neighborhood of $R \times \{0\}$ and replace it by Q_n. Refill the tube associated to each component of ∂R by an appropriate hyperbolic tube – this is done using the natural marking as in the previous examples; In this case ω_n for each torus will have large imaginary part, corresponding to the height of Q_n.

Thus as $n \to \infty$, each tube of ∂R will become a cusp, but a rank-1 cusp now instead of a rank-2 cusp. If we take a geometric limit with basepoint, say, on the boundary of the convex core, then the distance to the top and bottom of Q_n stays bounded, but its middle grows ever farther, and hence in the geometric limit Q_n will become two eventually periodic geometrically infinite ends of the form $R \times [0, \infty)$.

6.6. Other constructions and limits

It is clear from these examples that many other variations are possible. For example a combination of the iterated "drill holes" construction of §6.4 and Brock's partial pseudo-Anosov construction should yield sequences that develop many different kinds of Margulis tubes and almost-periodic regions, and whose geometric limits can be the complement of many, even infinitely many, level surfaces and curves. Indeed the homeomorphism type of a general geometric limit of quasi-Fuchsian groups can be quite intricate, a fact which is expressed in the structure

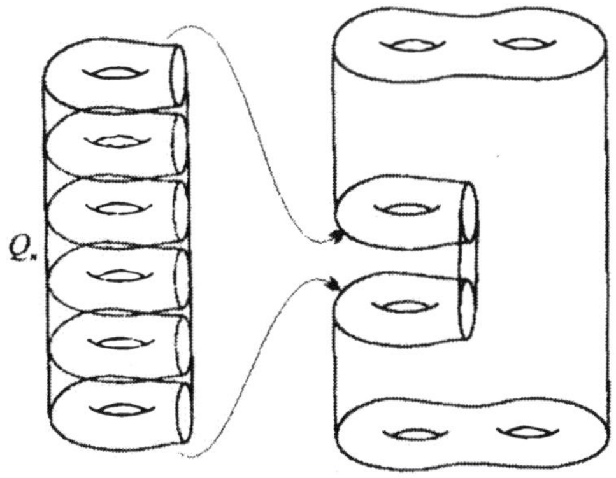

FIGURE 10. The bilipschitz model of the partial pseudo-Anosov example

of the "model manifolds" that we will discuss in Section 9.2. See also Soma [**97**], and the discussion in §10.

6.7. Bounded and unbounded geometry

It is worth dwelling a bit more on the example N from §6.2 which cyclically covers the pseudo-Anosov mapping torus N_φ, and its relation to the general case. First, since N admits a compact quotient, its geometry is bounded in a fairly strong sense. In particular its injectivity radii are bounded away from 0. Let us say that N in $AH(M)$ has *bounded geometry* if it admits a positive lower bound on injectivity radius (in the case with cusps we say that there is a positive infimum on the lengths of closed geodesics). This case is considerably simpler than the general case. In [**76, 77**] we showed that the Ending Lamination Conjecture holds in the setting of manifolds with bounded geometry (and no cusps).

Furthermore a bounded-geometry manifold homeomorphic to $S \times \mathbb{R}$ has structure similar to that of the pseudo-Anosov example, but not quite periodic: Part of the proof in [**76, 77**] involved showing that such a manifold is modeled, up to bilipschitz homeomorphism, by the "universal curve" over a Teichmüller geodesic. That is, there is a geodesic g in $\mathcal{T}(S)$ and a metric on $S \times \mathbb{R}$ such that $S \times \{t\}$ is isometric to the hyperbolic surface parameterized by $g(t)$ (where t is an arclength parameter for g). This metric on $S \times \mathbb{R}$ is within uniform bilipschitz distortion of the hyperbolic metric. In the pseudo-Anosov case, g is the axis of φ (as constructed by Bers in [**10**]). See also Mosher [**85**] and Bowditch [**17**] for new alternative proofs.

The examples in §§6.3–6.5 illustrate what happens when the bounded-geometry condition is dropped: the Margulis tubes cut the manifold into a number of different types of pieces, the universal curve model for the geometry no longer holds, and sequences of such manifolds can have exotic geometric limits (whereas it follows from Thurston's work that the geometric limit of a sequence of surface groups with uniformly bounded geometry is always a surface group). On the other hand the manifold decomposes into pieces that are either tubes, or bounded-geometry manifolds of lower genus (the Q_n in §6.5), or have bounded complexity. This behavior, properly formalized, turns out to hold in the general case.

McMullen [**73**] showed that unbounded-geometry manifolds are generic, in an appropriate sense, in the boundary of a Bers slice (a particular middle-dimensional slice of $AH(S \times [-1,1])$. This was generalized by Canary-Culler-Hersonsky-Shalen [**26**] to other deformation spaces.

Uniform control of the geometry of elements of $AH(S \times [-1,1])$ can only be possible if a way is found to predict (in terms of end invariants) the occurrence and configuration of Margulis tubes of short geodesics. In the sections that follow we will outline the mechanism for deducing, from the structure of the end invariants of N, the set of short geodesics in N and their topological arrangement.

7. Pleated surfaces

The geometry of a hyperbolic surface is rather simple and explicitly controlled by its topology. The Gauss-Bonnet theorem fixes its area; injectivity radius is bounded above; the thick-thin decomposition has a simple description in terms of collars and pieces of bounded complexity. On the other hand the hyperbolic metrics vary in a moduli or Teichmüller space so there is great flexibility. Thurston's idea of pleated surfaces harnesses this control coupled with flexibility by considering useful isometric maps of hyperbolic surfaces into hyperbolic 3-manifolds.

A pleated surface is something like a piecewise-geodesic surface, except that there are no "corner points" and the bending lines are typically infinitely many. More formally, it is a map $f : S \to N$ where N is a hyperbolic 3-manifold, together with a hyperbolic metric σ_f on the surface S, such that path lengths are preserved by f, and every point in S either has a neighborhood that is mapped totally geodesically or is on a leaf of a *geodesic lamination* λ_f whose leaves are mapped totally geodesically. Fairly detailed accounts of the theory of pleated surfaces can be found in [**30**] and [**36**].

Spinning. Typically we will have a fixed homotopy class of maps $[h : S \to N]$ in mind, and will look for pleated surfaces within this class. Starting with any essential simple closed curve γ on S for which $h(\gamma)$ is not homotopic to a point or into a cusp, there are pleated surfaces $f \sim h$ such that $\gamma \subset \lambda_f$. Suppose for simplicity that γ is part of a pants decomposition P of S. We can complete P to a lamination of S by adding leaves in each pair of pants that spiral toward the boundary (see Figure 11).

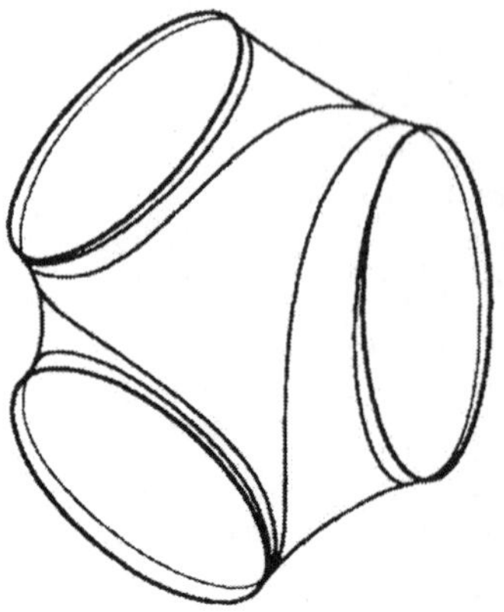

FIGURE 11. A lamination in a pair of pants with leaves spinning toward the boundary.

We can deform h to get a map taking each component of P to its unique geodesic representative. Assuming appropriate topological conditions on h (being an isomorphism on π_1 suffices) each of the infinite spiraling leaves can be realized in N as a leaf that spirals around the geodesic representatives of P. The remainder is a collection of ideal triangles that can only be mapped in one way. It is not hard to see that the resulting surface inherits a hyperbolic metric.

Other examples can be obtained from this one by a process of taking limits – given a sequence P_n converging to some lamination μ the associated sequence of pleated surfaces, if convergent, will yield a pleated surface which maps μ geodesically. This convergence is made possible by a nice *compactness property* of pleated surfaces – a sequence of pleated surfaces to a fixed manifold N, after taking a subsequence, either leaves every compact set, or converges in a suitable sense to a pleated surface, or degenerates in a predictable way, developing thin parts that enter cusps of N.

Pleated surfaces also arise naturally as boundaries of convex hulls of hyperbolic 3-manifolds. Thurston pointed out that these boundaries are embedded pleated surfaces with the added condition of convexity.

A lamination μ is *realizable* in a homotopy class $[h]$ if it is mapped geodesically by some pleated surface in $[h]$. If h is the inclusion map of an incompressible component of the compact core, non-realizability of μ turns out to be equivalent to the statement that μ has a component which is either homotopic to a cusp or is an ending lamination of N.

Pleated surfaces and ends. If E is an end of N facing an incompressible component S of the compact core, then we can begin to see the consequences of Thurston's simple degeneracy condition using pleated surfaces. A sequence α_i of simple curves in S whose geodesic representatives α_i^* exit E gives rise, via the spinning construction, to a sequence of pleated surfaces $f_i : S \to N$ (homotopic to the inclusion) whose images exit E (one must deal with the possibility that part of $f_i(S)$ is far out in E and part remains near the core – this is understood in terms of the thick-thin decomposition of the surface). Part (2) of Theorem 4.1, the existence of a sequence of *bounded* curves exiting the end, is now a consequence of the fact that the shortest curve in each hyperbolic metric σ_{f_i} has uniformly bounded length.

Because laminations can be organized in a continuum, namely in the sphere $\mathcal{PML}(S)$ (§4.1), it becomes possible to *interpolate continuously* between any two pleated surfaces. Not all the intermediate surfaces in such an interpolation are pleated – one must slightly generalize the category while keeping most of the geometric control. With this, Thurston showed that a simply degenerate end E can be "swept" by a family of these almost-pleated surfaces going all the way to infinity. This implies for example upper bounds on injectivity radius anywhere in E, as well as the topological tameness of E (part (4) of Theorem 4.1). Thurston also used this to control the growth of harmonic functions on N, thus obtaining a proof of *Ahlfors' Measure Conjecture* for the case that the core has incompressible boundary (see §10).

Injectivity and Efficiency. One can get finer geometric control from a pleated surface if one can control the extent to which it is "folded". The presence of a pleated surface $f : S \to N$ easily yields upper bounds: if a curve α has length L in σ_f then L bounds the length of the geodesic representative of $f(\alpha)$. In his proof of the Double Limit Theorem [**101**], which yielded the geometrization of manifolds that fibre over S^1, Thurston needed to obtain *lower bounds* of the form

$$\ell_N(f(\alpha)^*) \geq \ell_{\sigma_f}(\alpha) - C(\lambda_f, \alpha)$$

where $C(\lambda_f, \alpha)$ is a combinatorial constant determined by α and the lamination λ_f, and independent of the geometric data. That is, he needed to find in $f(S)$ an "efficient" representative of the curve α.

The main issue turned out to be preventing different leaves of λ_f from having images in N that are too close together and nearly parallel. Thurston proved the *Uniform Injectivity Theorem*, which provides such control under suitable topological restrictions (again f being an isomorphism on π_1 suffices). This theorem is proved by recourse to geometric limits: If it fails then there is a sequence of examples $f_i : S \to N_i$, in which leaves that remain apart in S get closer and more parallel in N_i. Extracting a suitable limit of the manifolds and maps Thurston obtained a map in which two leaves have images that coincide, which he then showed contradicted the topological restrictions.

The Uniform Injectivity Theorem also plays a major role in the proof of the Ending Lamination Conjecture; see Section 9.

8. The complex of curves

We have already seen, in the examples of §6 and in the discussion of pleated surfaces in §7, the important geometric role played by simple closed curves. A central ingredient in a complete analysis of the structure of Kleinian surface groups is the intrinsic combinatorial structure of the set of all simple closed curves on a surface (up to homotopy), as encoded by Harvey's *complex of curves*.

In this section let S be any compact surface with genus $g \geq 0$ and $n \geq 0$ boundary components. Let $\xi(S) = 3g + n$.

We define the complex of curves $\mathcal{C}(S)$ as follows when $\xi(S) > 4$: the vertices of $\mathcal{C}(S)$ are the essential homotopy classes of simple closed curves in S, and k-simplices are $(k+1)$-tuples $[\alpha_0, \ldots, \alpha_k]$ that have simultaneously disjoint representatives.

If $\xi(S) = 4$ then $(g, n) = (0, 4)$ or $(1, 1)$, and no two nonhomotopic essential simple curves can be disjoint. We redefine $\mathcal{C}(S)$ so that $[\alpha_0, \alpha_1]$ is an edge whenever α_0 and α_1 have geometric intersection number 1 (for $(1,1)$) or 2 (for $(0,4)$). The same can be done when $(g, n) = (1, 0)$. Other than that, $\mathcal{C}(S)$ is empty when $\xi(S) \leq 3$.

These complexes were formulated by Harvey in [**44**], and later applied by Harer [**42, 43**], Ivanov [**49, 51, 50**] and more recently Luo [**65, 64**] Korkmaz [**60**] and Hempel [**45**] to the study of the mapping class group of a surface, and to Heegaard decompositions of 3-manifolds.

In joint work with Masur, we studied $\mathcal{C}(S)$ as a metric space, by endowing each edge with length 1. We showed that

THEOREM 8.1. [**70**] *Whenever $\xi(S) \geq 4$, $\mathcal{C}(S)$ is an infinite-diameter δ-hyperbolic space.*

(See Bowditch [**16**] for a new and much improved proof). δ-hyperbolicity is a "coarse geometry" notion, introduced by Gromov [**41**] and Cannon [**31**], to capture some of the large-scale properties of spaces such as $\mathbb{H}^n$, and metric trees. Geodesics in a δ-hyperbolic space X tend to diverge exponentially fast, and there is a robust notion of *boundary at infinity* $\partial_\infty X$, which roughly speaking is the set of asymptotic classes of infinite quasi-geodesic rays. The union $X \cup \partial_\infty X$ admits a natural topology. It is natural to ask what the boundary at infinity of $\mathcal{C}(S)$ would be. Because $\mathcal{C}(S)$ is not locally finite it is not a proper metric space, and $\partial_\infty \mathcal{C}(S)$ does not have as many nice properties as less exotic examples. In particular it is not compact.

It is not unreasonable to expect that $\partial_\infty \mathcal{C}(S)$ should have something to do with laminations on S. After all, any infinite sequence of simple closed curves should accumulate on some set of laminations. A few moments of thought indicate that not all laminations should correspond to endpoints at infinity. For example, if λ is a lamination in S which is disjoint from a curve α, we can approximate λ (say in $\mathcal{UML}(S)$) by a sequence $\{\gamma_i\}$ of vertices in $\mathcal{C}(S)$, which are themselves disjoint from α. Thus $d(\alpha, \gamma_i) = 1$ and hence $\{\gamma_i\}$ are *not* a sequence going to infinity in $\mathcal{C}(S)$. If λ were *filling* then this particular phenomenon couldn't occur, and this suggests we should direct our attention to the filling laminations in S.

An additional issue that arises is whether points in $\partial\mathcal{C}(S)$ should correspond to *measured* or *unmeasured* laminations. This is a subtle point, since there are filling laminations which support more than one projective class of measures. Klarreich [**58**] resolved this question in a nice way:

THEOREM 8.2. [**58**] *The boundary $\partial_\infty \mathcal{C}(S)$ is homeomorphic to $\mathcal{EL}(S)$, in such a way that convergence to a point at infinity in $\mathcal{C}(S) \cup \partial_\infty \mathcal{C}(S)$ corresponds to convergence in $\mathcal{UML}(S)$.*

Recall that $\mathcal{EL}(S)$ are the filling laminations in $\mathcal{UML}(S)$, and by Theorem 4.3 are exactly those laminations that occur as ending laminations for manifolds without cusps. Thus our ending laminations appear as points at infinity in the complex of curves.

8.0.1. The torus and the Farey graph. When S is the torus (or one-holed torus or 4-holed sphere), the complex of curves and its boundary recapitulate the notion of approximation of irrational numbers by rationals via continued fraction expansions.

As indicated in §4.1, the vertices of $\mathcal{C}(S)$ in this case are the extended rationals $\widehat{\mathbb{Q}} = \mathbb{Q} \cup \{\infty\}$, associating to each curve its slope in $H_1(S)$. It is a nice exercise to check that the definition of $\mathcal{C}(S)$ in this case yields the "Farey Graph," which is embedded in the disk with vertices at the rational points of the circle. (See Figure 12, and [**79, 78**] for more details.)

In this case, the boundary of $\mathcal{C}(S)$ is just the irrational points in the circle, with convergence at infinity being the normal one in $\mathbb{R}$. Furthermore, there is a natural class of quasi-geodesic rays: Any irrational $r \in \mathbb{R} \setminus \mathbb{Q}$ has a unique continued-fraction expansion of the form

$$r = [n_0, n_1, n_2, \ldots] = n_0 + \cfrac{1}{n_1 + \frac{1}{n_2 + \frac{1}{n_3 + \cdots}}}$$

where $n_0 \in \mathbb{Z}$ and $n_i \in \mathbb{N}$ for $i > 0$. The sequence of partial expansions $\{[n_0, \ldots, n_k]\}$ forms a quasigeodesic ray converging to r.

The role of the coefficients n_i is played, for general surfaces, by the "subsurface coefficients" discussed in §9.2.

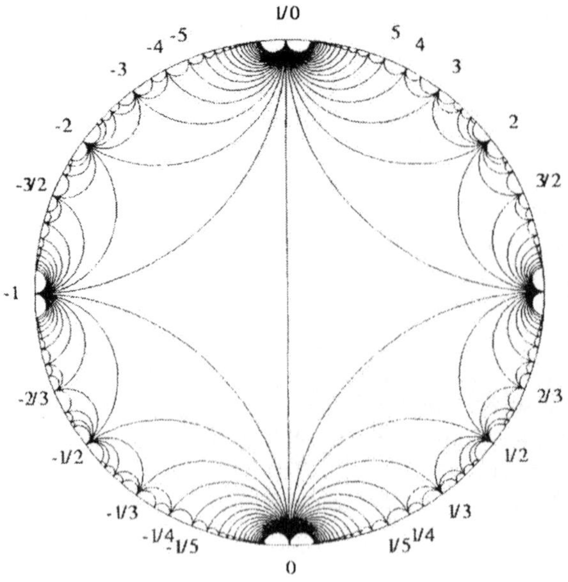

FIGURE 12. The Farey graph in the disk is the complex of curves of the torus

9. The proof of the Ending Lamination Conjecture

We will discuss now the main ideas that go into the proof of Thurston's conjecture. For another expository account of part of this argument (particularly §9.1 and §9.2) see [**83**].

To simplify this discussion, we will restrict to the case of ***doubly degnerate surface groups*** (although we will also discuss some examples from a more general class). That is, let us fix an element $[\rho] \in AH(S \times [-1,1])$, and let $N = N_\rho$ denote the quotient manifold. Suppose also, for simplicity, that N has no cusps, and both ends of N are simply degenerate, with ending laminations $\nu_\pm$.

Our goal is to determine the geometry of N uniquely from $\nu_\pm$, and this can be divided more or less into these steps:

1. Quasiconvexity and a-priori length bounds: Using hyperbolicity of $\mathcal{C}(S)$ and Thurston's Uniform Injectivity theorem for pleated surfaces, we analyze the set of vertices of $\mathcal{C}(S)$ whose length in N is bounded. In particular we obtain an interpolation between ν_- and ν_+ via a "hierarchy of geodesics" passing through this bounded-length vertex set.
2. Lipschitz Model: We apply the hierarchical structure of $\mathcal{C}(S)$ to build a *model* for N which admits a Lipschitz map into N.
3. Embeddings and Bi-Lipschitz bounds: we improve the Lipschitz map to a bi-Lipschitz homeomorphism. This together with Sullivan's rigidity theorem is sufficient to establish the conjecture.

9.1. Length functions and quasiconvexity

Part (2) of Thurston's Theorem 4.1 on simply degenerate ends tells us that the set of vertices

$$\mathcal{C}(\rho, L_0) \equiv \{\alpha \in \mathcal{C}_0(S) : \ell_N(\alpha^*) \leq L_0\},$$

for suitable L_0, accumulates onto the ending laminations $\nu_\pm$, which we now think of as points in $\partial_\infty \mathcal{C}(S)$. Because our goal is to control the geometry of N, we would like to know that these accumulation points at infinity for $\mathcal{C}(\rho, L_0)$ suffice to determine $\mathcal{C}(\rho, L_0)$, at least roughly.

Our first result in this direction, from [**80**], is:

THEOREM 9.1. *The set $\mathcal{C}(\rho, L_0)$ is r-quasiconvex in $\mathcal{C}(S)$ where r depends only on the topological type of S.*

A set $C \subset X$ is r-quasiconvex if any geodesic of X with endpoints in C remains in an r-neighborhood of C. If X is δ-hyperbolic it is relatively easy to prove that a subset C is quasi-convex: all that is needed is a (coarse) Lipschitz retraction, i.e. a map

$$\Pi : X \to C$$

which is a uniformly bounded distance from the identity on C, and satisfies

$$d(\Pi(x), \Pi(y)) \leq Kd(x,y) + D$$

with uniform K, D (see Lemma 3.3 of [**80**]).

Such a map

$$\Pi : \mathcal{C}(S) \to \mathcal{C}(\rho, L_0)$$

is constructed by the use of pleated surfaces. For any vertex α we consider the set of pleated surfaces $f : S \to N$ in the homotopy class of ρ which take α to α^*. The image $\Pi(\alpha)$ is obtained by selecting some curve of length bounded by L_0 (one always exists) on one of these surfaces. In order to obtain the coarse Lipschitz property for Π, it turns out to be sufficient to show that two different pleated surfaces mapping α to α^* yield curves of length $\leq L_0$ that are not too far apart in $\mathcal{C}(S)$. This is done using the Uniform Injectivity result discussed in §7. A bit more precisely: if f and g are both pleated surfaces taking α to α^*, then one can find a curve β in S which is formed of an arc along α of bounded length (in both σ_f and σ_g) composed with a "jump" of bounded length in σ_f. The Uniform Injectivity theorem is applied here to show that the jump is small also in σ_g.

Digression: motivation from harmonic maps. The idea for the map Π came from a natural construction in the *analytical* setting: Given a homotopy class $[h : S \to N]$ (with mild topological restrictions) where N is a hyperbolic 3-manifold, there exists a map

$$\Phi : \mathcal{T}(S) \to \mathcal{T}(S)$$

where $\mathcal{T}(S)$ is the Teichmüller space of S, defined as follows: Fixing a Riemannian metric on S, the Dirichlet energy of each $f : S \to N$, defined as

$$\mathcal{E}(f) = \tfrac{1}{2} \int_S |df|^2 \, dA,$$

is minimized in the homotopy class by a unique map, called a *harmonic map*. Since $\mathcal{E}(f)$ is in fact invariant under conformal changes of the domain metric, this harmonic map actually depends only on the choice of a point $\sigma \in \mathcal{T}(S)$. If we then pull back the metric of N via the harmonic map to S, we obtain a new point in $\mathcal{T}(S)$, and this is $\Phi(\sigma)$.

It is not hard to see that the Dirichlet energy is bounded on the image of Φ, and one may attempt to study the geometry of N by understanding the Dirichlet energy function on $\mathcal{T}(S)$ and the self-map Φ. In the case of manifolds with *bounded geometry* (lower bounds on injectivity radius) this was somewhat successful, and this was the basis of the solution of the Ending Lamination Conjecture in this case [**76, 77**]. However, in the presence of very short curves in N the analytical approach runs into trouble. It is hard to control the map Φ because when a harmonic map enters a very thin part its dependence on the domain structure becomes very loose. The combinatorics of the thin parts themselves begin to play an important role, and this is why the complex of curves must be considered.

The map Π, then, is a combinatorial analogue of Φ, in which we have replaced harmonic maps with pleated surfaces, relinquished precise analytic control, and obtained instead the coarse structure of δ-hyperbolicity and quasiconvexity.

The quasiconvexity theorem is not enough, however, to control the set $\mathcal{C}(\rho, L_0)$. Since $\mathcal{C}(S)$ is locally infinite, distance bounds are not easy to use. Finer control is obtained by looking at subsurfaces of S. If W is an essential subsurface of S, then there is a projection

$$\pi_W : \mathcal{C}(S) \to \mathcal{C}(W) \cup \{\emptyset\}$$

defined roughly as follows. If α intersects W essentially, select an essential arc of intersection and a curve γ in W disjoint from it. This is $\pi_W(\alpha)$ (the choices in this construction yield a bounded-diameter set in $\mathcal{C}(W)$). If α has no essential intersection with W then $\pi_W(\alpha) = \emptyset$. Let $d_W(\alpha, \beta)$ denote $d_{\mathcal{C}(W)}(\pi_W(\alpha), \pi_W(\beta))$ whenever these projections are nonempty.

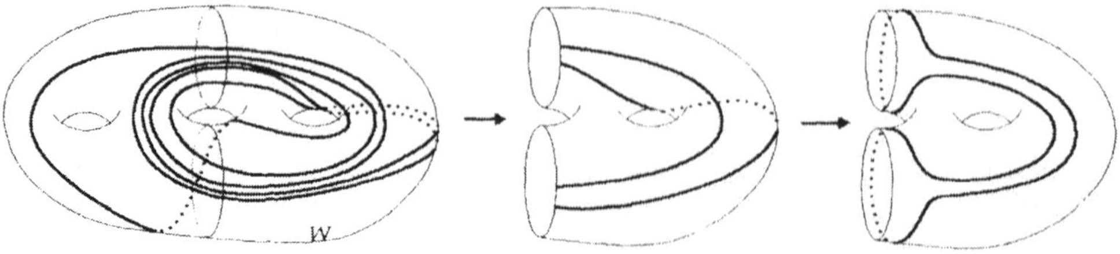

FIGURE 13. The subsurface projection π_W

The "relative quasiconvexity" theorem is the following:

THEOREM 9.2. [**82**] *The set* $\pi_W(\mathcal{C}(\rho, L))$ *is* r*-quasiconvex for each subsurface* W *of* S*, and a uniform* r.

This theorem does produce enough control to get a priori bounds on the set $\mathcal{C}(\rho, L_0)$. We will describe some of this in the next section.

9.2. Hierarchies and model manifolds

We can use the combinatorial structure of $\mathcal{C}(S)$ and the points at infinity $\nu_\pm$ to construct a certain manifold M_ν, equipped with a piecewise-Riemannian metric, and meant to be a bilipschitz model for the geometry of N.

We begin by letting g be an infinite geodesic in the 1-skeleton $\mathcal{C}_1(S)$ of $\mathcal{C}(S)$, such that the vertices of g converge in one direction to ν_- and in the other to ν_+. The existence of such a geodesic for two points on the boundary of a δ-hyperbolic space is easy to obtain by a limiting argument when the space is proper (bounded balls are compact). In the case of $\mathcal{C}_1(S)$ the limiting step is not automatic, but the machinery developed in Masur-Minsky [**71**] provides sufficient control to carry out such a limit.

We then build up g to obtain a "hierarchy of geodesics", the main construction of **[71]**. To give a hint of this procedure, let u, v, w be three succesive vertices of g, and let Y be the component of $S \setminus v$ containing u and w (if they were in different components they would have distance 1, contradicting the assumption that g is a geodesic). Then u and w can be joined by a new geodesic h in $C_1(Y)$, and all the vertices in h are in the link of v. We can repeat this at every vertex, obtaining a "thickening" of g into a collection of geodesics in links of vertices, and then repeat this inductively for each of those geodesics. The resulting object, a collection of geodesics supported in curve complexes of subsurfaces of S, is called a "hierarchy of geodesics". The construction is in fact considerably complicated by considerations of how to deal with endpoints. In the case of a 5-holed sphere the construction is especially simple to describe, and we refer to the expository article **[83]** for more details.

The structure of this hierarchy is controlled by $\nu_\pm$. In particular, for any essential subsurface $Y \subset S$ the projections $\pi_Y(\nu_\pm)$ (which are defined similarly to projections of simple closed curves) control whether or not Y appears as the support of one of the geodesics in the hierarchy: we show in **[71]** that if $d_Y(\nu_+, \nu_-)$ is sufficiently large then Y must be the support of a geodesic in the hierarchy with length roughly estimated by $d_Y(\nu_+, \nu_-)$.

On the hyperbolic geometry side, we showed in **[80]** that, if $d_Y(\nu_+, \nu_-)$ is large, then $\ell_N(\partial Y^*)$ is very small.

These results, together with the relative quasiconvexity theorem 9.2, allow us to show that

LEMMA 9.3. *If W is an essential subsurace of S then*

$$d_W(\alpha, \Pi(\alpha)) \leq D \tag{9.1}$$

over all vertices α appearing in H, for which $\pi_W(\alpha)$ and $\pi_W(\Pi(\alpha))$ are non-empty.

This roughly means that α and the bounded-length curves in $\Pi(\alpha)$ are not too far apart in a combinatorial sense. We can in fact use this to deduce an a priori length bound on α itself:

THEOREM 9.4. **[82]** *Given $[\rho] \in AH(S \times [-1,1])$ with ending laminations $\nu_\pm$, let H be a hierarchy constructed from a geodesic g in $C_1(S)$ joining ν_- and ν_+. Then all the vertices that appear in H are contained in $C(\rho, L)$, where L depends only on the topology of S.*

The hierarchy and the a priori bounds allow us to construct our model manifold M_ν and its Lipschitz map to N_ρ. M_ν, which can be identified with $S \times \mathbb{R}$, is a union of pieces called "blocks" and "tubes". The tubes are solid tori of the form (annulus)$\times$(interval), where the homotopy classes of the annuli are exactly the vertices of $C(S)$ that occur in the hierarchy construction. Each tube is given the structure of a hyperbolic tube as in §5, which is controlled by a Teichmüller coefficient $\omega \in \mathbb{H}^2$, defined with respect to a natural marking for ∂U. In fact these coefficients can be estimated from the data of the hierarchy, and hence from the subsurface projection distances, $\{d_Y(\nu_-, \nu_+)\}$.

The ***blocks*** are pieces that can be identified with $W \times [0,1]$ where W denotes a subsurface of S, and their geometric structures fall into a finite set of isometry types. Different blocks are glued along subsurfaces of the boundary in $W \times \{0\}$ and $W \times \{1\}$, so that the unglued parts of the boundary are annuli; these annuli form the boundaries of the tubes.

Theorem 9.4 gives us enough control to map the blocks of M_ν with Lipschitz bounds into N. The tubes U with large ω are shown to correspond to short curves in N.

Furthermore the the map can be adjusted so that each U with large ω maps properly onto the corresponding Margulis tube in N, and conversely that the complement of this set of tubes maps to the complement of the corresponding Margulis tubes.

In summary, we have a "Lipschitz model" for the thick part of N. It gives upper bounds for the lengths of closed geodesics in N, as well as a topological description of the arrangement of the components of the thin part (collars of short geodesics). However, it is not yet a bilipschitz model.

9.3. Examples

Let us consider what the model manifold would look like in some of the examples from §6.

The periodic case. If N_ρ is the cyclic cover of a fibred manifold, as described in §6.2, then $\nu_\pm$ are the fixed points of the pseudo-Anosov monodromy φ, and in the construction of the hierarchy we start with a geodesic g which is a "pseudo-axis" for φ: $\varphi(g)$ and g are, if not identical, at least within bounded distance of each other. The hierarchy in this case has a bounded structure: All geodesics that occur in its construction have uniformly bounded lengths (essentially because of the φ-periodicity), and as a result the model manifold has bounded geometry: all the tubes have bounded ω coefficients, and it follows that there is some positive lower bound on the length of the geodesics at their cores. The deck translation of the covering acts on the model by a bilipschitz homeomorphism, showing that the model structure is almost periodic. Of course we know that N_ρ itself is exactly periodic, but the hierarchy and model constructions are only rough approximations. In view of Sullivan's rigidity theorem (§9.4), this rough approximation suffices for our needs.

The drilling example. If ρ_n is the manifold $qf(X, D^n_\gamma(X))$ where D_γ is a Dehn twist, as described in §6.3, then the model manifold contains a bounded number of blocks and tubes of bounded ω, and one tube associated with γ, with coefficient $\omega \approx n + i$. This is essentially the model described in §6.3.

In Brock's example in §6.5, where $\rho_n = qf(X, \phi^n(X))$ with ϕ a partial pseudo-Anosov supported in $R \subset S$, the hierarchy will contain a geodesic in the curve complex $\mathcal{C}(R)$, and this will give an approximately periodic sequence of blocks corresponding to the piece Q_n described in §6.5. Again the tubes associated to ∂R will arise in the model.

In all the rest of the examples of §6 something similar happens: the structure of $qf(X,Y)$, in particular those tubes with large coefficient and small core length, is detected by the hierarchy construction and thus appears explicitly in the model manifold.

The punctured torus. The case that S is a one-holed torus, treated separately in [**79**], is especially simple. In this case the hierarchy corresponds to a geodesic in the Farey graph (§8.0.1), and the data for the model manifold is carried in the associated continued-fraction coefficients. For example if $\nu_- = \infty$ and ν_+ has continued-fraction expansion $[n_0, n_1, n_2, \ldots]$, then the continued-fraction approximants $[n_0, \ldots, n_k]$ represent the slopes of the curves associated to the tubes in the model, whose tube coefficients ω_k are estimated by $n_k + i$.

9.4. Bilipschitz bounds

In Brock-Canary-Minsky [**23**] we show that the Lipschitz model of the previous section can be converted into a bilipschitz homeomorphism

$$f : M_\nu \to N_\rho,$$

with uniform bounds. This is done by successively adjusting the map on pieces of the model, and maintaining geometric control by a variety of geometric limit arguments.

Embedding of horizontal subsurfaces. Let $f_0 : M_\nu \to N$ be the Lipschitz map described in §9.2. Within M_ν there are many "horizontal slices", which are surfaces F isotopic to level subsurfaces $Z \times \{t\} \subset S \times \mathbb{R}$, which have bounded geometry in the model metric, and are properly embedded in $M_\nu \setminus \mathcal{U}$. For example, in Brock's construction in §6.5, The subpiece Q_n contains many bounded-geometry surfaces isotopic to $R \times \{t\}$ with boundary in the hyperbolic tube(s) associated to ∂R. There are also the bottom and top boundaries of the convex core, which can be represented by horizontal slices of the form $S \times \{t\}$.

The restriction of f_0 to such a surface F is uniformly Lipschitz and homotopic to an embedding, and the trick is to deform it to a ***uniformly bilipschitz embedding.*** Using the techniques of Anderson-Canary [**5**] and Anderson-Canary-Culler-Shalen [**6**], we are able to show that the only obstruction to this is the possibility that $f_0|_F$ is "wrapped" around a Margulis tube of N: If $\mathbb{T}$ is a standard Margulis tube in N we say that a map $g : F \to N$ is wrapped around $\mathbb{T}$ if it cannot be deformed to $+\infty$ or $-\infty$ in $N \cong S \times \mathbb{R}$ in the complement of $\mathbb{T}$.

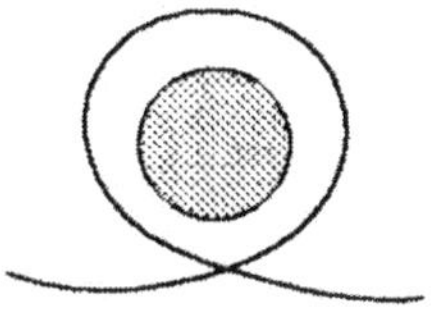

FIGURE 14. A surface wrapped around a Margulis tube. Here $\mathbb{T}$ is represented by a disk and the surface by a curve; crossing this figure by S^1 yields a correct configuration in a neighborhood of $\mathbb{T}$.

The argument is by contradiction. If there is a sequence of successively worse examples, i.e. Lipschitz maps $g_i : F_i \to N_i$ which are homotopic to embeddings and satisfy the topological condition of not being wrapped around Margulis tubes of N_i, but which do not admit uniformly bounded homotopies to embeddings, we can extract a geometric limit $g : F \to N$ in which some of the Margulis tubes have become parabolic cusps, and the limit map is not homotopic to an embedding at all. At this point we consider the way in which the subgroup $g_*(\pi_1(F))$ and its conjugates lie inside $\pi_1(N)$, and the techniques of [**5, 6**] produce a contradiction to the assumption that the original maps were not wrapped.

In the case of our model map, it is possible to show directly that for all of the horizontal slices F, $f_0|_F$ is not wrapped around any Margulis tubes. This is because we know that the Margulis tubes are all images of tubes in $\mathcal{U}$, and the complement of the tubes in $\mathcal{U}$ (with large ω) maps to the complement of the Margulis tubes. Since in the model it is easy to show that each of the slices can be pushed to ∞ or to $-\infty$ in the complement of the model tubes, the same follows for its image in N.

We conclude that f_0 on any horizontal slice can be made an embedding, in a uniform way.

Disjoint surfaces and partial orders. Thus fixing any set of disjoint horizontal slices, the map f_0 can be deformed in a uniform way to a map f_1 which is an embedding on each of these slices. However, there is nothing preventing these individual embeddings from intersecting, and it is not possible to fix these intersections by the simple expedient of performing surgeries on them: such surgeries increase the bilipschitz constants, and since there may be infinitely many horizontal slices a bad intersection pattern could cause the constants to grow without bound.

The solution to this is indirect. By the Lipschitz bounds on f_0 (and hence f_1) and the fact that each block in M_ν contains a bounded-length curve in a distinct homotopy class, it is not hard to see that there is a uniform upper bound on the number of horizontal slices whose images intersect the image of a given one. We use this fact to obtain a decomposition of M_ν with *local* disjointness properties. That is, we can carefully select among all the horizontal slices in M_ν to get a subset that cuts $M_\nu \setminus \mathcal{U}$ into pieces of bounded diameter and geometry, and such that the slices in the boundary of each piece have disjoint images. This requires careful use of the structure of the hierarchy to organize these slices, and an analysis of a "topological partial order" among embedded surfaces in $S \times \mathbb{R}$.

In Brock's example of §6.5, the selected slices would be some set of disjoint horizontal slices of type $R \times \{t\}$ in Q_n, chosen with appropriate spacing, as well as the top and bottom surfaces of the convex hull. This cuts M_ν into a large number of pieces of type $R \times [0,1]$ (with bounded geometry), which decompose Q_n, and a piece that can be identified with $M_\nu \setminus Q_n$ (and of course the Margulis tubes associated to ∂R).

The next step is to show that the topological partial order among the slices in M_ν is preserved by the map f_1, at least among those that are in the boundary of a particular complementary piece. The constants used in the selection of the subsets of slices must be chosen so that this additional condition of order preservation is satisfied. The proof that this is possible is, again, an argument by contradiction and passage to a geometric limit, using the structure of the lipschitz model map.

The order preservation property implies that the images of slices from our selected subset cut the hyperbolic manifold N into regions that are homeomorphic (preserving orientation) to the complementary regions of the cuts in M_ν. A final argument by contradiction and geometric limit shows that these homeomorphisms can be made uniformly bilipschitz. Thus they piece together to a bilipschitz map of $M_\nu \setminus \mathcal{U}$ to the thick part of N. Bilipschitz extension of the map to the tubes of $\mathcal{U}$ is now a fairly simple matter.

Applying Sullivan rigidity. Now suppose that ρ_1 and ρ_2 have the same pair of ending laminations $\nu_\pm$. Then there are bilipschitz homeomorphisms f_1 and f_2 from the same model manifold M_ν to N_1 and N_2, respectively. Thus $f_2 \circ f_1^{-1} : N_1 \to N_2$ is a bilipschitz homeomorphism. Since $\partial_\infty N_1$ and $\partial_\infty N_2$ are empty in the case we are considering, Sullivan's rigidity theorem implies that f must be homotopic to an isometry. This establishes the Ending Lamination Conjecture. (The case with nonempty ∂_∞ or with cusps is handled with standard modifications to this argument).

10. Other theorems and conjectures

The Ending Lamination Conjecture has long been known to imply the Bers-Sullivan-Thurston Density Conjecture, which can be stated this way:

Density Conjecture. *$AH(M)$ is the closure of its interior.*

In other words, this conjecture states that every hyperbolic 3-manifold with finitely generated fundamental group can be obtained as a limit of geometrically finite manifolds.

Interestingly, a new and unexpected proof was recently found, at least in the non-cusped cases (and with incompressible $\partial\overline{M}$), by Bromberg [**24**] and Brock-Bromberg [**21**]. This argument uses a branched-cover construction to convert a degenerate manifold with very short geodesics into a geometrically finite cone-manifold, and then deforms this to a smooth geometrically finite hyperbolic manifold using machinery developed by Hodgson-Kerckhoff [**46, 47**]. If there are infinitely many sufficiently short curves then this can be repeated to obtain a sequence of elements in the interior converging to the desired manifold. If not, then the manifold has bounded geometry, and this case is dealt with in [**81**].

The proof of the Density Conjecture via the Ending Lamination Theorem uses Thurston's Double Limit Theorem together with the Ahlfors-Bers quasiconformal deformation theory. These imply, as stated in Theorem 4.3 for the case without cusps, that any "legal" combination of end invariants can be obtained, and in fact constructed as a limit of points from the interior of $AH(M)$. Thus given an arbitrary point $[\rho]$ of $AH(M)$, a point $[\rho']$ on the closure of the interior of $AH(M)$ can be obtained with the same invariants. The Ending Lamination Theorem implies that $[\rho] = [\rho']$.

The Density Conjecture remains open in the compressible-boundary case. The main hurdle appears to be the Tameness Conjecture, which continues to resist attempts at its solution.

Another application of the solution to the ending lamination conjecture is the following further extension of the rigidity theory:

Topological Rigidity. *If $[\rho_1], [\rho_2] \in AH(M)$ where $\partial\overline{M}$ is incompressible, and the actions of ρ_1 and ρ_2 on $\widehat{\mathbb{C}}$ are topologically conjugate, then their actions are in fact quasiconformally conjugate.*

The point here is that a topological conjugacy at infinity implies that the two representations have homeomorphic compact cores, and the same ending laminations. The solution to the ending lamination conjecture then implies the quotient manifolds admit a bilipschitz homeomorphism, which extends to a quasiconformal conjugacy at infinity. The significance of this, as we saw in Section 9.4, is that it puts us in a position to apply Sullivan's Rigidity Theorem.

Geometric limits of hyperbolic 3-manifolds play a central role in the proof of our theorem, and indeed in the entire theory. Conversely, the existence of a bilipschitz model manifold gives us the kind of uniform control that allows us to describe fairly well the geometry of geometric limits. Here is a theorem about these geometric limits:

Topology of geometric limits. *Let $\overline{M}$ have incompressible boundary, and let N_i be a sequence of hyperbolic 3-manifolds homeomorphic to M, which converge geometrically to N_∞. Then N_∞ is homeomorphic to an open subset of M.*

Although this seems to be rather a weak statement, it does not follow from any kind of general considerations. In fact the topological type of a general geometric limit can be quite wild. Soma [**97**] has also proved this theorem, and additionally given a precise characterization of the possible topological types that occur as geometric limits

One way to measure the complexity of a hyperbolic 3-manifold is to consider volume growth functions. Fixing $\epsilon > 0$, let CN_ϵ be the ϵ-thick part of N, intersected with its convex core C_N. Let d_ϵ denote the metric induced by lengths of paths in CN_ϵ. Fix a basepoint x of CN_ϵ and define

$$g_N(r) = vol\{y \in CN_\epsilon : d_\epsilon(x,y) \leq r\}.$$

When N is geometrically finite, CN_ϵ is compact, so $g_N(r)$ is bounded. If N has a degenerate end with bounded geometry, then $g_N(r)$ grows linearly with r. An analysis of the geometry of the model manifolds verifies a conjecture of McMullen:

Polynomial volume growth. *Let N be any element of $AH(M)$, where $\partial\overline{M}$ is incompressible. Then*

$$g_N(r) = O(r^k)$$

where k can be computed from $\partial\overline{M}$.

For instance if $M = S \times [-1,1]$ for a closed surface S of genus g then $k = 3g-3$. This exponent is sharp, as can be shown by direct constructions.

Structure of the limit set. Ahlfors realized the importance of the limit set for the study of deformations of Kleinian groups. He formulated

Ahlfors' Measure Conjecture. *The limit set of a finitely generated Kleinian group is either the whole sphere, or has zero Lebesgue measure.*

Ahlfors proved this conjecture for geometrically finite groups [**1**]. The original motivation was the knowledge that no quasiconformal deformations can be supported on a set of zero measure. Although this application is now superseded by Sullivan's rigidity theorem (which applies even when the limit set is the whole sphere), the conjecture remains of interest. Thurston proved that the conjecture follows from geometric tameness, and hence Bonahon's theorem implies it in the incompressible-boundary setting. Later on, Canary [**29**] showed that in fact the assumption of topological tameness, by a clever branched-cover argument, suffices to make Thurston's techniques apply even in the case of compressible boundary, and Ahlfors' measure conjecture holds in this case also. Thus, this conjecture and a number of others would now follow from a solution of Marden's tameness conjecture. Recently, Brock-Bromberg-Evans-Souto [**22**] established the measure conjecture for all limits of geometrically finite manifolds, hence reducing it to the Density Conjecture. This proof relies on theorems establishing the Tameness Conjecture for certain limits of geometrically finite manifolds, extending results of Canary-Minsky [**27**], Ohshika [**88, 90**] and Evans [**37**].

A stronger version of Ahlfors' conjecture states that, when the limit set has full measure, the group acts ergodically on it. This was verified as part of the techniques of the previous paragraph, in all the topologically tame cases.

The measure-theoretic understanding of the limit set has also been further refined – by Patterson and Sullivan [**98**] (see also Nicholls [**87**]), who constructed invariant conformal densities on the limit set and related its Hausdorff dimension to the spectral theory of the quotient manifold in the geometrically finite case. These results were generalized by Canary [**28**] to the topologically tame category, and by others. Of particular note is the result of Bishop-Jones [**12**], who show that, for manifolds of infinite volume and finitely generated fundamental group, the Hausdorff dimension of the limit set is 2 if and only if the manifold is geometrically infinite.

In a different direction, it is of interest to know if the limit set is *locally connected*. This is true for example for quasifuchsian groups, whose limit sets are Jordan curves, but is a particularly difficult question for degenerate surface groups, i.e. geometrically infinite elements of $AH(S \times [-1, 1])$. More generally, local connectivity would be a consequence of a conjectural topological model for Kleinian group actions on the sphere. Cannon-Thurston [**32**] established the correctness of this model for groups closely related to the pseudo-Anosov examples of §6.2, and the author [**77**] extended this to bounded-geometry groups without cusps. Klarreich [**59**] extended this, still in the bounded geometry setting without cusps, to other manifolds M, and Bowditch [**18**] treated the bounded-geometry case with cusps.

McMullen proved local connectivity for surface groups where S is a one-holed torus, with an argument completely different from Cannon-Thurston's original approach, and utilizing the model manifold constructed in Minsky [**79**] in an earlier solution of the Ending Lamination Conjecture for that case. Now that there is a good description of a model manifold for the general case, it is interesting to see if McMullen's argument can be generalized. This looks like a fairly hard problem.

Topological structure of $AH(M)$. The Ending Lamination Conjecture provides a unique description of every point in $AH(M)$ (when $\partial\overline{M}$ is incompressible), but it does *not* give a description of the topology of $AH(M)$. This is due to two related facts: The ending invariants do not vary continuously in $AH(M)$ in any of the standard topologies (see Brock [**19**]), and there is a phenomenon of "bumping": the intersection of closures of components of the interior of $AH(M)$, which is not accounted for by ending laminations (see Anderson-Canary [**4**], Anderson-Canary-McCullough [**7**], Holt [**48**]). There is also a phenomenon of "self-bumping" of single components (see McMullen [**75**], Bromberg-Holt [**25**]): A component of the interior self-bumps at a boundary point p if its intersection with all sufficiently small neighborhoods of p is disconnected. Figure 1 suggests self-bumping of a quasifuchsian deformation space, although it has not to the author's knowledge been shown rigorously that the features of this figure are in fact true self-bumping, and not artifacts of the choice of 2-dimensional slice.

All known bumping phenomena involve essential annuli in $\overline{M}$. Given such an annulus A a sequence of elements of $AH(M)$ can be found whose geometric limit is homeomorphic to M minus a core curve of A in the interior – this gives rise to a rank 2 cusp. Then an exotic immersion of $\overline{M}$ into this cusped manifold yields a sequence of elements of $AH(M)$ which converge to the closure of a different component, or to the same component but from the "wrong" direction.

Although many different bumping and self-bumping points have been found, a topological model for all of $AH(M)$ remains out of reach. It is an indication of the incomplete nature of our knowledge that the following question is still without an answer in general:

Question. *Is the closure of a Bers slice homeomorphic to a closed ball?*

A Bers slice is the part of $int(AH(S \times [-1,1]))$ parameterized by $\{X\} \times \mathcal{T}(S)$ in the Ahlfors-Bers coordinates. Hence it is homeomorphic to the interior of a ball. A weaker question is whether there is no bumping of a Bers slice with itself, which seems quite plausible. The answer to this question is also known to be affirmative when $\mathcal{T}(S)$ is one complex dimensional, e.g. when S is a once-punctured torus [**79**].

References

[1] L. Ahlfors, *Fundamental polyhedrons and limit point sets for Kleinian groups*, Proc. Nat. Acad. Sci. **55** (1966), 251–254.

[2] L. Ahlfors and L. Bers, *Riemann's mapping theorem for variable metrics.*, Ann. of Math. **72** (1960), 385–404.

[3] Lars V. Ahlfors, *Finitely generated Kleinian groups*, Amer. J. Math. **86** (1964), 413–429.

[4] J. Anderson and R. Canary, *Algebraic limits of Kleinian groups which rearrange the pages of a book*, Invent. Math. **126** (1996), 205–214.

[5] ———, *Cores of hyperbolic 3-manifolds and limits of Kleinian groups*, Amer. J. Math. **118** (1996), 745–779.

[6] J. Anderson, R. Canary, M. Culler, and P. Shalen, *Free Kleinian groups and volumes of hyperbolic* 3*-manifolds*, J. Differential Geom. **43** (1996), no. 4, 738–782.

[7] J. Anderson, R. Canary, and D. McCullough, *The topology of deformation spaces of Kleinian groups*, Ann. of Math. (2) **152** (2000), no. 3, 693–741.

[8] L. Bers, *Simultaneous uniformization*, Bull. Amer. Math. Soc. **66** (1960), 94–97.

[9] ———, *Spaces of Kleinian groups*, Maryland conference in Several Complex Variables I, Springer-Verlag Lecture Notes in Math, No. 155, 1970, pp. 9–34.

[10] ———, *An extremal problem for quasiconformal mappings and a theorem by Thurston*, Acta Math. **141** (1978), 73–98.

[11] L. Bers and B. Maskit, *On a class of Kleinian groups*, Contemporary Problems in Theory Anal. Functions (Internat. Conf., Erevan, 1965) (Russian), Izdat. "Nauka", Moscow, 1966, pp. 44–47.

[12] C. J. Bishop and P. W. Jones, *Hausdorff dimension and Kleinian groups*, Acta Math. **179** (1997), no. 1, 1–39.

[13] F. Bonahon, *Bouts des variétés hyperboliques de dimension 3*, Ann. of Math. **124** (1986), 71–158.

[14] ———, *Geodesic laminations on surfaces*, Laminations and Foliations in Dynamics, Geometry and Topology (Stony Brook, NY) (M. Lyubich, J. Milnor, and Y. Minsky, eds.), Contemporary Mathematics, vol. 269, AMS, 2001, pp. 1–37.

[15] F. Bonahon and J. P. Otal, *Varietes hyperboliques a geodesiques arbitrairement courtes*, Bull. London Math. Soc. **20** (1988), 255–261.

[16] B. Bowditch, *Intersection numbers and the hyperbolicity of the curve complex*, Preprint, Southampton.

[17] ———, *Stacks of hyperbolic spaces and ends of 3-manifolds*, Preprint, Southampton.

[18] ———, *The Cannon-Thurston map for punctured-surface groups*, Preprint, Southampton.

[19] J. Brock, *Boundaries of Teichmüller spaces and end-invariants for hyperbolic* 3*-manifolds*, Duke Math. J. **106** (2001), no. 3, 527–552.

[20] ———, *Iteration of mapping classes and limits of hyperbolic 3-manifolds*, Invent. Math. **143** (2001), no. 3, 523–570.

[21] J. Brock and K. Bromberg, *On the density of geometrically finite Kleinian groups.*, Preprint 2002.

[22] J. Brock, K. Bromberg, R. Evans, and J. Souto, *Boundaries of deformation spaces and Ahlfors' measure conjecture.*, in preparation.

[23] J. Brock, R. Canary, and Y. Minsky, *Classification of Kleinian surface groups II: the ending lamination conjecture*, in preparation.

[24] K. Bromberg, *Projective structures with degenerate holonomy and the Bers density conjecture*, preprint.

[25] K. Bromberg and J. Holt, *Self-bumping of deformation spaces of hyperbolic 3-manifolds*, J. Differential Geom. **57** (2001), no. 1, 47–65.

[26] R. Canary, M. Culler, S. Hersonsky, and P. Shalen, *Density of maximal cusps in boundaries of quasiconformal deformation spaces*, preprint.

[27] R. Canary and Y. Minsky, *On limits of tame hyperbolic 3-manifolds*, J. Differential Geom. **43** (1996), 1–41.

[28] R. D. Canary, *On the laplacian and the geometry of hyperbolic 3-manifolds*, J. Differential Geom. **36** (1992), 349–367.

[29] ———, *Ends of hyperbolic 3-manifolds*, J. Amer. Math. Soc. **6** (1993), 1–35.

[30] R. D. Canary, D. B. A. Epstein, and P. Green, *Notes on notes of Thurston*, Analytical and Geometric Aspects of Hyperbolic Space, Cambridge University Press, 1987, London Math. Soc. Lecture Notes Series no. 111, pp. 3–92.

[31] J. Cannon, *The theory of negatively curved spaces and groups*, Ergodic theory, symbolic dynamics, and hyperbolic spaces (Trieste, 1989), Oxford Univ. Press, 1991, pp. 315–369.

[32] J. Cannon and W. Thurston, *Group invariant peano curves*, preprint, 1989.

[33] A. J. Casson and S. A. Bleiler, *Automorphisms of surfaces after Nielsen and Thurston*, Cambridge University Press, 1988.

[34] D. Wright D.Mumford, C.Series, *Indra's Pearls*, Cambridge Univ. Press, 2002.

[35] A. Douady and C.J. Earle, *Conformally natural extension of homeomorphisms of the circle*, Acta Math. **157** (1986), 23–48.

[36] D. B. A. Epstein and A. Marden, *Convex hulls in hyperbolic space, a theorem of Sullivan, and measured pleated surfaces*, Analytical and Geometric Aspects of Hyperbolic Space, Cambridge University Press, 1987, London Math. Soc. Lecture Notes Series no. 111, pp. 113–254.

[37] R. Evans, *Tameness persists in type-preserving limits*, Preprint.

[38] ———, *Deformation spaces of hyperbolic 3-manifolds: strong convergence and tameness*, Ph.D. thesis, U. of Michigan, 2000.

[39] A. Fathi, F. Laudenbach, and V. Poenaru, *Travaux de Thurston sur les surfaces*, vol. 66-67, Asterisque, 1979.

[40] L. Greenberg, *Fundamental polyhedra for kleinian groups*, Ann. of Math. (2) **84** (1966), 433–441.

[41] M. Gromov, *Hyperbolic groups*, Essays in Group Theory (S. M. Gersten, editor), MSRI Publications no. 8, Springer-Verlag, 1987.

[42] J. Harer, *Stability of the homology of the mapping class group of an orientable surface*, Ann. of Math. **121** (1985), 215–249.

[43] ———, *The virtual cohomological dimension of the mapping class group of an orientable surface*, Invent. Math. **84** (1986), 157–176.

[44] W. J. Harvey, *Boundary structure of the modular group*, Riemann Surfaces and Related Topics: Proceedings of the 1978 Stony Brook Conference (I. Kra and B. Maskit, eds.), Ann. of Math. Stud. 97, Princeton, 1981.

[45] J. Hempel, *3-manifolds as viewed from the curve complex*, Topology **40** (2001), no. 3, 631–657.

[46] C. D. Hodgson and S. P. Kerckhoff, *Universal bounds for hyperbolic Dehn surgery*, preprint, arXiv:math.GT/0204345.

[47] C. D. Hodgson and S. P. Kerckhoff, *Rigidity of hyperbolic cone-manifolds and hyperbolic Dehn surgery*, J. Differential Geom. **48** (1998), no. 1, 1–59.

[48] J. Holt, *Multiple bumping of deformation spaces of hyperbolic 3-manifolds*, preprint.

[49] N. V. Ivanov, *Complexes of curves and the Teichmüller modular group*, Uspekhi Mat. Nauk **42** (1987), 55–107.

[50] ———, *Complexes of curves and Teichmüller spaces*, Math. Notes **49** (1991), 479–484.

[51] ———, *Automorphisms of complexes of curves and of Teichmüller spaces*, Internat. Math. Res. Notices (1997), no. 14, 651–666.

[52] T. Jørgensen, *On cyclic groups of Möbius tranformations*, Math. Scand. **33** (1973), 250–260.

[53] ———, *On discrete groups of Möbius transformations*, Amer. J. of Math. **98** (1976), 739–49.

[54] M. Kapovich, *On the dynamics of pseudo-Anosov homeomorphisms on representation varieties of surface groups*, Ann. Acad. Sci. Fenn. Math. **23** (1998), no. 1, 83–100.

[55] ———, *Hyperbolic manifolds and discrete groups*, Progress in Mathematics, vol. 183, Birkhäuser Boston Inc., Boston, MA, 2001.

[56] L. Keen and Ç. Series, *Pleating coordinates for the Maskit slice of Teichmüller space*, Topology **32** (1993), 719–749.

[57] S. Kerckhoff and W. P. Thurston, *Noncontinuity of the action of the modular group at Bers' boundary of Teichmüller space*, Invent. Math. **100** (1990), 25–47.

[58] E. Klarreich, *The boundary at infinity of the curve complex and the relative Teichmüller space*, preprint.

[59] ———, *Semiconjugacies between Kleinian group actions on the Riemann sphere*, Amer. J. Math. **121** (1999), no. 5, 1031–1078.

[60] M. Korkmaz, *Automorphisms of complexes of curves on punctured spheres and on punctured tori*, Topology Appl. **95** (1999), no. 2, 85–111.

[61] I. Kra, *On spaces of Kleinian groups*, Comment. Math. Helv. **47** (1972), 53–69.

[62] R. S. Kulkarni and P. B. Shalen, *On Ahlfors' finiteness theorem*, Adv. Math. **76** (1989), no. 2, 155–169.

[63] G. Levitt, *Foliations and laminations on hyperbolic surfaces*, Topology **22** (1983), 119–135.

[64] F. Luo, *On Heegaard diagrams*, Math. Res. Lett. **4** (1997), no. 2-3, 365–373.

[65] ———, *Automorphisms of the complex of curves*, Topology **39** (2000), no. 2, 283–298.

[66] A. Marden, *The geometry of finitely generated Kleinian groups*, Ann. of Math. **99** (1974), 383–462.

[67] A. Marden and B. Maskit, *On the isomorphism theorem for Kleinian groups*, Invent. Math. **51** (1979), 9–14.

[68] B. Maskit, *On boundaries of Teichmüller spaces and on Kleinian groups II*, Ann. of Math. **91** (1970), 607–639.

[69] Bernard Maskit, *Classification of Kleinian groups*, Proceedings of the International Congress of Mathematicians (Vancouver, B. C., 1974), Vol. 2, Canad. Math. Congress, Montreal, Que., 1975, pp. 213–216.

[70] H. A. Masur and Y. Minsky, *Geometry of the complex of curves I: Hyperbolicity*, Invent. Math. **138** (1999), 103–149.

[71] ———, *Geometry of the complex of curves II: Hierarchical structure*, Geom. Funct. Anal. **10** (2000), 902–974.

[72] D. McCullough, *Compact submanifolds of 3-manifolds with boundary*, Quart. J. Math. Oxford **37** (1986), 299–306.

[73] C. McMullen, *Cusps are dense*, Ann. of Math. **133** (1991), 217–247.

[74] ———, *Renormalization and 3-manifolds which fiber over the circle*, Princeton University Press, Princeton, NJ, 1996.

[75] ———, *Complex earthquakes and Teichmüller theory*, J. Amer. Math. Soc. **11** (1998), no. 2, 283–320.

[76] Y. Minsky, *Teichmüller geodesics and ends of hyperbolic 3-manifolds*, Topology **32** (1993), 625–647.

[77] ———, *On rigidity, limit sets and end invariants of hyperbolic 3-manifolds*, J. Amer. Math. Soc. **7** (1994), 539–588.

[78] ———, *A geometric approach to the complex of curves*, Proceedings of the 37th Taniguchi Symposium on Topology and Teichmüller Spaces (S. Kojima et. al., ed.), World Scientific, 1996, pp. 149–158.

[79] ———, *The classification of punctured-torus groups*, Annals of Math. **149** (1999), 559–626.

[80] ———, *Kleinian groups and the complex of curves*, Geometry and Topology **4** (2000), 117–148.

[81] ———, *Bounded geometry in Kleinian groups*, Invent. Math. **146** (2001), 143–192, arXiv:math.GT/0105078.

[82] ———, *Classification of Kleinian surface groups I: models and bounds*, preprint, 2002.

[83] ———, *Combinatorial and geometrical aspects of hyperbolic 3-manifolds*, Kleinian Groups and Hyperbolic 3-Manifolds (V. Markovic Y. Komori and C. Series, eds.), London Math. Soc. Lec. Notes, vol. 299, Cambridge Univ. Press, 2003, pp. 3–40.

[84] H. Miyachi, *On cusps in the boundary of the Maskit slice for once punctured torus groups*, Sūrikaisekikenkyūsho Kōkyūroku (2000), no. 1153, 20–28, Comprehensive research on complex dynamical systems and related fields (Japanese) (Kyoto, 1999).

[85] L. Mosher, *Stable quasigeodesics in Teichmüller space and ending laminations*, preprint.

[86] G. D. Mostow, *Quasiconformal mappings in n-space and the rigidity of hyperbolic space forms*, Publ. I.H.E.S. **34** (1968), 53–104.

[87] P.J. Nicholls, *The ergodic theory of discrete groups*, London Math. Soc. Lecture Notes no. 143, Cambridge University Press, 1989.

[88] K. Ohshika, *On limits of quasi-conformal deformations of Kleinian groups*, Math. Z. **201** (1989), 167–176.

[89] ———, *Ending laminations and boundaries for deformation spaces of Kleinian groups*, J. London Math. Soc. **42** (1990), 111–121.

[90] ———, *Limits of geometrically tame Kleinian groups on boundaries for deformation spaces*, Invent. Math. **99** (1990), 185–203.

[91] J.-P. Otal, *Le théorèm d'hyperbolisation pour les variétés fibrées de dimension trois*, Astérisque No. 235, 1996.

[92] R. Penner and J. Harer, *Combinatorics of train tracks*, Annals of Math. Studies no. 125, Princeton University Press, 1992.

[93] G. Prasad, *Strong rigidity of* **Q**-*rank* 1 *lattices*, Invent. Math. **21** (1973), 255–286.

[94] G. P. Scott, *Compact submanifolds of 3-manifolds*, J. London Math. Soc. **7** (1973), 246–250.

[95] P. Scott, *The geometries of 3-manifolds*, Bull. London Math. Soc. **15** (1983), 401–487.

[96] P. B. Shalen, *Representations of 3-manifold groups*, Handbook of geometric topology, North-Holland, Amsterdam, 2002, pp. 955–1044.

[97] T. Soma, *Geometric limits of quasi-Fuchsian groups*, preprint.

[98] D. Sullivan, *The density at infinity of a discrete group of hyperbolic motions*, Publ. I.H.E.S. **50** (1979), 419–450.

[99] ———, *On the ergodic theory at infinity of an arbitrary discrete group of hyperbolic motions*, Riemann Surfaces and Related Topics: Proceedings of the 1978 Stony Brook Conference, Ann. of Math. Stud. 97, Princeton, 1981.

[100] ———, *Quasiconformal homeomorphisms and dynamics II: Structural stability implies hyperbolicity for Kleinian groups*, Acta Math. **155** (1985), 243–260.

[101] W. Thurston, *Hyperbolic structures on 3-manifolds, II: surface groups and manifolds which fiber over the circle*, preprint, arXiv:math.GT/9801045.

[102] ———, *The geometry and topology of 3-manifolds*, Princeton University Lecture Notes, online at http://www.msri.org/publications/books/gt3m, 1982.

[103] ———, *Three-Dimensional Geometry and Topology*, Princeton University Press, 1997, (S. Levy, ed.).

SUNY AT STONY BROOK

Current Developments in Mathematics, 2002
Pages 219 – 289

On the Cohomology of Locally Symmetric Spaces and of their Compactifications

Leslie Saper

Abstract. This expository article gives an introduction to the (generalized) conjecture of Rapoport and Goresky-MacPherson which identifies the intersection cohomology of a real equal-rank Satake compactification of a locally symmetric space with that of the reductive Borel-Serre compactification. We motivate the conjecture with examples and then give an introduction to the various topics that are involved: intersection cohomology, the derived category, and compactifications of a locally symmetric space, particularly those above. We then give an overview of the theory of $\mathcal{L}$-modules and micro-support which was developed to solve the conjecture but has other important applications as well. We end with sketches of the proofs of three main theorems on $\mathcal{L}$-modules that lead to the resolution of the conjecture. The text is enriched with many examples, illustrations, and references to the literature.

Contents

An expanded version of two talks at the Harvard-M.I.T. conference "Current Developments in Mathematics, 2002", November 15–17.

This material is based upon work supported in part by the National Science Foundation under grant DMS-9870162. The manuscript was typeset using AMS-LaTeX; the XY-pic package was used for figures and diagrams.

1. An example

We begin with the simplest example of a locally symmetric space. This is the moduli of complex elliptic curves $\mathbb{C}/\Lambda$, where Λ is a lattice in $\mathbb{C}$. The moduli space can be expressed as

$$\begin{aligned}\frac{\{\text{elliptic curves}\}}{\mathbb{C}\text{-isomorphism}} &\longleftrightarrow \{\text{lattices } \Lambda \text{ in } \mathbb{C}\}/\mathrm{SO}(2)\mathbb{R}_+^* \\ &\longleftrightarrow \mathrm{SL}_2(\mathbb{Z})\backslash \mathrm{GL}_2(\mathbb{R})^0 / \mathrm{SO}(2)\mathbb{R}_+^* \\ &\longleftrightarrow \mathrm{SL}_2(\mathbb{Z})\backslash H\end{aligned}$$

where $H = \{z \in \mathbb{C} \mid \operatorname{Im} z > 0\}$ is the upper half-plane and $\mathbb{R}_+^*$ denotes the group of positive real scalar matrices. Here we let $\left(\begin{smallmatrix} a & b \\ c & d\end{smallmatrix}\right) \in \mathrm{GL}_2(\mathbb{R})^0$ parametrize the lattice $\Lambda = \mathbb{Z}(a+ib) + \mathbb{Z}(c+id)$; the quotient by $\mathrm{SL}_2(\mathbb{Z})$ removes the dependence on the choice of an oriented basis. For future purposes, we will actually view H as $\mathrm{GL}_2(\mathbb{R})/\mathrm{O}(2)\mathbb{R}_+^*$.

A fundamental domain for the action of $\mathrm{SL}_2(\mathbb{Z})$ is pictured in Figure 1; the action of $\mathrm{SL}_2(\mathbb{Z})$ further identifies $z \leftrightarrow -\bar{z}$ for points in the boundary of the domain. Note the "cusp" at infinity which is not properly part of $X = \mathrm{SL}_2(\mathbb{Z})\backslash H$ but may be added to yield a compactification X^*. One may also consider $X = \Gamma\backslash H$ where Γ is a finite-index subgroup of $\mathrm{SL}_2(\mathbb{Z})$. Then X^* is formed by adding possibly several cusp points. Any representation (σ, E) of $\mathrm{GL}_2(\mathbb{R})$ defines a locally constant sheaf $\mathbb{E}$ on X, namely $\Gamma\backslash(H \times E) = H \times E/\!\sim$, where $(x,v) \sim (\gamma x, \sigma(\gamma)v)$ for all $\gamma \in \Gamma$. The classical Eichler-Shimura theorem **[71]** states

THEOREM 1.1. *Let E_k be the representation of* $\mathrm{GL}_2(\mathbb{R})$ *with highest weight k. Then $H^1_P(X^*;\mathbb{E}_k) \cong \mathcal{S}_{k+2}(\Gamma) \oplus \overline{\mathcal{S}_{k+2}(\Gamma)}$.*

The group appearing on the left-hand side is the ***parabolic cohomology***:

$$H^1_P(X^*;\mathbb{E}_k) \equiv \operatorname{Ker}\Big(H^1(X;\mathbb{E}_k) \to \bigoplus_{x\in \text{cusps}} H^1(U_x;\mathbb{E}_k)\Big),$$

where U_x is a punctured neighborhood of the cusp point x. It may also be defined by a modification of the usual complex computing the group cohomology $H(\Gamma; E_k)$. One should think of parabolic cohomology as a modification of ordinary cohomology in which the local cohomology near a cusp in degree 1 has been killed or truncated.

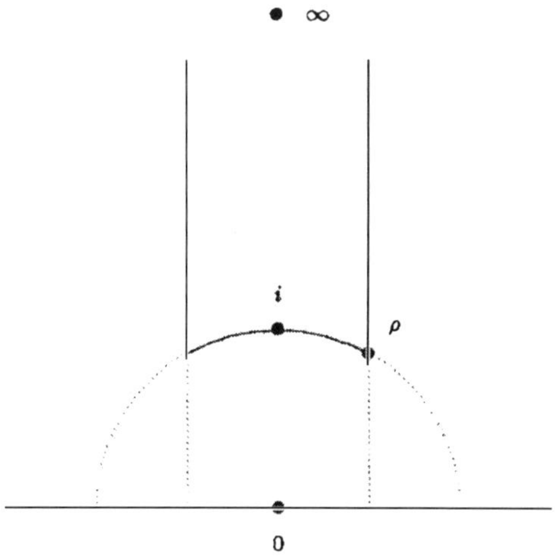

FIGURE 1. The fundamental domain for the action of $\mathrm{SL}_2(\mathbb{Z})$ on the upper half-plane

The space $\mathcal{S}_{k+2}(\Gamma)$ on the right-hand side is the space of *modular cusp forms of weight* $k+2$ *for* Γ:

$$f \in \mathcal{S}_{k+2}(\Gamma) \iff \begin{cases} f\colon H \to \mathbb{C} \text{ holomorphic}, \\ f\left(\frac{az+b}{cz+d}\right) = (cz+d)^{k+2} f(z) \text{ for } \begin{pmatrix} a & b \\ c & d \end{pmatrix} \in \Gamma, \\ f \text{ vanishes at the cusps.} \end{cases}$$

Modular forms (and their generalization, automorphic forms) play a central role in number theory. For example, note that the space X^* is actually an algebraic variety defined over a number field. An important application of Theorem 1.1 is that, for Γ a congruence subgroup, it allows one to relate the Hasse-Weil zeta function of X^* (which encodes the number of points of X^* defined over all finite fields) to the L-functions associated to modular forms. We will not discuss this further here; for details see Shimura's book [**71**].

2. L^2-cohomology

We now interpolate a group within the Eichler-Shimura isomorphism:

$$H^1_P(X^*; E_k) \cong H^1_{(2)}(X; \mathbb{E}_k) \cong \mathcal{S}_{k+2}(\Gamma) \oplus \overline{\mathcal{S}_{k+2}(\Gamma)} \tag{1}$$

The new group $H_{(2)}(X; \mathbb{E}_k)$ is the L^2-*cohomology*.

The L^2-cohomology is a variant of de Rham cohomology formed by imposing an L^2-growth condition on the differential forms. We will briefly review the theory in a general context; good references are the papers of Cheeger [**26**], Cheeger, Goresky, and MacPherson [**27**], and Zucker [**81**]. Suppose that M is an oriented Riemannian manifold (a V-manifold [**66**] or orbifold is sufficient) and that $\mathbb{E}$ is a metrized locally constant sheaf. In other words, $\mathbb{E}$ is the sheaf of sections of a flat vector bundle on M equipped with a fiber-wise Hermitian metric (which is not necessarily flat). Let d denote the exterior derivative on $A(M; \mathbb{E})$, the smooth $\mathbb{E}$-valued differential forms on M. For every $x \in M$, the norm on the tangent space $T_x M$ and the norm on the fiber $\mathbb{E}_x$ induce a norm on $\bigwedge T_x^* M \otimes \mathbb{E}_x$. Thus for $\phi \in A(M;\mathbb{E})$, we have a function $x \mapsto |\phi(x)|$. On the other hand, the Riemannian metric also induces a volume form dV. We define the L^2-norm of ϕ by

$$\|\phi\|^2 \equiv \int_M |\phi|^2 dV \tag{2}$$

and set

$$A_{(2)}(M;\mathbb{E}) \equiv \{\, \phi \in A(M;\mathbb{E}) \mid \|\phi\| < \infty \text{ and } \|d\phi\| < \infty \,\}. \tag{3}$$

The L^2-cohomology $H_{(2)}(M;\mathbb{E})$ is defined to be the cohomology of the complex $(A_{(2)}(M;\mathbb{E}), d)$; it has a topology (which may not be Hausdorff) given by the semi-norm

$$\|\,[\phi]\,\| \equiv \inf_{\eta \in A_{(2)}(M;\mathbb{E})} \|\phi + d\eta\| \qquad \text{for } [\phi] \in H_{(2)}(M;\mathbb{E}).$$

The L^2-cohomology is a *quasi-isometry* invariant of the metrics on M and $\mathbb{E}$: two metrics h and h' are said to be quasi-isometric if $c^{-1}h \le h' \le ch$ uniformly for some $c > 0$.

Sometimes it is useful to consider forms which are not necessarily smooth, but only measurable. The measurable forms ϕ which are L^2 and which have exterior derivatives in the distribution sense represented by L^2 forms yield a complex which we denote $(\operatorname{Dom} \bar{d}, \bar{d})$; this is indeed the closure in the sense of unbounded operators on Hilbert space of the d operator acting on the domain $A_{(2)}(M,\mathbb{E})$. A smoothing argument [**26**], [**28**] shows that this extended complex computes the same cohomology,

$$H_{(2)}(M;\mathbb{E}) = H(\operatorname{Dom} \bar{d}) = \operatorname{Ker} \bar{d} \,/ \operatorname{Im} \bar{d}\,.$$

It follows that

$$\begin{aligned} H_{(2)}(M;\mathbb{E}) &= \left(\operatorname{Ker} \bar{d} \,/\, \overline{\operatorname{Im} \bar{d}}\right) \oplus \left(\overline{\operatorname{Im} \bar{d}} \,/\, \operatorname{Im} \bar{d}\right) \\ &\cong \left(\operatorname{Ker} \bar{d} \cap \operatorname{Ker} \bar{d}^*\right) \oplus \left(\overline{\operatorname{Im} \bar{d}} \,/\, \operatorname{Im} \bar{d}\right). \end{aligned}$$

This is Hodge theory: the first term is a certain space of L^2 harmonic forms (in fact all L^2 harmonic forms if the Riemannian metric is complete) and the second term is either 0 or infinite dimensional. Note that $H_{(2)}(M;\mathbb{E})$ is Hausdorff if and only if the second term is 0.

In the example of §1, we give H the hyperbolic metric $(dx^2 + dy^2)/y^2$ under which H becomes a symmetric space. Since the metric is invariant under the action of $\mathrm{SL}_2(\mathbb{Z})$, it descends to a metric on $X = \Gamma \backslash H$, a locally symmetric space. Furthermore, any finite dimensional representation (σ, E) of $\mathrm{GL}_2(\mathbb{R})$ admits an ***admissible inner product***, that is, one for which $\sigma(g)^* = \sigma(\theta g)^{-1}$, where θ denotes the Cartan involution $g \mapsto {}^t g^{-1}$. Such an inner product induces a Hermitian metric on $\mathbb{E} = \Gamma \backslash (H \times E)$, namely

$$\|(g\,\mathrm{SO}(2), v)\| = \|\sigma(g^{-1})v\| \qquad \text{for } g \in \mathrm{SL}_2(\mathbb{R}). \tag{4}$$

A variant of this definition applies to $g \in \mathrm{GL}_2(\mathbb{R})$. These metrics on X and $\mathbb{E}$ are unique up to quasi-isometry (in fact up to scalar multiples if E is irreducible) and thus the L^2-cohomology $H_{(2)}(X;\mathbb{E}_k)$ is well-defined.

The first isomorphism of (1) is heuristically true since the metric near a cusp, say the cusp at ∞, may be expressed as

$$dr^2 + e^{-2r} d\theta^2, \qquad \text{where } r = \log y \in [b, \infty) \text{ and } \theta = x \text{ modulo } \mathbb{Z}, \tag{5}$$

and hence $d\theta$ is not L^2. The second isomorphism of (1) follows from Hodge theory and the Hodge decomposition: the coefficients of harmonic forms representing type $(1,0)$ cohomology classes can easily be seen to be modular forms of the appropriate weight; the L^2 condition implies the form is cuspidal.

Other moduli problems yield locally symmetric spaces and in view of the application mentioned in §1 and others, one would like a generalization of (1). In this paper, we will focus on the first isomorphism of (1).

3. Zucker's conjecture

In general a *locally symmetric space* has the form

$$X = \Gamma \backslash G / K A_G = \Gamma \backslash D \tag{6}$$

where

$G \equiv$ the real points of a connected reductive algebraic group defined over $\mathbb{Q}$,

$K \equiv$ a maximal compact subgroup of G,

$A_G \equiv$ the identity component $(\mathbb{R}_+^*)^s$ of a maximal $\mathbb{Q}$-split torus in the center of G,

$D \equiv G/KA_G$, the corresponding symmetric space,

$\Gamma \equiv$ an arithmetic subgroup of G.

We always give D a G-invariant Riemannian metric; this induces a Riemannian metric on X. Although the locally symmetric space X may have finite quotient singularities, for simplicity we will treat it and related spaces as if they were smooth; the correct treatment involves the language of V-manifolds or orbifolds. As in the case of $\mathrm{GL}_2(\mathbb{R})$, any representation E of G induces a metrized locally constant sheaf $\mathbb{E}$ on X.

NOTATION. In this paper we will indicate Lie groups, often defined as the real points of an algebraic group, by uppercase roman letters; their Lie algebras will be denoted by the corresponding lowercase fraktur letter. In order to lighten the exposition and the notation, we will take some liberties in terminology in this paper. Specifically if H is the Lie group of the real points (or the identity component of the real points) of an algebraic group defined over $\mathbb{Q}$, we may refer to H when properly we should refer to the underlying algebraic group. For example we might speak of a rational character of H defined over $\mathbb{Q}$. Or we will speak of a parabolic $\mathbb{Q}$-subgroup P of H, when we properly should be referring to the real points of a parabolic $\mathbb{Q}$-subgroup of the algebraic group underlying H.

An important special case is a *Hermitian locally symmetric space*, where D has a G-invariant complex structure. The generalization of the first isomorphism in (1) is

THEOREM 3.1 (Zucker's Conjecture). *Let X be a Hermitian locally symmetric space. There is a natural isomorphism $I_pH(X^*; \mathbb{E}) \cong H_{(2)}(X; \mathbb{E})$, where I_pH denotes middle-perversity intersection cohomology.*

This result was first conjectured by Zucker in [**81**] where some simple cases were proven. Other special cases of this result were proven by Borel [**10**], Borel and Casselman [**14**], and Zucker [**83**]. The theorem in general was proved by Looijenga [**49**] and independently by the author and Stern [**65**].

There are two new ingredients in this theorem we need to explain: the middle-perversity intersection cohomology $I_pH(X^*;\mathbb{E})$, which plays the role of parabolic cohomology, and the Baily-Borel-Satake compactification X^*, which generalizes adjoining cusp points to $\Gamma\backslash H$. Along the way we will also introduce the Borel-Serre compactification $\overline{X}$ and the reductive Borel-Serre compactification $\widehat{X}$ which will be important for the conjecture of Rapoport and Goresky-MacPherson.

4. Intersection cohomology

Intersection cohomology, developed by Goresky and MacPherson [**32**], [**33**], is a cohomology theory for singular spaces Z which shares many of the properties of ordinary cohomology theory applied to smooth spaces (e.g. Poincaré duality). It depends on a locally constant sheaf $\mathbb{E}$ on the smooth locus and a *perversity*:

$$p\colon \{2,\dots,d\} \to \mathbb{Z}, \quad p(2) = 0, \quad p(k) \le p(k+1) \le p(k)+1.$$

We will mainly be interested in p being one of the *middle perversities*:

$$n(k) = \left\lfloor \frac{(k-1)}{2} \right\rfloor \qquad \text{or} \qquad m(k) = \left\lfloor \frac{(k-2)}{2} \right\rfloor; \tag{7}$$

this is what is required in Zucker's conjecture and later in the conjecture of Rapoport and Goresky-MacPherson. Besides the original papers, a useful reference is Borel's book [**15**].

4.1. Simplicial intersection cohomology. The type of singular space for which $I_pH(Z;\mathbb{E})$ is usually defined is a pseudomanifold (see [**37**], [**43**], and [**55**] for more general contexts). A *d-dimensional stratified pseudomanifold* Z is a topological space with a filtration by closed subsets

$$Z = Z_d \supset Z_{d-1} = Z_{d-2} \supset Z_{d-3} \supset \cdots \supset Z_0 \supset Z_{-1} = \emptyset$$

such that

(i) For all $k \ge 0$, the subspace $S^k = Z_{d-k} \setminus Z_{d-k-1}$ is a topological $(d-k)$-manifold, called the *codimension-k stratum*.
(ii) The stratum S^0 is dense.
(iii) Any point $x \in S^k$ has an open neighborhood U in Z which is homeomorphic (as a stratified pseudomanifold) to $B_{d-k} \times c(L_{k-1})$, where B_{d-k} is a ball neighborhood of x in S^k and $c(L_{k-1})$ is a cone on a $(k-1)$-dimensional stratified pseudomanifold called the *link*.

The coefficient system $\mathbb{E}$ is a locally constant sheaf defined on S^0, the nonsingular stratum of Z. Examples of pseudomanifolds include complex analytic varieties as well as the reductive Borel-Serre compactification and the Satake compactifications of locally symmetric spaces to be discussed later.

The first definition of intersection cohomology [**32**], however, used a piecewise-linear structure on Z. Briefly, assume that Z is a piecewise-linear stratified pseudomanifold (that is, all spaces and maps in the above definition are taken in the piecewise-linear category). A locally finite *geometric chain* ξ is a locally finite simplicial chain for some triangulation of Z, however we identify two chains if they agree after passing to a common refinement of the triangulations; let $|\xi|$ denote the support of the chain. Let $\mathcal{O}$ denote the orientation sheaf of S^0 and let $I_pC^j(Z;\mathbb{E})$ denote the locally finite geometric $(d-j)$-chains ξ with values in $\mathbb{E}\otimes\mathcal{O}$ that satisfy the ***allowability*** conditions

$$\dim(|\xi| \cap Z_{d-k}) \leq d - j - k + p(k),$$
$$\dim(|\partial\xi| \cap Z_{d-k}) \leq d - j - 1 - k + p(k),$$

for all $k \geq 2$. Even though $\mathbb{E}$ is not defined on all of Z, the equations above for $k = 2$ ensure that the interiors of the $(d-j)$-simplices and the $(d-j-1)$-simplices in ξ lie in S^0, so it makes sense to say that ξ has "values in $\mathbb{E}\otimes\mathcal{O}$". The role of the perversity $p(k)$ in these equations is to bound the amount of intersection allowed beyond that which would occur in the case of transversal intersection.

Under the usual boundary operator, $I_pC(Z;\mathbb{E})$ becomes a complex whose cohomology is the intersection cohomology $I_pH(Z;\mathbb{E})$. The key thing to keep in mind about intersection cohomology is its local characterization. If $U = B_{d-k} \times c(L_{k-1})$ is a local neighborhood of $x \in S^k$ as above, then

$$I_pH^j(B_{d-k} \times c(L_{k-1});\mathbb{E}) \cong \begin{cases} I_pH^j(L_{k-1};\mathbb{E}) & \text{for } j \leq p(k), \\ 0 & \text{for } j > p(k). \end{cases} \tag{8}$$

Indeed this calculation and the facts that the Mayer-Vietoris sequence holds for intersection cohomology and that intersection cohomology agrees with ordinary cohomology on a manifold, suffice to inductively determine $I_pH(Z;\mathbb{E})$. Since $m(k) = n(k)$ for k even, it follows that both $p = m$ and $p = n$ yield the same theory for Z with only even codimension strata (for example, a complex analytic variety); certain spaces with possibly odd codimension strata will become important later. In our example of a quotient of the upper half-plane, a cusp x has codimension 2 and so the local intersection cohomology at x vanishes in degrees $j > p(2) = 0$ for $p = m$ or n; it is easy to see then that $I_pH^1(X^*;\mathbb{E}_k) \cong H^1_P(X^*;\mathbb{E}_k)$.

The constraints on a ordinary perversity and the condition in the definition of a stratified pseudomanifold that $S^1 = \emptyset$ ensure that $I_pH(Z;\mathbb{E})$ is independent of the stratification; it is not apparent however with this combinatorial definition that intersection cohomology is independent of the piecewise-linear structure.

4.2. The derived category. It has proven useful to take a sheaf-theoretic approach to intersection cohomology [**33**]; in particular, this allows an easy proof of topological invariance. In this approach, intersection cohomology is the hypercohomology of a certain object $\mathcal{I}_p\mathcal{C}(Z;\mathbb{E})$ in the constructible derived category of sheaves. We begin by briefly describing this category; a good reference is the book of Kashiwara and Schapira [**42**].

The (bounded) ***derived category of sheaves*** $\mathbf{D}^b(Z)$ has as objects bounded complexes of sheaves

$$\mathcal{S} \equiv \qquad \cdots \xrightarrow{d_{n-2}} \mathcal{S}^{n-1} \xrightarrow{d_{n-1}} \mathcal{S}^n \xrightarrow{d_n} \cdots$$

on Z. A single sheaf will always be treated as a complex which is zero except in degree 0. The definition of morphism is more subtle. Before giving it, recall that the ***cohomology sheaf*** $H(\mathcal{S})$ of $\mathcal{S}$ is the complex of sheaves (with zero differential) defined by $H^n(\mathcal{S}) \equiv \operatorname{Ker} d_n / \operatorname{Im} d_{n-1}$. Any morphism $\psi\colon \mathcal{S} \to \mathcal{S}'$ between complexes of sheaves induces a morphism $H(\psi)\colon H(\mathcal{S}) \to H(\mathcal{S}')$ between the cohomology sheaves; ψ is called a ***quasi-isomorphism*** if $H(\psi)$ is an isomorphism. Finally a morphism $\phi \in \mathrm{Mor}_{\mathbf{D}^b(Z)}$ is a certain equivalence class of diagrams $\mathcal{S} \xleftarrow{\phi_1} \mathcal{S}'' \xrightarrow{\phi_2} \mathcal{S}'$ of sheaf complex morphisms such that ϕ_1 is a quasi-isomorphism. (We will not define the equivalence relation here; to begin with, ϕ_1 and ϕ_2 should be taken to be homotopy classes of sheaf complex morphisms.) One defines a notion of composition of such morphisms.

It is clear that any sheaf complex morphism ψ induces a morphism of the corresponding objects in the derived category of sheaves; one sets $\mathcal{S}'' = \mathcal{S}$ with ϕ_1 the identity and $\phi_2 = \psi$. The definition is arranged so that a quasi-isomorphism becomes an isomorphism in the derived category and in fact we say that two complexes of sheaves are quasi-isomorphic if they are isomorphic in the derived category. Loosely put, the derived category is a localization of the ordinary category of complexes of sheaves obtained by "inverting" the quasi-isomorphisms. The cohomology sheaf $H(\mathcal{S})$ is well-defined for an object in the derived category and the ***constructible derived category*** is the full subcategory consisting of objects $\mathcal{S}$ such that $H(\mathcal{S})|_{S^k}$ is a locally constant sheaf with finitely generated stalks for every stratum S^k.

4.3. Functors on the derived category. Any bounded complex of sheaves $\mathcal{S}$ has an ***injective resolution***: a quasi-isomorphism $\mathcal{S} \to \mathcal{I}$ where $\mathcal{I}$ is a complex of injective sheaves; this is unique up to isomorphism in the derived category. Thus for any left exact functor F on sheaves one may define a corresponding functor RF (the ***right derived functor***) on the derived category of sheaves:

$$RF(\mathcal{S}) \equiv F(\mathcal{I}).$$

We will write RF simply as F since we will never consider the original functor F on complexes of sheaves. In this way, for any continuous map $k : Z \to W$ we obtain functors k_*, $k_!$, and k^* on the derived category. (The functor k^* is actually exact so one does not even need to pass to an injective resolution.) If k is a stratified map [**33**, §1.2] these are in fact functors on the constructible derived category.

The ***hypercohomology*** of an object $\mathcal{S}$ of the derived category is the right derived functor of the global sections functor $\Gamma(Z;\cdot)$. Thus $H(Z;\mathcal{S}) \equiv H(\Gamma(Z;\mathcal{I}))$.

The functor k^* is a left adjoint to k_*:

$$\mathrm{Mor}_{\mathbf{D}^b(Z)}(k^*\mathcal{S}, \mathcal{S}') = \mathrm{Mor}_{\mathbf{D}^b(W)}(\mathcal{S}, k_*\mathcal{S}').$$

It follows that there is a natural ***adjunction morphism***

$$\mathcal{S} \longrightarrow k_*k^*\mathcal{S}$$

characterized by the condition that $k^*\mathcal{S} \longrightarrow k^*k_*k^*\mathcal{S} = k^*\mathcal{S}$ is the identity morphism.

With slightly more work, one may also define a functor $k^!$ on the derived category which is a right adjoint to $k_!$. We will only need $k^!$ when k is the inclusion of a locally closed subset; in this case $k^! = k^* \circ R\Gamma_Z$, the derived "sections supported by Z" functor. Concretely, $k^!(\mathcal{S})$ is the sheaf defined by

$$k^!(\mathcal{S})(U) \equiv \mathrm{Ker}\big(\Gamma(\widetilde{U};\mathcal{I}) \to \Gamma(\widetilde{U} \setminus (Z \cap \widetilde{U});\mathcal{I})\big),$$

where $\widetilde{U} \subseteq W$ is an open subset such that $U = Z \cap \widetilde{U}$ is closed in $\widetilde{U}$.

It is not always necessary to use injective resolutions. For example, for hypercohomology it suffices to use a resolution by a complex of soft or fine sheaves provided the space is paracompact (as we will assume).

4.4. Sheaf-theoretic intersection cohomology. We need one more functor on the derived category. This is the truncation functor $\tau^{\leqslant n}$ defined by

$$\tau^{\leqslant n}\mathcal{S} \equiv \qquad \cdots \xrightarrow{d_{n-2}} \mathcal{S}^{n-1} \xrightarrow{d_{n-1}} \ker d_n \xrightarrow{d_n} 0 \longrightarrow \cdots ;$$

it is characterized by the fact there is a natural morphism $\tau^{\leqslant n}\mathcal{S} \to \mathcal{S}$ which induces an isomorphism on cohomology sheaves in degrees $\leq n$ and the zero map on cohomology sheaves in degrees $> n$.

For all $k \geq 2$, let $U_k \equiv Z \setminus Z_{d-k}$, and let $j_k \colon U_k \hookrightarrow U_{k+1} = U_k \cup S^k$ and $i_k \colon S^k \hookrightarrow U_{k+1}$ be the inclusions. (We will find it convenient in situations like this to use the same symbols to denote the inclusions of the same sets into larger subspaces.) The local calculation (8) is encoded into an object of the constructible derived category by Deligne's formula [**7**], [**33**]:

$$\mathcal{I}_pC(Z;\mathbb{E}) \equiv \tau^{\leqslant p(d)} j_{d*} \cdots \tau^{\leqslant p(3)} j_{3*} \tau^{\leqslant p(2)} j_{2*}\mathbb{E}. \tag{9}$$

This object is characterized up to isomorphism in the derived category by the following conditions:

(i) $i_2^*\mathcal{S} \cong \mathbb{E}$.
(ii) $H^j(i_k^*\mathcal{S}) = 0$ for $j > p(k)$.
(iii) The *attaching morphism* $i_k^*\mathcal{S} \to i_k^* j_{k*} j_k^*\mathcal{S}$ induces an isomorphism on cohomology sheaves in degrees $\leq p(k)$.

By definition, the intersection cohomology $I_pH(Z;\mathbb{E})$ is the hypercohomology of $\mathcal{I}_pC(Z;\mathbb{E})$. It is possible to recast these conditions in an equivalent form that does not make explicit use of the stratification [**15**], [**33**]; this proves that intersection cohomology is a topological invariant.

In the piecewise-linear case, the allowable locally finite geometric chains define a complex of sheaves $U \mapsto I_pC(U;\mathbb{E})$ which satisfies the above conditions (essentially by (8)) and hence is quasi-isomorphic to $\mathcal{I}_pC(Z;\mathbb{E})$. These sheaves are also soft, so their hypercohomology may be computed as the cohomology of global sections; it follows that the two definitions of intersection cohomology agree.

As in [**7**], we will sometimes consider a more general notion of perversity which is simply an integer-valued function $p \colon \mathcal{S} \to \mathbb{Z}$ on the set of strata $\mathcal{S}$. The formula (9) may be generalized to such p and the corresponding intersection cohomology may be defined; in general such a cohomology group depends upon the stratification.

4.5. Sheaf-theoretic L^2-cohomology. If the nonsingular stratum S^0 of Z has a Riemannian metric, one may localize the L^2-cohomology of S^0 at points of Z to define the *L^2-cohomology sheaf* $\mathcal{L}_{(2)}(Z;\mathbb{E})$. This is the complex of sheaves defined by the presheaf

$$U \mapsto A_{(2)}(U \cap S^0;\mathbb{E}).$$

In general this sheaf may not be fine; the difficulty is finding a partition of unity for a cover of Z with functions η such that $\eta|_{S^0}$ has bounded differential. However when this sheaf is fine, the L^2-cohomology of S^0 may be expressed as a hypercohomology group on Z: $H(Z;\mathcal{L}_{(2)}(Z;\mathbb{E})) = H_{(2)}(S^0;\mathbb{E})$.

Clearly the L^2-cohomology sheaf on the compactification X^* of $\mathrm{SL}_2(\mathbb{Z})\backslash H$ considered in §1 is fine. If instead one considers the compactification $\overline{X}$ obtained by adjoining $S^1 = \mathbb{Z}\backslash\mathbb{R}$ (an example of the Borel-Serre compactification to be considered later), the sheaf would not be fine since $|d\theta| \to \infty$ near the cusp by (5). It is not difficult to show that $\mathcal{L}_{(2)}(Z;\mathbb{E})$ is fine [**81**, (4.4)], [**83**] when Z is the reductive Borel-Serre compactification $\widehat{X}$ or one of the Satake compactifications X^* of a locally symmetric space to be introduced later.

The more precise form of Zucker's conjecture and Theorem 3.1 is that there is an isomorphism $\mathcal{I}_p\mathcal{C}(X^*;\mathbb{E}) \cong \mathcal{L}_{(2)}(X^*;\mathbb{E})$ in the derived category.

5. Compactifications of locally symmetric spaces

We need to consider three compactifications of a locally symmetric space X which fit into a diagram like this:

$$\begin{array}{ccccc} \overline{X} & \longrightarrow & \widehat{X} & \xrightarrow{\pi} & X^* \\ \text{Borel-Serre} & & \text{reductive Borel-Serre} & & \text{Satake} \end{array}$$

- The Borel-Serre compactification $\overline{X}$ [**16**] is "topological": it corresponds to adding a boundary (or more generally corners) at infinity and does not change the homotopy type of X.
- The reductive Borel-Serre compactification $\widehat{X}$ (first constructed by Zucker in [**81**, (4.1)]) is "geometric": the boundary strata of $\overline{X}$ become collapsed so that the Riemannian metric on X extends to a nondegenerate metric on each stratum at infinity.
- The compactification X^* is one of the various Satake compactifications [**67**], [**68**]. Zucker [**82**] has shown that it may be viewed as a quotient of $\widehat{X}$. In the special case where X is Hermitian locally symmetric, one of these is the Baily-Borel-Satake compactification. This is a complex-analytic compactification: the strata at infinity are further collapsed so that they also become Hermitian locally symmetric. In fact it is also an algebraic compactification: Baily and Borel [**4**] show in this case that X^* is a projective algebraic variety defined over a number field.

We will give the constructions of the three compactifications carefully in §§6–8 together with examples. In view of their importance to intersection cohomology, we will highlight for each compactification the boundary strata and their links. We will not however discuss the *toroidal compactifications* $\widetilde{X}$ [**3**]; these are defined in the Hermitian case and map to the Baily-Borel-Satake compactification X^*.

In the remainder of this section, we will describe the three compactifications qualitatively. First we give two simple examples.

5.1. Example: $X = \mathrm{SL}_2(\mathbb{Z})\backslash H$. In this case there is only one non-proper stratum for all three compactifications and the situation is quite simple. A basic result of reduction theory shows that for $y \gg 0$, the identifications induced by $\mathrm{SL}_2(\mathbb{Z})$ are already induced by its parabolic subgroup $\{\, \left(\begin{smallmatrix}1 & n\\ 0 & 1\end{smallmatrix}\right) | \, n \in \mathbb{Z}\,\}$. This group preserves the lines $y = b$ and acts on them by translation by elements of $\mathbb{Z}$. Thus in X for sufficiently large y these lines wrap up into a family of circles; for the compactification $\overline{X}$ we adjoin yet another circle at $y = \infty$; see Figure 2. However the norm of the tangent vector to the circles in this family decays as $y \to \infty$: $\left|\frac{\partial}{\partial x}\right| = \frac{1}{y} \to 0$. Thus the circle at infinity is collapsed to a point in $\widehat{X}$ and X^*. In summary:

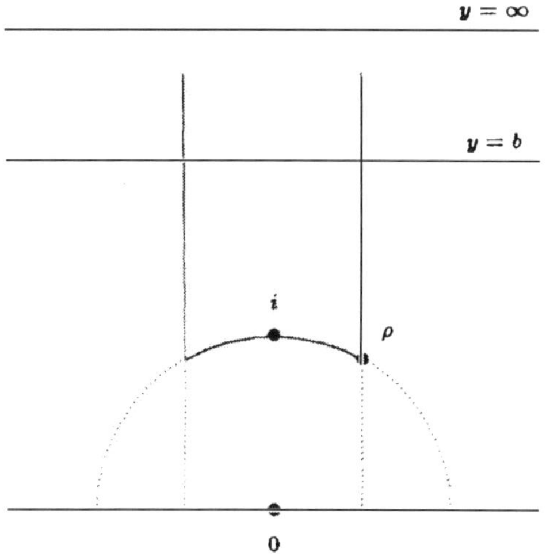

FIGURE 2. The upper half-plane with an additional line at $y = \infty$

	Boundary stratum	Link
$\overline{X}$	S^1	point
$\widehat{X}$	point	S^1
X^*	point	S^1

5.2. Example: Hilbert modular surface. The simplest Hermitian example where all three compactifications are different is the Hilbert modular surface. (See [**3**, §I.5] for a nice short exposition and [**39**] for more details.) This is the moduli space of complex tori $\mathbb{C}^2/\Lambda$ with "real multiplication" by $\mathcal{O}_k$, the ring of integers of a real quadratic extension $k = \mathbb{Q}(\sqrt{d})$, $d > 0$, of $\mathbb{Q}$. The symmetric space is simply $H \times H$; the arithmetic group $\Gamma = \mathrm{SL}_2(\mathcal{O}_k)$ acts on $H \times H$ via the embedding $\mathrm{SL}_2(\mathcal{O}_k) \hookrightarrow \mathrm{SL}_2(\mathbb{R}) \times \mathrm{SL}_2(\mathbb{R})$ given by the two real embeddings of k in $\mathbb{R}$. Again, reduction theory shows that "near infinity" corresponding to $y_1 y_2 \gg 0$, the group $\Gamma = \mathrm{SL}_2(\mathcal{O}_k)$ acts via its parabolic subgroup

$$\{\, \left(\begin{smallmatrix} u & 0 \\ 0 & u^{-1} \end{smallmatrix}\right) \mid u \in \mathcal{O}_k^\times \,\} \ltimes \{\, \left(\begin{smallmatrix} 1 & a \\ 0 & 1 \end{smallmatrix}\right) \mid a \in \mathcal{O}_k \,\} \tag{10}$$

and there is only one non-proper stratum for all three compactifications.

Write $\mathcal{O}_k = \mathbb{Z} + \mathbb{Z}\delta$; the image of the map $\mathcal{O}_k \to \mathbb{R}^2$ given by the two real embeddings of k in $\mathbb{R}$ is a lattice $\mathbb{Z}(1,1) + \mathbb{Z}(\delta, \bar{\delta})$. The second factor of (10) acts in the x_1x_2-plane via translation by this lattice. Let u_0 be a generator of the group of positive units $\mathcal{O}_{k+}^\times$. The first factor of (10) preserves the hyperbolas $y_1 y_2 = b$ and acts on them via $(y_1, y_2) \mapsto (u_0^{2k}, u_0^{-2k})$ for all $k \in \mathbb{Z}$. Furthermore it effects an automorphism of the lattice in the x_1x_2-plane. Thus in X for sufficiently large y_1y_2, we obtain a family of flat T^2-bundles (from the x_1x_2-plane) over S^1's (from the hyperbolas); for the compactification $\overline{X}$ we adjoin another copy of this bundle at $y_1 y_2 = \infty$; see Figure 3.

To understand the non-proper stratum in the other two compactifications first note that the product metric on $H \times H$ may be rewritten as

$$dr^2 + dt^2 + e^{-r}(e^{-t}dx_1^2 + e^t dx_2^2) \qquad \text{where } r = \log y_1 y_2 \text{ and } t = \log y_1/y_2;$$

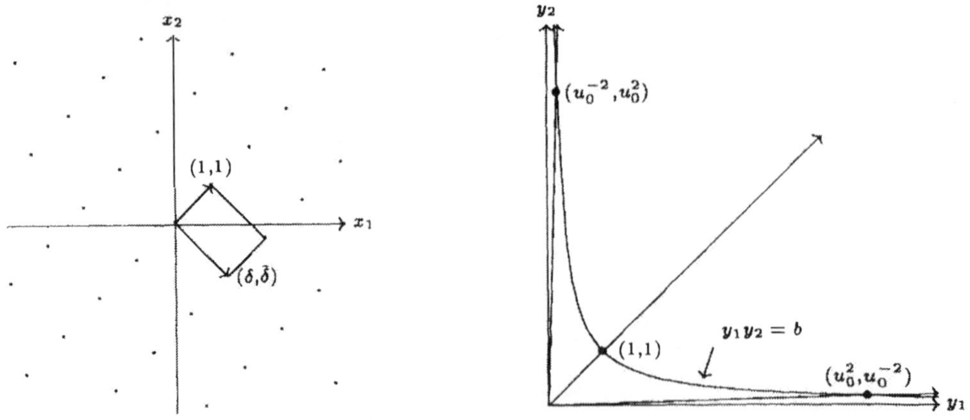

FIGURE 3. $\mathrm{SL}_2(\mathcal{O}_k)$ acting on $H \times H$

up to quasi-isometry this induces the metric

$$dr^2 + ds^2_{S^1} + e^{-r} ds^2_{T^2}$$

on X where $r \gg 0$. We see that the norm of a tangent vector to the T^2-fibers decays to 0 as $r \to \infty$ while the norm of a tangent vector to the base S^1 does not. Thus the T^2-fibers are collapsed in $\widehat{X}$ while the S^1 remains; since S^1 cannot have a complex structure, however, it is collapsed in X^*. In summary:

	Boundary stratum	Link
$\overline{X}$	flat T^2-bundle over S^1	point
$\widehat{X}$	S^1	T^2
X^*	point	flat T^2-bundle over S^1

5.3. The general case. In the general case the strata are no longer compact. For both the Borel-Serre compactification $\overline{X}$ and the reductive Borel-Serre compactification $\widehat{X}$ they are indexed by the set $\mathcal{P}$ of Γ-conjugacy classes of parabolic $\mathbb{Q}$-subgroups P of G; for every $P \in \mathcal{P}$, we let Y_P (respectively X_P) denote the corresponding stratum of $\overline{X}$ (respectively $\widehat{X}$). (Note that we will use the same symbol, say P, to denote both a parabolic $\mathbb{Q}$-subgroup and its Γ-conjugacy class.)

Associated to $G \in \mathcal{P}$ we have $X_G = Y_G = X$, the top-dimensional open stratum, while associated to a minimal parabolic $\mathbb{Q}$-subgroup $P_0 \in \mathcal{P}$ is a minimal-dimensional stratum. In fact the correspondence between the strata and $\mathcal{P}$ is order-preserving if we define the partial order on the set of strata by $S < T$ if and only if $S \subseteq \mathrm{cl}(T)$ (topological closure) and the partial order on $\mathcal{P}$ by $P < Q$ if and only if there exists $\gamma \in \Gamma$ so that $P \subset {}^{\gamma}Q \equiv \gamma Q \gamma^{-1}$.

For $P \in \mathcal{P}$, let $r(P)$ denote the *parabolic rank*; this is the number of proper parabolic $\mathbb{Q}$-subgroups in a maximal chain $P = P_r < P_{r-1} < \cdots P_1 < G$. Let $|\Delta_P|^\circ$ denote an open simplex of dimension $r(P) - 1$. We can now crudely describe the strata and their links. Each stratum X_P of $\widehat{X}$ will be a locally symmetric space; in the above example this was S^1. Each stratum Y_P of $\overline{X}$ is a flat bundle over X_P with general fiber a compact nilmanifold $\mathcal{N}'_P$; in the above example the nilmanifold was T^2. The links are now themselves stratified.

The corresponding data for a Satake compactification of $X = \Gamma\backslash D$ is in general more complicated; we will give Zucker's realization of X^* as a quotient of $\widehat{X}$ [**82**]. The construction depends on the choice of a certain representation U of G which must satisfy a condition called "geometric rationality" [**25**]. This data induces a decomposition $X_P = X_{P,\ell} \times X_{P,h}$ (modulo a finite quotient) for each stratum X_P of $\widehat{X}$. The strata of X^* will be the various $X_{P,h}$, and the quotient map $\widehat{X} \xrightarrow{\pi} X^*$ is induced by projection, stratum by stratum. The situation is complicated by the fact that several $P \in \mathcal{P}$ may yield the same stratum in X^*; we index the boundary stratum $X_{P,h}$ of X^* by the maximal such P, denoted $P^\dagger$. Let $\mathcal{P}^* = \{\, P^\dagger \mid P \in \mathcal{P} \,\}$.

We can summarize the above discussion in the following table; for simplicity we indicate just the smooth locus of the links.

	Boundary stratum associated to P	Link°
$\overline{X}$	$Y_P =$ flat $\mathcal{N}'_P$-bundle over X_P	$\lvert\Delta_P\rvert^\circ$
$\widehat{X}$	X_P	$\mathcal{N}'_P \times \lvert\Delta_P\rvert^\circ$
X^*	$X_{P^\dagger,h}$	$(\mathcal{N}'_{P^\dagger} \times \lvert\Delta_{P^\dagger}\rvert^\circ)$-bundle over $X_{P^\dagger,\ell}$

6. The Borel-Serre compactification

As in §3, we consider a locally symmetric space $X = \Gamma\backslash D$ where $D = G/KA_G$. In this section we outline the actual construction of $\overline{X}$ due to Borel and Serre [**16**]. The construction uses the language of algebraic groups, in particular their parabolic subgroups, for which good references are the texts of Borel [**12**] and Humphreys [**40**], as well as the paper by Borel and Tits [**17**]. For the reduction theory of arithmetic groups, one may consult Borel's book [**9**]. We illustrate the discussion with the example of $G = \mathrm{GL}_n(\mathbb{R})$ and $\Gamma = \mathrm{GL}_n(\mathbb{Z})$. In this case A_G is the subgroup of positive scalar matrices and we may pick $K = \mathrm{O}(n)$. The symmetric space $D = \mathrm{GL}_n(\mathbb{R})/\mathrm{O}(n)A_G$ is the space of positive definite $n \times n$ symmetric matrices modulo homothety and $X = \Gamma\backslash D$ is the moduli of positive definite quadratic forms on $\mathbb{Z}^n$ modulo homothety and isometry.

Let P be a parabolic $\mathbb{Q}$-subgroup, let N_P be its unipotent radical and $L_P = P/N_P$ its Levi quotient. We can when necessary lift L_P to a subgroup $\widetilde{L}_P$ of P such that $P = \widetilde{L}_P N_P$. In the case $G = \mathrm{GL}_n(\mathbb{R})$, any such P is conjugate under $\mathrm{GL}_n(\mathbb{Q})$ to the group of block upper-triangular matrices associated to a partition $n = n_0 + \cdots + n_r$, $n_i \geq 0$. A lift of the Levi quotient is the group of block diagonal matrices, $\mathrm{GL}_{n_0}(\mathbb{R}) \times \cdots \times \mathrm{GL}_{n_r}(\mathbb{R})$, while the unipotent radical N_P consists of those block upper-triangular matrices with identity matrices along the diagonal; see Figure 4. Conjugation by elements of N_P would take one lift $\widetilde{L}_P$ to all possible such lifts. The stabilizer of a basepoint $x_0 \in D$ has the form KA_G where K is a maximal compact subgroup. Thus the choice of a basepoint determines a maximal compact subgroup K with a corresponding Cartan involution θ; there exists a unique lift $\widetilde{L}_P$ of L_P which is stable under θ. Such a lift however is not necessarily defined over $\mathbb{Q}$.

Let A_P be the identity component of the maximal $\mathbb{Q}$-split torus in the center of L_P; it has a natural complement $M_P = {}^0L_P$ defined as

$$ {}^0L_P \equiv \bigcap_{\alpha \in X(L_P)} \ker \alpha^2, $$

$$\begin{pmatrix} \mathrm{GL}_{n_0}(\mathbb{R}) & & & * \\ & \mathrm{GL}_{n_1}(\mathbb{R}) & & \\ & & \ddots & \\ 0 & & & \mathrm{GL}_{n_r}(\mathbb{R}) \end{pmatrix} = \begin{pmatrix} \mathrm{GL}_{n_0}(\mathbb{R}) & & & 0 \\ & \mathrm{GL}_{n_1}(\mathbb{R}) & & \\ & & \ddots & \\ 0 & & & \mathrm{GL}_{n_r}(\mathbb{R}) \end{pmatrix} \begin{pmatrix} I_{n_0} & & & * \\ & I_{n_1} & & \\ & & \ddots & \\ 0 & & & I_{n_r} \end{pmatrix}$$

FIGURE 4. A parabolic subgroup $P = \widetilde{L}_P N_P$ of $\mathrm{GL}_n(\mathbb{R})$

where $X(L_P)$ denotes the (rational) characters of L_P defined over $\mathbb{Q}$. Thus $L_P = M_P A_P$ and this decomposition lifts to $\widetilde{L}_P = \widetilde{M}_P \widetilde{A}_P$. In the example, $\widetilde{A}_P$ consists of diagonal matrices which are positive scalar matrices in each block and $\widetilde{M}_P$ consists of block diagonal matrices $\mathrm{SL}^{\pm}_{n_0}(\mathbb{R}) \times \cdots \times \mathrm{SL}^{\pm}_{n_r}(\mathbb{R})$, where $\mathrm{SL}^{\pm}_{n_i}(\mathbb{R})$ denotes matrices with determinant ± 1.

There is a natural inclusion $A_G \subseteq A_P$ and a complement $A_P^G \subseteq A_P$ given as the joint kernel of the characters in $X(G)$. In our standard example, $a = \mathrm{diag}(\lambda_0 \cdot I_{n_0}, \dots, \lambda_r \cdot I_{n_r}) \in A_P$ lies in A_G if and only if $\lambda_0 = \cdots = \lambda_r$ and a lies in A_P^G if and only if $\det a = 1$. Since A_P^G is $\mathbb{Q}$-split, it has coordinates given by characters of P defined over $\mathbb{Q}$. In the next paragraph we will define an important and canonical set of characters Δ_P so that we have an isomorphism

$$A_P^G \xrightarrow{\sim} (\mathbb{R}^*_+)^{\Delta_P}, \qquad a \mapsto (a^{\alpha_1}, \dots, a^{\alpha_r}). \tag{11}$$

There is an adjoint action of $z \in L_P$ on the Lie algebra $\mathfrak{n}_P$ of N_P given by lifting z to $\widetilde{L}_P$ and then acting by the adjoint action. Although this action depends on the lift, its A_P^G-weights (which are induced from characters of P) do not. For P a minimal parabolic $\mathbb{Q}$-subgroup these are the positive roots of a root system, namely the $\mathbb{Q}$-root system of G, but this does not hold true in general. However it is still true that if we let $\Delta_P = \{\alpha_1, \dots, \alpha_r\}$ be the indecomposable elements among the A_P^G-weights of $\mathfrak{n}_P$, then all A_P^G-weights of $\mathfrak{n}_P$ can be expressed uniquely as non-negative integral linear combinations of elements of Δ_P. We call the weights in Δ_P the ***simple roots*** even though they may not generate a root system. There is also a "dual" basis $\widehat{\Delta}_P = \{\beta_1, \dots, \beta_r\}$ which in the case of a minimal parabolic $\mathbb{Q}$-subgroup is simply the set of fundamental weights. In our example, we have

$$a^{\alpha_i} = \lambda_i/\lambda_{i-1} \quad \text{and} \quad a^{\beta_i} = 1/(\lambda_0 \cdots \lambda_{i-1})$$

for $a = \mathrm{diag}(\lambda_0 \cdot I_{n_0}, \dots, \lambda_r \cdot I_{n_r}) \in A_P^G$.

For later use we single out two subsets of A_P^G, the ***strictly dominant cone***

$$A_P^{G+} \equiv \{a \in A_P^G \mid a^\alpha > 1 \text{ for all } \alpha \in \Delta_P\},$$

and the ***negative codominant cone***,

$${}^-A_P^G \equiv \{a \in A_P^G \mid a^\beta \leq 1 \text{ for all } \beta \in \widehat{\Delta}_P\}.$$

The Lie algebra $\mathfrak{a}_P^G$ of A_P^G, endowed with a Weyl group invariant inner product, is illustrated in Figure 5(a) in the case of a minimal parabolic $\mathbb{Q}$-subgroup of $\mathrm{GL}_3(\mathbb{R})$ together with the subsets $\mathfrak{a}_P^{G+} \equiv \log A_P^{G+}$ and ${}^-\mathfrak{a}_P^G \equiv \log {}^-A_P^G$. The figure also includes Δ_P and $\widehat{\Delta}_P$, identified with elements of $\mathfrak{a}_P^{G*}$ and then transfered to $\mathfrak{a}_P^G$ via the inner product.

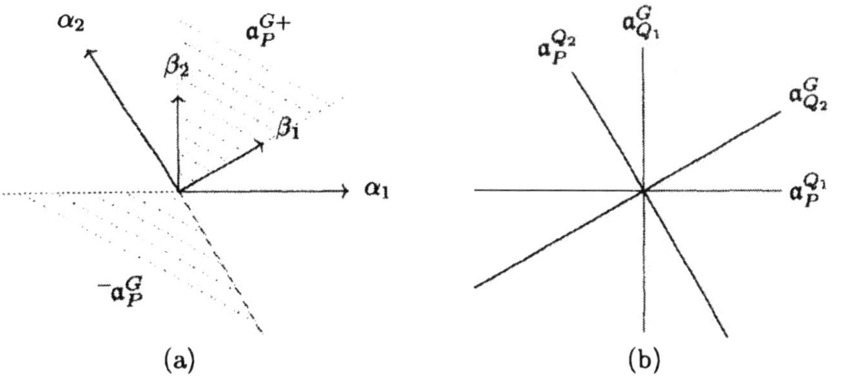

FIGURE 5. $\mathfrak{a}_P^G$, its subcones, and its subspaces for P a minimal parabolic subgroup of $\mathrm{GL}_3(\mathbb{R})$. (Here $\Delta_P = \{\alpha_1, \alpha_2\}$ and $\Delta_P^{Q_i} = \{\alpha_i\}$.)

The parabolic $\mathbb{Q}$-subgroups Q containing P are in one-to-one order preserving correspondence with subsets Δ_P^Q of Δ_P. Namely, consider $Q = \widetilde{L}_Q N_Q \supseteq P = \widetilde{L}_P N_P$; we can decompose $N_P = (N_P \cap \widetilde{L}_Q) N_Q$. Define Δ_P^Q to consist of those $\alpha \in \Delta_P$ whose weight space lies in $\mathfrak{n}_P \cap \widetilde{\mathfrak{l}}_Q$ as opposed to $\mathfrak{n}_Q$. For such a Q, there is a natural identification

$$A_Q^G \equiv \bigcap_{\alpha \in \Delta_P^Q} \ker \alpha \subseteq A_P^G.$$

The set of simple roots Δ_Q for Q is the set of restrictions $\alpha|_{A_Q^G}$ for $\alpha \in \Delta_P \setminus \Delta_P^Q$. Furthermore there is a complementary subgroup

$$A_P^Q \equiv \bigcap_{\alpha \in \Delta_P \setminus \Delta_P^Q} \ker \beta_\alpha \subseteq A_P^G$$

to A_Q^G; here $\beta_\alpha \in \widehat{\Delta}_P$ is the dual element to $\alpha \in \Delta_P$. The group A_P^Q may also be viewed as the analogue of A_P^G for the parabolic $\mathbb{Q}$-subgroup P/N_Q of L_Q. We have a direct decomposition

$$A_P^G = A_Q^G \times A_P^Q.$$

The corresponding decomposition of the Lie algebra in the example of $\mathrm{GL}_3(\mathbb{R})$ is illustrated in Figure 5(b). Note that the closure of the strictly dominant cone A_P^{G+} is stratified with strata $\{A_Q^{G+}\}_{Q \supseteq P}$, while $\{^{-}A_P^Q\}_{Q \supseteq P}$ are the closures of the strata of the negative codominant cone $^{-}A_P^G$.

The compactification $\overline{X}$ will be the quotient by Γ of a bordification $\overline{D}$ which will be constructed in three steps.

6.1. The first step is to form a partial bordification $\bar{A}_P^G$ of A_P^G, a "corner", by allowing the coordinates in (11) to attain infinity:

$$\bar{A}_P^G \xrightarrow{\sim} (\mathbb{R}_+^* \cup \{\infty\})^{\Delta_P}, \qquad a \mapsto (a^{\alpha_1}, \ldots, a^{\alpha_r}).$$

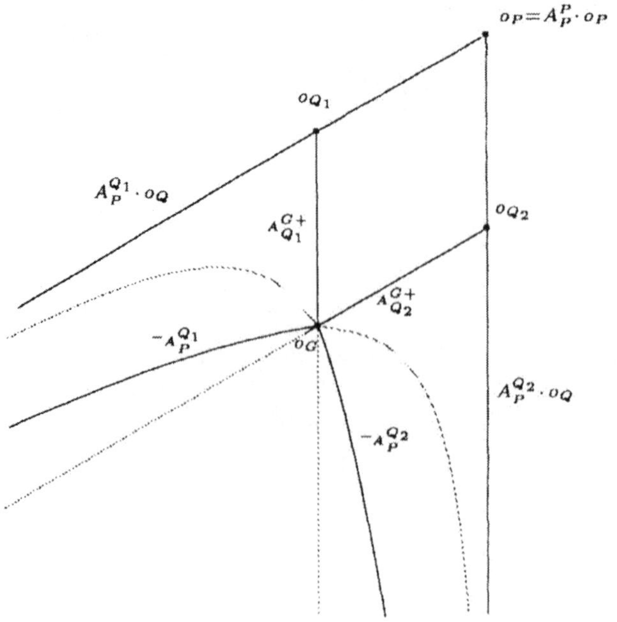

FIGURE 6. $\bar{A}_P^G$ for P a minimal parabolic subgroup of $\mathrm{GL}_3(\mathbb{R})$

The corner $\bar{A}_P^G$ is given a real-analytic structure by means of the inverses of these coordinates: $\bar{A}_P^G \to (\mathbb{R}_{\geq 0})^{\Delta_P}$, $a \mapsto (a^{-\alpha_1}, \dots, a^{-\alpha_r})$. For $Q \supseteq P$, the inclusion $A_Q^G \hookrightarrow A_P^G$ extends to the partial bordifications, $\bar{A}_Q^G \hookrightarrow \bar{A}_P^G$. The action of A_P^G on itself by multiplication extends naturally to an action of A_P^G on $\bar{A}_P^G$. In fact $\bar{A}_P^G$ is a semigroup. Set $\bar{A}_P^{G+}$ to be the subset where all coordinates are greater than 1.

Let $o_Q \in \bar{A}_P^G$ be the point with coordinates $\{\infty\}^{\Delta_P \setminus \Delta_P^Q} \times \{1\}^{\Delta_P^Q}$; this point has stabilizer A_Q^G so the orbit it generates is $A_P^Q \cdot o_Q$. The space $\bar{A}_P^G$ is stratified by these A_P^G-orbits:

$$\bar{A}_P^G = \coprod_{Q \supseteq P} A_P^Q \cdot o_Q .$$

Topologically $\bar{A}_P^G$ is homeomorphic to a cone $c(|\Delta_P|)$ on the simplex $|\Delta_P|$. (We denote by $|\Delta_P|$ the $(r-1)$-dimensional closed simplex whose vertices are identified with $\alpha \in \Delta_P$.) The vertex of the cone is $o_P = \{\infty\}^{\Delta_P}$ and the sub-cone $c(|\Delta_P^Q|)$ corresponds to the closure of the orbit $A_P^Q \cdot o_Q$.

All of the above illustrated in Figure 6 for P a minimal parabolic subgroup in $G = \mathrm{GL}_3(\mathbb{R})$. The figure is drawn using analytic coordinates based at o_P and chosen so that at o_G the diagram accurately reflects the geometry of a Weyl-group invariant inner product on the tangent space $\mathfrak{a}_P^G$. It is important to note that whereas the two 1-parameter subgroups $A_{Q_1}^G$ and $A_{Q_2}^G$ each intersect the boundary at infinity in 1 point, neither of the subgroups $A_P^{Q_1}$ or $A_P^{Q_2}$ intersect the boundary at all. This is due to the fact that the two simple roots are not orthogonal.

6.2. The second step in the construction of $\overline{D}$ is to extend the partial bordification of A_P^G to one of D. To do this, note that even though A_P is not naturally a subgroup of G, Borel and Serre define an action of A_P on D called the *geodesic action* [**16**, §3]. To describe it, fix a basepoint in D which determines a maximal compact subgroup $K \subseteq G$, a Cartan involution θ, and a θ-stable lift $\widetilde{L}_P$ of L_P. Note that $G = PK$, that is, P acts transitively on D; this is a generalization of Gram-Schmidt orthonormalization. Every point $x \in D$ can thus be written as $x = pKA_G$ for some $p \in P$, and the geodesic action of $a \in A_P$ on x is given by

$$a \circ x \equiv p\tilde{a}KA_G,$$

where $\tilde{a} \in \tilde{A}_P$ is the lift of a. The geodesic action is independent of the choice of basepoint and commutes with the usual action of P.

Let $e_P \equiv A_P^G \backslash D$ be obtained by collapsing the orbits of the geodesic action of A_P^G to points. We will further discuss the structure of e_P below. The quotient map $r_P \colon D \to e_P$ is a A_P^G-principal bundle; natural trivializing sections are given by the orbits of ${}^0P \equiv \bigcap_{\alpha \in X(P)} \ker \alpha^2$. Thus if we let $\mathcal{A}_P^G \equiv {}^0P \backslash D$ be the A_P^G-homogeneous space obtained by collapsing these sections to points, there is a natural product decomposition

$$D = \mathcal{A}_P^G \times e_P.$$

This generalizes the product decomposition of the upper half-plane into geodesics tending to ∞ (rays parallel to the positive y-axis) times horocycles at ∞ (lines parallel to the x-axis). For P a minimal parabolic $\mathbb{R}$-subgroup this is the decomposition induced from the Iwasawa decomposition of G.

The space $\mathcal{A}_P^G$ is an affine version of A_P^G; given a choice of basepoint in D, there is a canonical isomorphism of A_P^G-homogeneous spaces $\mathcal{A}_P^G \cong A_P^G$ such that the class of the basepoint maps to the identity. We can then define a bordification $\bar{\mathcal{A}}_P^G$ such that this isomorphism extends to an isomorphism $\bar{\mathcal{A}}_P^G \cong \bar{A}_P^G$. The resulting A_P^G-space (in fact $\bar{A}_P^G$-space) is independent of the choice of basepoint. Note that although the points $o_Q \in \bar{A}_P^G$ do not uniquely determine points in $\bar{\mathcal{A}}_P^G$ (except for $Q = P$), the orbits $A_P^Q \cdot o_Q \subseteq \bar{A}_P^G$ do indeed define well-defined orbits in $\bar{\mathcal{A}}_P^G$. We will again use the notation $A_P^Q \cdot o_Q$ to denote the corresponding orbit of $\bar{\mathcal{A}}_P^G$.

Borel and Serre construct a partial bordification $D(P)$ by replacing $\mathcal{A}_P^G$ in the above decomposition with $\bar{\mathcal{A}}_P^G$:

$$D(P) = \bar{\mathcal{A}}_P^G \times e_P. \tag{12}$$

We consider e_P as a stratum of $D(P)$ by identifying it with $o_P \times e_P = \{\infty\}^{\Delta_P} \times e_P$; the link of any point in e_P is $|\Delta_P|$. The projection map

$$D(P) = \bar{\mathcal{A}}_P^G \times e_P \longrightarrow \{\infty\}^{\Delta_P} \times e_P \equiv e_P \tag{13}$$

is called *geodesic retraction*.

6.3. The third and final step in the construction of $\overline{D}$ is to note that the construction above of the partial bordification is functorial in the sense that for $P \subseteq Q$ we can canonically identify $D(Q)$ with an open subset of $D(P)$. Indeed $e_Q \subseteq D(Q)$ corresponds to $(A_P^Q \cdot o_Q) \times e_P \subseteq D(P)$ and so $D(P)$ may be re-expressed as $\coprod_{Q \supseteq P} e_Q$. Borel and Serre then define a full bordification of D by setting

$$\overline{D} = \bigcup_P D(P) = \coprod_P e_P$$

where the union is over all parabolic $\mathbb{Q}$-subgroups P and the above identifications are taken into account. Let $\bar{e}_P$ denote the closure of e_P in $\overline{D}$.

The action of G on D extends to an action of $G(\mathbb{Q})$ on $\overline{D}$ and the *Borel-Serre compactification* is defined by

$$\overline{X} \equiv \Gamma\backslash\overline{D}.$$

Given a parabolic $\mathbb{Q}$-subgroup P, the various strata $e_{\gamma P}$ of $\overline{D}$ for $\gamma \in \Gamma$ become identified to one stratum Y_P in $\overline{X}$. Furthermore reduction theory shows that the identifications induced by Γ on e_P, and indeed even on a small neighborhood of $\overline{e}_P$, are already induced by $\Gamma_P \equiv \Gamma \cap P$. Thus there is a stratification

$$\overline{X} = \coprod_{P\in\mathcal{P}} Y_P, \qquad \text{where } Y_P \equiv \Gamma_P\backslash e_P.$$

The geodesic retraction (13) extends to a projection, also called geodesic retraction,

$$V \longrightarrow Y_P, \tag{14}$$

where V is a sufficiently small neighborhood of Y_P in $\overline{X}$. In fact, if O_P is a relatively compact open subset of Y_P and $\widetilde{O}_P$ is its inverse image in e_P, reduction theory shows that there exists $s_P \in \mathcal{A}_P^G$ such that the identifications induced by Γ on

$$\widetilde{W}_P \equiv (\bar{A}_P^{G+} \cdot s_P) \times \widetilde{O}_P \subsetneq D(P)$$

agree with those induced by Γ_P. However Γ_P acts only on the second factor. Thus

$$W_P \equiv \Gamma_P\backslash\widetilde{W}_P = (\bar{A}_P^{G+} \cdot s_P) \times O_P \subsetneq \Gamma_P\backslash D(P) \tag{15}$$

may be identified with a subset of $\overline{X}$; the geodesic retraction is projection on the second factor. The open subset W_P is called a *cylindrical set*; for appropriate choices of O_P and s_P (for all $P \in \mathcal{P}$) one obtains a *cylindrical cover*

$$\overline{X} = \bigcup_{P\in\mathcal{P}} W_P. \tag{16}$$

Let $x \in Y_P$. If O_P is a ball neighborhood B of x in Y_P the construction above yields a neighborhood in $\overline{X}$ of x which is homeomorphic (as a stratified space) to $B \times c(|\Delta_P|)$. (Here we have interchanged the factors in (15).) As B shrinks and s_P tends to o_P we obtain a basis of local neighborhoods of x. The link $|\Delta_P|$ is itself stratified by its intersections with the strata $Y_Q > Y_P$; this stratification is

$$|\Delta_P| = \coprod_{Q>P} |\Delta_P^Q|^\circ,$$

where $|\Delta_P^Q|^\circ$ denotes the open face with vertices Δ_P^Q.

7. The reductive Borel-Serre compactification

We outline the construction of $\widehat{X}$ which was introduced by Zucker [**81**]; see also [**30**].

Recall that e_P is a homogeneous space for ${}^0P$. Let $D_P = N_P\backslash e_P$ be obtained by collapsing the orbits of the unipotent radical; it a symmetric space for L_P, namely $D_P = L_P/K_PA_P$ where $K_P = K \cap P$. The quotient map $e_P \to D_P$ is a principal N_P-bundle. We can define a geodesic action of L_P on D [**61**, §1.1][1] which extends that of A_P; the orbits of the induced geodesic action of ${}^0L_P$ on e_P yield natural

[1]Precise references to numbered results in [**61**] refer to version 2 as posted on the **arXiv**. The numbering in [**61**] may change in the published version due to revision or a possible splitting of the paper.

trivializing sections of $e_P \to D_P$. We let $\mathcal{N}_P$ be obtained from e_P be collapsing these sections; thus there is a natural decomposition

$$e_P = D_P \times \mathcal{N}_P. \tag{17}$$

More concretely, the decomposition ${}^0P = \widetilde{M}_P \ltimes N_P$ (which depends on the choice of basepoint $x_0 \in D$) induces a decomposition

$$e_P \xrightarrow{\sim} D_P \times N_P, \qquad nmK_P \mapsto (mK_P, n), \qquad (n \in N_P,\, m \in \widetilde{M}_P). \tag{18}$$

This agrees with (17) under the canonical isomorphism of N_P-homogeneous spaces $\mathcal{N}_P \cong N_P$ which sends the class of x_0 to the identity. In the example considered in §6 (see Figure 4), D_P is the product of the spaces of positive definite $n_i \times n_i$ symmetric matrices modulo homothety, for $i = 0, \ldots, r$.

The action of $\gamma \in \Gamma_{N_P} \equiv \Gamma \cap N_P \subseteq \Gamma_P$ is solely on the second factor of (17); thus we obtain

$$\Gamma_{N_P} \backslash e_P = D_P \times \mathcal{N}'_P, \qquad \text{where } \mathcal{N}'_P \equiv \Gamma_{N_P} \backslash \mathcal{N}_P.$$

To achieve the full quotient $Y_P = \Gamma_P \backslash e_P$ we need to now take the quotient by $\Gamma_{L_P} \equiv \Gamma_P / \Gamma_{N_P}$. An element $\gamma \in \Gamma_{L_P}$ acts in the usual way on the first factor D_P but in view of (18) it also conjugates the second factor. Thus the trivial N_P-principal bundle $e_P \to D_P$ descends to a flat bundle of the arithmetic quotients, called the ***nilmanifold fibration***:

$$\begin{array}{c} Y_P \\ \Big\downarrow \text{flat } \mathcal{N}'_P\text{-bundle} \\ X_P \equiv \Gamma_{L_P} \backslash D_P \end{array}$$

The ***reductive Borel-Serre compactification*** $\widehat{X}$ is obtained by collapsing the fibers of the nilmanifold fibration on each stratum Y_P of $\overline{X}$. Thus there is a quotient map $\overline{X} \to \widehat{X}$ which is stratified with respect to the stratification

$$\widehat{X} = \coprod_{P \in \mathcal{P}} X_P.$$

The closure of a stratum is $\widehat{X}_P$, the reductive Borel-Serre compactification of X_P.

In order to understand the structure of $\widehat{X}$ near a point $x \in X_P$, we first look at the inverse image of x in $\overline{X}$. This is a full nilmanifold fiber $\mathcal{N}'_P$ in Y_P which has a basis of neighborhoods of the form

$$\widetilde{U} = B \times \mathcal{N}'_P \times c(|\Delta_P|),$$

where B maps diffeomorphically to a ball neighborhood of x in X_P. (Again we have interchanged the factors in (15).) The strata of $\widetilde{U}$ are

$$\widetilde{U} \cap Y_Q = \begin{cases} B \times \mathcal{N}'_P \times c(|\Delta_P|)^\circ & Q = G, \\ B \times \mathcal{N}'_P \times c(|\Delta_P^Q|)^\circ & P < Q < G, \\ B \times \mathcal{N}'_P \times c(\emptyset) & Q = P, \end{cases} \tag{19}$$

where $c(\emptyset)$ denotes the vertex of the cone and $c(|\Delta_P^Q|)^\circ \equiv c(|\Delta_P^Q|^\circ) \setminus c(\emptyset)$.

A neighborhood U of x is obtained by projecting $\widetilde{U}$ to $\widehat{X}$. Since the $\mathcal{N}'_Q$-fibers in Y_Q need to be collapsed for each $Q \geq P$, (19) is replaced by

$$U \cap X_Q = \begin{cases} B \times \mathcal{N}'_P \times c(|\Delta_P|)^\circ & Q = G, \\ B \times \mathcal{N}'^Q_P \times c(|\Delta^Q_P|)^\circ & P < Q < G, \\ B \times c(\emptyset) & Q = P, \end{cases} \tag{20}$$

where $\mathcal{N}'^Q_P \equiv (\Gamma_{N_P}/\Gamma_{N_Q})\backslash(N_Q\backslash\mathcal{N}_P)$. In particular, over $X = X_G$ no collapsing occurs, while over X_P the entire $\mathcal{N}'_P$ factor is collapsed to a point.

The link of the stratum $X_P \subseteq \widehat{X}$ is thus

$$(\mathcal{N}'_P \times |\Delta_P|)/\sim \,= \coprod_{Q>P} \mathcal{N}'^Q_P \times |\Delta^Q_P|^\circ, \tag{21}$$

where $(\Gamma_{N_P} n, a) \sim (\Gamma_{N_P} n', a)$ if $a \in |\Delta^Q_P|^\circ$ and $n' = un$ for some $u \in N_Q$.

8. Satake compactifications

The Satake compactification X^* depends on an irreducible representation (σ, U) of G. The construction proceeds in several stages. Satake [**67**] first constructs a compactification ${}_{\mathbb{R}}D^*$ of D depending on (σ, U). The compactification ${}_{\mathbb{R}}D^*$ is a disjoint union of subsets called *real boundary components*. The action of G on D extends to an action on ${}_{\mathbb{R}}D^*$ which permutes the boundary components. In [**68**], Satake defines $D^* \subseteq {}_{\mathbb{R}}D^*$ to be the union of certain of these boundary components, called *rational boundary components*, and shows that $G(\mathbb{Q})$ acts on D^*. He defines the *Satake topology* on D^* and sets $X^* \equiv \Gamma\backslash D^*$. These steps, and certain assumptions that have to be made on (σ, U), are described in more detail below. In addition to the above papers of Satake, references are the papers of Borel [**8**], Casselman [**25**], and Zucker [**82**].

8.1. Assume

(i) (σ, U) is nontrivial on each $\mathbb{R}$-simple component;
(ii) the maximal $\mathbb{Q}$-split torus and the maximal $\mathbb{R}$-split torus in the center of G coincide.

(This second assumption is not essential but it simplifies the exposition.) Give U an *admissible inner product*, that is, one for which

$$\sigma(g)^* = \sigma(\theta g)^{-1}, \qquad \text{for all } g \in G;$$

here θ is the Cartan involution associated to a choice of maximal compact subgroup $K \subseteq G$. There is an action of G on the real vector space $\operatorname{Herm}(U)$ of Hermitian endomorphisms of U given by

$$S \mapsto \sigma(g)S\sigma(g)^*, \qquad \text{for } S \in \operatorname{Herm}(U),\ g \in G.$$

In particular $g \in K$ acts by conjugation by $\sigma(g)$, while $g \in A_G$ acts by a real scalar (since the center of G acts by a rational character on an irreducible representation). Thus if we consider the induced projective representation, the class of the identity $[I] \in \mathbb{P}\operatorname{Herm}(U)$ is fixed by KA_G and there is a well-defined map

$$D = G/KA_G \longrightarrow \mathbb{P}\operatorname{Herm}(U), \qquad x = gKA_G \mapsto g\cdot[I] = \sigma(g)\sigma(g)^*.$$

This map is an embedding by the assumptions above and ${}_{\mathbb{R}}D^*$ is defined to be the closure of the image of D in $\mathbb{P}\operatorname{Herm}(U)$.

An important class of examples are the Hermitian symmetric spaces, in which D has a G-invariant complex structure. In this case the $\mathbb{R}$-root system is always type C_r or BC_r and there is a unique simple $\mathbb{R}$-root α_r at the end of the diagram with a double bond (if the type is C_2 (resp. BC_2), α_2 is the long (resp. short) simple root). If the highest restricted weight of (σ, U) is a multiple of the fundamental weight dual to α_r, the compactification ${}_{\mathbb{R}}D^*$ is homeomorphic to the closure of D in $\mathbb{C}^N$ under Harish-Chandra's embedding as a bounded symmetric domain [**38**]. This is called the *natural compactification*.

8.2. A *real boundary component* is by definition the set of points of ${}_{\mathbb{R}}D^*$ fixed by the unipotent radical of a parabolic $\mathbb{R}$-subgroup of G; if P is a parabolic $\mathbb{R}$-subgroup, we let $D_{P,h} \equiv \operatorname{Fix}(N_P) \cap {}_{\mathbb{R}}D^*$ denote the associated real boundary component. The compactification ${}_{\mathbb{R}}D^*$ is the disjoint union of its real boundary components. The action of P on $D_{P,h}$ descends to an action of the Levi quotient L_P and there is an almost direct product decomposition (direct product of subgroups modulo finite subgroup) of reductive $\mathbb{R}$-subgroups

$$L_P = \widetilde{L}_{P,h} \times L_{P,\ell}, \tag{22}$$

where $L_{P,\ell}$ is the subgroup fixing $D_{P,h}$ pointwise and the induced action of $L_{P,h} \equiv L_P/L_{P,\ell}$ on $D_{P,h}$ is that of a real reductive group on its associated symmetric space. This decomposition yields a product decomposition of the associated symmetric spaces,

$$D_P = D_{P,h} \times D_{P,\ell}. \tag{23}$$

In the case of the natural compactification of a bounded symmetric domain in $\mathbb{C}^N$, the real boundary components, which are again Hermitian symmetric spaces, can also be described as the *holomorphic arc components*: two points are in the same holomorphic arc component if there are a finite number of holomorphic maps of $\{z \in \mathbb{C} \mid |z| < 1\}$ into the closure of D for which the union of the images is connected and contains the two points.

8.3. We can be more specific about the subgroups $L_{P,\ell}$ and $\widetilde{L}_{P,h}$, in particular, their root systems. First some notation. Choose a maximal $\mathbb{R}$-split torus ${}_{\mathbb{R}}A$ of L_P; via an appropriate lift we may identify it with a maximal $\mathbb{R}$-split torus of G. Let ${}_{\mathbb{R}}\Delta$ be the simple $\mathbb{R}$-roots of G with respect to a fixed positive system of $\mathbb{R}$-roots which contains the roots in $\mathfrak{n}_P$. To any set of $\mathbb{R}$-weights ψ is associated a graph with vertex set ψ and an edge between α, $\beta \in \psi$ if $(\alpha, \beta) \neq 0$. (Here we view weights as belonging to ${}_{\mathbb{R}}\mathfrak{a}^*$ endowed with a Weyl group invariant inner product.) The set ψ is *connected* if the associated graph is connected.

Let ${}_{\mathbb{R}}\mu$ be the highest weight of U restricted to ${}_{\mathbb{R}}A$. The simple $\mathbb{R}$-roots ${}_{\mathbb{R}}\Delta^P \subseteq {}_{\mathbb{R}}\Delta$ of L_P may be orthogonally decomposed

$${}_{\mathbb{R}}\Delta^P = \kappa({}_{\mathbb{R}}\Delta^P) \sqcup \zeta({}_{\mathbb{R}}\Delta^P),$$

where $\kappa({}_{\mathbb{R}}\Delta^P)$ is the largest subset of ${}_{\mathbb{R}}\Delta^P$ such that $\kappa({}_{\mathbb{R}}\Delta^P) \cup \{{}_{\mathbb{R}}\mu\}$ is connected and $\zeta({}_{\mathbb{R}}\Delta^P)$ is the complement of $\kappa({}_{\mathbb{R}}\Delta^P)$ in ${}_{\mathbb{R}}\Delta^P$. Alternatively, a root $\alpha \in {}_{\mathbb{R}}\Delta^P$ is in $\kappa({}_{\mathbb{R}}\Delta^P)$ (resp. $\zeta({}_{\mathbb{R}}\Delta^P)$) if it can (resp. cannot) be connected to ${}_{\mathbb{R}}\mu$ via a chain of weights $\chi_0 \equiv \alpha, \chi_1, \ldots, \chi_{k-1}, \chi_k \equiv {}_{\mathbb{R}}\mu$ satisfying $\chi_i \in {}_{\mathbb{R}}\Delta^P$ for $i < k$ and $(\chi_{i-1}, \chi_i) \neq 0$ for $i \leq k$.

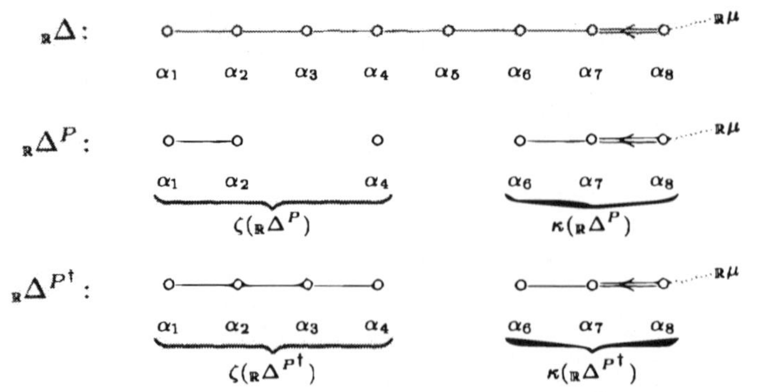

FIGURE 7. An example of the Dynkin diagrams of the $\mathbb{R}$-root systems of G, L_P, and $L_{P^\dagger}$, where $G = \mathrm{Sp}_{18}(\mathbb{R})$ and (σ, U) has highest weight $\mu = \beta_8$.

We can now describe the identity component of the centralizer $L_{P,\ell}$ of $D_{P,h}$ as the maximal normal connected $\mathbb{R}$-subgroup of L_P with simple $\mathbb{R}$-roots $\zeta({}_{\mathbb{R}}\Delta^P)$. Similarly the identity component of $\widetilde{L}_{P,h}$ is the minimal normal connected $\mathbb{R}$-subgroup of L_P with simple $\mathbb{R}$-roots $\kappa({}_{\mathbb{R}}\Delta^P)$.

Given a real boundary component $D_{P,h}$, the parabolic subgroup P is not uniquely determined: any parabolic $\mathbb{R}$-subgroup $P' \supseteq P$ such that $\kappa({}_{\mathbb{R}}\Delta^{P'}) = \kappa({}_{\mathbb{R}}\Delta^P)$ yields the same boundary component. The largest such parabolic $\mathbb{R}$-subgroup is denoted $P^\dagger$ and we set

$$\omega({}_{\mathbb{R}}\Delta^P) \equiv {}_{\mathbb{R}}\Delta^{P^\dagger}.$$

The parabolic $\mathbb{R}$-subgroups R for which ${}_{\mathbb{R}}\Delta^R = \omega(\psi)$ for some $\psi \subseteq {}_{\mathbb{R}}\Delta$ are called *saturated*; they are the normalizers of the real boundary components. An example for $G = \mathrm{Sp}_{18}(\mathbb{R})$ is pictured in Figure 7; here D is the Siegel generalized upper half-plane (a Hermitian symmetric space) and ${}_{\mathbb{R}}D^*$ is the natural compactification.

8.4. Let $A \subseteq {}_{\mathbb{R}}A$ be the identity components of a maximal $\mathbb{Q}$-split torus and a maximal $\mathbb{R}$-split torus respectively. The restriction of characters on ${}_{\mathbb{R}}A$ to characters on A yields a map on simple roots

$$\rho_{\mathbb{R}/\mathbb{Q}}\colon {}_{\mathbb{R}}\Delta \to \Delta \cup \{0\}$$

and we set

$$\epsilon_{\mathbb{R}/\mathbb{Q}}(\Upsilon) \equiv \rho_{\mathbb{R}/\mathbb{Q}}^{-1}(\Upsilon \cup \{0\}), \qquad \text{for any } \Upsilon \subseteq \Delta.$$

We wish now to define a Γ-invariant union of real boundary components $D^* \subseteq {}_{\mathbb{R}}D^*$ so that Γ acts discontinuously on D^* and $\Gamma\backslash D^*$ is compact. It is clear that not all real boundary components need to be included in D^* nor are desirable to be included. The following two points motivate the exact definition of D^*.

- D^* *should be big enough*: By reduction theory, a fundamental domain Ω for the action of Γ on D may be formed from a finite union of *Siegel sets* $\mathfrak{S} \equiv C\widetilde{A}^+_{P_0} b \cdot x_0$. Here P_0 is a minimal parabolic $\mathbb{Q}$-subgroup, $C \subseteq P_0$ is compact, $b \in \widetilde{A}_{P_0}$, and $x_0 \in D$ is a fixed basepoint. Call a real boundary component ***rationally visible*** if it occurs in the closure $\overline{\Omega} \subseteq {}_{\mathbb{R}}D^*$ for some fundamental domain as above. For the quotient $\Gamma\backslash D^*$ to be compact, we wish to include in D^* at least all rationally visible boundary components. One may check [**25**, Lemma 8.1] that the rationally visible boundary components are those whose normalizers R contain a minimal parabolic $\mathbb{Q}$-subgroup and satisfy

$$_{\mathbb{R}}\Delta^R = \omega(\epsilon_{\mathbb{R}/\mathbb{Q}}(\Upsilon)) \qquad \text{for some } \Upsilon \subseteq \Delta.$$

- D^* ***should not be too big***: On the other hand, call a real boundary component ***geometrically rational*** if its normalizer R is defined over $\mathbb{Q}$ and if there exists a normal subgroup $L'_{R,\ell} \subseteq L_{R,\ell}$ defined over $\mathbb{Q}$ such that $L_{R,\ell}/L'_{R,\ell}$ is compact. In this case, the decomposition in (22) can be modified to be defined over $\mathbb{Q}$, essentially by shifting certain compact factors from $L_{R,\ell}$ to $\widetilde{L}_{R,h}$; we denote these modified terms again by $L_{R,\ell}$ and $\widetilde{L}_{R,h}$. The geometrical rationality condition on a boundary component ensures that Γ_{N_R}, $\Gamma_{L_{R,\ell}} \equiv \Gamma_{L_R} \cap L_{R,\ell}$, and $\Gamma_{L_{R,h}} \equiv \Gamma_{L_R}/\Gamma_{L_{R,\ell}}$ are all arithmetic subgroups and hence act discontinuously on N_R, $D_{R,\ell}$, and $D_{R,h}$. To ensure that Γ acts discontinuously on D^*, we do not want to include in D^* boundary components that are not geometrically rational.

Any geometrically rational boundary component (in fact, any boundary component whose normalizer is defined over $\mathbb{Q}$) is rationally visible. The two points above suggest that we must assume conversely that

(iii) every rationally visible boundary component is geometrically rational.

A compactification $_{\mathbb{R}}D^*$ of D satisfying this assumption is called ***geometrically rational.*** Under this assumption, the geometrically rational boundary components will simply be called ***rational boundary components*** and are just those with normalizers defined over $\mathbb{Q}$. We define

$$D^* \equiv \coprod_{\substack{\text{rational} \\ \text{boundary components}}} D_{R,h}.$$

Note that under the assumption of geometric rationality, the Levi quotient L_P of any parabolic $\mathbb{Q}$-subgroup P will have a decomposition (possibly different from (22) due to shifts of compact factors) into an almost direct product of reductive $\mathbb{Q}$-subgroups

$$L_P = \widetilde{L}_{P,h} \times L_{P,\ell}, \tag{24}$$

where the $\mathbb{Q}$-root systems are $\kappa(\Delta^P)$ and $\zeta(\Delta^P)$ respectively. (The notation $\kappa(\Delta^P)$, $\zeta(\Delta^P)$, and $\omega(\Delta_P)$ is extended to subsets of Δ by using $_{\mathbb{Q}}\mu \equiv \mu|_A$ instead of $_{\mathbb{R}}\mu$.) The normalizers R of the rational boundary components will be the saturated parabolic $\mathbb{Q}$-subgroups; these are those with $\omega(\Delta^R) = \Delta^R$.

The Satake compactification is known to be geometrically rational if (σ, U) is strongly $\mathbb{Q}$-rational [**8**] or if $_{\mathbb{R}}D^*$ is the natural compactification of a Hermitian symmetric space [**4**]. In the latter case, as we have mentioned, all real boundary components are again Hermitian symmetric. Borel has suggested a natural generalization to consider. A real boundary component $D_{R,h}$ is ***equal-rank*** if $\mathbb{C}$-rank $L_{R,h}$ = rank $K_{R,h}$, where $K_{R,h}$ is a maximal compact subgroup of $L_{R,h}$. (A Hermitian symmetric space is automatically equal-rank.) Call $_{\mathbb{R}}D^*$ an ***real equal-rank Satake compactification*** if D and all its real boundary components are equal-rank. Zucker [**83**, (A.2)] has enumerated the possible examples when G is $\mathbb{R}$-simple; there are two new infinite families, $G = \mathrm{SO}(p,q)$ with $p+q$ odd and $G = \mathrm{Sp}(p,q)$, with ${}_{\mathbb{R}}\mu$ again connected to a simple $\mathbb{R}$-root at the end of the Dynkin diagram. In [**64**] we prove that a real equal-rank Satake compactification is geometrically rational, aside from some explicitly described $\mathbb{Q}$-rank 1 and 2 examples obtained by restriction of scalars.

8.5. The subspace topology on $D^* \subseteq {}_{\mathbb{R}}D^*$ is not the right one to use. Instead D^* is given the ***Satake topology***. To describe this topology, recall that $\overline{\Omega}$ is the closure in ${}_{\mathbb{R}}D^*$ of a fundamental domain Ω constructed from Siegel sets. We have $D^* = \Gamma \cdot \overline{\Omega}$. Then a basis of local neighborhoods of $x \in D^*$ in the Satake topology is given by subsets U such that

(i) If $x \in \gamma\overline{\Omega}$, then $U \cap \gamma\overline{\Omega}$ is open in the subspace topology on $\gamma\overline{\Omega} \subseteq {}_{\mathbb{R}}D^*$.
(ii) U is Γ_x-invariant, where $\Gamma_x = \{\gamma \in \Gamma \mid \gamma \cdot x = x\}$.

The *Satake compactification* X^* is defined by

$$X^* \equiv \Gamma \backslash D^*;$$

it has a stratification

$$X^* = \coprod_{R \in \mathcal{P}^*} X_{R,h}$$

where $\mathcal{P}^* \subseteq \mathcal{P}$ is the set of Γ-equivalence classes of saturated parabolic $\mathbb{Q}$-subgroups and

$$X_{R,h} \equiv \Gamma_{L_{R,h}} \backslash D_{R,h}.$$

The Satake compactification X^* associated to the natural compactification of a Hermitian symmetric space is known as the *Baily-Borel-Satake compactification*; Baily and Borel [**4**] show that it admits the structure of a normal projective variety defined over a number field. In this case $\mathcal{P}^*$ simply consists of Γ-equivalence classes of maximal parabolic $\mathbb{Q}$-subgroups. More generally, the Satake compactification X^* associated to a geometrically rational real equal-rank Satake compactification D^* is again called a *real equal-rank Satake compactification*.

8.6. There is an alternative description due to Zucker [**82**] of the topology on X^* that is quite useful. For $P \in \mathcal{P}$, we have the projection

$$D_P = D_{P,h} \times D_{P,\ell} \longrightarrow D_{P,h} = D_{P^\dagger,h}$$

onto the first factor of (23). This induces a map on the arithmetic quotients

$$\pi_P \colon X_P \longrightarrow X_{P^\dagger,h}.$$

In general this is a flat bundle with typical fiber $X_{P,\ell} \equiv \Gamma_{L_{P,\ell}} \backslash D_{P,\ell}$, however if we replace Γ_{L_P} by a certain finite index subgroup the corresponding bundle becomes trivial.

The maps π_P for the various $P \in \mathcal{P}$ fit together to define a surjective map

$$\pi \colon \widehat{X} \to X^*$$

and Zucker proves that the topology on X^* is the quotient topology induced by π. (See Figure 28 in §19 for a picture of the map π.) One can also define a continuous map $\widehat{D} \equiv \coprod_P D_P \to D^*$ before taking the arithmetic quotient, however this map is not necessarily open.

9. The Rapoport/Goresky-MacPherson conjecture

The preceding sections, and especially the table at the end of §5.3, explicitly illustrate how the links of strata become more complicated as we pass from $\overline{X}$ to $\widehat{X}$ to X^*. The intersection cohomology sheaf will likewise become more complicated. A hope therefore is that one might be able to transfer the study of the cohomology of various sheaves, in particular the intersection cohomology sheaf, from X^* to the simpler space $\widehat{X}$. (From now on in this paper, a "sheaf" means a complex of sheaves, viewed as an element of the constructible derived category.) In particular, the replacement of X^* by $\widehat{X}$ in Theorem 3.1, Zucker's conjecture, would be of considerable use in applications.

For example, if X^* is the Baily-Borel-Satake compactification of a Hermitian locally symmetric space, part of Langlands's program involves relating the Hasse-Weil zeta function of X^* to automorphic L-functions; see for example [**47**] where Picard modular surfaces are considered. One approach to establishing this relationship would be to compare the Arthur-Selberg trace formula [**2**] for a Hecke operator on $H_{(2)}(X;\mathbb{E})$ with a topological Lefschetz fixed point formula for $I_pH(X^*;\mathbb{E})$ such as Goresky and MacPherson give in [**35**]; this is reasonable since Zucker's conjecture shows that the groups are isomorphic. However Arthur's formula involves a sum over $P \in \mathcal{P}$ whereas a fixed point formula on X^* will involve a sum over $R \in \mathcal{P}^*$; since $\mathcal{P}$ indexes the strata of $\widehat{X}$, a fixed point formula on $\widehat{X}$ might be more amenable to a comparison with Arthur's formula.

These considerations were one motivation for the following conjecture. It was first posed by Rapoport in 1986 within a letter to Borel [**58**]; the conjecture was later independently posed by Goresky and MacPherson and described in an unpublished preprint [**34**].

CONJECTURE 9.1 (Rapoport/Goresky-MacPherson). *Let X be a Hermitian locally symmetric space and let p be a middle-perversity. There is a natural quasi-isomorphism $\mathcal{I}_p\mathcal{C}(X^*;\mathbb{E}) \cong \pi_*\mathcal{I}_p\mathcal{C}(\widehat{X};\mathbb{E})$ and hence $I_pH(X^*;\mathbb{E}) \cong I_pH(\widehat{X};\mathbb{E})$.*

Previously Zucker had noticed that the conjecture held for $G = \mathrm{Sp}_4(\mathbb{R})$ with $\mathbb{E} = \mathbb{C}$. In addition Goresky and MacPherson [**34**] announced the conjecture held for $G = \mathrm{Sp}_4(\mathbb{R})$, $\mathrm{Sp}_6(\mathbb{R})$, and (for $\mathbb{E} = \mathbb{C}$) $\mathrm{Sp}_8(\mathbb{R})$. In the case where $\mathbb{Q}$-rank $G = 1$ a proof was given by the author and Stern in an appendix to a paper of Rapoport's [**59**]. The full conjecture (in fact a generalization to the equal-rank setting) was finally proved by the author in 2001 [**61**] (see Theorem 11.1 below).

10. Weighted cohomology

With similar motivation, but in a different direction from the above conjecture, Goresky, Harder, and MacPherson have defined the weighted cohomology $W^\eta H(\widehat{X};\mathbb{E})$ and have proven [**30**]:

THEOREM 10.1 (Goresky-Harder-MacPherson). *Let X be a Hermitian locally symmetric space, let p be a middle-perversity, and let η be a middle-weight profile. There is a natural quasi-isomorphism $\mathcal{I}_p\mathcal{C}(X^*;\mathbb{E}) \cong \pi_*\mathcal{W}^\eta\mathcal{C}(\widehat{X};\mathbb{E})$ and hence an isomorphism $I_pH(X^*;\mathbb{E}) \cong W^\eta H(\widehat{X};\mathbb{E})$.*

Weighted cohomology depends on a locally constant sheaf $\mathbb{E}$ on X arising from a representation E of G, as well as on a *weight profile* $\eta \in \mathfrak{a}_{P_0}^{G*}$ where P_0 is a minimal parabolic $\mathbb{Q}$-subgroup. The *middle-weight profiles* are $\nu = -\rho$ and $\mu = -\rho + \epsilon\rho$, where $\epsilon > 0$ is very small and ρ is one-half the sum of the positive $\mathbb{Q}$-roots of G (counted with multiplicity). Like L^2-cohomology and intersection cohomology, there is a sheaf $\mathcal{W}^\eta\mathcal{C}(\widehat{X};\mathbb{E})$ whose hypercohomology is weighted cohomology.

We will give a local calculation of $W^\eta H(\widehat{X};\mathbb{E})$ analogous to the one for intersection cohomology in (8) of §4.1. Recall that the smooth part of the link of a stratum $X_P \subseteq \widehat{X}$ is $\mathcal{N}'_P \times |\Delta_P|^\circ$ (see (21) in §7). Van Est's theorem [**76**] shows that the cohomology $H(\mathcal{N}'_P \times |\Delta_P|^\circ;\mathbb{E}) \cong H(\mathcal{N}'_P;\mathbb{E})$ is naturally an A_P^G-module (in fact, an L_P-module); we will explain this in more detail in §12.1. Thus there is a decomposition $H(\mathcal{N}'_P;\mathbb{E}) = \bigoplus_\chi H(\mathcal{N}'_P;\mathbb{E})_\chi$ into A_P^G-weight spaces. If U is a small neighborhood of $x \in X_P$, the local weighted cohomology $W^\eta H^j(U;\mathbb{E})$ similarly has a decomposition into A_P^G-weight spaces and the local characterization is

$$W^\eta H(U;\mathbb{E})_\chi \cong \begin{cases} H(\mathcal{N}'_P;\mathbb{E})_\chi & \text{for } \chi \geq \eta_P, \\ 0 & \text{for } \chi \not\geq \eta_P; \end{cases} \tag{25}$$

here η_P is the restriction of η to $\mathfrak{a}_P^G$ and the inequality $\chi \geq \eta_P$ means that $\chi - \eta_P$ is a nonnegative linear combination of elements of Δ_P.

REMARKS. (i) Despite the similarity of (25) with the local calculation of intersection cohomology (8), an important difference is that (25) is not inductive. Thus in general local calculations involving weighted cohomology are easier than the corresponding local calculations for intersection cohomology.

(ii) There is a natural projection $U \cap X \to \mathcal{N}'_P$. If we represent a class in $H(\mathcal{N}'_P;\mathbb{E})_\chi \cong H(U \cap X;\mathbb{E})_\chi$ by a differential form ϕ which is the pullback of a form on $\mathcal{N}'_P$, then the condition $\chi \geq \mu_P$ is precisely the condition that ϕ is L^2. Thus middle-weight profile weighted cohomology may be viewed as an algebraic analogue of L^2-cohomology.

Goresky and MacPherson [**36**] have applied their fixed point formula [**35**] to calculate the Lefschetz number of a Hecke correspondence on $W^\eta H(\widehat{X};\mathbb{E})$. They prove in collaboration with Kottwitz [**31**] that the result agrees with Arthur's trace formula [**2**] on $H_{(2)}(X;\mathbb{E})$, in line with the motivation from the previous section.

A proof of Theorem 10.1 (different from the original one) can easily be sketched using Zucker's conjecture and later work of Nair [**54**]. First of all, Zucker's conjecture yields a quasi-isomorphism

$$\mathcal{I}_p\mathcal{C}(X^*;\mathbb{E}) \cong \mathcal{L}_{(2)}(X^*;\mathbb{E}). \tag{26}$$

However it is fairly immediate that

$$\mathcal{L}_{(2)}(X^*;\mathbb{E}) \cong \pi_*\mathcal{L}_{(2)}(\widehat{X};\mathbb{E}). \tag{27}$$

Finally Nair's result is that there is a quasi-isomorphism

$$\mathcal{L}_{(2)}(\widehat{X};\mathbb{E}) \cong \mathcal{W}^\eta\mathcal{C}(\widehat{X};\mathbb{E})$$

as suggested by Remark (ii) above and hence

$$\pi_*\mathcal{L}_{(2)}(\widehat{X};\mathbb{E}) \cong \pi_*\mathcal{W}^\eta\mathcal{C}(\widehat{X};\mathbb{E}). \tag{28}$$

Equations (26)–(28) yield Theorem 10.1.

11. Main theorem

Despite the success of weighted cohomology, Conjecture 9.1 is still of interest. The short argument given above for Theorem 10.1 does not apply to prove the conjecture since in general the analogue of Nair's theorem does not hold: $\mathcal{I}_p\mathcal{C}(\widehat{X};\mathbb{E})$ is not quasi-isomorphic to $\mathcal{L}_{(2)}(\widehat{X};\mathbb{E})$ or $\mathcal{W}^{\eta}\mathcal{C}(\widehat{X};\mathbb{E})$ on $\widehat{X}$. The only hope for the conjecture is that a quasi-isomorphism exists downstairs on X^* and indeed this is the case:

THEOREM 11.1. *Let X^* be a real equal-rank Satake compactification and let $p = m$ or n be a middle-perversity. There is a natural quasi-isomorphism $\mathcal{I}_p\mathcal{C}(X^*;\mathbb{E}) \cong \pi_*\mathcal{I}_p\mathcal{C}(\widehat{X};\mathbb{E})$ and hence an isomorphism $I_pH(X^*;\mathbb{E}) \cong I_pH(\widehat{X};\mathbb{E})$.*

This result is proved in [**61**] as a consequence of a number of results in the theory of $\mathcal{L}$-modules which we will outline in the remaining sections.

Note that the condition on X^*, that all real boundary components of ${}_{\mathbb{R}}D^*$ are equal-rank, always holds if D is Hermitian and X^* is the Baily-Borel-Satake compactification. Thus the theorem implies the conjecture of Rapoport and Goresky-MacPherson. Note that although the equal-rank condition implies that the strata of X^* have even dimension, the strata of $\widehat{X}$, on the other hand, may have odd codimension, even in the Hermitian case.

12. $\mathcal{L}$-modules

In the course of work on the Rapoport/Goresky-MacPherson conjecture, it was realized that many of the techniques could be applied to study other sheaves. Furthermore, many parts of the proof, including a crucial and difficult local property of $\mathcal{I}_p\mathcal{C}(\widehat{X};\mathbb{E})$, apparently held even when D was non-Hermitian or non-equal-rank. So it seemed desirable in view of other applications to separate the components of the proof and to work in a more general context. This was one motivation for us to develop the theory of $\mathcal{L}$-modules [**61**].

12.1. Overview. First recall the notation. The reductive Borel-Serre compactification has a stratification

$$\widehat{X} = \coprod_{P\in\mathcal{P}} X_P,$$

where $\mathcal{P}$ is the partially ordered set of Γ-conjugacy classes of parabolic $\mathbb{Q}$-subgroups of G. For a parabolic $\mathbb{Q}$-subgroup P of G, let N_P denote its unipotent radical and let $L_P = P/N_P$ denote its Levi quotient. The reductive Borel-Serre boundary stratum X_P is the locally symmetric space associated to L_P, namely

$$X_P = \Gamma_{L_P}\backslash L_P/K_PA_P = \Gamma_{L_P}\backslash D_P,$$

where A_P is the identity component of a maximal $\mathbb{Q}$-split torus in the center of L_P, $\Gamma_P = \Gamma\cap P$, $\Gamma_{N_P} = \Gamma\cap N_P$ and $\Gamma_{L_P} = \Gamma_P/\Gamma_{N_P}$. The link of $X_P \subseteq \widehat{X}$ has as its smooth open stratum $\mathcal{N}'_P \times |\Delta_P|^\circ$, where $\mathcal{N}'_P$ is a compact nilmanifold which is canonically diffeomorphic to $\Gamma_{N_P}\backslash N_P$ once a basepoint is chosen. The rest of the link is described in §7. In general, if $P \subseteq Q$, we extend all this notation to the parabolic $\mathbb{Q}$-subgroup P/N_Q of L_Q simply by adding a superscript Q; thus we have $N_P^Q = N_P/N_Q$, etc.

An $\mathcal{L}$-module

$$\mathcal{M} = (E_P, f_{PQ})_{P\leq Q\in\mathcal{P}}$$

on $\widehat{X}$ is a combinatorial analogue of a constructible complex of sheaves. (The precise definition will be given below.) Like the constructible derived category of sheaves on $\widehat{X}$, the category of $\mathcal{L}$-modules has the functors k_*, $k_!$, k^*, and $k^!$ as well as truncation functors. Furthermore there is a realization functor $\mathcal{M} \mapsto \mathcal{S}(\mathcal{M})$ from $\mathcal{L}$-modules to the constructible derived category of sheaves, commuting with these functors. Many familiar sheaves can be realized as $\mathcal{S}(\mathcal{M})$ (or "lifted to an $\mathcal{L}$-module"), for example, the intersection cohomology sheaf $\mathcal{I}_p\mathcal{C}(\widehat{X};\mathbb{E})$, the weighted cohomology sheaf $\mathcal{W}^\eta\mathcal{C}(\widehat{X};\mathbb{E})$, or the sheaf $i_{G*}\mathbb{E}$ whose hypercohomology is the ordinary cohomology $H(X;\mathbb{E})$. The global cohomology of an $\mathcal{L}$-module is defined to be the hypercohomology of its realization:

$$H(\widehat{X};\mathcal{M}) \equiv H(\widehat{X};\mathcal{S}(\mathcal{M})).$$

In fact the realization functor yields an explicit complex of fine sheaves so that the cohomology may be computed from the complex of their global sections; we will see this complex is easy to work with.

Let $i_P\colon X_P \hookrightarrow \widehat{X}$ and $\hat{\imath}_P\colon \widehat{X}_P \hookrightarrow \widehat{X}$ denote the inclusion of a stratum and of its closure. Recall that for a sheaf $\mathcal{S}$ to be in the constructible derived category, the local cohomology sheaf $H(i_P^*\mathcal{S})$ along X_P must be a finitely generated locally constant sheaf for all P. It is equivalent[2] to require that the local cohomology $H(i_P^!\mathcal{S})$ supported on X_P is a finitely generated locally constant sheaf for all P; we prefer this point of view for reasons that will become clear later. Consequently for a sheaf, $H(i_P^!\mathcal{S})$ may be identified with a graded representation of the arithmetic subgroup $\Gamma_{L_P} \subseteq L_P$. An important difference between a sheaf and an $\mathcal{L}$-module is that for an $\mathcal{L}$-module $\mathcal{M}$, $H(i_P^!\mathcal{M}) = H(E_P, f_{PP})$ is actually a graded representation of the full reductive group L_P.

The reason this is possible, namely that the usual functorial operations applied to representations of Levi quotients yield representations of Levi quotients (as opposed to merely representations of fundamental groups) is van Est's theorem [**76**]. We will illustrate this with the simplest example of such a functorial operation. Suppose we start with a representation E of $\Gamma \subseteq G$ and hence a locally constant sheaf $\mathbb{E}$ on $X = X_G$. Pushforward $\mathbb{E}$ to $i_{G*}\mathbb{E}$ on $\widehat{X}$, and then restrict it to $i_P^* i_{G*}\mathbb{E}$ on X_P. The local cohomology $H(i_P^* i_{G*}\mathbb{E})$ is simply the cohomology of the smooth open stratum of the link $\mathcal{N}'_P \times |\Delta_P|^\circ$, which is homotopic to $\mathcal{N}'_P$. Thus the local cohomology is $H(\mathcal{N}'_P;\mathbb{E})$ and since this is the cohomology of the fiber of the flat bundle $Y_P \to X_P$, it has an action on it of $\Gamma_{L_P} \subseteq L_P$. However van Est's theorem says that if E actually arose from a representation of G, there is an isomorphism of Γ_{L_P}-modules

$$H(\mathcal{N}'_P;\mathbb{E}) \cong H(\mathfrak{n}_P;E). \tag{29}$$

The Γ_{L_P}-action on the complex $\bigwedge \mathfrak{n}_P^* \otimes E$ is obtained by lifting Γ_{L_P} to G; the induced action on the cohomology $H(\mathfrak{n}_P;E)$ is independent of the lift. This Γ_{L_P}-action, however, extends to an L_P-action defined by the same procedure.

The idea of representing a sheaf by combinatorial and linear algebraic data is not new. In other contexts there is the theory of S-constructible sheaves [**41**], [**42**, Exercise VIII.1] and MacPherson's theory of cellular sheaves [**70**], [**78**], [**79**]. These two examples are based on a decomposition into contractible cells, unlike the decomposition of $\widehat{X}$ into strata. For examples with non-contractible cells, there is

[2]This follows by the methods of [**15**, V.3] though it is not explicitly stated there.

the theory of sheaves on fans [5], [6], [22], [23], [52] and Braden and MacPherson's moment graphs [21]. There is also forthcoming work of Braden [20] on Koszul duality, in which he constructs a combinatorial category of mixed sheaves on affine toric varieties. The forthcoming work of Pink and Wildeshaus on automorphic sheaves [80] applies in a context closest to that of $\mathcal{L}$-modules. The basic linear algebraic data in all these theories is the local cohomology along a cell or stratum, $i_P^*\mathcal{S}$. An $\mathcal{L}$-module uses instead the dual notion of the local cohomology supported on a stratum $E_P = i_P^!\mathcal{S}$ as the building block. This approach turns out to make the construction of a useful realization technically easier.

12.2. $\mathcal{L}$-modules. An *$\mathcal{L}$-module* $\mathcal{M} = (E_\cdot, f_{\cdot\cdot})$ on $\widehat{X}$ consists of

- for all $P \in \mathcal{P}$, a graded algebraic representation E_P of L_P;
- for all $P \leq Q$, morphisms $f_{PQ}\colon H(\mathfrak{n}_P^Q; E_Q) \xrightarrow{[1]} E_P$.

(The superscript [1] indicates that f_{PQ} is graded of degree 1, that is, it maps $H^i(\mathfrak{n}_P^Q; E_Q)$ into E_P^{i+1}.) This data must satisfy the condition that for all $P \leq R$,

$$\sum_{P\leq Q\leq R} f_{PQ} \circ H(\mathfrak{n}_P^Q; f_{QR}) = 0. \tag{30}$$

In particular, (E_P, f_{PP}) is a complex of L_P-modules. Here $H(\mathfrak{n}_P^Q; f_{QR})$ is viewed as a map $H(\mathfrak{n}_P^R; E_R) \to H(\mathfrak{n}_P^Q; E_Q)$ via the canonical isomorphism

$$H(\mathfrak{n}_P^R; E_R) \cong H(\mathfrak{n}_P^Q; H(\mathfrak{n}_Q^R; E_R)). \tag{31}$$

This is a good place to recall the standard notation regarding shifts of degree: if C is a complex, then $C[k]$ denotes the shifted complex defined by

$$C[k]^i \equiv C^{i+k}, \qquad d_{C[k]} \equiv (-1)^k d_C.$$

This applies to graded objects (viewed as complexes with zero differential) and ordinary objects (viewed as complexes with nonzero entries only in degree 0) as well. Thus if $P < Q$ and there are no elements of $\mathcal{P}$ strictly between P and Q, the condition (30) shows that f_{PQ} is a morphism of complexes,

$$(H(\mathfrak{n}_P^Q; E_Q), H(\mathfrak{n}_P^Q; f_{QQ})) \longrightarrow (E_P, f_{PP})[1].$$

A ***morphism*** $\phi = (\phi_{\cdot\cdot})$ of $\mathcal{L}$-modules $\mathcal{M} \to \mathcal{M}'$ consists of graded maps

$$\phi_{PQ}\colon H(\mathfrak{n}_P^Q; E_Q) \longrightarrow E'_P \qquad \text{for all } P \leq Q$$

satisfying

$$\sum_{P\leq Q\leq R} \phi_{PQ} \circ H(\mathfrak{n}_P^Q; f_{QR}) = \sum_{P\leq Q\leq R} f'_{PQ} \circ H(\mathfrak{n}_P^Q; \phi_{QR}) \qquad \text{for all } P \leq R.$$

A subspace $W \subseteq \widehat{X}$ is ***admissible*** if it is a locally closed union of strata; set $\mathcal{P}(W) = \{\, P \in \mathcal{P} \mid X_P \subseteq W \,\}$. In this case we define an $\mathcal{L}$-module on W as above but with P, Q, and R restricted to belong to $\mathcal{P}(W)$.

To illustrate the above definition, suppose that $\mathbb{Q}$-rank $G = 2$ and that there is only one Γ-conjugacy class of parabolic $\mathbb{Q}$-subgroups of each type; an example where this occurs is $G = \mathrm{Sp}_4(\mathbb{R})$ and $\Gamma = \mathrm{Sp}_4(\mathbb{Z})$. Then $\mathcal{P} = \{P, Q_1, Q_2, G\}$ with the covering relations $P < Q_1, Q_2 < G$. In Figure 8 we represent $\mathcal{P}$ by its Hasse diagram (which we invert so as to match with Figure 6). The data of an $\mathcal{L}$-module $\mathcal{M}$ for this example (but not condition (30)) is displayed in Figure 9.

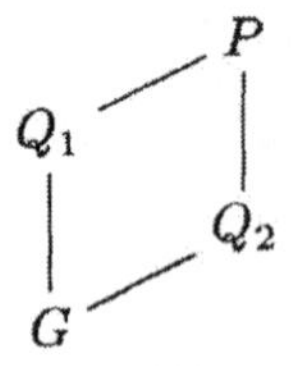

FIGURE 8. The (inverted) Hasse diagram for $\mathcal{P} = \{P < Q_1, Q_2 < G\}$

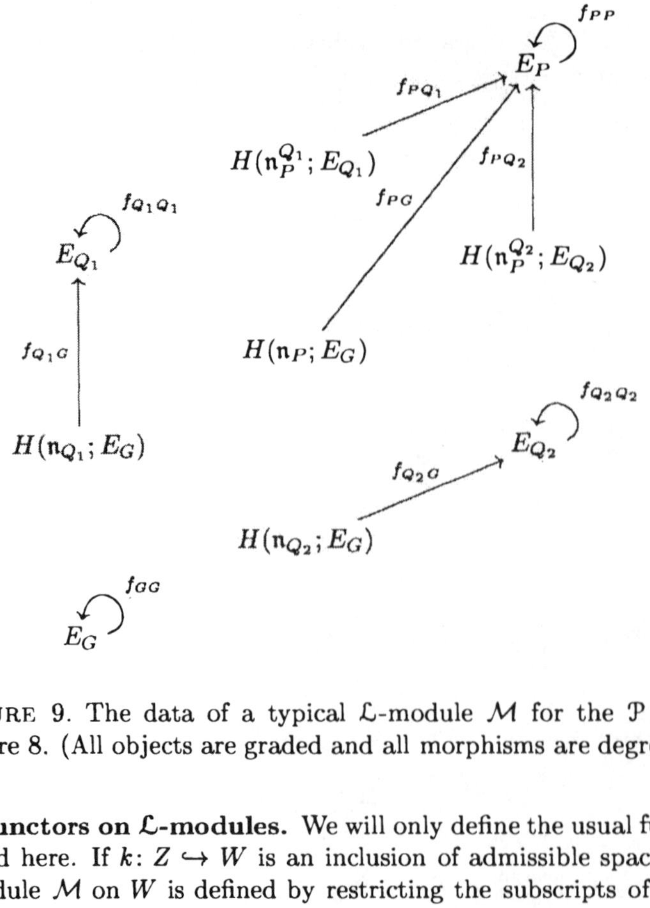

FIGURE 9. The data of a typical $\mathcal{L}$-module $\mathcal{M}$ for the $\mathcal{P}$ from Figure 8. (All objects are graded and all morphisms are degree 1.)

12.3. Functors on $\mathcal{L}$-modules. We will only define the usual functors in the cases we need here. If $k\colon Z \hookrightarrow W$ is an inclusion of admissible spaces, then $k^!\mathcal{M}$ for an $\mathcal{L}$-module $\mathcal{M}$ on W is defined by restricting the subscripts of E_P and f_{PQ} to belong to $\mathcal{P}(Z)$. In particular,

$$i_P^!\mathcal{M} = (E_P, f_{PP}).$$

for all $P \in \mathcal{P}(W)$. As we have noted, this is a complex of L_P-modules; its cohomology $H(i_P^!\mathcal{M})$ is the *local cohomology supported on* X_P. The less trivial example of $\hat{i}_{Q_1}^!\mathcal{M}$ for our example is represented in Figure 10.

On the other hand, for $P \in \mathcal{P}(W)$ we define

$$i_P^*\mathcal{M} \equiv \left(\bigoplus_{P \leq Q} H(\mathfrak{n}_P^Q; E_Q), \sum_{P \leq Q \leq R} H(\mathfrak{n}_Q^R; f_{QR}) \right).$$

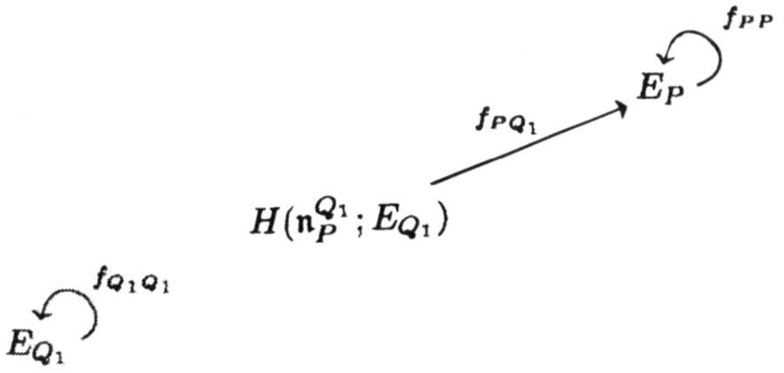

FIGURE 10. The $\mathcal{L}$-module $\hat{\imath}^!_{Q_1}\mathcal{M}$ ($\mathcal{M}$ as in Figure 9)

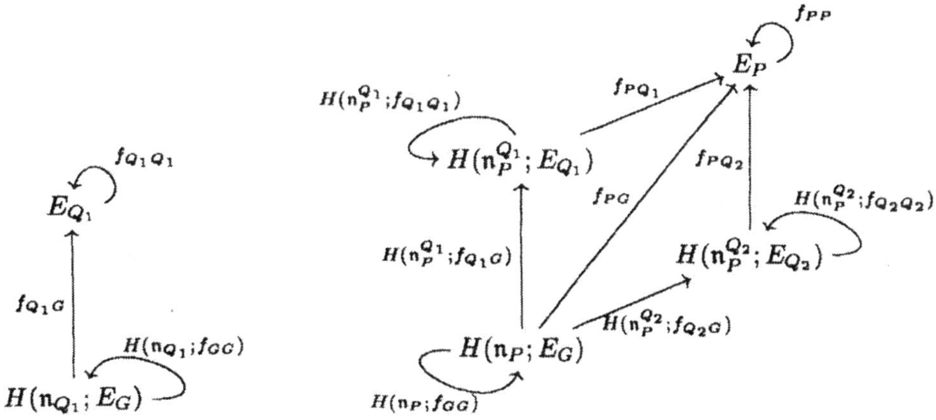

(a) $i^*_{Q_1}\mathcal{M}$ (b) $i^*_P\mathcal{M}$

FIGURE 11. The complexes computing local cohomology of $\mathcal{M}$ at Q_1 and at P ($\mathcal{M}$ as in Figure 9)

Again by (30) this is a complex of L_P-modules; its cohomology $H(i^*_P\mathcal{M})$ is the *local cohomology at* X_P (or *at* P). The complexes $i^*_{Q_1}\mathcal{M}$ and $i^*_P\mathcal{M}$ for our usual example are pictured in Figure 11.

If $R \in \mathcal{P}(W)$, the pullback $\hat{\imath}^*_R\mathcal{M} = (E'_\cdot, f'_{\cdot\cdot})$ to an entire closed face $\widehat{X}_R \cap W$ is defined by

$$E'_P \equiv \bigoplus_{\substack{Q \geq P \\ Q \cap \overline{R} = P}} H(\mathfrak{n}_P^Q; E_Q) \qquad \text{for all } P \leq R,$$

with the obvious induced morphisms for $f'_{\cdot\cdot}$.

Finally $k_*\mathcal{M}'$ for an $\mathcal{L}$-module $\mathcal{M}'$ on Z is defined by extending E_P and f_{PQ} to be zero if P or Q belongs to $\mathcal{P}(W) \setminus \mathcal{P}(Z)$.

In the cases we have defined, the usual adjoint relations hold among these functors. (For some of the other cases, it may be necessary to pass to the "homotopy category of $\mathcal{L}$-modules" which we do not consider here.) In particular, i^*_P is a left adjoint to i_{P*} on $\mathcal{L}$-modules and so there is a natural *adjunction morphism*

$$\mathcal{M} \longrightarrow i_{P*}i^*_P\mathcal{M}$$

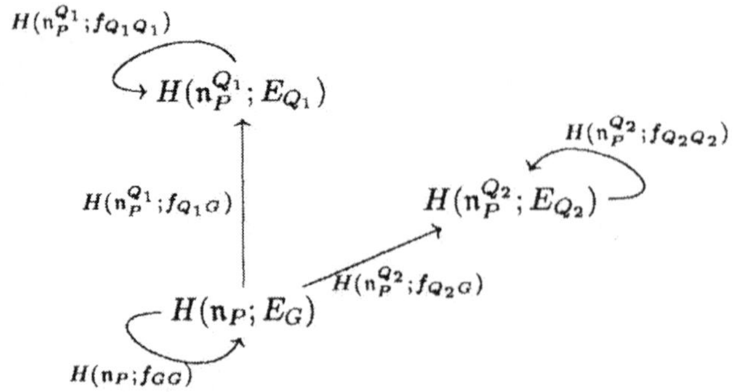

FIGURE 12. The complex computing the link cohomology of $\mathcal{M}$ at P ($\mathcal{M}$ as in Figure 9)

which is characterized by the condition that $i_P^*\mathcal{M} \longrightarrow i_P^* i_{P*} i_P^*\mathcal{M} = i_P^*\mathcal{M}$ is the identity map.

Besides $i_{P*}i_P^*\mathcal{M}$, certain other compositions of these functors are important. For example, let $j_P \colon W \setminus X_P \hookrightarrow W$ denote the inclusion of the complement of X_P. The functor $i_P^* j_{P*} j_P^*\mathcal{M}$ is a complex of L_P-modules whose cohomology is the *link cohomology of* $\mathcal{M}$ *at* P. An example is pictured in Figure 12.

The composed functor $i_P^*\hat{i}_Q^!\mathcal{M}$, where $P \leq Q$, will be important in §13 when we define the micro-support of an $\mathcal{L}$-module. Here are some examples. Since $\hat{i}_G^!$ is the identity, Figure 11 already gives two examples in the case $Q = G$. The other extreme is $Q = P$; in this case $i_P^*\hat{i}_Q^!\mathcal{M} = \hat{i}_P^!\mathcal{M} = (E_P, f_{PP})$. For an example with $P < Q < G$, consider

(32) $\quad i_P^*\hat{i}_{Q_1}^!\mathcal{M} =$

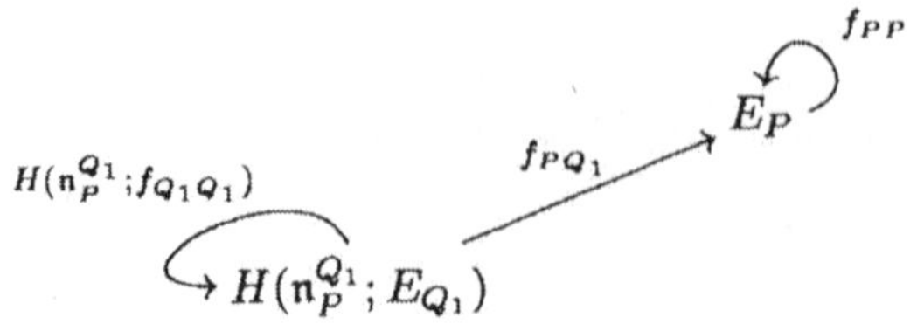

where $\mathcal{M}$ is as in Figure 9.

12.4. Realization of $\mathcal{L}$-modules. In order to define the realization sheaf of an $\mathcal{L}$-module, we need to recall the notion of special differential forms. The sheaf of *special differential forms* $\mathcal{A}_{\mathrm{sp}}(\widehat{X}; \mathbb{E})$ on $\widehat{X}$ for a G-module E was introduced by Goresky, Harder, and MacPherson [**30**]. It is a fine resolution of $i_{G*}\mathbb{E}$ whose sections on $U \subseteq \widehat{X}$ are smooth differential forms on the inverse image $\widetilde{U} \subseteq \overline{X}$ with coefficients in $\mathbb{E}$ which satisfy certain boundary conditions. Namely, in a neighborhood $\widetilde{U}$ of a boundary point in Y_P, the forms are locally pullbacks of forms on Y_P under the geodesic retraction $\widetilde{U} = B \times c(|\Delta_P|) \to B \subseteq Y_P$. Furthermore these boundary values are required to be constant along the nilmanifold fibers $\mathcal{N}_P'$. In the case that $E = \bigoplus_i E^i[-i]$ is graded, the differential d is the alternating sum of the corresponding de Rham exterior differentials. One may show, essentially by applying harmonic projection along the nilmanifold fibers and using van Est's

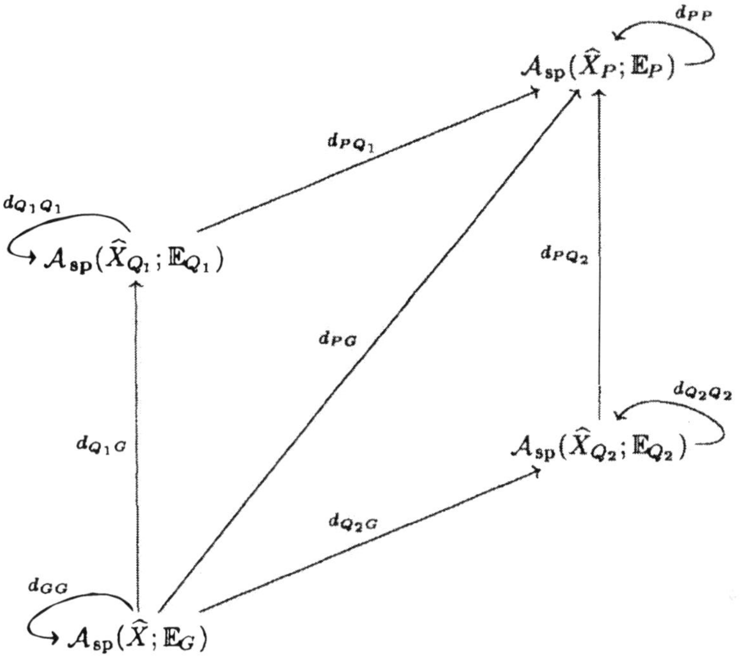

FIGURE 13. The realization sheaf $\mathcal{S}(\mathcal{M})$ ($\mathcal{M}$ as in Figure 9)

theorem (29), that there is a natural morphism, "restriction to a stratum",

$$k_P\colon \mathcal{A}_{sp}(\widehat{X}; \mathbb{E}) \longrightarrow \hat{\imath}_{P*}\mathcal{A}_{sp}(\widehat{X}_P; \mathbb{H}(\mathfrak{n}_P; E))$$

which restricts over $\widehat{X}_P$ to be a quasi-isomorphism. Given a graded L_Q-module E_Q we may likewise define the sheaf of special differential forms $\mathcal{A}_{sp}(\widehat{X}_Q; \mathbb{E}_Q)$, a differential d_Q, and a natural morphism

$$k_{PQ}\colon \mathcal{A}_{sp}(\widehat{X}_Q; \mathbb{E}_Q) \longrightarrow \hat{\imath}_{P*}\mathcal{A}_{sp}(\widehat{X}_P; \mathbb{H}(\mathfrak{n}_P^Q; E_Q)).$$

The ***realization*** $\mathcal{S}(\mathcal{M})$ of an $\mathcal{L}$-module $\mathcal{M}$ on $\widehat{X}$ can now be defined to be the complex of sheaves

$$\left(\bigoplus_{P\in\mathcal{P}} \hat{\imath}_{P*}\mathcal{A}_{sp}(\widehat{X}_P; \mathbb{E}_P), \sum_{P\leq Q\in\mathcal{P}} d_{PQ}\right)$$

where

$$d_{PQ} = \begin{cases} d_P + \mathcal{A}_{sp}(\widehat{X}_P; f_{PP}) & \text{if } P = Q \\ \mathcal{A}_{sp}(\widehat{X}_P; f_{PQ}) \circ k_{PQ} & \text{if } P < Q. \end{cases}$$

The realization sheaf $\mathcal{S}(\mathcal{M})$ for $\mathcal{M}$ as in Figure 9 is pictured in Figure 13. (We do not show the factorization of the arrows however.)

It is common to compute the cohomology of a sheaf by a complex consisting of differential forms satisfying boundary conditions or growth conditions depending on the particular sheaf. In view of the definition of $\mathcal{S}(\mathcal{M})$, however, we have a uniform type of complex to compute the cohomology of an $\mathcal{L}$-module: special differential forms with coefficients. The complication is that these "forms" are actually collections of forms on $\widehat{X}$ and on the various strata $\widehat{X}_P$, and there are interaction terms obtained by restricting to boundary strata and applying f_{PQ}. The boundary conditions mentioned above are translated into the choice of the coefficients and the f_{PQ}; see the example of intersection cohomology below.

It is easy to check that the realization functor commutes with the usual functors in the cases we have defined them, that is, there are natural quasi-isomorphisms $\mathcal{S}(k^!\mathcal{M}) \cong k^!\mathcal{S}(\mathcal{M})$, etc.

12.5. Reduced $\mathcal{L}$-modules. An $\mathcal{L}$-module $\mathcal{M} = (E_\cdot, f_{\cdot\cdot})$ is called *reduced* if $f_{PP} = 0$ for all P. One can show that for any $\mathcal{L}$-module $\mathcal{M}$, there exists a reduced $\mathcal{L}$-module $\mathcal{M}_0$ which is quasi-isomorphic to $\mathcal{M}$; quasi-isomorphic here means that there exists a third $\mathcal{L}$-module $\mathcal{M}'$ and morphisms $\mathcal{M} \leftarrow \mathcal{M}' \to \mathcal{M}_0$ which both induce isomorphisms on local cohomology. It follows that $H(\widehat{X};\mathcal{M}) \cong H(\widehat{X};\mathcal{M}_0)$. One could develop the theory by requiring $f_{PP} = 0$ in the definition of an $\mathcal{L}$-modules to begin with, however in that case the definitions of the usual functors, in particular the truncation functors defined below, would be more complicated.

12.6. Simplest example. The simplest sheaf that can be lifted to an $\mathcal{L}$-module is $i_{G*}\mathbb{E}$, where E is a G-module in degree 0; the cohomology is $H(\widehat{X}; i_{G*}\mathbb{E}) \cong H(X;\mathbb{E}) \cong H(\Gamma; E)$. The lifted $\mathcal{L}$-module $i_{G*}E$ is defined by

$$E_P \equiv \begin{cases} E & \text{if } P = G, \\ 0 & \text{if } P \neq G, \end{cases}$$
$$f_{PQ} \equiv 0.$$

The local cohomology at X_P is simply $i_P^* i_{G*}E = H(\mathfrak{n}_P; E)$.

12.7. Truncation functors. Recall that the formula for the intersection cohomology sheaf (9) involves the functor $\tau^{\leqslant n}$ which truncates local cohomology in degrees greater than n:

$$(33) \qquad \tau^{\leqslant n}\mathcal{S} \equiv \qquad \cdots \xrightarrow{d_{n-2}} \mathcal{S}^{n-1} \xrightarrow{d_{n-1}} \ker d_n \xrightarrow{d_n} 0 \longrightarrow \cdots .$$

In order to lift intersection cohomology to an $\mathcal{L}$-module, one needs an analogue of this truncation functor for $\mathcal{L}$-modules.

Consider the simplest $\mathcal{L}$-module $i_{G*}E$ on the simplest nontrivial $\widehat{X}$ where there is only one singular stratum corresponding to the parabolic subgroup P. We assume E is purely in degree 0 and $n \geq 0$. What should the $\mathcal{L}$-module $\tau^{\leqslant n} i_{G*}E$ be? The data $E_G = E$ cannot be changed; instead one must define E_P and f_{PG} so that the local cohomology $H(i_P^*\tau^{\leqslant n} i_{G*}E)$ at X_P changes from $H(\mathfrak{n}_P; E)$ to $\bigoplus_{i\leq n} H^i(\mathfrak{n}_P; E)[-i]$. Since the pullback $i_P^*\tau^{\leqslant n} i_{G*}E$ is

$$H(\mathfrak{n}_P; E) \xrightarrow{f_{PG}} E_P$$

and there is a natural decomposition

$$(34) \qquad H(\mathfrak{n}_P; E) = \Big(\bigoplus_{i\leq n} H^i(\mathfrak{n}_P; E)[-i]\Big) \oplus \Big(\bigoplus_{i>n} H^i(\mathfrak{n}_P; E)[-i]\Big),$$

the desired truncation of local cohomology can be arranged by setting

$$(35) \qquad E_P = \Big(\bigoplus_{i>n} H^i(\mathfrak{n}_P; E)[-i]\Big)[-1]$$

and letting f_{PG} be the natural degree 1 map (projection onto the second factor of (34)).

In essence the "internal" truncation functor $\tau^{\leqslant n}$ (33) has been realized as an "external" truncation functor. McConnell does something similar with his "flabby truncation" functor $\check{\tau}_{\leqslant n}$ [**52**]. That this is possible is not surprising. Recall that for a map of complexes $f\colon (C, d_C) \to (D, d_D)$ (where the objects lie in any additive category), the *mapping cone* $M(f)$ is the complex defined by

$$\left(C[1] \oplus D, \begin{pmatrix} d_{C[1]} & 0 \\ f & d_D \end{pmatrix} \right)$$

The truncation $\tau^{\leqslant n}\mathcal{S}$ of a sheaf is quasi-isomorphic to the mapping cone

$$M(\mathcal{S} \to \tau^{>n}\mathcal{S})[-1],$$

where

$$\tau^{>n}\mathcal{S} \equiv \qquad \cdots 0 \longrightarrow \mathcal{S}^{n+1}/\operatorname{Im} d_n \xrightarrow{d_{n+1}} \mathcal{S}^{n+2} \xrightarrow{d_{n+2}} \cdots \tag{36}$$

and

$$\mathcal{S} \to \tau^{>n}\mathcal{S} \tag{37}$$

is the natural map. With this notation, equation (35) may be re-expressed as

$$E_P = \tau^{>n} H(\mathfrak{n}_P; E)[-1] \tag{38}$$

(the degree shift is applied after the truncation).

In order to define $\tau^{\leqslant n}\mathcal{M}$ for a general $\mathcal{L}$-module $\mathcal{M}$, we proceed stratum by stratum. Namely assume that for a fixed stratum X_P, the local cohomology of $\mathcal{M}$ at all X_Q with $Q > P$ already vanishes in degrees greater than n. There is an obvious notion of a mapping cone for a morphism of $\mathcal{L}$-modules and to truncate the local cohomology at P we define

$$\tau_P^{\leqslant n}\mathcal{M} \equiv M(\mathcal{M} \to i_{P*}\tau^{>n}i_P^*\mathcal{M})[-1].$$

The morphism here is the composition

$$\mathcal{M} \longrightarrow i_{P*}i_P^*\mathcal{M} \longrightarrow i_{P*}\tau^{>n}i_P^*\mathcal{M}$$

of the adjunction morphism with one induced from (37). An example of $\tau_P^{\leqslant n}$ where P has parabolic rank 2 will be given in the next subsection. The functor $\tau^{\leqslant n}$ is defined to be the composition of all the functors $\tau_P^{\leqslant n}$, where one applies $\tau_R^{\leqslant n}$ before $\tau_Q^{\leqslant n}$ if $Q < R$.

It is easy to check that there is a natural quasi-isomorphism

$$\mathcal{S}(\tau^{\leqslant n}\mathcal{M}) \cong \tau^{\leqslant n}\mathcal{S}(\mathcal{M}).$$

12.8. Intersection cohomology. The formula (9) for the intersection cohomology sheaf $\mathcal{I}_p\mathcal{C}(\widehat{X}; \mathbb{E})$ involves pushforward and truncation; since both of these functors have been defined for $\mathcal{L}$-modules, commuting with the realization functor, we can lift intersection cohomology to an $\mathcal{L}$-module $\mathcal{I}_p\mathcal{C}(\widehat{X}; E)$. Specifically, enumerate the elements of $\mathcal{P}$ as $P_0 = G, P_1, \dots, P_N$ such that $\dim X_{P_i} \geq \dim X_{P_j}$ for $i < j$. For $k < N$, let

$$j_{P_k} : \coprod_{i<k} X_{P_i} \hookrightarrow \coprod_{i \leq k} X_{P_i}$$

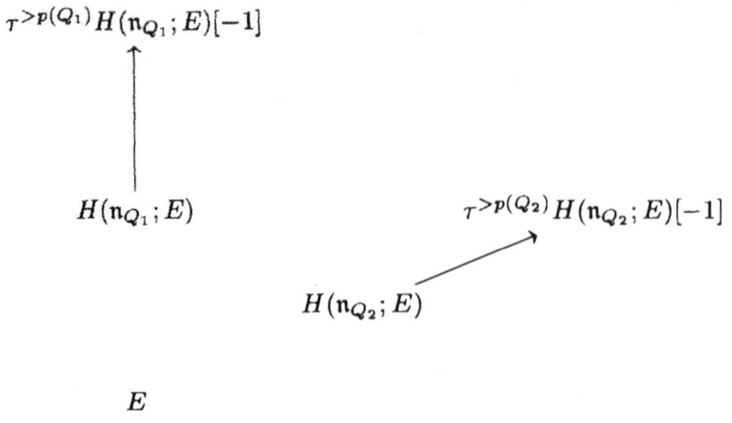

FIGURE 14. The $\mathcal{L}$-module $\mathcal{I}_p\mathcal{C}(X \sqcup X_{Q_1} \sqcup X_{Q_2}; E)$. (The unlabeled morphisms are the natural degree 1 maps.)

be the inclusion map. Define

$$p(Q) \equiv p(\dim \mathfrak{n}_Q + \#\Delta_Q) \qquad \text{for } Q \in \mathcal{P}; \tag{39}$$

this is reasonable since $\dim \mathfrak{n}_Q + \#\Delta_Q$ is the codimension of X_Q in $\widehat{X}$ (namely 1 more than the dimension of the link (21)). The intersection cohomology $\mathcal{L}$-module is defined as

$$\mathcal{I}_p\mathcal{C}(\widehat{X};E) \equiv \tau^{\leqslant p(P_N)} j_{P_N*} \cdots \tau^{\leqslant p(P_2)} j_{P_2*} \tau^{\leqslant p(P_1)} j_{P_1*} E,$$

where the G-module E is viewed as usual as an $\mathcal{L}$-module on $X = X_G$.

Let us illustrate this procedure for our standard $\mathbb{Q}$-rank 2 example with $\mathcal{P}$ as in Figure 8. Obtaining the $\mathcal{L}$-module over $X_G \sqcup X_{Q_1} \sqcup X_{Q_2}$ is straightforward; the result is pictured in Figure 14. The complex $i_P^* j_{P*}\mathcal{I}_p\mathcal{C}(X \sqcup X_{Q_1} \sqcup X_{Q_2}; E)$ computing the intersection cohomology of the link at P is then

$$(C, d_C) \equiv \begin{array}{ccc} H(\mathfrak{n}_P^{Q_1}; \tau^{>p(Q_1)}H(\mathfrak{n}_{Q_1};E))[-1] & & \\ \uparrow & & H(\mathfrak{n}_P^{Q_2}; \tau^{>p(Q_2)}H(\mathfrak{n}_{Q_2};E))[-1] \\ & \nearrow & \\ H(\mathfrak{n}_P;E) & & \end{array}; \tag{40}$$

the final $\mathcal{L}$-module $\mathcal{I}_p\mathcal{C}(\widehat{X};E)$ is pictured in Figure 15.

13. Micro-support of an $\mathcal{L}$-module

A key local invariant of an $\mathcal{L}$-module defined in [**61**] is its *micro-support*; the terminology reflects the rough analogy that exists between this concept and that introduced by Kashiwara and Schapira [**42**] for sheaves. The micro-support $\mathrm{SS}(\mathcal{M})$ of an $\mathcal{L}$-module $\mathcal{M}$ is a certain set of irreducible representations V of the various L_P which we now define.

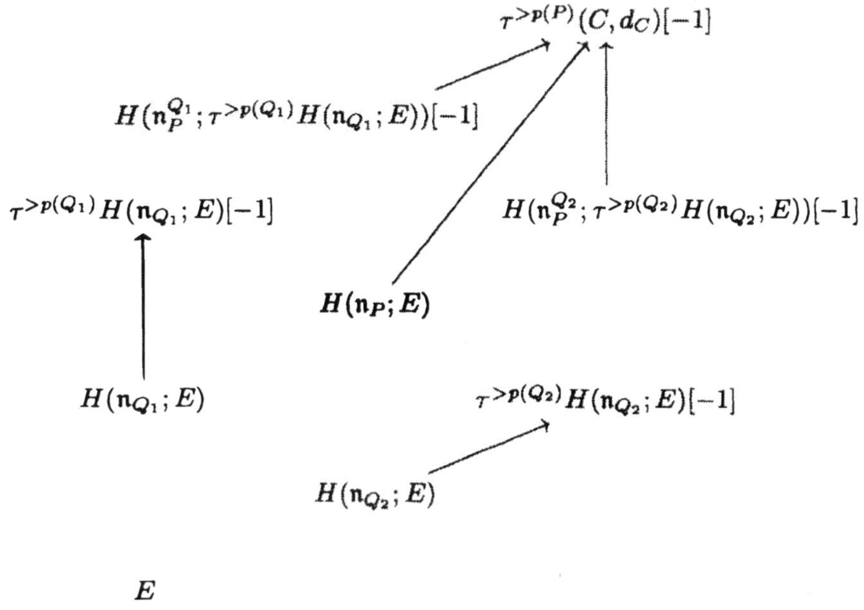

FIGURE 15. The $\mathcal{L}$-module $\mathcal{I}_p\mathcal{C}(\widehat{X}; E)$. (The unlabeled morphisms are the natural degree 1 maps.)

13.1. Definition of micro-support. Let $\mathfrak{Irr}(L_P)$ denote the irreducible algebraic representations of L_P. Recall we have decomposed $L_P = M_P A_P$, where A_P is the identity component of the maximal $\mathbb{Q}$-split torus in the center of L_P. For any $V \in \mathfrak{Irr}(L_P)$, let ξ_V be the character by which A_P acts on V and let $\mathbb{C}_{\xi_V}$ be the corresponding 1-dimensional representation. We can then write V as $V|_{M_P} \otimes \mathbb{C}_{\xi_V}$. Recall that Δ_P denotes the simple A_P-weights of the adjoint action of (a lift of) L_P on $\mathfrak{n}_P = \mathrm{Lie}(N_P)$, and that the parabolic $\mathbb{Q}$-subgroups $Q \geq P$ are parametrized by subsets Δ_P^Q of Δ_P. Consequently we can define parabolic $\mathbb{Q}$-subgroups $P \leq Q_V \leq Q'_V$ by setting

$$\Delta_P^{Q_V} = \{\, \alpha \in \Delta_P \mid (\xi_V + \rho, \alpha) < 0 \,\},$$
$$\Delta_P^{Q'_V} = \{\, \alpha \in \Delta_P \mid (\xi_V + \rho, \alpha) \leq 0 \,\};$$

here ρ is one-half the sum of the A_P-weights of $\mathfrak{n}_P$ (counted with multiplicity).

Let $\mathcal{M}$ be an $\mathcal{L}$-module on $\widehat{X}$. Define $\mathrm{SS}(\mathcal{M})$ to consist of those $V \in \mathfrak{Irr}(L_P)$ (for $P \in \mathcal{P}$) such that

(i) $(V|_{M_P})^* \cong \overline{V|_{M_P}}$, and

(ii) $H(i_P^* \hat{\imath}_Q^! \mathcal{M})_V \neq 0$ for some $Q_V \leq Q \leq Q'_V$.

Here $H(i_P^* \hat{\imath}_Q^! \mathcal{M})_V$ denotes the V-isotypical component of $H(i_P^* \hat{\imath}_Q^! \mathcal{M})$. Let

$$c(V; \mathcal{M}) \leq d(V; \mathcal{M})$$

be the least and greatest degrees for which (ii) holds.

Condition (i) says that $V|_{M_P}$ is *conjugate self-contragredient.* This condition is automatic if D_P is equal-rank, $\mathbb{C}$-rank $M_P = \operatorname{rank} K_P$ (see the discussion later in §13.2). Condition (ii) may be interpreted geometrically as follows. Consider the long exact sequence of the pair $(U, U \setminus (U \cap \widehat{X}_Q))$ where U is a small neighborhood of a point of X_P. Since $H(i_P^* \hat{\imath}_Q^! \mathcal{M})_V \cong H(U, U \setminus (U \cap \widehat{X}_Q); \mathcal{M})_V$, equation (ii) is equivalent to

$$H(U; \mathcal{M})_V \longrightarrow H(U \setminus (U \cap \widehat{X}_Q); \mathcal{M})_V$$

failing to be an isomorphism in some degree. Note that a functor like $i_P^* \hat{\imath}_Q^!$ was used by Goresky and MacPherson [**35**] to describe the local contributions to a Lefschetz fixed point formula; more recently it has been used in Braden's work [**19**] on hyperbolic localization.

We say that an element $V \in \mathrm{SS}(\mathcal{M})$ belongs to $\mathrm{SS}_{\mathrm{ess}}(\mathcal{M})$, the *essential micro-support*, if furthermore

(ii)′ $$H(i_P^* \hat{\imath}_{Q_V}^! \mathcal{M})_V \longrightarrow H(i_P^* \hat{\imath}_{Q'_V}^! \mathcal{M})_V \qquad \text{is nonzero.}$$

All of the above was for an $\mathcal{L}$-module on $\widehat{X}$. Consider more generally an admissible subspace $W \subseteq \widehat{X}$ which has a unique maximal stratum X_R. We define the micro-support and the essential micro-support of an $\mathcal{L}$-module on W similarly except that we restrict V to belong to $\coprod_{P \in \mathcal{P}(W)} \mathfrak{Irr}(L_P)$ and replace Q_V and Q'_V by parabolic $\mathbb{Q}$-subgroups Q_V^R and $Q_V^{\prime R}$ defined by

$$\Delta_P^{Q_V^R} = \{\, \alpha \in \Delta_P^R \mid (\xi_V + \rho, \alpha) < 0 \,\},$$
$$\Delta_P^{Q_V^{\prime R}} = \{\, \alpha \in \Delta_P^R \mid (\xi_V + \rho, \alpha) \leq 0 \,\}.$$

13.2. Example: Micro-support of $i_{G*}E$. We illustrate the definition by calculating the micro-support of $i_{G*}E$ [**62**]. This also gives us an opportunity to introduce Kostant's theorem which will be important later.

For $\mathcal{M} = i_{G*}E$, we have $E_P = 0$ for all $P \neq G$ and thus $\hat{\imath}_Q^! \mathcal{M} = 0$ unless $Q = G$; if $Q = G$ we have $\hat{\imath}_G^! \mathcal{M} = \mathcal{M}$. Thus for any $P \in \mathcal{P}$, we only need to consider irreducible L_P-modules V such that $Q'_V = G$, that is, such that

(41) $$(\xi_V + \rho, \alpha) \leq 0 \qquad \text{for all } \alpha \in \Delta_P;$$

such a V will actually be in $\mathrm{SS}(i_{G*}E)$ if $(V|_{M_P})^* \cong \overline{V|_{M_P}}$ and if V occurs in $H(i_P^* i_{G*}E) = H(\mathfrak{n}_P; E)$.

We therefore need to know the irreducible components of $H(\mathfrak{n}_P; E)$; in fact a theorem of Kostant [**45**] completely describes $H(\mathfrak{n}_P; E)$ as an L_P-module. Since this will be essential later as well, we review it here. First some notation. Let $\mathfrak{h} = \mathfrak{h}_{M_P} + \mathfrak{a}_P$ be a Cartan subalgebra of $\mathfrak{l}_P$. Via a lift we may view $\mathfrak{h}$ as a Cartan subalgebra of $\mathfrak{g}$ as well and we choose a set of simple $\mathbb{C}$-roots ${}_{\mathbb{C}}\Delta$ for $\mathfrak{g}_{\mathbb{C}}$ such that the roots in $\mathfrak{n}_{P\mathbb{C}}$ are positive; as usual, let ρ denote one-half the sum of the positive roots (its restriction to $\mathfrak{a}_P$ agrees with the previous definition of ρ). Let W be the Weyl group of $\mathfrak{g}_{\mathbb{C}}$ and let W^P denote the subgroup corresponding to the Weyl group of $\mathfrak{l}_{P\mathbb{C}}$. The length of $w \in W$ is denoted $\ell(w)$ and we let W_P be the set of minimal length coset representatives of $W^P\backslash W$. Assume E is irreducible with highest weight λ. Kostant's theorem says that

$$H(\mathfrak{n}_P; E) = \bigoplus_{w\in W_P} V_{w(\lambda+\rho)-\rho}[-\ell(w)], \tag{42}$$

where V_μ denotes the irreducible L_P-module with highest weight μ.

We also need to understand when $V|_{M_P}$ is conjugate self-contragredient. The involution $V|_{M_P} \mapsto (\overline{V|_{M_P}})^*$ on representations induces an involution on highest weights, $\mu|_{\mathfrak{h}_{M_P}} \mapsto \tau_P(\mu|_{\mathfrak{h}_{M_P}})$. It is always the case that $\tau_P(\rho|_{\mathfrak{h}_{M_P}}) = \rho|_{\mathfrak{h}_{M_P}}$, so $V_{w(\lambda+\rho)-\rho}|_{M_P}$ is conjugate self-contragredient if and only if

$$\tau_P(w(\lambda+\rho)|_{\mathfrak{h}_{M_P}}) = w(\lambda+\rho)|_{\mathfrak{h}_{M_P}}.$$

The involution τ_P is described in [**13**]; it is the composition of the opposition involution and the "$*$-action" of complex conjugation [**75**]. Concretely, let ${}_{\mathbb{C}}\Delta^P \subseteq {}_{\mathbb{C}}\Delta$ denote the simple $\mathbb{C}$-roots of $\mathfrak{l}_{P\mathbb{C}}$. Then

$$\tau_P = (-w_0^P) \circ (w_c^P c)$$

where $w_0^P \in W^P$ is the longest element, c is complex conjugation, and $w_c^P \in W^P$ is such that $w_c^P(c_{\mathbb{C}}\Delta^P) = {}_{\mathbb{C}}\Delta^P$. Even more concretely, if $\mathfrak{h}$ is a *fundamental* (that is, maximally compact) θ-stable Cartan subalgebra of $\mathfrak{l}_P$ and the positive system is chosen so that ${}_{\mathbb{C}}\Delta^P$ is θ-stable, then $\tau_P = \theta$.

The micro-support of $i_{G*}E$ is thus

$$\mathrm{SS}(i_{G*}E) = \coprod_P \{\, V_{w(\lambda+\rho)-\rho} \mid (w(\lambda+\rho),\alpha) \le 0 \text{ for all } \alpha \in \Delta_P,$$
$$\tau_P(w(\lambda+\rho)|_{\mathfrak{h}_{M_P}}) = w(\lambda+\rho)|_{\mathfrak{h}_{M_P}} \,\};$$

to obtain an expression for the essential micro-support $\mathrm{SS}_{\mathrm{ess}}(i_{G*}E)$, we impose instead the strict inequality $(w(\lambda+\rho),\alpha) < 0$. For $V = V_{w(\lambda+\rho)-\rho} \in \mathrm{SS}(i_{G*}E)$, we have

$$c(V; i_{G*}E) = d(V; i_{G*}E) = \ell(w). \tag{43}$$

14. A vanishing theorem for the cohomology of an $\mathcal{L}$-module

The justification for the definition of $\mathrm{SS}(\mathcal{M})$ is that it enters into a general vanishing theorem for the cohomology of an $\mathcal{L}$-module (see Theorem 14.1 below).

First some definitions. Let $\mathcal{M}$ be an $\mathcal{L}$-module on $\widehat{X}$ and let $V \in \mathrm{SS}(\mathcal{M})$ be an L_P-module. Let μ be the highest weight of V with respect to a fundamental Cartan subalgebra $\mathfrak{h}$ of $\mathfrak{l}_P$ and a θ-stable positive system; view μ as an element of $\mathfrak{l}_{P\mathbb{C}}^*$ by extending it to be 0 on all root spaces. Let $L_P(\mu) \subseteq L_P$ be the stabilizer of μ for the coadjoint action; this is a reductive $\mathbb{R}$-subgroup whose $\mathbb{C}$-roots are those $\mathbb{C}$-roots of L_P which are orthogonal to μ. Let

$$D_P(V) = L_P(\mu)/(K_P \cap L_P(\mu))A_P$$

be the associated symmetric space. The space $D_P(V)$ is not well-defined, even up to isomorphism, since it depends on the choice of a θ-stable positive system; we will assume however that the positive system has been chosen so that $D_P(V)$ has the maximum possible dimension.

Define

$$c(\mathcal{M}) = \inf_{V\in \mathrm{SS}(\mathcal{M})} \tfrac{1}{2}(\dim D_P - \dim D_P(V)) + c(V;\mathcal{M}),$$
$$d(\mathcal{M}) = \sup_{V\in \mathrm{SS}(\mathcal{M})} \tfrac{1}{2}(\dim D_P + \dim D_P(V)) + d(V;\mathcal{M}).$$

As we will see in Theorem 16.1, the significance of $(\dim D_P \pm \dim D_P(V))/2$ is that this is the range of degrees in which $H_{(2)}(X_P;\mathbb{V})$ can be nonzero. In these definitions, one can also consider just $V \in \mathrm{SS}_{\mathrm{ess}}(\mathcal{M})$ and redefine $c(V;\mathcal{M}) \le d(V;\mathcal{M})$ to be the least and greatest degrees in which (ii)$'$ is nonzero; one can prove that the same values of $c(\mathcal{M})$ and $d(\mathcal{M})$ are obtained.

The following vanishing theorem is proved in [**61**, Theorem 10.4]:

THEOREM 14.1. *Let $\mathcal{M}$ be an $\mathfrak{L}$-module on $\widehat{X}$. Then*

$$H^j(\widehat{X};\mathcal{M}) = 0 \qquad \text{for } j \notin [c(\mathcal{M}), d(\mathcal{M})].$$

In particular, if $\mathrm{SS}(\mathcal{M}) = \emptyset$ then $H(\widehat{X};\mathcal{M}) = 0$ identically.

We will sketch a proof of the theorem in §17. It has a number of ingredients, including the author's past work on tilings of locally symmetric spaces [**60**], various analytic estimates, and a vanishing theorem for L^2-cohomology which we will recall in §16.

15. Application: a vanishing theorem for ordinary cohomology

One can immediately apply Theorem 14.1 and our calculation of $\mathrm{SS}_{\mathrm{ess}}(i_{G*}E)$ to obtain a vanishing theorem for the ordinary cohomology $H(X;\mathbb{E})$. This applies to all X and E, however in special cases we can be more explicit regarding the degree bounds. For example, we have proved [**62**, Theorem 5]:

THEOREM 15.1. *If D is equal-rank and E has regular highest weight, then*

$$H^j(X;E) = 0 \qquad \text{for } j < \tfrac{1}{2}\dim X.$$

The same theorem has been announced by Li and Schwermer [**48**] in a slightly strengthened form. They drop the equal-rank condition and replace the vanishing range by

$$j < \tfrac{1}{2}(\dim X - (\mathbb{C}\text{-rank}\,{}^0G - \operatorname{rank} K));$$

it is clear that this strengthened form also follows from our proof in [**62**]. Theorem 15.1 answers a question of Tilouine [**53**, §8.7] in the Hermitian case; the case $G = R_{k/\mathbb{Q}}\,\mathrm{GSp}_4(\mathbb{R})$ where k is a totally real number field was proved in [**74**] using results of Franke [**29**]. For an application of the theorem see Mauger's thesis [**51**].

In order to deduce Theorem 15.1 from Theorem 14.1, equation (43) shows that we need an estimate on $\ell(w)$. The following basic lemma provides the needed estimate; it will also be crucial later in studying the micro-support of $\mathcal{I}_pC(\widehat{X};E)$ and its behavior under functorial operations, Theorems 18.1 and 19.1.

LEMMA 15.2. *Let E be an irreducible G-module with highest weight λ. Let P be a parabolic $\mathbb{Q}$-subgroup and let $w \in W_P$. Set $V = V_{w(\lambda+\rho)-\rho}$ and assume that $V|_{M_P}$ conjugate self-contragredient.*

(i) *If $(w(\lambda+\rho),\alpha) \le 0$ for all $\alpha \in \Delta_P$, then*

$$\ell(w) \ge \tfrac{1}{2}(\dim \mathfrak{n}_P + \dim \mathfrak{n}_P(V)).$$

(ii) *If* $(w(\lambda+\rho),\alpha) \geq 0$ *for all* $\alpha \in \Delta_P$, *then*

$$\ell(w) \leq \tfrac{1}{2}(\dim \mathfrak{n}_P - \dim \mathfrak{n}_P(V)).$$

The quantity $\mathfrak{n}_P(V)$ in the lemma is defined as follows; see [**61**, §24.1] for more details. Let μ be the highest weight of V for a fundamental θ-stable Cartan subalgebra of $\mathfrak{l}_P$ and θ-stable positive system. Let $\mathfrak{n}_P(\mu)$ be the sum of the irreducible $L_P(\mu)$-submodules of $\mathfrak{n}_{P\mathbb{C}}$ whose weights are stable under the action of $-\tau_P$; observe that $-\tau_P$ is complex conjugation in this situation so that $\mathfrak{n}_P(\mu)$ contains in particular all positive real root spaces. Let $\mathfrak{n}_P(V)$ be any of the $\mathfrak{n}_P(\mu)$ with maximal dimension as we vary the θ-stable positive system.

We can now deduce the theorem (in its strengthened form). The Kostant-Sugiura classification of Cartan subalgebras [**44**], [**72**], [**73**] shows that $\operatorname{rank} K - \operatorname{rank} K_P$ is at most the number of positive real roots. Thus

$$\dim \mathfrak{n}_P(V) \geq \operatorname{rank} K - \operatorname{rank} K_P. \tag{44}$$

The hypothesis that E has regular highest weight implies that $\dim D_P(V) = \mathbb{C}\text{-rank}\, M_P - \operatorname{rank} K_P$ for any $V \in \mathrm{SS}(i_{G*}E)$. These facts and Lemma 15.2(i) allow one to estimate

$$\begin{aligned}
&\tfrac{1}{2}(\dim D_P - \dim D_P(V)) + c(V; i_{G*}E) \\
&\quad \geq \tfrac{1}{2}\big(\dim D_P + \dim \mathfrak{n}_P - (\mathbb{C}\text{-rank}\, M_P - \operatorname{rank} K)\big) \\
&\quad = \tfrac{1}{2}\big(\dim X - (\mathbb{C}\text{-rank}\, {}^0G - \operatorname{rank} K)\big).
\end{aligned}$$

Sketch of the proof of Lemma 15.2. We will sketch the argument with $\dim \mathfrak{n}_P(V)$ replaced by $\operatorname{rank} K - \operatorname{rank} K_P$; see [**61**, Lemma 24.2] for the complete proof. Recall that $\ell(w)$ can be viewed as the number of positive $\mathbb{C}$-roots γ for which $w^{-1}\gamma$ is negative; for $w \in W_P$ all these roots are in $\mathfrak{n}_{P\mathbb{C}}$. Since $\lambda+\rho$ is strictly dominant, $(w(\lambda+\rho),\gamma) = (\lambda+\rho, w^{-1}\gamma)$ is never zero and will be negative precisely when $w^{-1}\gamma < 0$. Thus $\ell(w)$ is the number of roots γ in $\mathfrak{n}_{P\mathbb{C}}$ for which

$$(w(\lambda+\rho),\gamma) \leq 0. \tag{45}$$

However complex conjugation defines an involution $\gamma \mapsto \bar{\gamma}$ on the roots in $\mathfrak{n}_{P\mathbb{C}}$; we will prove (i) by showing that (45) holds for at least one of γ and $\bar{\gamma}$.

Assume that $\mathfrak{h} = \mathfrak{h}_{M_P} + \mathfrak{a}_P$ is a fundamental θ-stable Cartan subalgebra of $\mathfrak{l}_P$ and that the set of positive roots of $\mathfrak{l}_{P\mathbb{C}}$ are θ-stable. Decompose $\mathfrak{h} = \mathfrak{h}_{M_P,1} + (\mathfrak{h}_{M_P,-1} + \mathfrak{a}_P)$ according to the ±1-eigenspaces of the Cartan involution. In this case, the operator τ_P is simply θ [**13**], so the condition that $(V_{w(\lambda+\rho)-\rho}|_{M_P})^* \cong \overline{V_{w(\lambda+\rho)-\rho}|_{M_P}}$ is equivalent to $w(\lambda+\rho)|_{\mathfrak{h}_{M_P,-1}} = 0$. It follows that

$$(w(\lambda+\rho),\gamma) = (w(\lambda+\rho),\gamma|_{\mathfrak{h}_{M_P,1}}) + (w(\lambda+\rho),\gamma|_{\mathfrak{a}_P}).$$

Since $\gamma|_{\mathfrak{a}_P}$ is a linear combination with nonnegative coefficients of $\alpha \in \Delta_P$, the second term is nonpositive by hypothesis; on the other hand, the first term is negated when γ is replaced by $\bar{\gamma}$. Thus (45) holds for one of either γ or $\bar{\gamma}$ as desired. In particular it holds for all roots such that $\gamma = \bar{\gamma}$, that is, the positive real roots. Since, as we have already noted, the number of positive real roots is at least $\operatorname{rank} K - \operatorname{rank} K_P$, assertion (i) follows with $\dim \mathfrak{n}_P(V)$ substituted by $\operatorname{rank} K - \operatorname{rank} K_P$. Assertion (ii) is proven similarly. □

16. A vanishing theorem for L^2-cohomology

As we will see in the next section, the proof of Theorem 14.1 will be reduced to the study of the L^2-cohomology groups $H_{(2)}(X_P;\mathbb{V})$ for all $P \in \mathcal{P}$ and all L_P-modules $V \in \mathrm{SS}(\mathcal{M})$. Thus we need a vanishing theorem for L^2-cohomology. The theorem below was proved by the author and Stern in [**65**] though it is not stated explicitly in this form; it is based on work of Raghunathan [**56**], [**57**]. The statement and proof appeared in [**61**, Theorem 14.4].

THEOREM 16.1. *Let X be a locally symmetric space and let E be a representation of G with corresponding metrized locally constant sheaf $\mathbb{E}$.*

(i) *If $(E|_{{}^0G})^* \not\cong \overline{E|_{{}^0G}}$, then $H_{(2)}(X;\mathbb{E}) = 0$.*

(ii) *If $(E|_{{}^0G})^* \cong \overline{E|_{{}^0G}}$, then $H^j_{(2)}(X;\mathbb{E}) = 0$ for*

$$j \notin [\tfrac{1}{2}(\dim D - \dim D(E)), \tfrac{1}{2}(\dim D + \dim D(E))]$$

and $H^{(\dim D - \dim D(E))/2}_{(2)}(X;\mathbb{E})$ is Hausdorff.

The proof of the theorem is based on the following criterion which will be used in the proof of Theorem 14.1 later as well:

PROPOSITION 16.2. *Let M be a complete Riemannian manifold with a metrized locally constant sheaf $\mathbb{E}$. Then for every j, the following two conditions are equivalent.*

(i) *$H^j_{(2)}(M;\mathbb{E}) = 0$ and $H^{j+1}_{(2)}(M;\mathbb{E})$ is Hausdorff;*

(ii) *there exists $c > 0$ such that*

$$\|d\phi\|^2 + \|d^*\phi\|^2 \geq c\|\phi\|^2$$

for all compactly supported smooth forms ϕ of degree j.

The complex of compactly supported forms in $A_{(2)}(X;\mathbb{E})$ may be identified as in [**18**, VII.2.7], [**50**] with a Koszul complex $C({}^0\mathfrak{g}, K; A^0_c(\Gamma\backslash{}^0G) \otimes E)$. Thus the differential may be decomposed $d = d_0 + d_1$, where d_0 corresponds to the algebraic action of ${}^0\mathfrak{g}$ on E and d_1 corresponds to the action of ${}^0\mathfrak{g}$ on $A^0_c(\Gamma\backslash{}^0G)$ by differentiation. One calculates using integration by parts that

$$\|d\phi\|^2 + \|d^*\phi\|^2 \geq (\Delta_0\phi, \phi),$$

where Δ_0 is the nonnegative algebraic Laplacian corresponding to d_0; it is the Laplacian for the complex of invariant forms $C({}^0\mathfrak{g}, K; E)$. A careful analysis of the possible zero eigenvectors of Δ_0 yields Theorem 16.1.

Alternatively, note that a vanishing theorem for $(\mathfrak{g}, K)$-cohomology of unitary representations was proved by Vogan and Zuckerman [**77**] based on work by Kumaresan [**46**]; this implies a vanishing theorem for L^2-cohomology (at least the Hausdorff part) which one can show [**63**] is equivalent with Theorem 16.1.

17. Proof of the vanishing theorem for the cohomology of an $\mathcal{L}$-module

In the following pages we sketch a proof of Theorem 14.1, the vanishing theorem for the cohomology $H(\widehat{X};\mathcal{M})$ of an $\mathcal{L}$-module $\mathcal{M} = (E_\cdot, f_{\cdot\cdot})$. Most steps of the proof are accompanied by an illustrative figure.

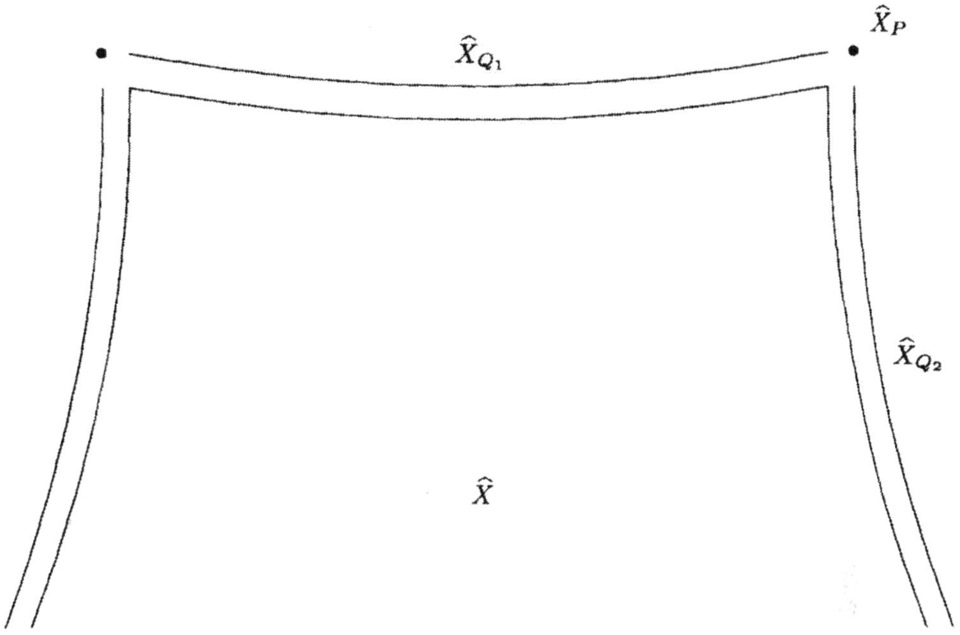

FIGURE 16. The complex $\Gamma(\widehat{X}, \mathcal{S}(\mathcal{M}))$ lives on the disjoint union $\coprod_P \widehat{X}_P$

The cohomology $H(\widehat{X}; \mathcal{M})$ may be computed as the cohomology of the complex of global sections of the fine realization sheaf $\mathcal{S}(\mathcal{M})$ defined in §12.4. Ignoring the differentials, this complex is

$$\Gamma(\widehat{X}, \mathcal{S}(\mathcal{M})) = \bigoplus_{P \in \mathcal{P}} A_{\mathrm{sp}}(\widehat{X}; \mathbb{E}_P).$$

Thus a "form" ϕ is really a collection $(\phi_P)_{P\in\mathcal{P}}$ of special differential forms on the various strata as pictured in Figure 16. The differential is

$$d\phi = d(\phi_P)_{P\in\mathcal{P}} = \Big(\sum_{P\leq Q} d_{PQ}\phi_Q\Big)_{P\in\mathcal{P}} = \Big(d_P\phi_P + \sum_{P\leq Q} f_{PQ}(\phi_Q|_{\widehat{X}_P})\Big)_{P\in\mathcal{P}}$$

where d_P denotes the de Rham differential and we write $\phi_Q|_{\widehat{X}_P}$ for $k_{PQ}(\phi_Q)$.

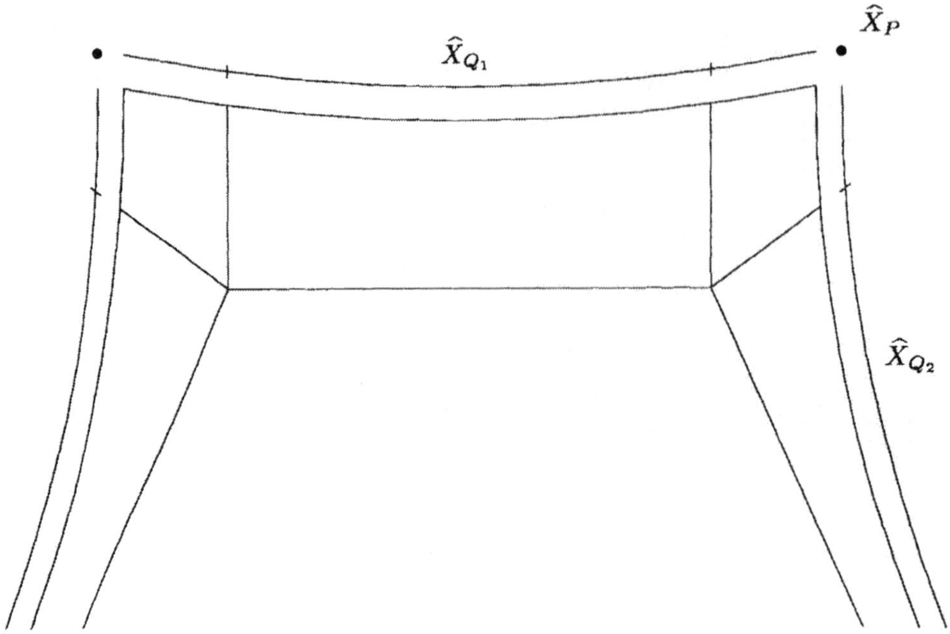

FIGURE 17. The Arthur-Langlands Partition

We would like to use Hodge theory to calculate the cohomology of this complex and for that we need to consider the L^2-norm

$$||\phi||^2 \equiv \sum_P ||\phi_P||^2.$$

However in the locally symmetric metric on X_P, the boundary is at infinite distance. Thus the condition $||\phi|| < \infty$ imposes nontrivial L^2-growth conditions on the ϕ_P which would change the cohomology.

The solution is to replace each X_P with a diffeomorphic copy which is embedded as a domain in X_P with compact closure. If the diffeomorphic embedding extended to the manifold with corners $\overline{X}_P$ the new boundary faces would now be at finite distance. To do this naturally, consider the Arthur-Langlands partition [**1**] of X. This was extended by the author to a partition of $\overline{X}$ (and hence $\widehat{X}$) in [**60**]; see Figure 17.

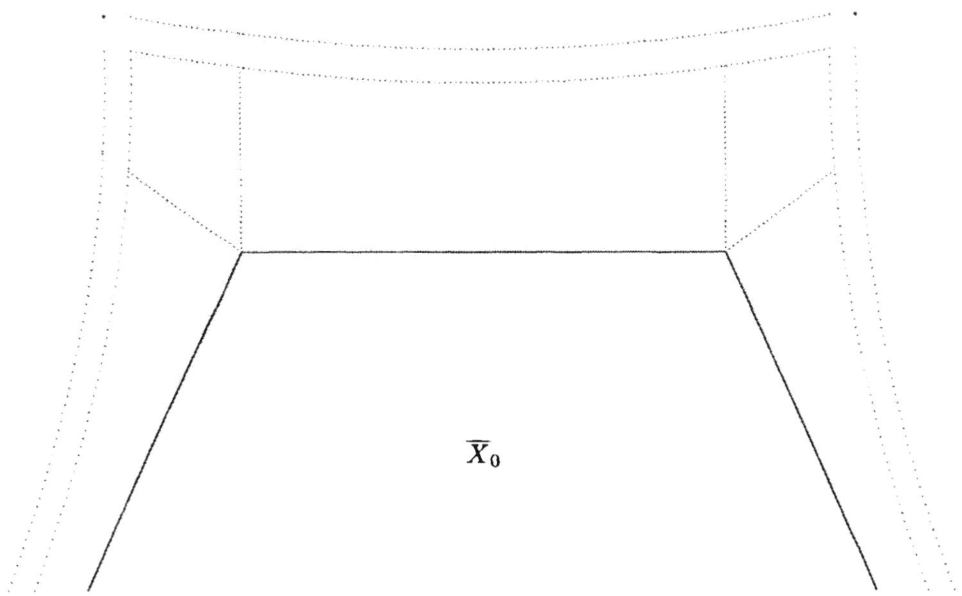

FIGURE 18. The Central Region

The Arthur-Langlands partition divides X into disjoint regions indexed by $\mathcal{P}$; it depends on a certain parameter $\mathbf{b} = (b_P)_{P\in\mathcal{P}}$, where $b_P \in \mathcal{A}_P^G$. We will only need here the central region $\overline{X}_0$, which is the one indexed by $G \in \mathcal{P}$. Recall that although the decomposition $D = \mathcal{A}_P^G \times e_P$ does not descend to a decomposition $\mathcal{A}_P^G \times Y_P$ of X, such a decomposition does hold within an appropriate cylindrical set W_P (see (15) in §6.3). We can use this decomposition to describe $\overline{X}_0$:

$$\overline{X}_0 \cap W_P = ((^{-}A_P^G \cdot b_P) \times Y_P) \cap W_P.$$

In [60] it was shown that there exists a natural piecewise analytic diffeomorphism (depending on another parameter) of $\overline{X}$ with $\overline{X}_0$. So the next step is to throw away everything outside of $\overline{X}_0$ and start over.

We endow the central region $\overline{X}_0$ with the metric induced by restriction from the locally symmetric metric on X. Since $\overline{X}_0$ is then a compact Riemannian manifold with corners, a smooth form on $\overline{X}_0$ or any of its boundary faces will automatically have finite L^2-norm.

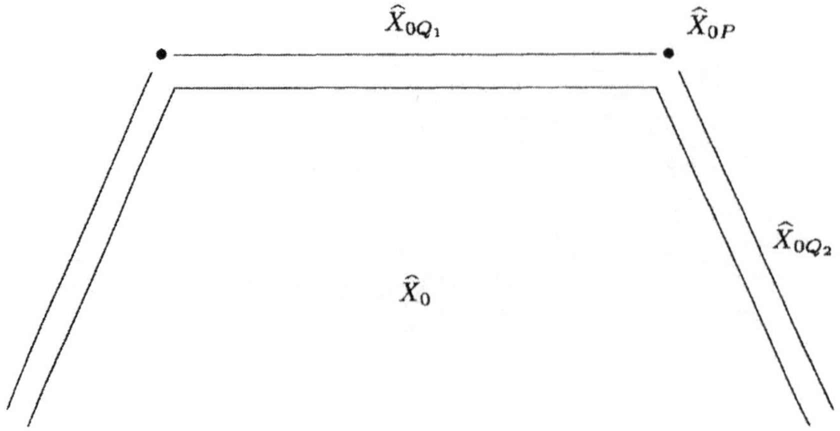

FIGURE 19. The complex $\Gamma(\widehat{X}_0, \mathcal{S}_0(\mathcal{M}))$ lives on the disjoint union $\coprod_P \widehat{X}_{0P}$

Since $\overline{X}_0$ is diffeomorphic with $\overline{X}$ we can mimic the construction of $\widehat{X}$ and $\mathcal{S}(\mathcal{M})$ to obtain a realization $\mathcal{S}_0(\mathcal{M})$ on $\widehat{X}_0$ whose global sections also compute $H(\widehat{X};\mathcal{M})$; see Figure 19. Since we have removed the difficulty with L^2-growth conditions, Hodge theory shows that classes in $H(\widehat{X};\mathcal{M})$ have harmonic representatives in $\Gamma(\widehat{X}_0, \mathcal{S}_0(\mathcal{M}))$. This is the analogue of the classical Hodge theorem for a compact Riemannian manifold.

We now follow an argument from [**65**] however there is an added combinatorial complication since our "forms" are actually collections of forms and there are interaction terms in the differential. The aim is to show that there exists $c > 0$ such that the estimate

$$\|d\phi\|^2 + \|d^*\phi\|^2 \geq c\|\phi\|^2 \tag{46}$$

holds for all ϕ with degree $j \notin [c(\mathcal{M}), d(\mathcal{M})]$. This will suffice to prove the theorem by (a generalization of) Proposition 16.2.

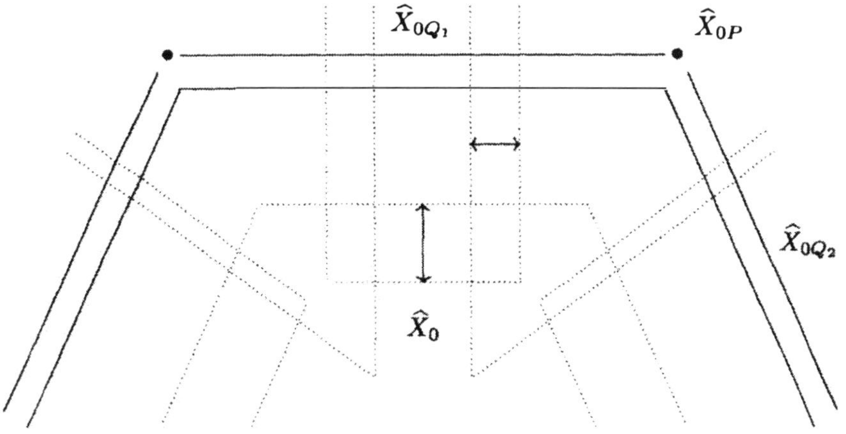

FIGURE 20. Covering by cylindrical sets W_P

Let $\{W_P\}_{P\in\mathcal{P}}$ be a cylindrical cover of $\overline{X}$ as in (16) of §6.3; we can arrange that for all $P \in \mathcal{P}$, the cylindrical set $W_P = (\bar{A}_P^{G+} \cdot s_P) \times O_P$ projects to a relatively compact subset of X_{0P}. Such a cover induces a cover $\{W_P\}_{P\leq Q}$ of each $\widehat{X}_{0Q}$; for simplicity we denote the elements of such a cover again by W_P. The situation is pictured in Figure 20.

Assume that for each $P \in \mathcal{P}$ we can establish an estimate

$$\|d\phi\|^2 + \|d^*\phi\|^2 \geq c_P\|\phi\|^2 \tag{47}$$

for all ϕ with compact support in $\widehat{X}_0 \cap W_P$ and degree $j \notin [c(\mathcal{M}), d(\mathcal{M})]$. Here $c_P > 0$ is to be a constant independent of O_P. If this is possible, and if there exists a partition of unity $\{\eta_P\}_{P\in\mathcal{P}}$ with $|d\eta_P| < \epsilon$ for all P (where $\epsilon > 0$ depends on $\{c_P\}_{P\in\mathcal{P}}$), then the above estimates may be patched together to yield (46). Such a partition of unity always exists. The point is that the magnitude $|d\eta_P|$ is roughly inversely proportional to the "width" of the overlaps $W_P \cap W_Q$ for all $Q \neq P$; two examples of these "widths", for $W_{Q_1}\cap W_P$ and $W_G \cap W_{Q_1}$, are indicated in Figure 20. However it is possible to choose $\{W_P\}_{P\in\mathcal{P}}$ so that these widths are arbitrarily large; this may involve enlarging $\widehat{X}_0$ through manipulation of the parameter $\mathbf{b}$ but there is plenty of room to do this since the boundary of $\widehat{X}$ is infinitely far away. Thus $\{\eta_P\}_{P\in\mathcal{P}}$ can be chosen with arbitrarily small derivative.

It remains to restrict our attention to a single W_P and prove (47).

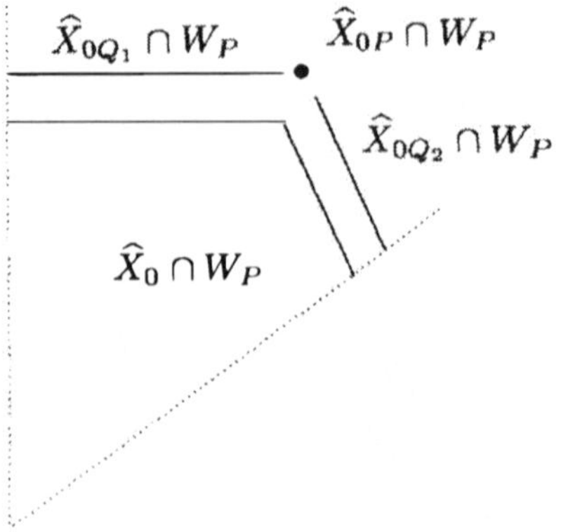

FIGURE 21. A single cylindrical set W_P

A section $\phi = (\phi_Q)_{Q \geq P}$ of $\mathcal{S}_0(\mathcal{M})$ over $\widehat{X}_0 \cap W_P$ is composed of $\mathbb{E}_Q$-valued forms ϕ_Q which live on $\widehat{X}_{0Q} \cap W_P$ as in Figure 21 above. We can apply harmonic projection to each ϕ_Q along the nilmanifold fibers $\mathcal{N}'^Q_P$ [**81**] and obtain an $(\mathbb{H}(\mathfrak{n}^Q_P; E_Q) \otimes \mathbb{C}_\rho)$-valued form on $((^-A^Q_P \cdot b_P) \times X_P) \cap W_P$; the factor $\mathbb{C}_\rho$ accounts for the volume of the nilmanifold fibers. However since we are seeking to prove the estimate (47) for compactly supported forms, it certainly suffices to consider more generally forms whose components have compact support anywhere in $(^-A^Q_P \cdot b_P) \times X_P$, which for simplicity we write as $^-A^Q_P \times X_P$. The result is that we have unfolded the situation of Figure 19, which was a quotient by Γ, to the following quotient by Γ_P:

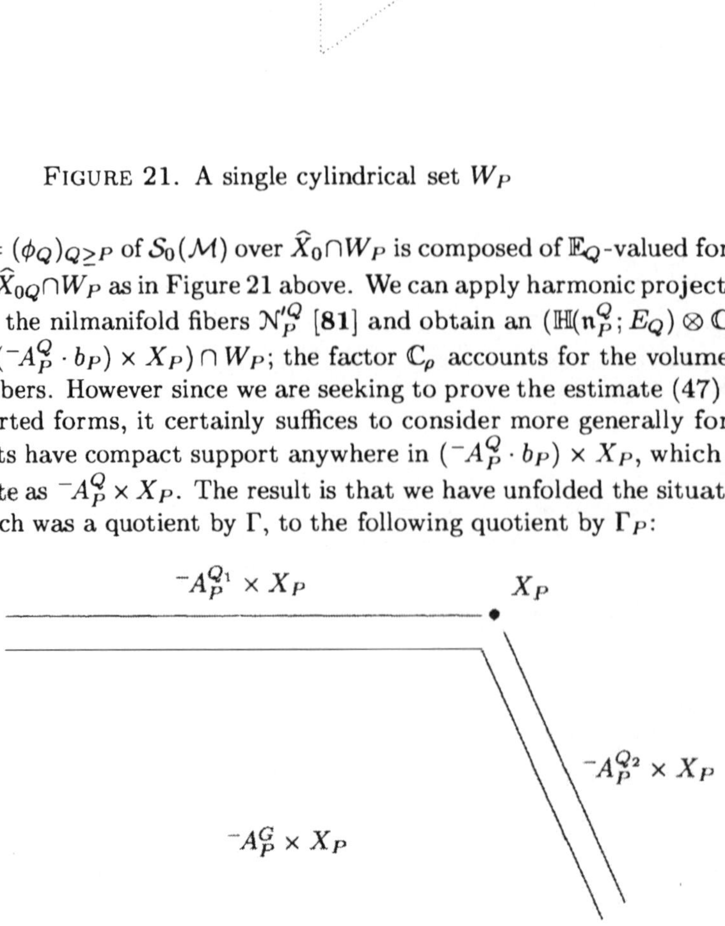

We need to establish the estimate (47) for compactly supported forms in the following unfolded complex:

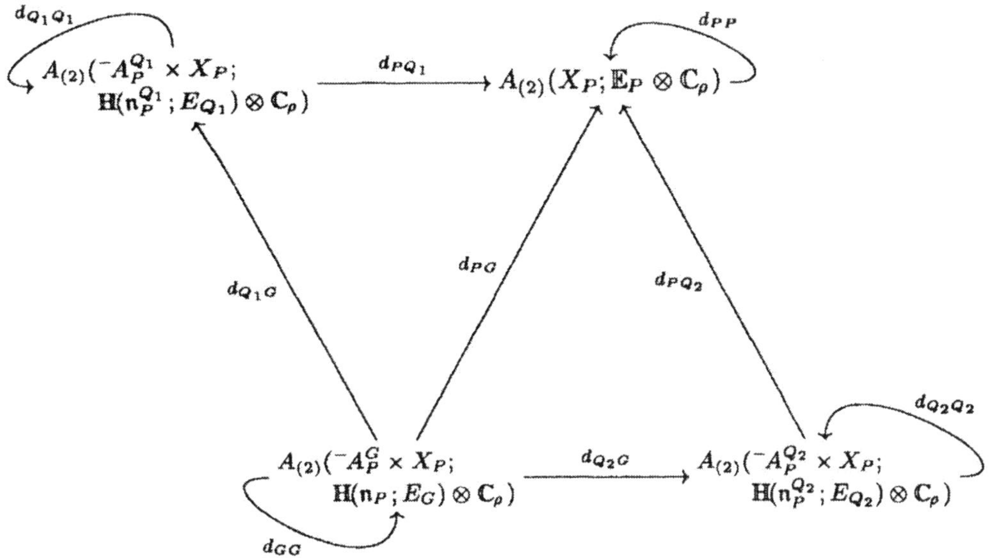

By Proposition 16.2 it suffices to show that the cohomology of this complex vanishes in degrees outside of $[c(\mathcal{M}), d(\mathcal{M})]$ and that the cohomology is Hausdorff in degree $c(\mathcal{M})$. It also suffices to do this with the coefficients replaced by their $V \otimes \mathbb{C}_\rho$-isotypical component for every irreducible representation V of L_P.

Begin the calculation of the cohomology by forming the cohomology of each summand (for simplicity we assume that $f_{QQ} = 0$ for all Q, that is, $\mathcal{M}$ is reduced as in §12.5):

(48)

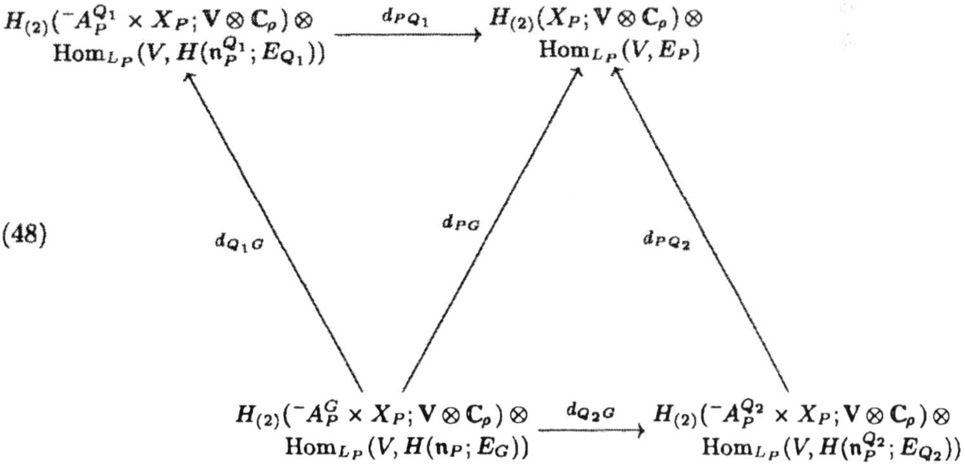

Consider a typical L^2-cohomology group that occurs in (48), say

$$H_{(2)}({}^-A_P^Q \times X_P; \mathbb{V} \otimes \mathbb{C}_\rho) \tag{49}$$

where $Q \geq P$. The locally constant sheaf $\mathbb{V} \otimes \mathbb{C}_\rho$ is constant along the ${}^-A_P^Q$ factor however its metric is not; the coefficients add an exponential weight of $a^{-2(\xi_V + \rho)}$

to the L^2-norm integral. This weight factors as

$$a^{-2(\xi_V+\rho)} = \prod_{\alpha\in\Delta_P^Q} (a^{-\beta_\alpha})^{2(\xi_V+\rho,\alpha)}$$

corresponding to the product decomposition

$${}^{-}A_P^Q \cong [1,\infty)^{\Delta_P^Q}, \qquad a \mapsto (a^{-\beta_{\alpha_1}},\dots,a^{\beta_{\alpha_r}}).$$

However the weighted L^2-cohomology $H_{(2)}([1,\infty);s^k)$ (where $s = a^{-\beta_\alpha}$ and $[1,\infty)$ has the multiplicatively invariant metric ds^2/s^2) vanishes for $k > 0$ and is $\mathbb{C}$ for $k < 0$. Thus whenever there exists $\alpha \in \Delta_P^Q$ such that $(\xi_V+\rho,\alpha) > 0$, a Künneth-type argument [**81**] shows that (49) vanishes. Similarly if $(\xi_V+\rho,\alpha) < 0$ for all α we can drop the ${}^{-}A_P^Q$ factor in (49). The situation where $(\xi_V+\rho,\alpha) = 0$ for some α can be handled by a slightly more delicate argument.

In order to illustrate the remainder of the proof, suppose that ξ_V satisfies

$$(\xi_V+\rho,\alpha_1) < 0, \qquad (\xi_V+\rho,\alpha_2) > 0, \tag{50}$$

where $\Delta_P = \{\alpha_1,\alpha_2\}$ and $\Delta_P^{Q_i} = \{\alpha_i\}$ for $i = 1, 2$. By the above arguments, equation (48) becomes

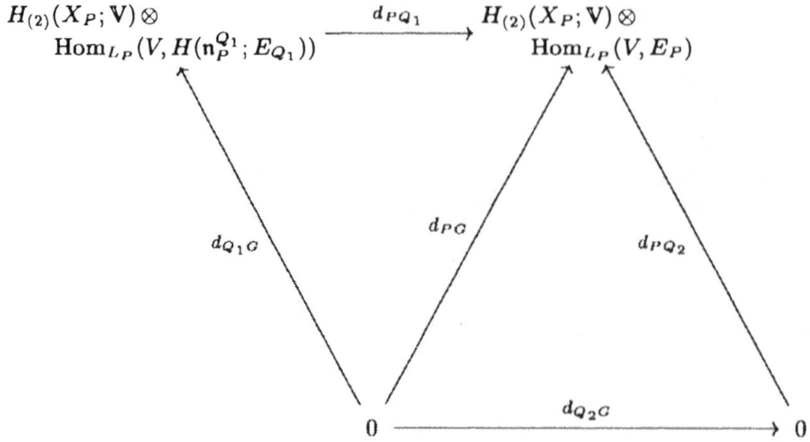

However $Q_V = Q'_V = Q_1$ in this case so the total cohomology is (compare (32) in §12.3)

$$H_{(2)}(X_P;\mathbb{V}) \otimes \mathrm{Hom}_{L_P}(V, H(i_P^* \hat{i}_{Q_V}^! \mathcal{M})). \tag{51}$$

The vanishing theorem for L^2-cohomology, Theorem 16.1, shows that the first factor of (51) vanishes unless $(V|_{M_P})^* \cong \overline{V|_{M_P}}$ and the degree is in

$$[\tfrac{1}{2}(\dim D_P - \dim D_P(V)), \tfrac{1}{2}(\dim D_P + \dim D_P(V))];$$

the cohomology is also Hausdorff in degree $\frac{1}{2}(\dim D_P - \dim D_P(V))$. On the other hand, if $(V|_{M_P})^* \cong \overline{V|_{M_P}}$ then the second factor of (51) vanishes unless $V \in \mathrm{SS}(\mathcal{M})$ and the degree is in

$$[c(V;\mathcal{M}), d(V;\mathcal{M})].$$

The sum of these two degree ranges gives the desired result.

This proves the vanishing theorem for the cohomology of an $\mathcal{L}$-module.

18. Micro-support of intersection cohomology

When an $\mathfrak{L}$-module is known explicitly the micro-support is not difficult to compute; for example, consider our calculation of $SS(i_{G*}E)$ in §13.2. Unfortunately $\mathcal{I}_p\mathcal{C}(\widehat{X};E)$ is defined inductively and we do not have an explicit closed formula for its local cohomology. Furthermore the conjugate-self-contragredient condition in the definition of micro-support does not behave well under induction. Nonetheless we can prove the following combinatorial theorem [**61**, Corollary 17.2]:

THEOREM 18.1. *Let E be an irreducible G-module and let $p = m$ or n be a middle-perversity. Assume the $\mathbb{Q}$-root system of G does not contain a factor of type D_n, E_n, or F_4. If $(E|_{{}^0G})^* \cong \overline{E|_{{}^0G}}$, then* $SS_{ess}(\mathcal{I}_p\mathcal{C}(\widehat{X};E)) = \{E\}$.

Actually we can describe the entire micro-support, not just the essential micro-support; see [**61**, Theorem 17.1] and (56), (57) below. In fact [**61**] treats the more general case in which $(E|_{{}^0G})^* \cong \overline{E|_{{}^0G}}$ is not assumed. We expect that the condition on the $\mathbb{Q}$-root system will be able to be removed; the condition is satisfied in the Hermitian case or in the case of a real equal-rank Satake compactification.

What does the theorem say? Recall that an L_P-module V belongs to $SS_{ess}(\mathcal{M})$ if and only if

(i) $\quad (V|_{M_P})^* \cong \overline{V|_{M_P}}$, and

(ii)$'$ $\quad H(i_P^* \hat{i}^!_{Q_V}\mathcal{M})_V \longrightarrow H(i_P^* \hat{i}^!_{Q'_V}\mathcal{M})_V$ is nonzero,

where we have defined Q_V and Q'_V by

$$\Delta_P^{Q_V} = \{\, \alpha \in \Delta_P \mid (\xi_V + \rho, \alpha) < 0 \,\},$$
$$\Delta_P^{Q'_V} = \{\, \alpha \in \Delta_P \mid (\xi_V + \rho, \alpha) \leq 0 \,\}.$$

Since $i_G^* \hat{i}^!_G \mathcal{I}_p\mathcal{C}(\widehat{X};E) = E$, we certainly have $E \in SS_{ess}(\mathcal{I}_p\mathcal{C}(\widehat{X};E))$ if $(E|_{{}^0G})^* \cong \overline{E|_{{}^0G}}$. The theorem is asserting that this is all: for every proper $P \in \mathcal{P}$ with parabolic rank $r(P) = 1, 2, \dots$, there is no irreducible L_P-module $V \in SS_{ess}(\mathcal{I}_p\mathcal{C}(\widehat{X};E))$.

The only candidates for V are the irreducible constituents of $H(\mathfrak{n}_P;E)$, namely $V_{w(\lambda+\rho)-\rho}$ for $w \in W_P$. The assertion of the theorem amounts to a subtle relationship between the combinatorics of the Weyl element w, namely the minimal lengths in the cosets $W^R w$ for all $R \geq P$, and the geometry of $w(\lambda+\rho)$ *vis à vis* the roots $\alpha \in \Delta_P$. Since these matters play an important role in Goresky and MacPherson's topological trace formula [**36**] and in the representation of cohomology classes by Eisenstein series [**69**], it is likely that Theorem 18.1 will have applications beyond the Rapoport/Goresky-MacPherson conjecture.

The use of micro-support and of intersection cohomology is an efficient way to encode this relationship; a description of the result without these tools becomes unmanageable rapidly as the parabolic rank increases. However it is not difficult for low parabolic rank. In the following subsections we will sketch proofs for the parabolic rank 1 and 2 cases that will give some indication of the above relationship. Unfortunately these arguments do not generalize to higher parabolic rank; in particular, the condition on the $\mathbb{Q}$-root system begins to play a role with parabolic rank 4. In the last subsection, we indicate some of the difficulties and outline how the general proof in [**61**] works.

From now on, we assume that E is irreducible with highest weight λ. We will use Kostant's theorem and the notation introduced in §13.2.

18.1. Parabolic rank 1. The simplest case is when P is a maximal parabolic $\mathbb{Q}$-subgroup. Write $\Delta_P = \{\alpha\}$. For any $\mathfrak{L}$-module $\mathcal{M}$, the expression $H(i_P^* \hat{\imath}_Q^! \mathcal{M})$ is either the local cohomology at P or the local cohomology supported at P depending on whether $Q = G$ or $Q = P$ respectively. Given the definition of intersection cohomology (see in particular (38) in §12.7), the cohomology of the link is $H(\mathfrak{n}_P; E)$ and we have

$$H(i_P^* \hat{\imath}_Q^! \mathcal{I}_p\mathcal{C}(\widehat{X}; E)) = \begin{cases} \tau^{\leq p(P)} H(\mathfrak{n}_P; E) & \text{if } Q = G, \\ \tau^{>p(P)} H(\mathfrak{n}_P; E)[-1] & \text{if } Q = P. \end{cases} \tag{52}$$

Recall that Kostant's theorem says that

$$H(\mathfrak{n}_P; E) = \bigoplus_{w \in W_P} H(\mathfrak{n}_P; E)_w = \bigoplus_{w \in W_P} V_{w(\lambda+\rho)-\rho}[-\ell(w)]. \tag{53}$$

Thus fix $V = V_{w(\lambda+\rho)-\rho}$ for some $w \in W_P$. Since $\xi_V + \rho = w(\lambda + \rho)$, the values of Q_V and Q'_V are given by

- $(w(\lambda + \rho), \alpha) < 0 \implies Q_V = Q'_V = G$,
- $(w(\lambda + \rho), \alpha) = 0 \implies Q_V = P,\ Q'_V = G$,
- $(w(\lambda + \rho), \alpha) > 0 \implies Q_V = Q'_V = P$.

On the other hand, it follows from (52) that the map in (ii)$'$ is 0 precisely in the following cases:

- when $\ell(w) > p(P)$ and $Q_V = Q'_V = G$,
- when $Q_V = P$, $Q'_V = G$, and
- when $\ell(w) \leq p(P)$ and $Q_V = Q'_V = P$.

By (7) the value of $p(P)$ is

$$p(P) = p(\dim \mathfrak{n}_P + \#\Delta_P) = \begin{cases} \lfloor (\dim \mathfrak{n}_P - 1)/2 \rfloor & p = m, \\ \lfloor (\dim \mathfrak{n}_P)/2 \rfloor & p = n. \end{cases}$$

From the above facts, it is easy to see that the theorem in this case is equivalent to following proposition, which is a generalization of an old observation of Casselman [**24**] in the $\mathbb{R}$-rank 1 case; our proof uses Lemma 15.2.

PROPOSITION 18.2. *Let E be an irreducible G-module with highest weight λ. Let P be a maximal parabolic $\mathbb{Q}$-subgroup with $\Delta_P = \{\alpha\}$ and let $w \in W_P$. Assume that $(V_{w(\lambda+\rho)-\rho}|_{M_P})^* \cong \overline{V_{w(\lambda+\rho)-\rho}|_{M_P}}$.*

(i) *If $(w(\lambda + \rho), \alpha) \leq 0$ then $\ell(w) \geq (\dim \mathfrak{n}_P)/2$.*
(ii) *If $(w(\lambda + \rho), \alpha) \geq 0$ then $\ell(w) \leq (\dim \mathfrak{n}_P)/2$.*

Furthermore assume that $(E|_{{}^\circ G})^* \cong \overline{E|_{{}^\circ G}}$. Then

$$\ell(w) = (\dim \mathfrak{n}_P)/2 \implies (w(\lambda + \rho), \alpha) = 0 \textit{ for all } \alpha \in \Delta_P.$$

PROOF. The first two parts are simply Lemma 15.2 with the $\dim \mathfrak{n}_P(V)$ term omitted. For the final assertion, note that the proof of Lemma 15.2 together with the hypothesis $\ell(w) = (\dim \mathfrak{n}_P)/2$ shows that for each root γ in $\mathfrak{n}_{P\mathbb{C}}$, exactly one of $w^{-1}\gamma$ and $w^{-1}\bar{\gamma} = -w^{-1}(\theta\gamma)$ is negative, and that there are no real roots. The first statement implies that the positive system $w\Phi^+$ is θ-stable [**61**, Lemma 8.6]. The second statement, together with the Kostant-Sugiura classification of Cartan subalgebras [**44**], [**72**], [**73**], implies that $\mathfrak{h}$ is a fundamental Cartan subalgebra for $\mathfrak{g}$. Thus the operator τ_G equals θ as well, and hence the hypothesis $(E|_{{}^\circ G})^* \cong \overline{E|_{{}^\circ G}}$ is equivalent to $w\lambda|_{\mathfrak{h}_{M_P,-1} + \mathfrak{a}_P^G} = 0$. (This is because $w\lambda$ is the highest weight of E

with respect to $w\Phi^+$.) Since $w\rho|_{\mathfrak{h}_{M_P,-1}+\mathfrak{a}_P^G} = 0$ due to $w\Phi^+$ being θ-stable, we find that $w(\lambda+\rho)|_{\mathfrak{a}_P^G} = 0$ as desired. □

18.2. Bidegree in $H(\mathfrak{n}_P; E)$. For P with parabolic rank ≥ 2 we need a refinement of Kostant's theorem. Recall that for $P \leq Q$ we have an isomorphism

$$H(\mathfrak{n}_P; E) \cong H(\mathfrak{n}_P^Q; H(\mathfrak{n}_Q; E)); \tag{54}$$

in particular $H(\mathfrak{n}_P; E)$ is bigraded given $Q \geq P$. We write the bidegree as (ℓ^Q, ℓ_Q) and define ***truncation by bidegree*** functors $\tau^{\ell_Q \leq n}$ and $\tau^{\ell_Q > n}$ on such bigraded modules. How is this structure reflected in the description of $H(\mathfrak{n}_P; E)$ given by Kostant's theorem (53)?

Write

$$W = W^P W_P \qquad \text{and} \qquad W = W^Q W_Q$$

where W^P is the Weyl group of L_P and W_P is the set of minimal length coset representatives of $W^P \backslash W$, and similarly for Q. Since P/N_Q is a parabolic subgroup of L_Q, we also have a decomposition

$$W^Q = W^P W_P^Q$$

and thus $W = W^P W_P^Q W_Q$. One can show in fact that

$$W_P = W_P^Q W_Q$$

and hence for $w \in W_P$ we may write $w = w^Q w_Q$ corresponding to this decomposition. It is easy then to verify that

$$H(\mathfrak{n}_P; E)_w \cong H(\mathfrak{n}_P^Q; H(\mathfrak{n}_Q; E)_{w_Q})_{w^Q}.$$

So the bidegree of $H(\mathfrak{n}_P; E)_w$ in (54) is $(\ell^Q(w), \ell_Q(w)) \equiv (\ell(w^Q), \ell(w_Q))$.

Recall that $\ell(w)$ may be interpreted as the number of positive $\mathbb{C}$-roots which w^{-1} sends to negative roots. For $w \in W_P$, all such roots occur in $\mathfrak{n}_{P\mathbb{C}}$; the bidegree $\ell_Q(w)$ is the number of those roots occurring in $\mathfrak{n}_{Q\mathbb{C}}$ and the bidegree $\ell^Q(w)$ is the number of roots occurring in (a lift of) $\mathfrak{n}_{P\mathbb{C}}^Q$. In Figure 22 this is illustrated for two examples in the case $G = \mathrm{GL}_n(\mathbb{R})$, where positive roots may be identified with matrix positions above the diagonal. As one compares the various $\ell_Q(w)$, pictures such as these give helpful insight even for other groups, however one must be aware of the implicit assumption in the pictures that the Dynkin diagram of the root system is linear.

18.3. Parabolic rank 2. Now assume $r(P) = 2$ and write $\Delta_P = \{\alpha_1, \alpha_2\}$ with $\Delta_P^{Q_i} = \{\alpha_i\}$ for $i = 1, 2$. The intersection cohomology $\mathcal{L}$-module is illustrated in Figure 15 of §12.8. In order to find a complex computing the link intersection cohomology at P, we start with $H(\mathfrak{n}_P; E)$ and truncate it via a mapping cone in bidegree $\ell_{Q_1} > p(Q_1)$ and independently in bidegree $\ell_{Q_2} > p(Q_2)$. This yields the complex (40). We find that $H(\mathfrak{n}_P; E)_w$ contributes a nonvanishing component to the link cohomology if neither truncation applies or if both truncations apply (in which case its degree is shifted up by 1):

$$\tau^{\ell_{Q_1} \leq p(Q_1)} \tau^{\ell_{Q_2} \leq p(Q_2)} H(\mathfrak{n}_P; E) \oplus \tau^{\ell_{Q_1} > p(Q_1)} \tau^{\ell_{Q_2} > p(Q_2)} H(\mathfrak{n}_P; E)[-1]. \tag{55}$$

See §18.4 below for a further discussion.

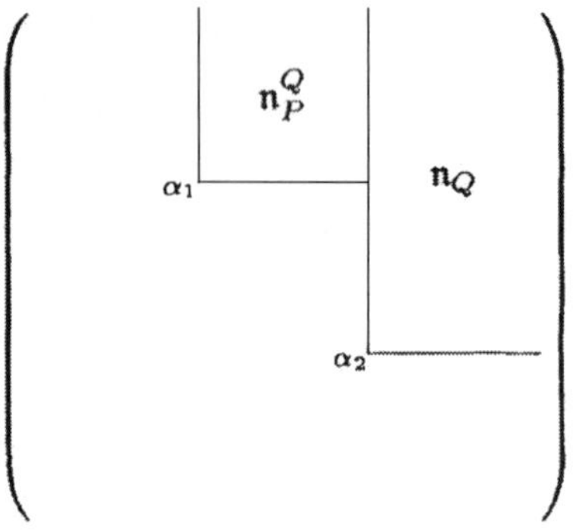

(a) $\Delta_P = \{\alpha_1, \alpha_2\}$, $\Delta_P^Q = \{\alpha_1\}$

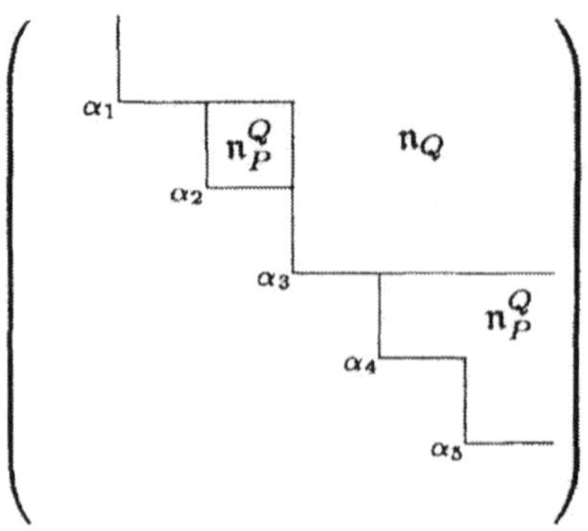

(b) $\Delta_P = \{\alpha_1, \dots, \alpha_5\}$, $\Delta_P^Q = \{\alpha_2, \alpha_4, \alpha_5\}$

FIGURE 22. Unipotent radicals for various $P \leq Q$ in $\mathrm{GL}_n(\mathbb{R})$

This enables us to calculate $H(i_P^* \hat{i}_Q^! \mathcal{I}_p\mathcal{C}(\widehat{X}; E))$ if $Q = G$ or $Q = P$. For $Q = G$ we obtain the local intersection cohomology at P,

$$\tau^{\leq p(P)}\big(\tau^{\ell_{Q_1} \leq p(Q_1)} \tau^{\ell_{Q_2} \leq p(Q_2)} H(\mathfrak{n}_P; E) \oplus \tau^{\ell_{Q_1} > p(Q_1)} \tau^{\ell_{Q_2} > p(Q_2)} H(\mathfrak{n}_P; E)[-1]\big).$$

For $Q = P$ we obtain the local intersection cohomology supported at P,

$$\tau^{> p(P)}\big(\tau^{\ell_{Q_1} \leq p(Q_1)} \tau^{\ell_{Q_2} \leq p(Q_2)} H(\mathfrak{n}_P; E) \oplus \tau^{\ell_{Q_1} > p(Q_1)} \tau^{\ell_{Q_2} > p(Q_2)} H(\mathfrak{n}_P; E)[-1]\big)[-1].$$

In order to calculate micro-support, we need to gather some calculations and a generalization of Proposition 18.2. First, given that we are working with a middle perversity, we have

$$p(P) = \begin{cases} \lfloor (\dim \mathfrak{n}_P)/2 \rfloor & p = m, \\ \lfloor (\dim \mathfrak{n}_P + 1)/2 \rfloor & p = n, \end{cases} \quad \text{and} \quad p(Q_i) = \begin{cases} \lfloor (\dim \mathfrak{n}_{Q_i} - 1)/2 \rfloor & p = m, \\ \lfloor (\dim \mathfrak{n}_{Q_i})/2 \rfloor & p = n. \end{cases}$$

Next it is helpful to note that if k and l are integers, then

$$l \leq \lfloor k/2 \rfloor \iff l \leq k/2 \qquad \text{and} \qquad l > \lfloor k/2 \rfloor \iff l > k/2;$$

the analogous formulas for $<$ and $\geq$ do not hold. Finally the generalization of Proposition 18.2 is:

PROPOSITION 18.3. *Let E be an irreducible G-module with highest weight λ. Let P be a parabolic $\mathbb{Q}$-subgroup and let $w \in W_P$. Assume that $(V_{w(\lambda+\rho)-\rho}|_{M_P})^* \cong \overline{V_{w(\lambda+\rho)-\rho}|_{M_P}}$.*

(i) *If $(w(\lambda + \rho), \alpha) \leq 0$ for all $\alpha \in \Delta_P$, then $\ell_Q(w) \geq (\dim \mathfrak{n}_Q)/2$ for all $Q \geq P$.*

(ii) *If $(w(\lambda + \rho), \alpha) \geq 0$ for all $\alpha \in \Delta_P$, then $\ell_Q(w) \leq (\dim \mathfrak{n}_Q)/2$ for all $Q \geq P$.*

Furthermore assume that $(E|_{{}^\circ G})^* \cong \overline{E|_{{}^\circ G}}$ and that the hypotheses of (i) *or* (ii) *hold. Then*

$$\ell(w) = (\dim \mathfrak{n}_P)/2 \implies (w(\lambda + \rho), \alpha) = 0 \text{ for all } \alpha \in \Delta_P.$$

SKETCH OF PROOF. For every A_P-weight α in $\mathfrak{n}_Q$ (not necessarily simple), the argument we gave for Lemma 15.2 may be applied to the $\mathbb{C}$-roots γ for which

$\gamma|_{\mathfrak{a}_P} = \alpha$. The result is an estimate on the contribution to $\ell_Q(w)$ from such roots. The sum of these estimates over all A_P-weights in $\mathfrak{n}_Q$ yields the proposition. The last assertion follows as in Proposition 18.2. □

Consider $V = V_{w(\lambda+\rho)-\rho}$ for some $w \in W_P$ and assume that $(V|_{M_P})^* \cong \overline{V|_{M_P}}$. In the case $Q_V = P$, we have $(w(\lambda+\rho), \alpha) \geq 0$ for all $\alpha \in \Delta_P$. It follows from the proposition and the preceding calculations that $H(i_P^* \hat{\imath}_{Q_V}^! \mathcal{I}_p\mathcal{C}(\widehat{X}; E))_V$ is 0 unless $\ell(w) = (\dim \mathfrak{n}_P)/2$, in which case $(w(\lambda+\rho), \alpha) = 0$ for all $\alpha \in \Delta_P$ and thus $Q'_V = G$. Conversely if $Q'_V = G$ then $(w(\lambda+\rho), \alpha) \leq 0$ for all $\alpha \in \Delta_P$ and it follows that either $H(i_P^* \hat{\imath}_{Q'_V}^! \mathcal{I}_p\mathcal{C}(\widehat{X}; E))_V = 0$ or $\ell(w) = (\dim \mathfrak{n}_P)/2$ and $Q_V = P$.

Thus for any V which satisfies Q_V or Q'_V equal to either P or G and which can conceivably contribute to micro-support, we have $Q_V = P$, $Q'_V = G$, $\ell(w) = (\dim \mathfrak{n}_P)/2$, and

$$H(i_P^* \hat{\imath}_{Q_V}^! \mathcal{I}_p\mathcal{C}(\widehat{X}; E))_V = \begin{cases} V[-\ell(w)-2] & \text{if } p = m, \\ 0 & \text{if } p = n. \end{cases} \tag{56}$$

$$H(i_P^* \hat{\imath}_{Q'_V}^! \mathcal{I}_p\mathcal{C}(\widehat{X}; E))_V = \begin{cases} 0 & \text{if } p = m, \\ V[-\ell(w)] & \text{if } p = n. \end{cases} \tag{57}$$

Such V belong to $\mathrm{SS}(\mathcal{I}_p\mathcal{C}(\widehat{X}; E))$ but obviously do not contribute to the essential micro-support; we call them ***fundamental*** since they occur if and only if P contains a fundamental parabolic $\mathbb{R}$-subgroup of G [**13**], [**61**, Lemma 8.8]. (Recall that a parabolic $\mathbb{R}$-subgroup is ***fundamental*** if a fundamental Cartan subalgebra for its Levi quotient lifts to a fundamental Cartan subalgebra for G.)

Unlike the parabolic rank 1 case, there are two more cases of $H(i_P^* \hat{\imath}_Q^! \mathcal{I}_p\mathcal{C}(\widehat{X}; E))$ to consider, namely when $Q = Q_1$ or $Q = Q_2$. Since the two cases are identical after relabeling, we focus on $Q = Q_1$. Use the complex $\hat{\imath}_P^* \hat{\imath}_Q^! \mathcal{M}$ of (32) in §12.3 applied to the $\mathcal{L}$-module $\mathcal{M} = \mathcal{I}_p\mathcal{C}(\widehat{X}; E)$ pictured in Figure 15. We compute that

$$\begin{aligned} H(i_P^* \hat{\imath}_{Q_1}^! \mathcal{I}_p\mathcal{C}(\widehat{X}; E)) =& \tau^{\ell_{Q_1} > p(Q_1)} \tau^{\ell_{Q_2} \leq p(Q_2)} H(\mathfrak{n}_P; E)[-1] \oplus \\ & \tau^{\leq p(P)} \big(\tau^{\ell_{Q_1} > p(Q_1)} \tau^{\ell_{Q_2} > p(Q_2)} H(\mathfrak{n}_P; E)[-1]\big) \oplus \\ & \tau^{> p(P)} \tau^{\ell_{Q_1} \leq p(Q_1)} \tau^{\ell_{Q_2} \leq p(Q_2)} H(\mathfrak{n}_P; E)[-1]. \end{aligned}$$

The first term are those classes in $\tau^{\ell_{Q_1} > p(Q_1)} H(\mathfrak{n}_P; E)[-1]$ which map to 0 in the link cohomology, the next term are those classes which map to nonzero elements in the link cohomology which are not truncated at P, and the third term are those link cohomology classes being truncated at P which do not come from $\tau^{\ell_{Q_1} > p(Q_1)} H(\mathfrak{n}_P; E)[-1]$.

We can now handle V such that $Q_V = Q'_V = Q_1$. In this case

$$H(i_P^* \hat{\imath}_{Q_1}^! \mathcal{I}_p\mathcal{C}(\widehat{X}; E))_V = 0$$

by the proposition below (which is actually slightly stronger than necessary). Consequently V cannot be in $\mathrm{SS}(\mathcal{I}_p\mathcal{C}(\widehat{X}; E))$, much less the essential micro-support. The case where $Q_V = Q'_V = Q_2$ is treated similarly.

PROPOSITION 18.4. *Let E be an irreducible G-module with highest weight λ. Let P be a parabolic $\mathbb{Q}$-subgroup with $\Delta_P = \{\alpha_1, \alpha_2\}$ and let $w \in W_P$. Assume that*

$(V_{w(\lambda+\rho)-\rho}|_{M_P})^* \cong \overline{V_{w(\lambda+\rho)-\rho}|_{M_P}}$ *and that*

$$(w(\lambda+\rho),\alpha_1) \le 0 \quad \textit{and} \quad (w(\lambda+\rho),\alpha_2) \ge 0.$$

(i) *If* $\ell_{Q_1}(w) \ge (\dim \mathfrak{n}_{Q_1})/2$ *then* $\ell(w) \ge (\dim \mathfrak{n}_P)/2$.
(ii) *If* $\ell_{Q_2}(w) \le (\dim \mathfrak{n}_{Q_2})/2$ *then* $\ell(w) \le (\dim \mathfrak{n}_P)/2$.

If either hypothesis is a strict inequality, the corresponding conclusion is also strict.

SKETCH OF PROOF. Apply Proposition 18.2 to the group L_{Q_1} with the representation $E_{Q_1} = H^{\ell_{Q_1}(w)}(\mathfrak{n}_{Q_1}; E)_{w_{Q_1}}$, the maximal parabolic subgroup P/N_{Q_1}, and the Weyl element $w^{Q_1} \in W_P^{Q_1}$. If μ is the highest weight of E_{Q_1}, the representation $H^{\ell^{Q_1}(w)}(\mathfrak{n}_P^{Q_1}; E_{Q_1})_{w^{Q_1}} = V_{w^{Q_1}(\mu+\rho)-\rho}$ is identical with $V_{w(\lambda+\rho)-\rho}$ and hence satisfies the conjugate self-contragredient condition. It follows from the proposition that $\ell^{Q_1}(w) \ge (\dim \mathfrak{n}_P^{Q_1})/2$ which proves (i) since $\ell(w) = \ell_{Q_1}(w) + \ell^{Q_1}(w)$. Part (ii) follows similarly. □

REMARK. Whereas the conjugate self-contragredience of $H(\mathfrak{n}_P; E)_w$ trivially implied the conjugate self-contragredience of $H(\mathfrak{n}_P^{Q_1}; E_{Q_1})_{w^{Q_1}}$ in the proof above, it does not in general imply the conjugate self-contragredience of $H(\mathfrak{n}_{Q_1}; E)_{w_{Q_1}}$. As we will indicate in §18.5 below, this is a fundamental difficulty in formulating a proof that applies to arbitrary parabolic rank. The natural approach would be to obtain information regarding the cohomology of the link by applying the theorem inductively to larger parabolic subgroups; this fails since the self-contragredience condition is not preserved under the induction which replaces w by w_R for $R \ge P$.

18.4. Geometric interpretation of link cohomology. Although the calculation of link intersection cohomology preceding (55) is trivial, it is helpful when considering higher parabolic rank to view it geometrically. Thus we digress to indicate this point of view. Recall that $|\Delta_P|$ denotes the closed $(r(P)-1)$-simplex with vertices $\alpha \in \Delta_P$; this is a stratified space with strata $|\Delta_P^Q|^\circ$ indexed by $Q > P$. For $w \in W_P$, define the *w-shifted perversity*

$$p_w(Q) \equiv p(Q) - \ell_Q(w) = p(\dim \mathfrak{n}_Q + \#\Delta_Q) - \ell_Q(w), \quad Q \ge P;$$

we view this as a generalized perversity for the space $|\Delta_P|$ and so the intersection cohomology $I_{p_w}H(|\Delta_P|;\mathbb{Z})$ is defined. The $H(\mathfrak{n}_P;E)_w$-isotypical component of the complex $i_P^* j_{P*} j_P^* \mathcal{I}_p\mathcal{C}(\widehat{X};E)$ computing link intersection cohomology can be expressed as

$$H(\mathfrak{n}_P;E)_w \otimes I_{p_w}C(|\Delta_P|;\mathbb{Z});$$

we take this as defining $I_{p_w}C(|\Delta_P|;\mathbb{Z})$. The corresponding isotypical component of the link intersection cohomology is

$$H(\mathfrak{n}_P;E)_w \otimes I_{p_w}H(|\Delta_P|;\mathbb{Z}). \tag{58}$$

The isotypical components of $H(i_P^*\mathcal{I}_p\mathcal{C}(\widehat{X};E))$, the local intersection cohomology, are expressed similarly using $I_{p_w}H(c(|\Delta_P|);\mathbb{Z})$. Here $c(|\Delta_P|)$ is the cone on $|\Delta_P|$, a stratified space with strata $c(|\Delta_P^Q|)^\circ$ indexed by $Q \ge P$. (The stratum associated to $Q = P$ is the vertex of the cone.)

For example, in the case of parabolic rank two, $|\Delta_P|$ is a closed 1-simplex with vertices $\{\alpha_1,\alpha_2\}$. Let $B_w \subseteq \{\alpha_1,\alpha_2\}$ denote the set of vertices where a degree 0

$I_{p_w}C(|\Delta_P|;\mathbb{Z})$:

B_w:	$\emptyset$	$\{\alpha_1\}$	$\{\alpha_2\}$	$\{\alpha_1, \alpha_2\}$
$I_{p_w}H(|\Delta_P|;\mathbb{Z})$:	$\mathbb{Z}$	0	0	$\mathbb{Z}[-1]$

FIGURE 23. The intersection cohomology $I_{p_w}H(|\Delta_P|;\mathbb{Z})$ in parabolic rank 2

class on the interior $|\Delta_P|^\circ$ would be truncated according to p_w, that is,

$$B_w \equiv \begin{cases} \emptyset & \text{if } \ell_{Q_1}(w) \leq p(Q_1),\ \ell_{Q_2}(w) \leq p(Q_2), \\ \{\alpha_1\} & \text{if } \ell_{Q_1}(w) > p(Q_1),\ \ell_{Q_2}(w) \leq p(Q_2), \\ \{\alpha_2\} & \text{if } \ell_{Q_1}(w) \leq p(Q_1),\ \ell_{Q_2}(w) > p(Q_2), \\ \{\alpha_1, \alpha_2\} & \text{if } \ell_{Q_1}(w) > p(Q_1),\ \ell_{Q_2}(w) > p(Q_2). \end{cases}$$

Then $I_{p_w}C(|\Delta_P|;\mathbb{Z})$ is the complex $\mathbb{Z} \to \mathbb{Z}^{\#B_w}$ (compare (40) in §12.8).

We graphically represent $I_{p_w}C(|\Delta_P|;\mathbb{Z})$ in Figure 23 along with the value of $I_{p_w}H(|\Delta_P|;\mathbb{Z})$ by placing a dot near α_i if truncation is taking place there. The corresponding diagrams for parabolic rank 3 are given in Figure 24; here $|\Delta_P|$ is a 2-simplex and we draw a line near an edge if truncation of the degree 0 class on the interior is occurring, and we draw a dot near a vertex if the degree 0 or 1 class on its link is being truncated.

Properly speaking, we should decorate the dots and lines in these figures with the degree $p_w(Q)$ above which we truncate. However the cohomology potentially being truncated occurs only in one degree (namely 0 or 1), so simply indicating whether or not it is truncated is sufficient. The same shortcut can be taken in parabolic rank 4, but can fail beginning with parabolic rank 5. The point is that if P has parabolic rank 5, a vertex in $|\Delta_P|$ corresponds to a subgroup of parabolic rank 4 and, as we will indicate in the next paragraph, the isotypical components of link intersection cohomology at a parabolic rank 4 subgroup can live in several degrees.

Incidentally, the figures show that in both parabolic rank 2 and 3, the isotypical components (58) of link intersection cohomology (55) occur in a single degree. This phenomenon does not persist however; it can fail for parabolic rank ≥ 4. For example, if P has parabolic rank 4 and thus $|\Delta_P|$ is a 3-dimensional simplex, there are two configurations that lead to cohomology in more one degree: one can have truncation at all 4 faces as well as at 3 non-coplanar edges or dually one can have truncation at all 4 vertices and at 3 edges bounding a face. The resulting complex in the first case is $0 \to \mathbb{Z} \to \mathbb{Z}^4 \to \mathbb{Z}^3 \to \mathbb{Z}$ which yields cohomology in degrees 1 and 2. These configurations can actually arise from Weyl group elements.[3]

[3]Among the real symplectic groups one must wait until $G = \mathrm{Sp}_{20}(\mathbb{R})$ to see them. For this group, the parabolic P with $\Delta^P = \{\alpha_2, \alpha_4, \alpha_5, \alpha_7, \alpha_8, \alpha_9\}$ has three elements $w \in W_P$ which yield the first configuration above; one of them is

$$w = s_3 s_2 s_1 s_6 s_5 s_4 s_3 s_2 s_7 s_6 s_5 s_4 s_3 s_8 s_{10} s_9 s_8 s_7 s_6 s_5 s_4$$
$$s_{10} s_9 s_8 s_7 s_6 s_5 s_{10} s_9 s_8 s_7 s_6 s_{10} s_9 s_8 s_7 s_{10} s_9 s_8 s_{10} s_9 s_{10}.$$

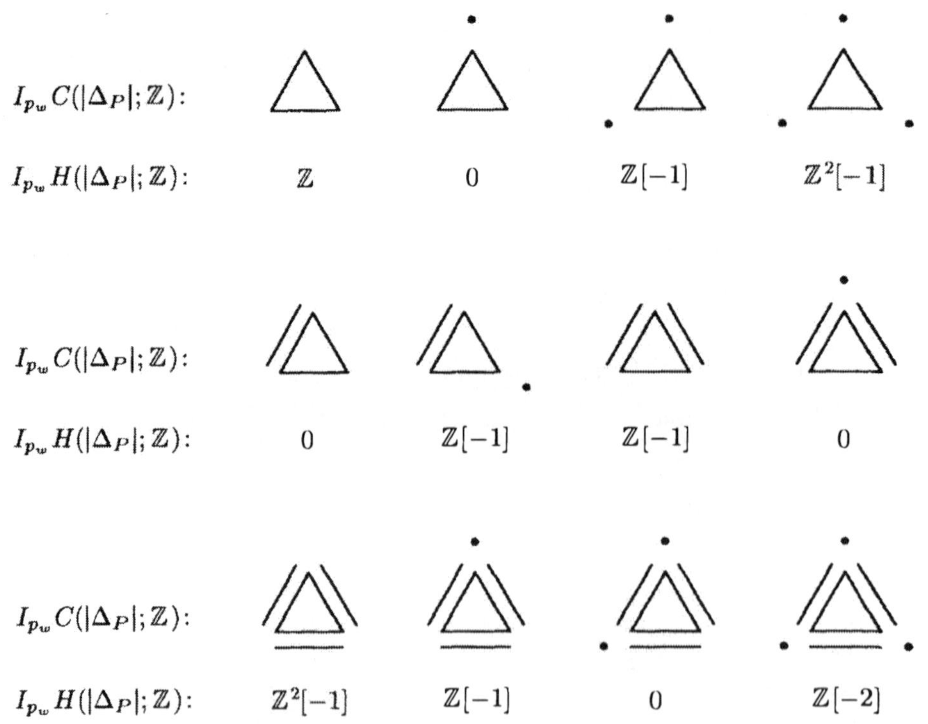

FIGURE 24. The intersection cohomology $I_{p_w}H(|\Delta_P|;\mathbb{Z})$ in parabolic rank 3. (Configurations differing from previous ones by a rotation are omitted.)

Finally, note that in all the illustrated examples, the intersection cohomology groups $I_{p_w}H(|\Delta_P|;\mathbb{Z})$ can be viewed simply as relative cohomology groups $H(|\Delta_P|, C_w)$ for a certain subset $C_w \subseteq \partial|\Delta_P|$. Namely, C_w is the union of the boundary strata at which truncation is taking place. Again this interpretation continues to hold in parabolic rank 4 but fails (at least with the above definition of C_w) beginning with parabolic rank 5.

18.5. Parabolic rank $\geq$ 3. If Q_V or Q'_V is equal to P or G, the argument already given using Proposition 18.3 applies to P with any parabolic rank to show that $V = H^{\ell(w)}(\mathfrak{n}_P; E)_w$ can only contribute to the micro-support when it is fundamental and can never contribute to the essential micro-support [**61**, Proposition 17.7].

It remains to consider V such that $P < Q_V \leq Q'_V < G$. In this case there is a direct though tedious proof for parabolic rank 3 that V is not in the micro-support; it uses Figure 24 and the same ideas as in the parabolic rank 2 case. Obviously this approach does not lend itself to a general argument; it requires a case-by-case calculation of $H(i_P^* \hat{\imath}_Q^! \mathcal{I}_p\mathcal{C}(\widehat{X}; E))_V$ for all the situations illustrated in Figure 24.

Unfortunately in this example, $\ell(w) = 42$, $\dim \mathfrak{n}_P = 90$, and hence $p_w(P) = 4$ or $4\frac{1}{2}$. This means that both of the link cohomology classes in degrees 1 and 2 will not be truncated. We expect that there are examples in higher rank where only part of the link cohomology is truncated.

We will now outline the ingredients for a general proof. In the case of parabolic rank 3, a complete proof can be constructed from what we present here; in general one must consult [**61**]. Recall that we wish to show that

$$H(i_P^* \hat{i}_Q^! \mathcal{I}_p \mathcal{C}(\widehat{X};E))_V = 0 \tag{59}$$

under the hypotheses

(i) $V = H^{\ell(w)}(\mathfrak{n}_P; E)_w$ is conjugate self-contragredient when restricted to M_P.

(ii) $P < Q_V \le Q \le Q'_V < G$.

We can re-express

$$H(i_P^* \hat{i}_Q^! \mathcal{I}_p \mathcal{C}(\widehat{X};E))_V \cong H(\mathfrak{n}_P;E)_w \otimes I_{p_w} H_{c(|\Delta_P^Q|)}$$

[**61**, Proposition 17.4], where we use the convenient shorthand

$$I_{p_w} H_{c(|\Delta_P^Q|)} \equiv I_{p_w} H\big(c(|\Delta_P|), c(|\Delta_P|) \setminus c(|\Delta_P^Q|); \mathbb{Z}\big)$$

for the generalized perversity intersection cohomology supported on $c(|\Delta_P^Q|)$. (We will henceforth omit the coefficients $\mathbb{Z}$ from the notation.) Thus (59) may be replaced by

$$I_{p_w} H_{c(|\Delta_P^Q|)} = 0. \tag{60}$$

We exploit the local calculation of intersection cohomology (8) at the vertex of the cone,

$$I_{p_w} H^j(c(|\Delta_P|)) = \begin{cases} I_{p_w} H^j(|\Delta_P|) & \text{for } j \le p_w(P), \\ 0 & \text{for } j > p_w(P). \end{cases} \tag{61}$$

From this and the exact sequence of the pair, one can show that (60) is equivalent to

$$I_{p_w} H^{j-1}(|\Delta_P| \setminus |\Delta_P^Q|) = 0 \qquad \text{for } j > p_w(P)+1, \tag{62a}$$

$$\operatorname{Im}\Big(I_{p_w} H^{p_w(P)}(|\Delta_P| \setminus |\Delta_P^Q|) \longrightarrow I_{p_w} H^{p_w(P)+1}_{|\Delta_P^Q|}\Big) = 0 \quad \text{for } j = p_w(P)+1, \tag{62b}$$

$$I_{p_w} H^j_{|\Delta_P^Q|} = 0 \qquad \text{for } j < p_w(P)+1. \tag{62c}$$

We will focus on proving (62a). It turns out that the argument also shows that the domain of the map in (62b) vanishes in half of the cases. Then a dual argument proves (62c) and also that the range of the map in (62b) vanishes in the other half of the cases.

There are two spectral sequences [**61**, Lemma 3.7] we use that abut to

$$I_{p_w} H^{j-1}(|\Delta_P| \setminus |\Delta_P^Q|). \tag{63}$$

First some notation. Let $S > P$ be the parabolic $\mathbb{Q}$-subgroup *complementary* to Q relative to P, that is, such that $\Delta_P^S = \Delta_P \setminus \Delta_P^Q$. For each $\alpha \in \Delta_P$, let U_α be the open star neighborhood of the corresponding vertex of $|\Delta_P|$; note that $U_\alpha = \bigcup_{\alpha\in\Delta_P^R \subseteq \Delta_P} |\Delta_P^R|^\circ$. Set $U_R \equiv \bigcap_{\alpha\in\Delta_P^R} U_\alpha$ for $R > P$, the open star neighborhood of $|\Delta_P^R|^\circ$, and define a parabolic $\mathbb{Q}$-subgroup $Q \vee R$ by $\Delta_P^{Q\vee R} = \Delta_P^Q \cup \Delta_P^R$. The

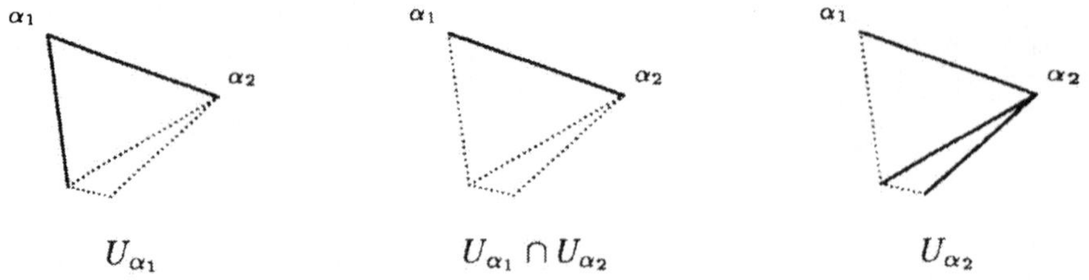

FIGURE 25. The Mayer-Vietoris spectral sequence for $I_{p_w}H(|\Delta_P| \setminus |\Delta_P^Q|)$, where $\Delta_P^Q = \{\alpha_3, \alpha_4\}$

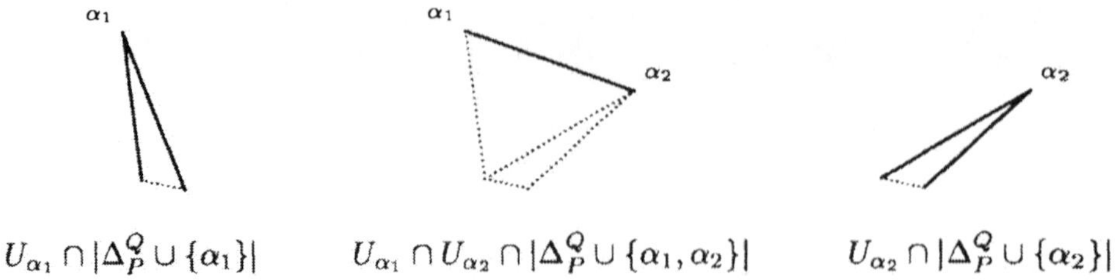

FIGURE 26. The Fary spectral sequence for $I_{p_w}H(|\Delta_P| \setminus |\Delta_P^Q|)$, where $\Delta_P^Q = \{\alpha_3, \alpha_4\}$

Mayer-Vietoris spectral sequence for the open cover $\{U_\alpha\}_{\alpha \in \Delta_P^S}$ of $|\Delta_P| \setminus |\Delta_P^Q|$ has[4]

$$E_1^{p,j-1-p} = \bigoplus_{\substack{P<R\leq S \\ \#\Delta_P^R = p+1}} I_{p_w}H^{j-1-p}(U_R) \cong \bigoplus_{\substack{P<R\leq S \\ \#\Delta_P^R = p+1}} I_{p_w}H^{j-1-p}(c(|\Delta_R|)). \tag{64}$$

See Figure 25. On the other hand, the projection from $|\Delta_P^Q|$ defines a fibration $|\Delta_P| \setminus |\Delta_P^Q| \to |\Delta_P^S|$. For $P < R \leq S$, the inverse image of the stratum $|\Delta_P^R|^\circ \subseteq |\Delta_P^S|$ is $U_R \cap |\Delta_P^{Q\vee R}| \subseteq |\Delta_P| \setminus |\Delta_P^Q|$. The associated *Fary spectral sequence* has

$$E_1^{-p,j-1+p} = \bigoplus_{\substack{P<R\leq S \\ \#\Delta_P^R = p}} I_{p_w}H^{j-1}_{U_R \cap |\Delta_P^{Q\vee R}|} \cong \bigoplus_{\substack{P<R\leq S \\ \#\Delta_P^R = p}} I_{p_w}H^{j-1}_{c(|\Delta_R^{Q\vee R}|)}. \tag{65}$$

See Figure 26.

The term of (64) associated to $R > P$ vanishes by the truncation condition (61) at the vertex when $j-1-p > p_w(R)$, that is, when

$$j > p_w(R) + \#\Delta_P^R. \tag{66}$$

Since p is a middle perversity, it is easy to verify that

$$p_w(P) = p_w(R) + \lfloor(\dim \mathfrak{n}_P^R + \#\Delta_P^R)/2\rfloor - \ell^R(w) + \delta,$$

where $\delta = 0$ or 1 depending on the choice of p and the parity of $(\dim \mathfrak{n}_P^R + \#\Delta_P^R)$ and $(\dim \mathfrak{n}_P + \#\Delta_P)$. However we know by Proposition 18.3 (applied to w^R and

[4]We are using p both as an index and a perversity; this should not cause confusion.

the parabolic P/N_R of L_R) that

$$\ell^R(w) \leq (\dim \mathfrak{n}_P^R)/2. \tag{67}$$

Thus the Mayer-Vietoris spectral sequence yields the vanishing of (63) for

$$j > p_w(P) + \lceil \#\Delta_P^R/2 \rceil - \delta. \tag{68}$$

Assume that $r(P) = 3$. Then $\#\Delta_P^R \leq 2$ and thus (63) vanishes for $j > p_w(P)+1$ and even for $j = p_w(P) + 1$ if $\delta = 1$. Thus the Mayer-Vietoris spectral sequence argument yields a complete proof for the parabolic rank 3 case.

For $r(P) \geq 4$ the argument leading to (68) is not sufficient. It would be sufficient, however, if (67) were far enough away from equality. It could also be improved if we had more precise knowledge about $I_{p_w} H(|\Delta_R|)$, which suggests the utility of an inductive argument. On the other hand, the Fary spectral sequence explicitly calls out for an inductive argument. In fact, if we could apply (60) by induction, with P replaced by R, w by w_R, and Q by $Q \vee R$, all the terms in (65) would vanish. Of course this is not possible (and would lead to too strong a result) since the hypotheses (i) and (ii) are not preserved under such an induction.

The approach in [**61**] is to replace (i) and (ii) by other hypotheses that are more suited to induction and use a combination of the two spectral sequences. Recall that (i) and (ii) were used as input to Proposition 18.3; more precisely they imply that for all A_P-weights α occurring in $\mathfrak{n}_P$,

$$\begin{cases} \ell_\alpha(w) \geq (\dim \mathfrak{n}_\alpha)/2 & \text{if } \alpha \text{ occurs in } \mathfrak{n}_P^Q, \\ \ell_\alpha(w) \leq (\dim \mathfrak{n}_\alpha)/2 & \text{if } \alpha \text{ occurs in } \mathfrak{n}_P^S. \end{cases} \tag{69}$$

Here $\dim \mathfrak{n}_\alpha$ is the number of $\mathbb{C}$-roots γ such that $\gamma|_{\mathfrak{a}_P} = \alpha$ and $\ell_\alpha(w)$ is the number of such roots which in addition satisfy $w^{-1}\gamma < 0$. The conditions (69) are still not preserved by induction from P to R. However we can prove [**61**, Lemma 17.8][5] that there exists a parabolic $\mathbb{Q}$-subgroup $T \geq P$ (depending on w) with $r(T) = 1$ or 2 and $r(Q \vee T)$, $r(S \vee T) \leq 1$ such that for all A_P-weights α occurring in $\mathfrak{n}_P$,

$$\begin{cases} \ell_\alpha(w) \geq (\dim \mathfrak{n}_\alpha)/2 & \text{if } \alpha \text{ occurs in } \mathfrak{n}_T^{Q\vee T}, \\ \ell_\alpha(w) \leq (\dim \mathfrak{n}_\alpha)/2 & \text{if } \alpha \text{ occurs in } \mathfrak{n}_T^{S\vee T}. \end{cases} \tag{70}$$

We illustrate the difference between conditions (69) and (70) in Figure 27 (here $P \leq Q$ are as in Figure 22(b) and only one possible value of T is considered). It is clear from the figure that significant information is gained in passing from (69) to (70) even though some information is lost as well. Clearly the conditions (70) are preserved by induction from P to R provided $R \leq T$.

The proof of (62a)–(62c), and hence $I_{p_w} H_{c(|\Delta_P^Q|)} = 0$, in [**61**, Theorem 17.9] uses the existence of T satisfying (70) (plus one additional hypothesis). The proof of (62a) is separated into two parts. The vanishing of $I_{p_w} H^{j-1}_{|\Delta_P^{Q\vee T}|\setminus|\Delta_P^Q|}$, is proved using the Fary spectral sequence and induction from P to $R \leq T$. The other part,

[5]This is where the condition on the $\mathbb{Q}$-root system in Theorem 18.1 comes into play. Recall that a ***quasi-minuscule representation*** is one in which the non-zero weights form a single Weyl group orbit. The root systems of type A_n, B_n, C_n, and G_2 are precisely the irreducible root systems such that the corresponding simply connected $\mathbb{Q}$-split group has a quasi-minuscule representation whose weights under the usual ordering are totally ordered; thus our hypothesis allows us to express the $\mathbb{Q}$-roots of G as differences of weights $\varepsilon_i - \varepsilon_j$ analogously to the usual expression of roots in GL_n. We expect that a further study of quasi-minuscule representations will allow us to treat the cases D_n, E_n, and F_4, although we may have to allow $r(T) = 3$ or 4.

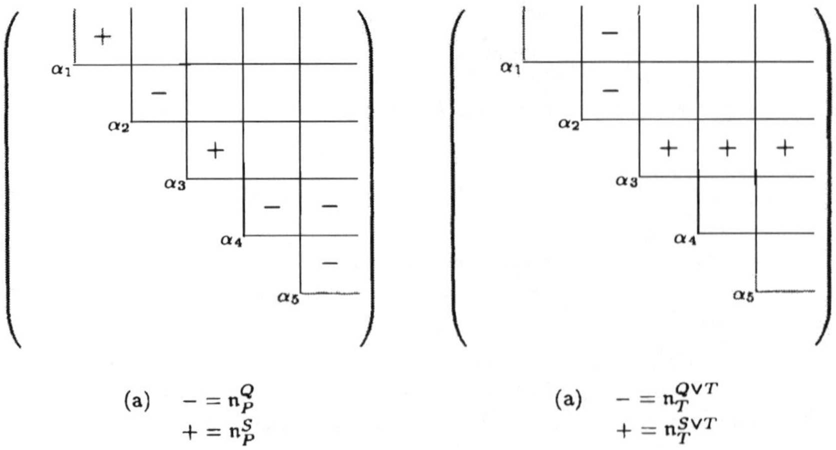

FIGURE 27. An example of $\mathfrak{n}_P^Q$ versus $\mathfrak{n}_T^{Q\vee T}$ and $\mathfrak{n}_P^S$ versus $\mathfrak{n}_T^{S\vee T}$ in $GL_n(\mathbb{R})$. Here $\Delta_P = \{\alpha_1,\dots,\alpha_5\}$, $\Delta_P^Q = \{\alpha_2,\alpha_4,\alpha_5\}$, and $\Delta_P^T = \{\alpha_1,\alpha_4,\alpha_5\}$.

the vanishing of $I_{p_w}H^{j-1}(|\Delta_P|\setminus|\Delta_P^{Q\vee T}|)$, is proved with the Mayer-Vietoris spectral sequence (note that $\#\Delta_P^R \leq 1$ in this case).

19. Functoriality of micro-support

It is important to understand how the micro-support of an $\mathcal{L}$-module $\mathcal{M}$ changes when the various functors k_*, k^*, and $k^!$ are applied to $\mathcal{M}$. An example of this for the case of a direct image functor has already been illustrated in §13.2 with the calculation of $SS(i_{G*}E)$. It remains to consider inverse image functors. Specifically one would like understand how $SS(k^*\mathcal{M})$, for example, is derived from $SS(\mathcal{M})$. Furthermore, in view of Theorem 14.1, it is important also to estimate how $c(k^*\mathcal{M})$ and $d(k^*\mathcal{M})$ relate to the corresponding quantities for $\mathcal{M}$. A number of general theorems of this sort are found in [**61**]; here we will instead simply illustrate these ideas by sketching the proof of the following result needed for Theorem 11.1 and illustrated in Figure 28. The complete proof is found in [**61**, Corollary 26.2].

THEOREM 19.1. *Let $\mathcal{M}$ be an $\mathcal{L}$-module on $\widehat{X}$ with $SS_{ess}(\mathcal{M}) = \{E\}$ and $c(E;\mathcal{M}) = d(E;\mathcal{M}) = 0$. Let $x \in X_{R,h}$ be a point on a stratum of a real equal-rank Satake compactification X^*. Let $\pi\colon \widehat{X} \to X^*$ be Zucker's projection and let $k\colon \pi^{-1}(x) \cong \widehat{X}_{R,\ell} \hookrightarrow \widehat{X}$ denote the inclusion. Then*

$$d(k^*\mathcal{M}) \leq \tfrac{1}{2}\operatorname{codim} X_{R,h} - \#\Delta_R \qquad \text{and} \qquad c(k^!\mathcal{M}) \geq \tfrac{1}{2}\operatorname{codim} X_{R,h} + \#\Delta_R.$$

We will make several simplifying assumptions along the way in order to highlight the basic idea. To begin with though, we do not make any assumption about $SS(\mathcal{M})$ nor about the equal-rank nature of the boundary components of X^*; these will be added later to emphasize where they are needed.

19.1. Consider the triple of subsets,

$$\{x\} \subseteq X_{R,h} \subseteq U_{R,h} \subseteq X^*,$$

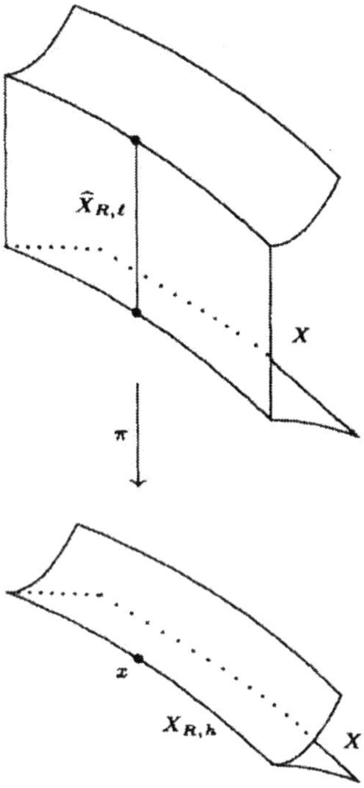

FIGURE 28. The projection $\pi\colon \widehat{X} \to X^*$ over a point $x \in X_{R,h}$ (adapted from [**62**])

where

$$U_{R,h} \equiv \coprod_{\mathrm{cl}(X_{S,h}) \supseteq X_{R,h}} X_{S,h}$$

is the open star neighborhood of $X_{R,h}$. The inclusion k may thus be factored into the composition of three inclusions,

$$\begin{array}{ccccccc} \pi^{-1}(x) & \xhookrightarrow{i_{R,\ell}} & \pi^{-1}(X_{R,h}) & \xhookrightarrow{i_R} & \pi^{-1}(U_{R,h}) & \xhookrightarrow{j} & \widehat{X}\ , \\ & & \| & & \||| & & \\ & & \widehat{X}_R \cap W & & W & & \end{array}$$

and we will examine (in reverse order) how each affects micro-support. For simplicity we will only consider k^*; the functor $k^!$ is treated similarly.

19.2. An open embedding. The map

$$j\colon W = \coprod_{\substack{P \le R \\ P^\dagger = R}} \coprod_{\substack{Q \ge P \\ Q \cap R = P}} X_Q \hookrightarrow \widehat{X}$$

is an embedding of an open admissible subspace. The effect on micro-support is simple [**61**, Proposition 22.2]:

$$\mathrm{SS}(j^*\mathcal{M}) = \mathrm{SS}(\mathcal{M}) \cap \Big(\coprod_{Q \in \mathcal{P}(W)} \mathfrak{Irr}(L_Q) \Big) \tag{71}$$

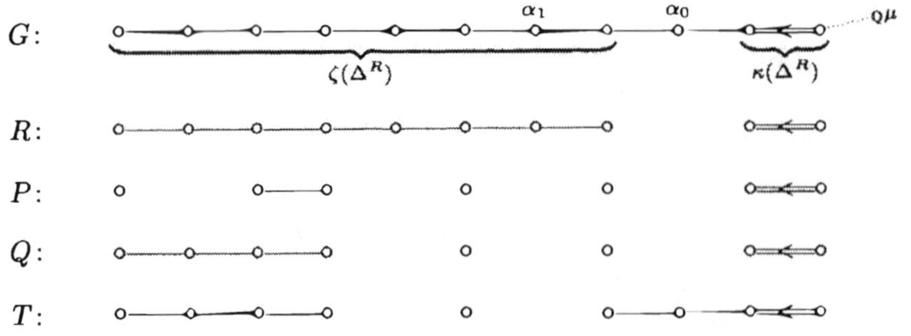

FIGURE 29. An example of the Dynkin diagrams for G and the Levi quotients of various parabolic $\mathbb{Q}$-subgroups considered in the text. (Here $G = \mathrm{Sp}_{22}(\mathbb{R})$ and X^* is the Baily-Borel-Satake compactification.)

and for $V \in \mathrm{SS}(\jmath^*\mathcal{M})$, $c(V;\jmath^*\mathcal{M}) = c(V;\mathcal{M})$ and $d(V;\jmath^*\mathcal{M}) = (V;\mathcal{M})$.

19.3. The inclusion of the closure of a stratum. The next map

$$\hat{\imath}_R \colon \widehat{X}_R \cap W = \coprod_{\substack{P \leq R \\ P^\dagger = R}} X_P \hookrightarrow W$$

is the main issue; it is the embedding of a (relatively) closed stratum. To treat it we outline a simplified version of the arguments of [**61**, Propositions 22.6 and 23.3]. Set $\mathcal{M}_1 = \jmath^*\mathcal{M}$ whose micro-support we understand by (71). Let $P \in \mathcal{P}(\widehat{X}_R \cap W)$. If an L_P-module V belongs to $\mathrm{SS}(\hat{\imath}_R^*\mathcal{M}_1)$, then $V|_{M_P}$ is conjugate self-contragredient and

$$H^j(i_P^*\hat{\imath}_Q^!(\hat{\imath}_R^*\mathcal{M}_1))_V \neq 0$$

for some $Q \leq R$ such that $Q_V^R \leq Q \leq Q_V'^R$ and some j. Note that $\hat{\imath}_Q^!\hat{\imath}_R^*\mathcal{M}_1 = \hat{\imath}_T^!\mathcal{M}_1$ where $T \geq Q$ is the unique parabolic $\mathbb{Q}$-subgroup complementary to R relative to Q, that is, such that $\Delta_Q^T = \Delta_Q \setminus \Delta_Q^R$. (See Figure 29 for an example of P, Q, R, and T.) Thus we have

$$H^j(i_P^*\hat{\imath}_T^!\mathcal{M}_1)_V \neq 0. \tag{72}$$

We will show that if V is not already in $\mathrm{SS}(\mathcal{M}_1)$, then it is explicitly related to an element $\tilde{V} \in \mathrm{SS}(\hat{\imath}_{\tilde{R}}^*\mathcal{M}_1)$, where $X_{\tilde{R},h}$ is a larger stratum in $U_{R,h}$. A repetition of the argument will then lead us to an element in $\mathrm{SS}(\mathcal{M}_1)$.

For simplicity now assume:

(i) The Dynkin diagram of the $\mathbb{Q}$-root system for G is linear.
(ii) The highest $\mathbb{Q}$-weight ${}_{\mathbb{Q}}\mu$ of the representation defining the Satake compactification is orthogonal to all simple $\mathbb{Q}$-roots except one at the end of the Dynkin diagram.
(iii) All rational boundary components are equal-rank.

These conditions are satisfied for practically all the real equal-rank Satake compactifications (including the Baily-Borel-Satake compactification) considered in Theorem 11.1, and in particular for the example illustrated in Figure 29. Assumptions (i) and (ii) are simply for ease of exposition; they ensure that

- All saturated parabolic $\mathbb{Q}$-subgroups R are maximal.

Assumption (iii) will be needed to ensure that $\tilde{V}$ below is actually conjugate self-contragredient.

Since R is maximal by the above assumptions, write $\Delta_P^R = \Delta_P \setminus \{\alpha_0\}$ and let $\tilde{P} > P$ have $\Delta_P^{\tilde{P}} = \{\alpha_0\}$. We need to understand how T relates to Q_V and Q'_V. By definition, $T = Q \vee \tilde{P}$, that is, $\Delta_P^T = \Delta_P^Q \cup \{\alpha_0\}$. For Q_V and Q'_V there are three cases:

(a) $(\xi_V + \rho, \alpha_0) < 0$ and thus $Q_V = Q_V^R \vee \tilde{P}$ and $Q'_V = Q_V'^R \vee \tilde{P}$.
(b) $(\xi_V + \rho, \alpha_0) = 0$ and thus $Q_V = Q_V^R$ and $Q'_V = Q_V'^R \vee \tilde{P}$.
(b) $(\xi_V + \rho, \alpha_0) > 0$ and thus $Q_V = Q_V^R$ and $Q'_V = Q_V'^R$.

In cases (a) and (b), $Q_V \leq T \leq Q'_V$ and thus equation (72) implies that $V \in \mathrm{SS}(\mathcal{M}_1)$. In case (c) on the other hand, $T \not\leq Q'_V$.

We investigate case (c) further. Consider the long exact sequence of the triple $(U, U \setminus (U \cap \widehat{X}_Q), U \setminus (U \cap \widehat{X}_T))$, where U is a small neighborhood of a point of X_P. The V-isotypical part of this sequence is

$$\cdots \longrightarrow H^j(i_P^* \hat{\imath}_Q^! \mathcal{M}_1)_V \longrightarrow H^j(i_P^* \hat{\imath}_T^! \mathcal{M}_1)_V \longrightarrow H^j(i_P^* i_{\tilde{P}*}(i_{\tilde{P}}^* \hat{\imath}_T^! \mathcal{M}_1))_V \longrightarrow \cdots .$$

By (72), the middle term is nonzero for some j, so either the first or last term must also be nonzero. If the first term is nonzero, again $V \in \mathrm{SS}(\mathcal{M}_1)$ since $Q_V \leq Q \leq Q'_V$. On the other hand, it is easy to see that the last term can be rewritten as $H^\ell(\mathfrak{n}_P^{\tilde{P}}; H^{j-\ell}(i_{\tilde{P}}^* \hat{\imath}_T^! \mathcal{M}_1))_V$, so if it is nonzero we must have

$$H^{j-\ell(w)}(i_{\tilde{P}}^* \hat{\imath}_T^! \mathcal{M}_1)_{\tilde{V}} \neq 0 \tag{73}$$

for some irreducible $L_{\tilde{P}}$-module $\tilde{V}$ such that $V = H^{\ell(w)}(\mathfrak{n}_P^{\tilde{P}}; \tilde{V})_w$ for some $w \in W_P^{\tilde{P}}$. It is not difficult to check [**61**, Lemma 23.2] that $\tilde{V}|_{M_{\tilde{P}}}$ is conjugate self-contragredient from (iii) and the fact that any representation of an equal-rank group is conjugate self-contragredient.

By assumptions (i) and (ii), the roots in Δ_P have a linear ordering induced from that of Δ and α_0 will occur at one end; this means that α_0 is orthogonal to all $\alpha \in \Delta_P \setminus \Delta_P^{\tilde{P}}$ except for one, say α_1. Set $\tilde{\alpha} = \alpha|_{A_{\tilde{P}}}$ for $\alpha \in \Delta_P \setminus \Delta_P^{\tilde{P}}$. One can show [**61**, Lemmas 22.5 and 23.1] that

$$(\xi_{\tilde{V}} + \rho, \tilde{\alpha}) \begin{cases} = (\xi_V + \rho, \alpha) & \text{for } \alpha \neq \alpha_1, \\ > (\xi_V + \rho, \alpha_1) & \text{for } \alpha = \alpha_1 \end{cases} \tag{74}$$

(the inequality is since we are in case (c)). Thus if $\alpha_1 \notin \Delta_P^Q$ (as in Figure 29), then $Q_{\tilde{V}} \leq T \leq Q'_{\tilde{V}}$ and hence (73) implies $\tilde{V} \in \mathrm{SS}(\mathcal{M}_1)$.

If $\alpha_1 \in \Delta_P^Q$, we want to apply to (73) the reverse of the reasoning that led to (72). Set $\tilde{R}$ to be the maximal $\mathbb{Q}$-parabolic subgroup with $\Delta_P^{\tilde{R}} = \Delta_P \setminus \{\alpha_1\}$ and set $\tilde{Q} \equiv \tilde{R} \cap T$ (see Figure 30). Then $\hat{\imath}_T^! \mathcal{M}_1 = \hat{\imath}_{\tilde{Q}}^! \hat{\imath}_{\tilde{R}}^* \mathcal{M}_1$ and equation (74) implies that

$$Q_{\tilde{V}}^{\tilde{R}} \leq \tilde{Q} \leq Q_{\tilde{V}}'^{\tilde{R}}.$$

Thus we have

$$H^{j-\ell(w)}(i_{\tilde{P}}^* \hat{\imath}_{\tilde{Q}}^! (\hat{\imath}_{\tilde{R}}^* \mathcal{M}_1))_{\tilde{V}} \neq 0$$

and hence $\tilde{V} \in \mathrm{SS}(\hat{\imath}_{\tilde{R}}^* \mathcal{M}_1)$. In this case we may repeat the argument with V replaced by $\tilde{V}$ to eventually obtain an element of $\mathrm{SS}(\mathcal{M}_1)$.

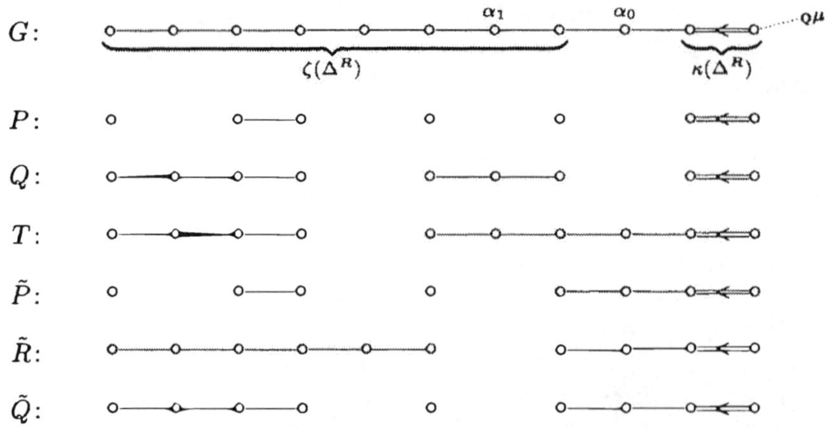

FIGURE 30. The case $\alpha_1 \in \Delta_P^Q$

In conclusion, we see that for every $V \in \mathrm{SS}(\hat{\imath}_R^* \mathcal{M}_1)$, either

(i) $V \in \mathrm{SS}(\mathcal{M}_1)$, or

(ii) $V = H^{\ell(w)}(\mathfrak{n}_P^{\tilde{P}}; \tilde{V})_w$ for an $L_{\tilde{P}}$-module $\tilde{V} \in \mathrm{SS}(\mathcal{M}_1)$ satisfying

$$\tilde{P} \cap R = P,$$

$$(\xi_V + \rho, \alpha) > 0 \text{ for all } \alpha \in \Delta_P^{\tilde{P}}, \text{ and}$$

$$d(V; \hat{\imath}_R^* \mathcal{M}_1) \le d(\tilde{V}; \mathcal{M}_1) + \ell(w),$$

or

(iii) V is obtained from an element of $\mathrm{SS}(\mathcal{M}_1)$ by a sequence of steps similar to (ii).

19.4. The inclusion of a fiber. The set $\widehat{X}_R \cap W = \pi^{-1}(X_{R,h})$ is a flat bundle over $X_{R,h}$ with typical fiber $\widehat{X}_{R,\ell}$. Our final map

$$\hat{\imath}_{R,\ell} \colon \pi^{-1}(x) \cong \widehat{X}_{R,\ell} \hookrightarrow \widehat{X}_R \cap W$$

is the inclusion of such a fiber. Thus $\widehat{X}_{R,\ell}$ is not an admissible subspace of $\widehat{X}_R \cap W$ and the functors $\hat{\imath}_{R,\ell}^*$ and $\hat{\imath}_{R,\ell}^!$ are defined differently from those in §12.3.

Instead, note that the spaces $\widehat{X}_{R,\ell}$ and $\widehat{X}_R \cap W$ have strata indexed by the same set: $P \in \mathcal{P}$ such that $P \le R$ with $P^\dagger = R$. For such P, the corresponding stratum of $\widehat{X}_{R,\ell}$ is $X_{P,\ell}$ while the corresponding stratum of $\widehat{X}_R \cap W$ is X_P. Let $\mathcal{M}_2 \equiv \hat{\imath}_R^* j^* \mathcal{M} = (E_\cdot, f_{\cdot\cdot})$ be an $\mathcal{L}$-module on $\widehat{X}_R \cap W$ (whose micro-support we understand by the preceding two subsections). The $\mathcal{L}$-module $\hat{\imath}_{R,\ell}^* \mathcal{M}_2 = (E'_\cdot, f'_{\cdot\cdot})$ on $\widehat{X}_{R,\ell}$ is defined [**61**, §3.5] by

$$E'_{P_\ell} \equiv \mathrm{Res}_{L_{P,\ell}}^{L_P} E_P,$$

$$f'_{P_\ell Q_\ell} \equiv \mathrm{Res}_{L_{P,\ell}}^{L_P} f_{PQ};$$

the $\mathcal{L}$-module $\hat{\imath}_{R,\ell}^! \mathcal{M}_2$ is defined similarly but with a degree shift of $-\dim D_{R,h}$.

The effect of $\hat{\imath}^*_{R,\ell}$ (and likewise $\hat{\imath}^!_{R,\ell}$) on micro-support is simple [**61**, Proposition 2.8]:

$$\mathrm{SS}(\hat{\imath}^*_{R,\ell}\mathcal{M}_2) = \{\operatorname{Res}^{L_P}_{L_{P,\ell}} V \mid V \in \mathrm{SS}(\mathcal{M}_2)\}.$$

(This is an abuse of notation since $\operatorname{Res}^{L_P}_{L_{P,\ell}} V$ is not actually an irreducible $L_{P,\ell}$-module, merely isotypical. In the above formula we mean the unique irreducible $L_{P,\ell}$-module that occurs in $\operatorname{Res}^{L_P}_{L_{P,\ell}} V$.)

19.5. Sketch of the proof of Theorem 19.1. Since $k^*\mathcal{M} = \hat{\imath}^*_{R,\ell}\hat{\imath}^*_R j^*\mathcal{M}$ and we assume that $\mathrm{SS}_{\mathrm{ess}}(\mathcal{M}) = \{E\}$ with $c(E;\mathcal{M}) = d(E;\mathcal{M}) = 0$, Theorem 19.1 may be deduced from the 3 preceding subsections. The only issue is to estimate $\ell(w)$ which is accomplished by the following proposition. This is the point at which we need to assume that all real (as opposed to rational) boundary components are equal-rank.

PROPOSITION 19.2. *Let ${}_{\mathbb{R}}D^*$ be a real equal-rank Satake compactification. Let P be a parabolic $\mathbb{Q}$-subgroup and let $w \in W_P$. Let $V = H^{\ell(w)}(\mathfrak{n}_P; E)_w$ be the corresponding irreducible L_P-module and assume that $(V|_{M_P})^* \cong \overline{V|_{M_P}}$.*

(i) *If $(\xi_V + \rho, \alpha) \le 0$ for all $\alpha \in \Delta_P$, then*

$$\ell(w) \ge (\dim \mathfrak{n}_P + \#\Delta_P + \dim D_{P,\ell}(V))/2.$$

(ii) *If $(\xi_V + \rho, \alpha) \ge 0$ for all $\alpha \in \Delta_P$, then*

$$\ell(w) \le (\dim \mathfrak{n}_P - \#\Delta_P - \dim D_{P,\ell}(V))/2.$$

The proposition follows from Lemma 15.2 and the estimate [**61**, Lemma 25.2]

$$\dim \mathfrak{n}_P(V) \ge \#\Delta_P + \dim D_{P,\ell}(V). \tag{75}$$

A similar estimate was proved in [**11**, §5.5] which only requires that D and $D_{R,h}$ are equal-rank and that a certain condition (B) holds which has to be verified case-by-case (it is satisfied for most real equal-rank Satake compactifications.) The proof of the estimate in [**61**] is free from case-by-case analysis but strongly uses the assumption that all real boundary components are equal-rank.

20. Proof of the Rapoport/Goresky-MacPherson conjecture

To prove Theorem 11.1 we need to show that $\pi_*\mathcal{I}_p\mathcal{C}(\widehat{X};\mathbb{E}) \cong \mathcal{I}_p\mathcal{C}(X^*;\mathbb{E})$. This is done by verifying that $\pi_*\mathcal{I}_p\mathcal{C}(\widehat{X};\mathbb{E})$ satisfies the local characterization of intersection cohomology on X^*, conditions (i)–(iii) from §4.4. Condition (i) is obvious. Let $i_x\colon \{x\} \hookrightarrow X^*$ denote the inclusion of a point $x \in X^*$. The local vanishing condition (ii) amounts to

$$H^j(i_x^*\pi_*\mathcal{I}_p\mathcal{C}(\widehat{X};\mathbb{E})) = 0 \qquad \text{for } x \in X_{R,h},\ j \ge \tfrac{1}{2}\operatorname{codim} X_{R,h} \tag{76}$$

since $p(k) = k/2 - 1$ for either middle perversity when k is even. Given the other conditions and constructibility, the attaching condition (iii) is equivalent [**15**, V.4], [**33**] to a local covanishing condition:

$$H^j(i_x^!\pi_*\mathcal{I}_p\mathcal{C}(\widehat{X};\mathbb{E})) = 0 \qquad \text{for } x \in X_{R,h},\ j \le \tfrac{1}{2}\operatorname{codim} X_{R,h}. \tag{77}$$

Let $k\colon \pi^{-1}(x) \hookrightarrow \widehat{X}$ be the inclusion. We may re-express

$$H^j(i_x^*\pi_*\mathcal{I}_p\mathcal{C}(\widehat{X};\mathbb{E})) \cong H^j(\pi^{-1}(x); k^*\mathcal{I}_p\mathcal{C}(\widehat{X};\mathbb{E}));$$

since $\mathcal{I}_p\mathcal{C}(\widehat{X};\mathbb{E}) = \mathcal{S}(\mathcal{I}_p\mathcal{C}(\widehat{X};E))$, this is isomorphic to the $\mathcal{L}$-module cohomology $H^j(\widehat{X}_{R,\ell};k^*\mathcal{I}_p\mathcal{C}(\widehat{X};E))$. Thus we can use Theorem 14.1 to see this vanishes for $j > d(k^*\mathcal{I}_p\mathcal{C}(\widehat{X};E))$. By Theorem 18.1, we know that $\mathrm{SS}_{\mathrm{ess}}(\mathcal{I}_p\mathcal{C}(\widehat{X};E)) = \{E\}$ and hence $d(k^*\mathcal{I}_p\mathcal{C}(\widehat{X};E)) \leq \frac{1}{2}\operatorname{codim} X_{R,h} - 1$ by Theorem 19.1. This establishes (76); an analogous argument treats (77).

21. A generalization of Goresky-Harder-MacPherson's theorem

The weighted cohomology $\mathcal{L}$-module $\mathcal{W}^\eta\mathcal{C}(\widehat{X};E)$ may be constructed [**61**, §6] by a weight truncation functor similar to the degree truncation functor that is used for $\mathcal{I}_p\mathcal{C}(\widehat{X};E)$. The following analogue of Theorem 18.1 is proved in [**61**, Theorem 16.3]. Note that there is no assumption on the $\mathbb{Q}$-root system. Unlike Theorem 18.1, this theorem is not difficult to prove; this is because there is an explicit non-inductive formula for the local weighted cohomology.

THEOREM 21.1. *Let E be an irreducible G-module and let $\eta = \mu$ or ν be a middle-weight profile. If $(E|_{{}^0G})^* \cong \overline{E|_{{}^0G}}$, then* $\mathrm{SS}_{\mathrm{ess}}(\mathcal{W}^\eta\mathcal{C}(\widehat{X};E)) = \{E\}$.

Consequently, the argument in the preceding section also proves the following generalization of Goresky, Harder, and MacPherson's result [**30**], Theorem 10.1:

THEOREM 21.2. *Let X^* be a real equal-rank Satake compactification, let p be a middle-perversity, and let η be a middle-weight profile. There is a natural quasi-isomorphism $\mathcal{I}_p\mathcal{C}(X^*;\mathbb{E}) \cong \pi_*\mathcal{W}^\eta\mathcal{C}(\widehat{X};\mathbb{E})$ and hence an isomorphism $I_pH(X^*;\mathbb{E}) \cong W^\eta H(\widehat{X};\mathbb{E})$.*

References

1. J. Arthur, *A trace formula for reductive groups. I. Terms associated to classes in $G(\mathbf{Q})$*, Duke Math. J. **45** (1978), 911–953.
2. ———, *The L^2 Lefschetz numbers of Hecke operators*, Inv. Math. **97** (1989), 257–290.
3. A. Ash, D. Mumford, M. Rapoport, and Y. Tai, *Smooth compactification of locally symmetric varieties*, Math Sci Press, Brookline, 1975.
4. W. Baily and A. Borel, *Compactification of arithmetic quotients of bounded symmetric domains*, Ann. of Math. **84** (1966), 442–528.
5. G. Barthel, J.-P. Brasselet, K.-H. Fieseler, and L. Kaup, *Equivariant intersection cohomology of toric varieties*, Algebraic geometry: Hirzebruch 70 (Warsaw, 1998), Contemp. Math., vol. 241, Amer. Math. Soc., Providence, RI, 1999, pp. 45–68.
6. ———, *Combinatorial intersection cohomology for fans*, Tohoku Math. J. (2) **54** (2002), no. 1, 1–41.
7. A. Beilinson, J. Bernstein, and P. Deligne, *Faisceaux pervers*, Analyse et topologie sur les espaces singuliers (I), CIRM, 6–10 juillet 1981, Astérisque, vol. 100, 1982.
8. A. Borel, *Ensembles fondamentaux pour les groupes arithmétiques*, Colloque sur la Théorie des Groupes Algébriques, Tenu à Bruxelles du 5 au 7 juin 1962, Centre Belge de Recherches Mathématiques, 1962, pp. 23–40.
9. ———, *Introduction aux groupes arithmétiques*, Hermann, Paris, 1969.
10. ———, *L^2-cohomology and intersection cohomology of certain arithmetic varieties*, Emmy Noether in Bryn Mawr (Bryn Mawr, Pa., 1982), Springer, New York, 1983, pp. 119–131.
11. ———, *A vanishing theorem in relative Lie algebra cohomology*, Algebraic Groups: Utrecht 1986 (Berlin) (A. M. Cohen, W. H. Hesselink, W. L. J. van der Kallen, and J. R. Strooker, eds.), Lecture Notes in Mathematics, vol. 1271, Springer-Verlag, 1987, pp. 1–16.
12. ———, *Linear algebraic groups*, Springer-Verlag, Berlin, 1991.
13. A. Borel and W. Casselman, *L^2-cohomology of locally symmetric manifolds of finite volume*, Duke Math. J. **50** (1983), 625–647.
14. ———, *Cohomologie d'intersection et L^2-cohomologie de variétés arithmétiques de rang rationnel 2*, C. R. Acad. Sci. Paris Sér. I Math. **301** (1985), no. 7, 369–373.

15. A. Borel et al., *Intersection cohomology*, Birkhäuser, Boston, 1984.
16. A. Borel and J.-P. Serre, *Corners and arithmetic groups*, Comment. Math. Helv. **48** (1973), 436–491.
17. A. Borel and J. Tits, *Groupes réductifs*, I.H.E.S. Publ. Math. **27** (1965), 55–151.
18. A. Borel and N. Wallach, *Continuous cohomology, discrete subgroups, and representation of reductive groups*, Ann. of Math. Stud., vol. 94, Princeton University Press, Princeton, 1980.
19. T. Braden, *Hyperbolic localization of intersection cohomology*, preprint, arXiv:math.AG/0202251, 9 pages, to appear in Transform. Groups, 2002.
20. ———, *Koszul duality for toric varieties*, preprint, 2003.
21. T. Braden and R. MacPherson, *From moment graphs to intersection cohomology*, Math. Ann. **321** (2001), no. 3, 533–551.
22. P. Bressler and V. A. Lunts, *Toric varieties and minimal complexes*, preprint, arXiv:alg-geom/9712007, 11 pages, 1997.
23. ———, *Intersection cohomology on nonrational polytopes*, Compositio Math. **135** (2003), no. 3, 245–278.
24. W. Casselman, *L^2-cohomology for groups of real rank one*, Representation Theory of Reductive Groups (P. Trombi, ed.), Birkhäuser, Boston, 1983, pp. 69–84.
25. ———, *Geometric rationality of Satake compactifications*, Algebraic groups and Lie groups, Austral. Math. Soc. Lect. Ser., **9**, Cambridge Univ. Press, Cambridge, 1997, pp. 81–103.
26. J. Cheeger, *On the Hodge theory of Riemannian pseudomanifolds*, Geometry of the Laplace Operator (Providence), Proc. Symposia Pure Math., vol. 36, Amer. Math. Soc., 1980.
27. J. Cheeger, M. Goresky, and R. MacPherson, *L^2-cohomology and intersection homology of singular algebraic varieties*, Seminar on Differential Geometry, Ann. of Math. Stud., vol. 102, Princeton Univ. Press, Princeton, N.J., 1982, pp. 303–340.
28. G. de Rham, *Variétés différentiables*, Hermann, Paris, 1973.
29. J. Franke, *Harmonic analysis in weighted L_2-spaces*, Ann. Sci. École Norm. Sup. (4) **31** (1998), 181–279.
30. M. Goresky, G. Harder, and R. MacPherson, *Weighted cohomology*, Invent. Math. **116** (1994), 139–213.
31. M. Goresky, R. Kottwitz, and R. MacPherson, *Discrete series characters and the Lefschetz formula for Hecke operators*, Duke Math. J. **89** (1997), 477–554.
32. M. Goresky and R. MacPherson, *Intersection homology theory*, Topology **19** (1980), no. 2, 135–162.
33. ———, *Intersection homology II*, Invent. Math. **72** (1983), 77–129.
34. ———, *Weighted cohomology of Satake compactifications*, Centre de recherches mathématiques, preprint #1593, 1988.
35. ———, *Local contribution to the Lefschetz fixed point formula*, Invent. Math. **111** (1993), no. 1, 1–33.
36. ———, *The topological trace formula*, preprint, arXiv:math.RT/0105038, 74 pages, 2001.
37. N. Habegger and L. Saper, *Intersection cohomology of cs-spaces and Zeeman's filtration*, Invent. Math. **105** (1991), no. 2, 247–272.
38. Harish-Chandra, *Representations of semisimple Lie groups VI: Integrable and square-integrable representations*, Amer. J. Math. **78** (1956), no. 3, 564–628.
39. F. Hirzebruch, *Hilbert modular surfaces*, Enseignement Math. (2) **19** (1973), 183–281.
40. J. Humphreys, *Linear algebraic groups*, Springer-Verlag, New York, 1975, Graduate Texts in Mathematics, No. 21.
41. M. Kashiwara, *The Riemann-Hilbert problem for holonomic systems*, Publ. Res. Inst. Math. Sci. **20** (1984), no. 2, 319–365.
42. M. Kashiwara and P. Schapira, *Sheaves on manifolds*, Springer-Verlag, Berlin, 1990.
43. H. King, *Topological invariance of intersection homology without sheaves*, Topology Appl. **20** (1985), no. 2, 149–160.
44. B. Kostant, *On the conjugacy of real Cartan subalgebras. I.*, Proc. Nat. Acad. Sci. U. S. A. **41** (1955), 967–970.
45. ———, *Lie algebra cohomology and the generalized Borel-Weil theorem*, Ann. of Math. **74** (1961), 329–387.
46. S. Kumaresan, *On the canonical k-types in the irreducible unitary g-modules with nonzero relative cohomology*, Invent. Math. **59** (1980), no. 1, 1–11.

47. R. P. Langlands and D. Ramakrishnan (eds.), *The zeta functions of Picard modular surfaces*, Les Publications CRM, Montréal, 1992.
48. J.-S. Li, *A vanishing result for the cohomology of arithmetic groups*, Talk at the Conference "Representation Theory of Lie Groups (July 2002 – January 2003)", National University of Singapore, August 2002, http://www.ims.nus.edu.sg/Programs/liegroups/abstracts.htm#lijs2.
49. E. Looijenga, *L^2-cohomology of locally symmetric varieties*, Compositio Math. **67** (1988), 3–20.
50. Y. Matsushima and S. Murakami, *On vector bundle valued harmonic forms and automorphic forms on symmetric Riemannian manifolds*, Ann. of Math. **78** (1963), 365–416.
51. D. Mauger, *Algèbre de Hecke quasi-ordinaire universelle d'un groupe réductif*, Thèse de doctorat, Université Paris-Nord, 2000.
52. M. McConnell, *Intersection cohomology of toric varieties*, preprint, http://www.math.okstate.edu/ mmcconn/toric2.ps, 23 pages, 1999.
53. A. Mokrane and J. Tilouine, *Cohomology of Siegel varieties with p-adic integral coefficients and applications*, Astérisque (2002), no. 280, 1–95, Cohomology of Siegel varieties.
54. A. Nair, *Weighted cohomology of arithmetic groups*, Ann. of Math. **150** (1999), 1–31.
55. F. Quinn, *Intrinsic skeleta and intersection homology of weakly stratified sets*, Geometry and topology (Athens, Ga., 1985), Lecture Notes in Pure and Appl. Math., vol. 105, Dekker, New York, 1987, pp. 233–249.
56. M. S. Raghunathan, *Vanishing theorems for cohomology groups associated to discrete subgroups of semisimple Lie groups*, Osaka J. Math. **3** (1966), 243–256.
57. ———, *Corrections to "Vanishing theorems ... "*, Osaka J. Math. **16** (1979), 295–299.
58. M. Rapoport, 1986, letter to A. Borel.
59. ———, *On the shape of the contribution of a fixed point on the boundary: The case of $\mathbb{Q}$-rank 1 (with an appendix by L. Saper and M. Stern)*, The Zeta functions of Picard modular surfaces (R. P. Langlands and D. Ramakrishnan, eds.), Les Publications CRM, Montréal, 1992, pp. 479–488.
60. L. Saper, *Tilings and finite energy retractions of locally symmetric spaces*, Comment. Math. Helv. **72** (1997), 167–202.
61. ———, *$\mathcal{L}$-modules and micro-support*, submitted, arXiv:math.RT/0112251, v2, 88 pages, 2001.
62. ———, *$\mathcal{L}$-modules and the conjecture of Rapoport and Goresky-MacPherson*, to appear in *Proceedings of the Automorphic Semester at Institute Henri Poincaré, February - June 2000*, arXiv:math.RT/0112250, v2, 16 pages, 2001.
63. ———, September 27, 2002, letter to W. Casselman.
64. ———, *Geometric rationality of equal-rank Satake compactifications*, submitted, arXiv:math.RT/0211138, 14 pages, 2002.
65. L. Saper and M. Stern, *L_2-cohomology of arithmetic varieties*, Ann. of Math. **132** (1990), 1–69.
66. I. Satake, *On a generalization of the notion of a manifold*, Proc. Nat. Acad. Sci. U. S. A. **42** (1956), 359–363.
67. ———, *On representations and compactifications of symmetric Riemannian spaces*, Ann. of Math. **71** (1960), 77–110.
68. ———, *On compactifications of the quotient spaces for arithmetically defined discontinuous groups*, Ann. of Math. **72** (1960), 555–580.
69. J. Schwermer, *Eisenstein series and cohomology of arithmetic groups: the generic case*, Invent. Math. **116** (1994), 481–511.
70. A. Shepard, *A cellular description of the derived category of a stratified space*, Ph.D. dissertation, Brown University, 1985.
71. G. Shimura, *Introduction to the arithmetic theory of automorphic functions*, Publications of the Mathematical Society of Japan, vol. 11, Princeton University Press, Princeton, NJ, 1994, Reprint of the 1971 original, Kanô Memorial Lectures, 1.
72. M. Sugiura, *Conjugate classes of Cartan subalgebras in real semisimple Lie algebras*, J. Math. Soc. Japan **11** (1959), 374–434.
73. ———, *Correction to my paper: "Conjugate classes of Cartan subalgebras in real semisimple Lie algebras"*, J. Math. Soc. Japan **23** (1971), 379–383.

74. J. Tilouine and E. Urban, *Several variable p-adic families of Siegel-Hilbert cusp eigensystems and their Galois representations*, Ann. Sci. École Norm. Sup. (4) **32** (1999), 499–574.
75. J. Tits, *Classification of algebraic semisimple groups*, Algebraic Groups and Discontinuous Subgroups (Providence), Proc. Symposia Pure Math., vol. 9, Amer. Math. Soc., 1966, pp. 33–62.
76. W. T. van Est, *A generalization of the Cartan-Leray sequence, I, II*, Indag. Math. **20** (1958), 399–413.
77. D. Vogan, Jr. and G. Zuckerman, *Unitary representations with nonzero cohomology*, Compositio Math. **53** (1984), no. 1, 51–90.
78. M. Vybornov, *Constructible sheaves on simplicial complexes and Koszul duality*, Math. Res. Lett. **5** (1998), no. 5, 675–683.
79. ______, *Sheaves on triangulated spaces and Koszul duality*, preprint, `arXiv:math.AT/9910150`, v2, 41 pages, 2000.
80. J. Wildeshaus, *Degeneration of automorphic sheaves*, Talk at the Conference "Shimura Varieties and Automorphic Forms", Japan-U.S. Mathematics Institute, Johns Hopkins University, March 2001.
81. S. Zucker, *L_2 cohomology of warped products and arithmetic groups*, Invent. Math. **70** (1982), 169–218.
82. ______, *Satake compactifications*, Comment. Math. Helv. **58** (1983), 312–343.
83. ______, *L_2-cohomology and intersection homology of locally symmetric varieties, II*, Compositio Math. **59** (1986), 339–398.

DEPARTMENT OF MATHEMATICS, DUKE UNIVERSITY, BOX 90320, DURHAM, NC 27708, U.S.A.
E-mail address: `saper@math.duke.edu`
URL: `http://www.math.duke.edu/faculty/saper`

Made in United States
Orlando, FL
04 November 2026

90206104R00162

Made in United States
Orlando, FL
04 November 2026

90206109R00100